Plant Molecular Biology

Molecular Genetic Analysis of Plant Development and Metabolism

NATO ASI Series

Advanced Science Institutes Series

A series presenting the results of activities sponsored by the NATO Science Committee, which aims at the dissemination of advanced scientific and technological knowledge, with a view to strengthening links between scientific communities.

The Series is published by an international board of publishers in conjunction with the NATO Scientific Affairs Division

A Life Sciences B Physics	Plenum Publishing Corporation London and New York
C Mathematical and Physical Sciences D Behavioural and Social Sciences E Applied Sciences	Kluwer Academic Publishers Dordrecht, Boston and London
F Computer and Systems Sciences G Ecological Sciences H Cell Biology I Global Environmental Change	Springer-Verlag Berlin Heidelberg New York London Paris Tokyo Hong Kong Barcelona Budapest

NATO-PCO DATABASE

The electronic index to the NATO ASI Series provides full bibliographical references (with keywords and/or abstracts) to more than 30000 contributions from international scientists published in all sections of the NATO ASI Series. Access to the NATO-PCO DATABASE compiled by the NATO Publication Coordination Office is possible in two ways:

- via online FILE 128 (NATO-PCO DATABASE) hosted by ESRIN, Via Galileo Galilei, I-00044 Frascati, Italy.
- via CD-ROM "NATO Science & Technology Disk" with user-friendly retrieval software in English, French and German (© WTV GmbH and DATAWARE Technologies Inc. 1992).

The CD-ROM can be ordered through any member of the Board of Publishers or through NATO-PCO, Overijse, Belgium.

Series H: Cell Biology, Vol. 81

Plant Molecular Biology

Molecular Genetic Analysis of Plant Development and Metabolism

Edited by

Gloria Coruzzi

Department of Biology
New York University
1009 Main Building, Washington Square
New York, NY 10003, USA

Pere Puigdomènech

CID-CSIC
Jordi Girona, 18–26
08034 Barcelona, Spain

Springer-Verlag
Berlin Heidelberg New York London Paris Tokyo
Hong Kong Barcelona Budapest
Published in cooperation with NATO Scientific Affairs Division

Proceedings of the NATO Advanced Study Institute on Plant Molecular Biology – Molecular Genetic Analysis of Plant Development and Metabolism, held at Cala Viñas, Mallorca, Spain, May 9–19, 1993

ISBN-13: 987-3-642-78854-3 e-ISBN-13: 987-3-642-78852-9
DOI: 10.1007/987-3-642-78852-9

CIP data applied for

Softcover reprint of the hardcover 1st edition 1994

Typesetting: Camera ready by authors
31/3130 - 5 4 3 2 1 0 - Printed on acid-free paper

PREFACE

The VIIth NATO Advanced Study Institute on Plant Molecular Biology was held in Cala Vinas, Mallorca, Spain, from May 10-19, 1993. The topic, "Molecular-genetic Analysis of Plant Metabolism and Development" brought together at one meeting two "camps", the developmental biologists and the plant biochemists, who many times fail to see the connection between plant physiology and development.

The talks focussed on studies involving cases where genetics and molecular biology have helped to elucidate fundamental issues in plant development and metabolism. Many of these were previously intractable questions to answer using standard biochemical or developmental techniques. The talks covered a range of model genetic organisms including most notably Arabidopsis and Maize. The value of concentrating efforts to characterize genes and mutants in a model plant is that eventually a molecular-genetic "blueprint" of plant development and metabolism may be constructed. The reports at the meeting demonstrated that not only are mutants in development and metabolism found, in many cases there has been excellent progress in identifying the affected gene. The power of genetics is to identify as yet unknown genes, and to uncover their function. In many of the reports, a combination of molecular biology, cell biology, and phenotypic analysis of mutants has been used to uncover a role for these novel gene products. The contributions in this book summarize lectures of invited speakers. They have not been subjected to any major cutting or editing, in order to preserve the individual nature of each contribution.

Many of the presentations highlighted the value of conducting basic research in plants and showed examples where basic studies have led to applications in agricultural biotechnology. The most notable example of this was the use of pollen-specific promoters to engineer male sterile plants. The molecular-genetic research on metabolism presented also has had a large impact on the engineering of quality traits in crop plants. We are pleased to acknowledge the generous contributions of the following companies and organizations who recognize the fact that basic plant research fuels agricultural biotechnology; NATO, The International Society of Plant Molecular Biology, Monsanto, CIBA-GEIGY, DuPont, Pioneer Hi-Bred, DeKalb Plant Genetics, ICI Seeds, Sanofi Agro-Veterinaire, Rhone-Poulenc, Sandoz Agro, Agrigenetics, and the following Spanish Agencies and Companies; Pharmacia Iberica, Direccio General de Recerca, Generalitat de Catalunya, Consejo Superior de Investigaciones Cientificas, Direccion General de Investigacion Cientifica y Tecnica, Plan Nacional de Investigacion Cientifica y Techica. Without the support of the above organizations, this meeting would not have taken place.

We would also like to thank our colleagues in Barcelona and New York for their enthusiastic participation in managing the logistics of the conference. Finally we thank all the partcipants in this meeting for their valuable contributions and for the many stimulating discussions which ensued.

Pere Puigdomenech,
Barcelona, Spain

Gloria Coruzzi
New York, USA

CONTENTS

PLANT DEVELOPMENT

PLANT METABOLISM

RESPONSES TO HORMONES AND STRESS

GENE REGULATION

POST-TRANSCRIPTIONAL REGULATION

MAPPING AND TRANSFORMATION

The Genetic and Molecular Basis of Flower Development in Arabidopsis

Elliot M. Meyerowitz
Division of Biology 156-29
California Institute of Technology
Pasadena, California 91125
U.S.A.

My laboratory studies pattern development in flowers, concentrating on two basic patterns that develop in parallel. The first is the pattern of organ number and position. The flowers of the plant we study, the mustard *Arabidopsis thaliana,* have a total of 16 organs: four organs in the first organ whorl, with a 90-degree angle between adjacent organs, four organs inside and alternate to the first whorl organs, six third-whorl organs with their own characteristic spacing and divergence angles, and two fourth-whorl organs (Müller, 1961; Smyth et al., 1990). This number and position of floral organs is specific to members of the mustard family and close relatives; other families have quite different patterns of organ number and position. The second pattern, which develops to a large degree independently of the first, is the pattern of organ identity. From the outside to the inside of the flower, one finds the four organ types sepal, petal, stamen, then carpel occurring in this order in successive organ whorls. This radial order of organ identities is not at all family-specific, being almost completely uniform in flowering plants. We would like to understand how the activities of specific genes produce both patterns in developing floral primordia, to understand how these patterns have arisen in the evolution of flowering plants, and to know what differences in the genes that control these patterns give the diversity of flower forms that exists in nature.

The radial pattern of organ identities depends on a set of floral organ identity genes. Mutations in these genes fall into three predominant phenotypic classes, each class affecting the

NATO ASI Series, Vol. H 81
Plant Molecular Biology
Edited by G. Coruzzi and P. Puigdomènech

organ types that appear in two adjacent whorls of the flower. By a series of genetic experiments, we have demonstrated that these three classes of genes specify the four types of floral organ found in *Arabidopsis* in a combinatorial fashion. Class A genes such as *APETALA2* act in whorls 1 and 2, class B genes such as *PISTILLATA* and *APETALA3* act in whorls 2 and 3, and class C genes such as *AGAMOUS* act in whorls 3 and 4 in early flower development. Sepals form in whorl 1 because in this region only class A genes are active, petals form in whorl 2 as a result of the combined activity of A and B class genes, stamens form in whorl 3, where class B and class C genes are coexpressed, and carpels develop in whorl 4 due to the activity of class C genes (Bowman et al., 1988; 1989; 1991; Meyerowitz et al., 1991; Coen and Meyerowitz, 1991; Jack et al., 1993). Molecular cloning of *Arabidopsis* organ identity genes from all of the three classes has been accomplished, by my lab and by others (Yanofsky et al., 1990; Jack et al., 1992; Mandel et al., 1992; Goto and Meyerowitz, 1993; Okamuro et al., 1993). The DNA sequences of the cloned organ identity genes indicate that most but not all of them are DNA-binding transcription factors of the MADS-box family. Ectopic expression of these genes in transgenic plants supports our organ identity model, and has allowed us to design and produce new types of flowers.

Since the radial pattern of organ identity in flowers is regulated by the spatial pattern of activity of these organ identity genes, the activity of these genes themselves must be regulated by prepattern (cadastral) genes, whose function is to regulate the spatial expression pattern of the organ identity genes. An example of such a gene is *SUPERMAN*, whose function is to regulate the initial pattern of expression of the class B organ identity genes *APETALA3* and *PISTILLATA*. When *SUP* is mutant, class B genes are expressed in the fourth whorl of developing flowers. As a consequence, the carpels are reduced or eliminated, and there are additional stamens near the center of the flower (Schultz et al., 1991; Bowman et al., 1992). A more complex interaction between genes that regulates the activity patterns of organ identity genes is the negative interaction between C class and A class genes. When the A class gene *APETALA2* is mutant, the C class gene *AGAMOUS* (and perhaps others, as well) is active in whorls 1 and 2, causing development of carpels in whorl 1, and stamens in whorl 2. Conversely, *AGAMOUS* mutants have A activity in whorls 3 and 4, and thus have petals in whorl 3, and sepals in whorl 4, at least in one interpretation of the mutant flowers (Bowman et al., 1991; Meyerowitz et al., 1991). Additional cadastral genes must exist, since the known genes and gene interactions do not fully explain the activity pattern of the organ identity genes.

An even earlier-acting set of genes are the meristem identity genes, whose mutant phenotype is the partial replacement of flowers with inflorescence shoots. Among these genes are *LEAFY,* *APETALA1*, *CAULIFLOWER*, and others (Irish and Sussex, 1990; Schultz and Haughn, 1991; Bowman, 1991; Weigel et al.., 1992; Huala and Sussex, 1992; Bowman et al., 1993). The evidence to date indicates that the meristem identity genes are responsible for the initiation of expression of the later-acting organ identity genes, and thus are the genes responsible for telling a lateral meristem to develop into a flower rather than a leaf with an axillary stem (Weigel and Meyerowitz, 1993). The meristem identity genes may therefore be the earliest set of genes active in developing flowers, and may be the genes that respond to the environmental sigmals that induce flowering.

We know less about the first pattern established in developing flowers, the pattern of organ number and position, though we have recently started the genetic and molecular analysis of a several genes whose mutant phenotypes are either an increase in number of organs (as in the first and second whorls of *PERIANTHIA* mutants, and in the inner whorls of *CLAVATA1*, *CLAVATA2* and *CLAVATA3* mutants, Running and Meyerowitz, unpublished; Clark and Meyerowitz, unpublished), or a specific decrease in organ number (as in the third whorl of *PERIANTHIA* mutants, or in mutants for the gene *HANABA TARANU*, Sakai and Meyerowitz, unpublished). Our initial analysis of these mutants, which relies upon laser scanning confocal microsopy to count the cells in different areas in developing flowers, indicates that control of the timing and position of cell division in the earlier stages of floral primordium formation is a key process in the establishment of appropriate floral organ number. We do not yet have a simple and predictive model for the control of the number and position of floral organs, however.

There are additional patterns and processes in flower development that we also do not understand at a mechanistic level. One is the growth of flowers to achieve their final size. Different members of the mustard family have flowers with essetially identical structures, but with very different sizes. There must therefore be mechanisms for changing flower size without changing organ number, position, or identity. We have never seen a single gene mutation that mimics this evolutionary difference. Another pattern in flowers develops later than those discussed so far; it is the shape of individual floral organs. This also varies widely in evolution, and varies considerably even within the mustard family. This variation must be due to differential patterns of cell division and cell expansion, since these are the only processes that contribute to organ shape in plants. We do have mutations that alter organ shape, but it is difficult to know which of these act trivially, by generally limiting the ability of cells to divide at appropriate rates, or to properly expand, and which might specifically affect individual cell divisions, or the regulated expansion of particular cells. One mutation that indicates the possibility that cell division in flowers is regulated differently than the similar process in vegetative plant parts is *TSO*, in which cell division seems to cease shortly after the specification of a lateral meristem as floral. DNA replication may still occur, though, since the cells of the mutant flower continue to expand, and their nuclei grow larger (Liu, Running and Meyerowitz, unpublished).

References:

Bowman, J.L., Yanofsky, M.F., and Meyerowitz, E.M. (1988). *Arabidopsis thaliana*: A review. Oxford Surv. Plant Mol. Cell. Biol. 5, 57-87.

Bowman, J.L., Smyth, D.R., and Meyerowitz, E.M. (1989). Genes directing flower development in *Arabidopsis*. The Plant Cell. 1, 37-52.
Bowman, J.L. (1991) Molecular genetics of flower development in *Arabidopsis thaliana*. Ph.D. thesis, California Institute of Technology.
Bowman, J.L., Smyth, D.R. and Meyerowitz, E.M (1991) Genetic interactions among floral homeotic genes of *Arabidopsis*. Development 112, 1-20.
Bowman, J.L., Sakai, H., Jack, T., Weigel, D., Mayer, U. and Meyerowitz, E.M (1992) *SUPERMAN*, a regulator of floral homeotic genes in *Arabidopsis*. Development 114, 599-615.
Bowman, J.L., Alvarez, J., Weigel, D., Meyerowitz, E.M and Smyth, D.R. (1993) The specification of floral meristem identity in *Arabidopsis thaliana*. submitted for publication.
Coen, E.S. and Meyerowitz, E.M (1991) The war of the whorls: genetic interactions controlling flower development. Nature 353, 31-37.
Goto, K. and Meyerowitz, E.M. (1993) Molecular cloning, expression and biochemical activity of the Arabidopsis homeotic gene PISTILLATA. submitted for publication.
Huala, E. and Sussex, I.M. (1992) *LEAFY* interacts with floral homeotic genes to regulate Arabidopsis floral development. Plant Cell 4, 901-913.
Irish, V.F. and Sussex, I.M. (1990) Function of the *apetala-1* gene during *Arabidopsis* floral development Plant Cell 2, 741-753.
Jack, T., Brockman, L.L. and Meyerowitz, E.M (1992) The homeotic gene *APETALA3* of *Arabidopsis thaliana* encodes a MADS box and is expressed in petals and stamens. Cell 68, 683-697.
Jack, T., Sieburth, L.E. and E.M. Meyerowitz (1993) Genes that control flower development in *Arabidopsis*. Seminars in Developmental Biology 4, 51-63.
Mandel, M.A., Gustafson-Brown, C., Savidge, B. and Yanofsky M.F. (1992) Molecular characterization of the *Arabidopsis* floral homeotic gene *APETALA1*. Nature 360, 273-277.
Meyerowitz, E.M., Bowman, J.L., Brockman, L.L., Drews, G.N., Jack, T., Sieburth, L.E. and Weigel, D. (1991) A genetic and molecular model for flower development in *Arabidopsis thaliana*. Development 112, suppl. 1, 157-168.
Müller, A. (1961). Zur Charakterisierung der Blüten und Infloreszenzen von *Arabidopsis thaliana* (L.) Heynh. *Kulturpflanze* **9**, 364-393.
Okamuro, J., den Boer, B., Van Montagu, M. and Jofuku, D. (1993) *APETALA2*: Characterization of a novel plant homeotic regulatory gene from Arabidopsis. J. Cell. Biochem. suppl. 17B, 16.
Schultz, E. A., and Haughn, G. W. (1991). Leafy, a homeotic gene that regulates inflorescence development in Arabidopsis. Plant Cell 3, 771-781.
Schultz, E. A., Pickett, F. B., and Haughn, G. W. (1991). The flo10 gene-product regulates the expression domain of homeotic genes ap3 and pi in Arabidopsis flowers. Plant Cell 3, 1221-1237.
Smyth, D.R., Bowman, J.L., and Meyerowitz, E.M. (1990) Early flower development in *Arabidopsis*. Plant Cell 2, 755-767.

Weigel, D., Alvarez, J., Smyth, D.R., Yanofsky, M.F. and Meyerowitz, E.M (1992) *LEAFY* controls floral meristem identity in *Arabidopsis*. Cell 69, 843-859.

Weigel, D.W. and Meyerowitz, E.M. (1993) Activation of *Arabidopsis* floral homeotic genes. submitted for publication.

Yanofsky, M.F., Ma, H., Bowman, J.L., Drews, G.N., Feldmann, K.A., and Meyerowitz, E.M. (1990) The protein encoded by the *Arabidopsis* homeotic gene *agamous* resembles transcription factors. Nature 346, 35-39.

Genetic and Molecular analyses of pattern formation genes in maize

B. Veit, D. Jackson, R. Kerstetter, L. Smith, E. Vollbrecht J. Yamaguchi and Sarah Hake, Plant Gene Expression Center, 800 Buchanan Street, Albany, CA 94710

Introduction

One of the most salient features of pattern formation in plants is the extended period over which it occurs. Unlike animal development, in which the many of the large scale patterns are established during embryogenesis, many key aspects of a plant's architecture are determined much later. In most seed plants, virtually the entire shoot system forms from a small group of cells contained in the shoot apical meristem (Steeves and Sussex, 1989). Although this meristem forms during embryogenesis, the initials it contains that will proliferate and provide the basis for tissues and organs of the shoot system are still largely undetermined at this time. It is only as the mitotic products of these initials are displaced away from the apical meristem that they take on their characteristic identities.

Cells produced by the apical meristem can be divided into at least 3 distinct lineages (Derman, 1953) that are differentiated from each other by their postition within the shoot and characteristic patterns of cell division. Cells in the outermost layer, the L1, divide primarily in an anticlinal plane, thus producing a relatively uniform layer that encases the shoot. More internal lineages tend to divide in a less polarized manner to generate less stratified internal tissues. As these clonally distinct layers are displaced away from the growing shoot tip, they are partitioned into the various anatomical elements that comprise the shoot. On the flanks of a growing shoot, leaves are initiated a short distance from the growing tip with a angular displacement between successive leaves that is characteristic of the species (reviewed by Smith, 1992). In most angiosperms, groups of cells in close proximity to leaf primordia become organized into axillary buds, while the remaining cells contribute to stem tissue to form an internode that separates successive leaves. The

NATO ASI Series, Vol. H 81
Plant Molecular Biology
Edited by G. Coruzzi and P. Puigdomènech

regular pattern by which these anatomical elements are formed produces a shoot system which can be thought of as being composed of metameric units termed phytomers.

Relatively little is known of the mechanisms responsible for partitioning the various elements of the phytomer from the growing shoot tip. Several lines of evidence suggest that the fates of cells produced by the apical meristem are not fixed from their inception, but instead are influenced by their position within the developing shoot and in relation to previously formed anatomical elements (Poethig, 1987). . For example, the shared identity of cells within a specific layer appears to reflect their common position within that layer rather than a common cell lineage; isolated cells that are introduced from adjacent layers by occassional out of plane cell divisions assume an identity consistent with their new position. At the level of organ initiation events, incisions that separate the meristem from newly formed leaves shift the normal pattern of phyllotaxis (Snow and Snow, 1933). These experiments suggest that the position of leaf initiation is influenced by older leaves, and that cells can be recruited to form leaf primordia at new sites well after they are set off by the apical meristem.

To better understand the underlying molecular mechanisms which partition the developing shoot, we are studying several classes of genes in maize whose expression patterns or mutant phenotypes suggest a role in this process. We describe here progress towards understanding the function of these genes through a combination of genetic and molecular stategies.

Plant homeobox genes

For several years, it was widely assumed that plants lacked a class of genes common to animals which share a conserved DNA binding motif known as the homeobox and which have in many cases been shown to play an important role in determining cell fate. Their absence in plants had been rationalized partly on the basis of the very different developmental strategies employed by plants and animals. It was therefore a surprise to many when the <u>Knotted1</u> (<u>Kn1</u>) gene from maize was found to contain this DNA binding motif

(Vollbrecht et al., 1991). More recently, additional classes of homeobox genes from other plant species have been described (Schena and Davis, 1992; table 1).

Plant Homeodomains

```
                     helix I             helix II  turn  helix III
Kn1             SKKKKKGKLPKEARQQLLSWWDQHYKWPYPSETQKVALAESTGLDLKQINNWFINQRKRHWKPS
Rs1             ---------------K--H--EL----------E-I---------Q------------------
KNAT1           ---------------K--T--EL---------SE-------N---Q------------------
R18             --R------------K--H--EL----------E-M----T----P-----------------A
P11             ---------------K--H--EL----------E-I----A----Q-------------D----
T6a             M--R-------------VD--LR-I-------S--L---------Q------------------
Sbh1            M--R--------------E--NR---------S--L---------Q------------------
ECA             C--R--D--------K-----EL--R------ME-I--
R4              L--R-------D--TV--E--NT--R----T-ED--R--AM----P------------------
B15-1           L--R-------D--SA-MD--NT--R----T--D--T

ZMH1            LR-RRA----GDTTSI-KQ--QE-S-----T-DD-AK-V-----Q--------------N-HNN
ZMH2            LR-RRA----GDTAST-KA--QA-S-----T-ED-AR-VQE---Q--------------N-HNN
R7              MR-RRA----GDTASV-KA--QA-S-----T-DD-AR-VQE---Q--------------N-HSN
R6              MR-RRA----GDTASV-KA--QA-S-----T-DD-AR-VQE---Q--------------N-HSN
T12             LR-RRA----GDTTSV-KA--QS-S-----T-ED-AR-VQE---Q--------------N-HSN

ZMHOX1A         NSTAR--HFGPVIN-K-HEHF...KTQ----RSV-ES---EL--TFR-V-K--ETR-HSARVA-
ZMHOX1B         NIKDR--HFGPVIS-K-HEHF...KTQ----RSL-ES---EL--TFH-V-R--E-R-HFARLA-

ATHB1           QLPE--RR-TT-QVHL-EKSF...ETENKLEPER-TQ--KKL--QPR-VAV--Q-R-A-WKTKQ
ATHB2           DNSR--LR-S-DQSAI-EETF...KDHSTLNPKQ-Q---KQL--RAR-VEV--Q-R-A-TKLKQ

                     .       .  .   .                      .    .  ** * * . .
Animal               R       Q  L   F        Y     R  A  L L   QIKIWFQNRR K K
consensus                                                       V

Plant                K       A  L   W        Y     K  A  T L   QINNWFINQR R W
consensus            A       T      F        K        V  N      VAK  ETR  X R
                             I               T           L       EK   Q   S K
                             Q                                    V       F
```

Table 1. Summary of described plant homeobox genes grouped according to similarity. Homeodomains designated with T are from tomato, (N. Sinha, unpublished results); ATHB homeoboxes are from Arabidopsis. The remainder of the plant homeodomains were isolated from maize.

Given the important role of homeobox genes in animal development, it was especially interesting that the Kn1 mutant should have a "developmental" phenotype. The Kn1 mutation appears to interfere with determinant patterns of growth in the lateral veins of leaves, leading to the formation of extra cells and associated outpocketings or "knots" of tissue (fig.1).

Figure 1. Comparison of leaf blade organization from normal (a) vs. Kn1-O (b) plants

Dosage analysis of Kn1 had suggested that the mutation's dominant effect on leaf development reflects a neomorphic gene activity (i.e. the production of a novel gene product, or the vast overexpression of a normal gene product) (Freeling and Hake, 1985). Molecular analyses of Kn1 have borne out the latter alternative. Of more than a dozen Kn1 alleles that have been characterized to date, none appear to significantly alter the coding region of the gene. Instead, these dominant alleles all contain insertions or rearrangements that appear to alter the regulation of a normal gene product (Hake, 1992). In wild type plants, Kn1 transcripts and protein are undetectable in leaves, but are readily observed in meristematic tissues such as vegetative meristems or immature inflorescences. In mutant plants, the pattern of Kn1 expression in meristematic tissues appears unchanged; however, expression of what appear to be normal Kn1 transcript and protein is also observed in leaves.

The ectopic expression of Kn1 conditioned by mutant alleles is localized to the lateral veins over which knots typically form (Smith et al, 1992). Interestingly, this expression appears more uniform than the sporadic distribution of the knots themselves. Similar patterns of sporadic knotting are also caused by placing the Kn1 transcription unit under the control of a 35S promoter in what are otherwise wild type plants (Williams and Lemaux, unpublished observations). A similar tendency towards indeterminant growth has also been seen in tobacco plants containing a 35S-Kn1 construct in which expression of Kn1 is correlated with the formation of abnormal shoots from the normally determinant leaves (Sinha et al, in press). Thus, in several contexts, the expression of Kn1 acts in some way to maintain cells in an undetermined state.

The notion that Kn1 expression might normally maintain indeterminancy has been supported by careful examination of its expression in normal vegetative and floral meristems. Although Kn1 is expressed at high levels in both meristem types, it is absent at sites on the flanks of the apical meristem where lateral organs such as leaves are normally initiated (Smith et al., 1992). The possibility of this correspondence is especially compelling in the vegetative meristem where the arrangement of the 100 or so cells in which this down-regulation is observed matches quite well to the position and number of leaf founder cells inferred from clonal analysis studies (Poethig, 1984). This down-regulation of Kn1 preceeds any overt morphological sign of leaf initiation, thus providing one of the earliest markers yet defined for this event. Given the previously discussed association of Kn1 expression with indeterminant patterns of growth, it seems plausible that the down-regulation of Kn1 could be a precondition for the initiation of determinant organs such as leaves.

Although this characteristic pattern of down-regulation has been described for both the Kn1 protein (Smith et al, 1992) and transcript (D. Jackson, unpublished results), their expression patterns do not precisely overlap. The most striking difference is the lack of detectable Kn1 transcripts in the L1 layers of the meristem; by contrast, Kn1 protein is consistently detected in this layer by immunolocalization methods. The possibility that this L1 signal is an artifact resulting from cross-reaction of antibody with non-Kn1 species is argued against by the similar patterns of expression that are detected by several independent monoclonal anti-sera raised against the Kn1 protein. On the other hand, the absence of signal detected by *in situ* hybridization can not be explained simply by the L1 being inacessible to probes since signal was detected in this layer with other probes. While other explanations are possible, one that merits consideration is intercellular movement of the Kn1 protein into the L1 from adjacent layers. Alternatively, the Kn1 transcript may turn over more rapidly than its protein product in L1 tissues.

Recent efforts to determine the normal function of the Kn1 gene have focused on the isolation of loss of function alleles. Given that Kn1 is expressed from very early stages of embryo formation, it is possible that a loss of function mutation would result in an embryo lethal phenotype. With this possibility in mind, screens have been designed in which the loss of function alleles would be recovered as heterozygotes with a wild type alleles as follows: Stocks that are homozygous for dominant Kn1 allele would first be mutagenized, then outcrossed to normal plants. Any normal progeny that resulted would either be heterozygous for derivatives of the dominant allele, or carry dominant suppressors of the Kn1 phenotype. A number of such progeny have been isolated and are now being analyzed.

Other genes that contain Kn1-like homeoboxes have been isolated from genomic and cDNA libraries of maize by low stringency hybridization. Through the use of recombinant inbreds lines, these have been mapped to various regions of the maize genome. Though these genes could be described as a family based on their sequence similarity, their map positions show no evidence of clustering similar to the Bithorax or Antenapedia complexes of Drosophila. However, several of these genes show meristematic patterns of expression that are similar, though not identical to Kn1, suggesting these structurally related genes may perform similar functions. ECA (ear clone A) for example, one of the homeobox genes most similar in sequence to Kn1, is preferentially expressed in shoot apical meristems, is down regulated at sites of leaf initiation and appears absent at the transcriptional level in L1 layers. ECA expression does not completely overlap with Kn1 expression, especially in regions of developing vasculature.

ECB, another clone whose sequence is very similar to Kn1, has recently been shown to correspond to the dominant leaf mutant Rough Sheath (Rs) by Freeling and colleagues (Freeling, 1992). To the extent that this gene, like ECA and Kn1, shows meristematic patterns of expression, it seems possible that Kn1 like homeobox genes form a gene family that play an important role in patterning the meristem.

terminal ear: altered partitioning of phytomer elements

In addition to characterizing various members of the Kn1 homeobox gene family, we have pursued other classes of patterning genes in maize through on-going transposon mutagenesis screens. One mutant that has been tagged by this approach, terminal ear (te) (Matthews,1972) appears to specifically perturb the phytomer organization of the primary shoot axis (B. Veit, unpublished results). The formation of abnormally short internodes on the upper part of the plant (fig. 2) causes the upper leaves to enclose the terminally positioned tassel, much like husk leaves normally enclose axillary positioned ears. Less consistently, the basal branches of the normally staminate tassel develop as pistillate structures. A second phenotype of terminal ear concerns the abnormal shape of leaves initiated on the primary plant axis, especially in the more severe alleles. Viewed face on, these leaves have a more angular shape than usual, with the blade portion of the leaf smaller and projecting at more of a right angle from the main axis than is normal.

The altered form of both vegetative leaves and their associated internodes suggests that the te mutant disturbs mechanisms by which different elements of the phytomer are partitioned from the shoot apical meristem. This interpretation is consistent with data that show a substantially increased number of internodes formed in terminal ear plants compared to their normal siblings.

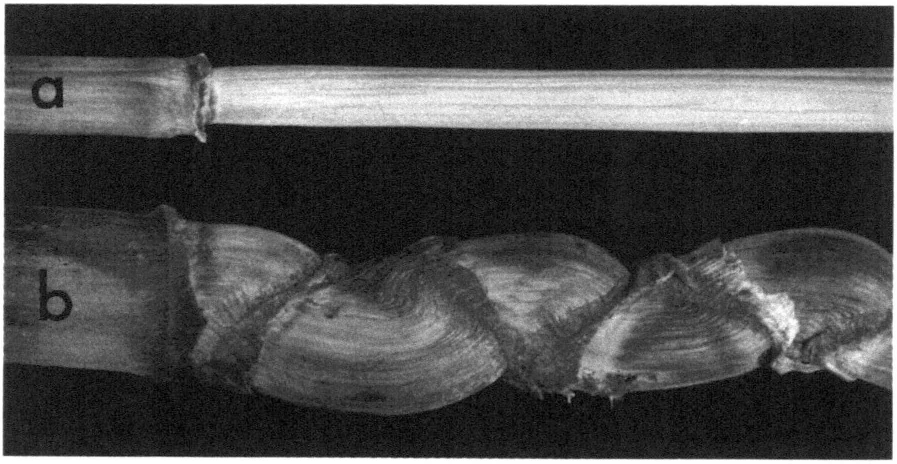

Figure 2. Comparison of upper stalk of normal (a) vs. terminal ear (b) plant.

Since mutant plants appear to flower no later than their wild type sibs, the mutant shoot apical meristem appears to set off phytomer units more frequently than its normal counterpart. Perhaps as a consequence, the extra internodes that do form tend to be shorter than normal.

How the te mutant would alter the pattern by which phytomer units are set off is not clear. A simple model supposes that mechanisms that normally restrict the initiation of leaf primordia are compromised, causing abnormally frequent leaf initiation events. These precocious initiation events could lead to the formation of misshapen leaves and internodes if an insufficient number of cells were available for recruitment into these structures at the time of their initiation. For example, if some portion of the cells normally allocated to the internode portion of the phytomer were instead recruited into a leaf, an assymmetric pattern of internode development might result. In the shortest internodes of te plants, precisely this pattern is observed. A pronounced curvature is evident in the internodes, with the shortest side of each internode always positioned below the midrib of the leaf above. This pattern of assymmetry is consistent with the clonal relationship of the internode to the leaf above which has been described for corn (Johri and Coe, .

Progress towards a molecular understanding of te function is being made through the analysis of 6 independent te alleles that have been recovered from untargetted transposon mutagenesis screens of lines carrying active Mutator (in collaboration with S. Briggs, Pioneer Hybrid). Two of these alleles have been characterized in detail. Both contain an insertion of a Mu8 element, the insertion sites of the element in the two alleles differing by approximately 200 bp (B. Veit, unpublished results. Attempts to define a transcription unit by using cloned flanking genomic sequence as probes are in progress.

Although the remaining alleles have not been characterized in as much detail on a molecular level, this collection displays a striking range of phenotypes. The most severe of these, te1-4, is typically less than half the height of normal sibs, produces strongly modified leaves, has uniformly shortened nodes, and typically develops a tassel which is completely enclosed in upper leaves. By contrast, te1-8 is relatively normal, with the upper nodes most severely affected but with the tassel relatively free of the upper leaves. All of these mutants are currently being introgressed into more uniform genetic backgrounds to more clearly delineate allele specific differences.

Prospects

It will be interesting to determine what sort of functional interactions might exist between various classes of meristem patterning mutants. If, for example, the down-regulation of the Kn1 gene is truly a precondition for the initiation of leaves, this pattern might be expected to be altered in te, a mutant in which the frequency and nature of leaf intiation events appears altered. Analysis of the expression patterns the Kn1 gene and its relatives in te plants and other mutants that alter specific aspects of meristem function should greatly enhance our understanding of how patterns are established in the shoot apical meristem.

References

Dermen, H. (1953). Periclinal cytochimeras and origin of tissues in stem and leaf of peach. Am. J. Bot. 57, 737-746

Freeling, M. and Hake, S. (1985). Developmental genetics of mutants that specify knotted leaves in maize. Genetics 111, 617-634

Freeling, M. (1992). A conceptual framework for maize leaf development. Dev. Biol. 153, 44-58.

Hake, S. (1992). Unravelling the knots in plant development. Trends Genet. 8, 109-114

Johri, M. and Coe, E. (1983). Clonal Analysis of Corn Plant Development. Devel. Biol. 97, 154-172

Matthews, D. L., Grogan, C. O. and Manchester, C. E. (1974). Terminal ear mutant of maize (*Zea maize L.*). J. Agric. Sci., Camb. 82, 433-435.

Poethig, R.S. (1984). Cellular parameters of leaf morphogenesis in maize and tobacco. In "contemporary Problems of Plant Anatomy" (R.A. White and W. C. Dickson, Eds.) pp.235-259. Academic Press, New York

Poethig, R.S. (1987). Clonal analysis of cell lineage patterns in plant development. Amer. J. Bot.74, 581-594

Smith, L., Greene, B., Veit, B. and Hake, S. (1992). A dominant mutation in the maize homeobox gene, Knotted-1, causes its ectopic expression in leaf cells with altered fates. Development 116, 21-30

Smith, L. and Hake, S. (1992). The initiation and determination of leaves. Plant Cell 4, 1017-1027

Shena, M. and Davis, R.W. (1992). HD-zip proteins: members of an Arabidopsis homeodomain protein superfamily. Proc. Natl. Acad. Sci. 89:3894-3898

Snow, M. and Snow, R. (1933). The effect of displacing a primordium. Phil. Trans. Roy. Soc. B.222, 353-400

Steeves, T. and Sussex, I. (1989). Patterns in Plant Development, 2nd ed. (Cambridge: Cambridge University Press)

Vollbrecht, E., Veit, B., Sinha, N. and Hake, S. (1991). The developmental gene Knotted1 is a member of a maize homeobox gene family. Nature 350, 241-243

The Cellular and Molecular Biology of the Quiescent Centre in Relation to Root Development

Peter W. Barlow
Department of Agricultural Science
University of Bristol
AFRC Institute of Arable Crops Research
Long Ashton Research Station
Bristol BS18 9AF
England, UK

Introduction

The cells at the summit of an actively growing primary root, derived from the anti-apical pole of the embryo, are the source of all of the other cells of that organ. Some of these latter cells then generate new growth centres which initiate new roots, again with a new set of 'source' cells at their summit. The source cells thus comprise a minimal set of structural initials for the root. In some ferns, the structural initial is reduced to a single 'apical' cell which divides frequently, whereas in gymnosperms and angiosperms there are usually a number of initials which have a low division frequency. Experimentally, it has been shown that, in the latter case, cells immediately neighbouring the structural initials share their proliferative properties and, together with them, constitute a quiescent centre (QC). The slow rate of proliferation in the QC is associated with a prolonged G_1 phase of the mitotic cycle which, in turn, is linked to a slow rate of cell growth. These properties, together with its special position within the root, characterize the QC as a stem-cell population, a view reinforced by the ability of QC cells to grow and proliferate more rapidly in circumstances that interfere with the proliferative behaviour of cell in the bulk of the meristem (Barlow 1978). The interrelationship of the various types of cells and their different behaviours within the root tip are summarized in Figure 1.

NATO ASI Series, Vol. H 81
Plant Molecular Biology
Edited by G. Coruzzi and P. Puigdomènech

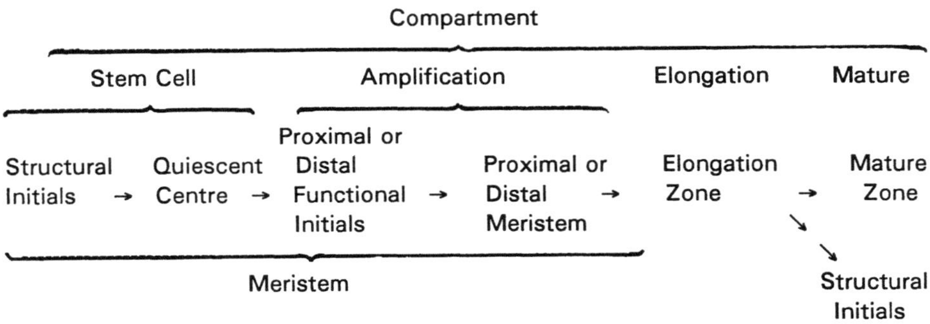

Figure 1. Structure of a root tip in terms of the behaviour of its cells in proliferation and differentiation.

In view of the differential behaviour of cells within the meristem, explanations, in terms of cellular and molecular activities, are required for the following features of the QC which set it apart from other meristematic cells:

- The low rate of cell growth and proliferation within the QC;
- The means by which QC cells can be switched to rapid proliferation, and, consequently, the homeostatic relationship between the QC and other zones of the meristem and beyond;
- The presence of the QC at a particular location within the apex and the means whereby it maintains its size.

Each of these topics will be dealt with as far as present knowledge or concepts permit. Discussion will draw on already published work, as well as work in progress with two model systems afforded by primary root apices of maize (*Zea mays* L.) seedlings and root apices of tomato (*Lycopersicon esculentum* Mill.) grown *in vitro*.

Cell Growth

Control over the rate of cell growth is the key to understanding the behaviour of the QC, as indeed it is in the remainder of the meristem where cell proliferation is more active. This is because cell growth drives the mitotic cycle and not *vice versa*. In yeast, such a linkage may be effected through a 'sizer' control point (Wheals 1982, Tyson 1987); this may be true of plant cells too, although it has not so far been successfully demonstrated (Demchenko and Ivanov 1977). Also, cell growth is intrinsic to meristematic zones and continues even if their mitotic cycle is inhibited. It is difficult to estimate directly the rate of cell growth in the QC because its cells are hidden from view. However, using the pole of the root as a reference point and recording the position of the end walls of cellular packets in the epidermis lineage of germinating maize roots, it was possible to estimate (Barlow unpublished) a mean relative elemental elongation rate (strain rate) of 0.6 % h^{-1} for the boundary wall of the root on the distal surface of the QC. This value compares with 40 % h^{-1} for the root boundary walls in the zone of elongation 4 mm proximal to QC. Probably other end walls in the QC could be similarly tracked to obtain estimates of strain rates orthogonal to the boundary wall. Indirect estimates of cellular growth within the QC as a whole can be calculated from the rate of cell proliferation in the QC if it is assumed that cells divide at double their birth size (V_B) and that the heterogeneous proliferative behaviour of the cell population can be satisfactorily taken into account. In the case of the QC of maize, a cell number doubling time (DT_N) of 170 h translates into a relative elemental rate of cell volume growth (RER_V) of 0.4 % h^{-1} for cells with a mean V_B of ~490 μm^3 (Barlow 1973, 1974). In the QC of tomato, where DT_N is 214 h, the estimated RER_V is 0.3 % h^{-1} (V_B ~320 μm^3). These values compare with RER_Vs of about 3.6 % h^{-1} (V_B ~850 μm^3) and 4.3 % h^{-1} (V_B ~500 μm^3) for meristematic cells in the stele of the two species, respectively, indicating a 10-fold difference in cell growth rate between QC and meristem (Barlow 1991).

Mutation at the *Gib-1* locus of tomato affects the pattern of cell proliferation in the QC (Barlow 1991). In *in-vitro*-grown roots of *gib-1* plants, DT_N for the QC was found to be less than that in the wild-type; thus, a RER_V of about 0.9 % h^{-1} (V_B ~415 μm^3) was estimated, which is about three times the RER_V found in the wild-type QC. Moreover, the RER_V of meristematic stelar cells is about 5.1 % h^{-1} (V_B ~600 μm^3), which is about 18 % faster than similar cells in wild-type roots. A growth differential also exists between meristematic cortical cells of mutant and wild-type. These results suggests that the *gib-1* mutation relaxes the control over cell enlargement, particularly in the QC, where the differential between the growth of its cells and those in the meristem is less marked than it is in the wild-type.

The *gib-1* mutation affects gibberellin biosynthesis by depressing the activity of *ent*-kaurene synthetase A (Bensen and Zeevaart 1990). While pleiotropic effects are possible, it is logical to suppose that the altered pattern of cell growth in the QC of the mutant is somehow the result of a decreased level of gibberellin. One way in which this might be effected is through a gibberellin-mediated alteration of the chemical composition of the cell walls that makes them more extensible. When ancymidol, a synthetic inhibitor of gibberellin biosynthesis, is applied to pea roots, they show a response (Tanimoto 1988) similar to that caused by the *gib-1* mutation in tomato roots. There is also an increased ratio of galactose to arabinose in extracts of cell walls. However, if longitudinal and transversal cell walls have different compositions with respect to these sugars, then the altered galactose/arabinose ratio may have a trivial explanation based on the fact that, in ancymidol-treated roots, the proportion of cell wall material attributable to either the longitudinal or the transversal walls is changed in favour of the latter type of wall. Gibberellins are also reported to affect other cell wall components (such as peroxidases) which may in turn modulate organ extensibility (reviewed by Métraux 1987). But most of these last-mentioned reports relate to shoots and it remains to be seen whether similar findings apply to roots.

The orientation of cell growth throughout the tomato root, including the QC, is also affected by the *gib-1* mutation so that cells in roots of the mutant tend to be wider and shorter than cells in roots of the wild-type (Barlow 1991). This is why there is an increased proportion of anticlinal walls within the total wall population, as mentioned above. It is possible that the increased RER_V in the QC of the mutant may relate to altered orientation of cell enlargement if it were shown that anticlinal walls in the mutant root increase in area faster than they do in the wild-type. That this may be so is suggested by the more rapid increase in cell width in the QC of the mutant during a seven-day growth period. However, it is not known to what extent comparison of cell wall dimensions are rendered unreliable by the continual interpolation of new walls formed at cytokinesis, since these walls continually fractionate the length or width of any previously-formed cell wall. Clearly, what is needed is a measure of wall growth in the absence of cell division (but if this is to be achieved experimentally polarity of growth must remain unaltered) to gain insight into the allometry of growth within the QC. Such an approach has been applied to the QC of maize using hydroxyurea as a division inhibitor (Barlow 1969), but this work only established a correspondence between RER_V and DT_N and did not examine QC growth in closer detail.

By studying the microtubular cytoskeleton of meristematic cells of gibberellin-deficient roots, insight was gained into the control of wall growth and its relationship with endogenous gibberellin status (Baluška, Parker and Barlow 1993). Cortical microtubules (CMTs) are intimately concerned with the oriented deposition of cellulose microfibrils which, in turn, regulate the orientation of cell growth and hence the rate of wall growth in any given direction. CMTs in meristematic cells of *d-5* mutant maize roots were appreciably less well aligned under longitudinal walls than CMTs in wild-type roots. Similarly, maize roots grown in the presence of *2S,3S* paclobutrazol, an inhibitor of gibberellin biosynthesis, also showed a less ordered arrangement of CMTs; this feature could be normalized by added gibberellic acid (GA_3) (Baluška *et al.* 1993). The tendency to a more random orientation of CMTs in roots with low gibberellin status coincided with a decreased length of the cells and a corresponding increase in cell width. Cell walls of tomato are associated with less numerous arrays of CMTs and so alterations to their disposition are less easy to discern, particularly in the QC where there may be gibberellin-induced disturbances to CMTs that correlate with the altered pattern of cell growth. Nevertheless, it was possible to categorize CMT patterns in the QC and show that in *gib-1* mutant roots there were slightly fewer cells with randomly oriented CMTs than in the wild-type. On the other hand, more cells in the *gib-1* QC had CMTs with transverse and oblique orientations. However, these differences were scored in QCs with already established patterns of development. What may be more important from the point of view of explaining the contrasting patterns of cell growth, are the orientations of CMTs in cells that are precursors of the QC in the youngest stages of root primordium development.

Cell Proliferation

Rates of cell proliferation are characteristically low in the QC of all species so far examined; in fact, it is this feature that led to the adoption of the term 'quiescent centre' (Clowes 1956). In general, the most elastic phase of the mitotic cycle is G_1. The significance of the G_1 phase is questionable (Cooper 1982); in some proliferating systems G_1 is absent (DNA synthesis commencing in telophase), in others G_1 is prolonged almost indefinitely in some of the cells. This latter feature has given rise to the notion of a G_0 phase which replaces G_1, or of G_1 being split into two compartments, an A state followed by a B phase which precedes the S phase; the duration of G_1 is therefore determined by a transition probability which regulates exit from the A state (Smith and Martin 1973). So far, it has not been possible to discover whether or not the prolonged G_1 phase in the QC is subdivided into discrete sub-phases which require distinct transition events (such as alterations to $p34^{cdc2}$ or to a homologue of a human tumour repressor, e.g. p53) for the cell to pass from one sub-phase to the next. *In situ* hybridization or antibody probes making use of specific molecular markers applied to QC cells may resolve this problem. However, certainly in maize, and possibly in tomato, the QC is heterogeneous with respect to the proliferative behaviour of its cells. There is evidence that not all cells are participating in the mitotic cycle (i.e. some cells are temporarily arrested in G_0 or have a low probability of passing from A state), but this is based on comparisons of DT_N using different methods of analysis which estimate cell cycle parameters that pertain to different fractions of the QC population as a whole (Clowes 1971). Unfortunately, no cytological markers are known for G_0 cells of plant meristems as are claimed to exist in animal cell populations (Cohen *et al.* 1981). However, long-term labelling of the QC with ^3H-thymidine does reveal a group of cells at the pole of the stele having a longer pre-DNA-replication phase than cells of the cortical portion of the QC (Barlow 1978). Interestingly, the nuclei of these same stelar cells have more dispersed chromatin than other QC nuclei. These particular cells may be true G_0 cells. Cells in this location are also interesting from a theoretical point of view as they constitute a null growth point in tensorial analyses of the cellular organization of roots (Hejnowicz and Karczewski 1993).

Some attention has been given to the possible hormonal control of the mitotic cycle in the QC of maize. Exposure of the QC by removal of the overlying root cap immediately stimulates its cells to enter S phase and then divide (Barlow 1974). Local application of either 0.1 μM zeatin (Z) or 1 μM zeatin riboside (Zr) to the exposed QC abolish this stimulation, whereas 50 μM abscisic acid (ABA) suppresses almost all DNA synthesis (Barlow and Pilet 1984, Müller, Pilet and Barlow 1993), though with this latter hormone the response is also conditioned by the way in which the roots are handled. On the other hand, 0.14 μM indole acetic acid (IAA) applied to intact roots stimulates the entry of QC cells into S phase and also shortens their mitotic cycle, but lengthens it elsewhere in the meristem (Barlow unpublished). In the respective experimental systems, ABA inhibits cell growth (Barlow and Pilet 1984) whereas IAA stimulates it (Barlow 1969). Thus, the results again point to cell growth as the primary regulator of mitotic activity in the QC, and that, if hormones are genuinely implicated in the proliferative behaviour of the QC, their effect might be mediated at the cell wall.

At a finer level of resolution, microtubules (MTs) have been found to participate in regulating mitotic cycle activities in the QC of maize roots (Baluška and Barlow 1993). Colchicine, oryzalin and cold conditions (4 °C) all cause disintegration of MTs within the QC. This coincides with a stimulation of DNA synthesis (i.e. passage from G_1 to S). Simultaneously, nuclei swell and their chromatin disperses in response to each treatment. The control experiment of stabilizing MTs and then applying each treatment is in hand, but no results are yet available. Nevertheless, the suspicion is that endoplasmic MTs (EMTs), which span the cytoplasm between MT-organizing centres (MTOCs) on the nuclear envelope and the plasmalemma, somehow 'lock' the nucleus and its chromatin in a pre-replicative mode. When the EMTs are disassembled, proteins then enter the nucleus, disperse the chromatin, and initiate DNA synthesis. When QC cells are stimulated to grow by decapping or IAA treatment, the EMTs are forced to remake their connections with the MTOCs to accomodate the increasing distances which they now have to bridge between the nuclear surface and plasmalemma. The enhanced dynamism of the EMTs leads them to relax their hold on the nucleus thus enabling it to initiate S phase. In undisturbed QCs, where there is only slow growth of the cells, the MTs are correspondingly less dynamic and hence the relative stability of the EMTs maintains a repressive control over nuclear activity.

Maintenance of the QC

Many circumstances, some of which have been described above, alleviate the quiescence of the QC and allow its cells to cycle rapidly (Clowes 1972). This switch in proliferative behaviour seems always to involve an impairment of proliferation in cells neighbouring the QC suggesting that, in undisturbed roots, there is a repressive growth control imposed on the QC from without (R in Figure 2). It is also possible that there could be a 'quiescence factor' that regulates the behaviour of the QC from within (Q in Figure 2). This 'factor' could take the form of either the absence of a promotor of growth (needed to initiate 'start' in the mitotic cycle) or the presence of a repressor of growth (which interferes with wall loosening, for example) that is made within the QC itself. Moreover, an extrinsic factor, R, might even induce a quiescence factor (Q' in Figure 2), as would seem probable in an interactive system such as a meristem. Thus, the problem is to identify which, if any, of these possible types of constraints actually operate, and then, of what they consist.

Repressive factors, R, may be chemical (R_c) or physical (R_p). The former may be mobile hormones. Consideration of their known effects and locations in roots was behind the attempts, mentioned earlier, to identify the effects of Z, Zr and ABA on the QC. ABA, for example, is known to be synthesised in the root cap and to be

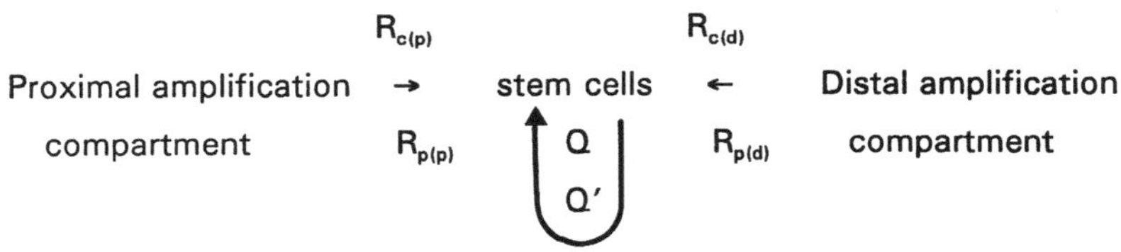

Figure 2. Various types of regulatory controls over the low proliferative rate of a stem cell population such as the QC. R, extrinsic repressive control; this may be either chemical, c, or physical, p, and originate in a site proximal (p) or distal (d) to the QC. Intrinsic repressive control may be either autonomous, Q, or induced, Q'. In the latter case, Q' is induced by R_c or R_p.

transported into the proximal portion of the root (Jaunin, Hofer and Pernet 1985), whereas Z and Zr have been proposed as regulators of quiescence, though the evidence for this is rather slender (Feldman 1975). Moreover, removing distal portions of the cap stimulates the QC to proliferate (Clowes 1972), as does decreasing the length of the proximal meristem by cold treatment (Barlow and Adam 1989) or radiation (Clowes 1972). In both cases, the greater the number of meristematic cells that are lost, the stronger the response of the QC, suggesting a direct relationship between proximal and distal meristems (sources of R_c) and the QC (a sink for R_c). However, immunocytochemical methods for detecting cytokinins (Z and Zr) in roots have not given any evidence suggesting that they are especially abundant in the QC of either maize or tomato (Zavala and Brandon 1983, Sossountzov *et al.* 1988).

Physical factors (R_p) include constraints to cell growth imposed by the outer boundary wall at the pole of the root, where the elemental rate of extension is low. Constraints may also be due to the physical presence of the cap whose major plane of growth tends to be antagonistic to the preferred orientation of growth of the distal cells of the QC (Barlow 1974). However, a system of interacting physical factors is easier to envisage as regulating the QC of 'closed' meristems (such as found in maize), where there is a clear boundary wall separating QC from the cap, than it is in 'open' meristems (such as in pea), where no such QC/cap boundary is obvious because the cell files seem to run uninterruptedly from the proximal meristem, through the QC, into the cap. In an 'open' meristem, chemical signals would seem to have precedence over physical signals and hence would serve as a relatively unsophisticated default mechanism for quiescence. When present, physical constraints would provide an additional control, and hence indicate a redundancy of control mechanisms; such redundancy is a regular feature of developmental systems (Malacinski and Neff 1990). A highly redundant control system may be of vital necessity, given the importance of the QC as a stem-cell population: it is required not only to maintain the integrity of the root during normal growth but also to serve as a source of new cells during the regenerative growth that follows a perturbation.

If the QC of open meristems represents a system uncomplicated by physical constraints but, nevertheless, is one that still possesses redundancy of control, then this redundancy may also include intrinsic repressors of quiescence (Q or Q'). The prospect of there being such repressors (or, alternatively, the absence of activators of growth, such as recently described by McQueen-Mason, Durachko and Cosgrove 1992) has initiated the search for patterns of gene activity unique to the QC. The experimental approach has been to extract total RNA from QCs that have been manually isolated from maize root tips (Müller *et al.* 1993, Sabelli *et al.* 1993), isolate and reverse-transcribe the RNA into cDNA, amplify the cDNA by polymerase chain reaction (PCR), and compare the library of PCR-generated cDNAs with other such libraries constructed from elsewhere in the meristem as well as from QCs caused to become proliferative by decapping (Sabelli *et al.* 1993). This differential screening method has revealed certain RNA species, and hence patterns of gene activity, that are specific to the QC. However, since QCs of young primary roots from germinating grains are used as the source material, it is necessary to rule out that the apparently QC-specific gene products are not the remnants of gene activities associated with the quiescence of the embryonic radicle present in the pre-germinated, dry grain: that is, the QC, or a fraction of it, may be a zone of the radicle that has not yet fully germinated. Since a QC is present in primary roots much longer than those studied by the above-mentioned methods, and can even increase in size as roots grow, this argument cannot be carried too far. It does have some weight, though, because cells in different zones of the germinating root meristem enter S phase and mitosis at different times after imbibition (Barlow unpublished) and there are genes (which presumably have products) known to be associated with embryonic root quiescence (e.g. genes at the *vp* locus); such genes would need to be repressed and their products eliminated at germination before an autonomously functioning root apex and QC could be said to exist.

When a new cap is regenerated following decapping, part of the regeneration process involves the development of a new QC (Barlow 1974). If physical factors are set aside as a control of quiescence, as perhaps they can be in open meristems, then it seems most likely that one aspect of cap regeneration is the restoration of the chemical repressor that induces the quiescence factor, i.e. $R_{c(d)} \rightarrow Q'$ (see Figure 2). Thus, the minimal determinant of the QC is a group of cells at the distal end of the root that produce $R_{c(d)}$. The position of a potential QC region in the regenerating apex is a function of the distance of suitably responsive cells from the $R_{c(d)}$-producing cells. These two factors, distance and $R_{c(d)}$ concentration, comprise the minimal positional

Table 1. Presence (+) or absence (-) of putative physical and chemical influences that regulate the presence or absence of a QC and the type of meristem in which the QC is found.

Degree of redundancy	$R_{c(p)}$	$R_{p(p)}$	$R_{c(d)}$	$R_{p(d)}$	Q	Q′	Type of Meristem
0	-	-	-	-	-	-	With apical cell (some ferns)
1	-	-	-	-	+	-	Hypothetical (non-dividing apical cell)
2	-	-	-	+	-	+	Open meristem
2a	-	-	-	+	(+)	+	Open meristem
3	-	-	+	+	-	+	Closed meristem
3a	-	(+)	+	+	-	+	Closed meristem

information needed to specify both the presence and the location of a QC. A set of controls, shown in Figure 2 and discussed above, can therefore be proposed Table 1) for the regulation of QC behaviour. This set of controls is also ordered to reflect the increasing degrees of redundancy that correspond with roots of different types of construction.

In undisturbed roots, the QC may actively maintain its size in accordance with the constraints proposed in Table 1. For example, the QC of *in-vitro*-grown roots of *gib-1* tomato undergoes cycles of expansion and shrinkage (Barlow and Parker unpublished). Because cell growth and division in the QC is relatively rapid in the mutant root, this cyclical activity is more evident than it is in the wild-type root (though there are indications that this cycle exits here too, but has a longer period). During the expansion phase, cells divide and remain within the QC, suggesting that the quiescent condition is inherited by daughter cells (Barlow 1991). However, an upper limiting size is reached and a group of cells in the cortex complex of the QC, distal to the cortex-stele boundary, begins to proliferate more rapidly. Consequently, the QC shrinks. A further consequence is that new cap initials are formed so that, in addition to cycles of QC formation and reformation, there are parallel cycles of cap generation and regeneration. Periods of apparently natural cap regeneration have been recorded in maize roots which may indicate similar expansion/shrinkage cycles in the QC (Barlow and Rathfelder 1985, Clowes and Wadekar 1989). Thus, the QC is a dynamic system, responding to a size control residing within the cellular milieu of the root which must be additional to those already suggested.

Conclusion

A fair amount is known about the cell biology of the QC - probably enough to invite sensible questions about its control that can be answered with currently available techniques of molecular biology. To follow, in the present context, the jig-saw puzzle metaphor that is often used in biological discussions, the root can be regarded as the whole puzzle, the cells as its pieces, and the molecular activities as the patterns on each piece. The QC comprises a few pieces of the whole; so far, they lack much hint of what their pattern might be. When part of its pattern is revealed, the rest might be easier to establish later on. Moreover, sometimes a pattern in one area of a puzzle predicts the pattern in another area, assuming the general theme is known. Hopefully, as more and more information is revealed about the pieces (the cells) and their inter-relationships within the vicinity of the QC, progress will accelerate towards comprehending not only its molecular details but also the significance of the QC in relation to the function of the root as a whole.

References

Baluška F, Barlow PW (1993) The role of the microtubular cytoskeleton in determining nuclear chromatin structure and passage of maize root cells through the cell cycle. Europ J Cell Biol 61 (in press)

Baluška F, Parker JS, Barlow PW (1993) A role for gibberellic acid in orienting microtubules and regulating cell growth polarity in maize root cortex. Planta (in press)

Barlow PW (1969) Cell growth in the absence of division in a root meristem. Planta 88: 215-223

Barlow PW (1973) Mitotic cycles in root meristems. In: The cell cycle in development and differentiation, eds M Balls and FS Billett, pp 133-165. Cambridge University Press, Cambridge

Barlow PW (1974) Regeneration of the cap of primary roots of *Zea mays*. New Phytol 73: 937-954

Barlow PW (1978) The concept of the stem cell in the context of plant growth and development. In: Stem cells and tissue homeostasis, eds BI Lord, CS Potten and RJ Cole, pp 87-113. Cambridge University Press, Cambridge

Barlow PW (1991) The meristem and quiescent centre in cultured root apices of the *gib-1* mutant of tomato (*Lycopersicon esculentum* Mill.). Ann Bot 69: 533-543

Barlow PW, Adam JS (1989) The response of the primary root meristem of *Zea mays* L. to various periods of cold. J Exp Bot 40: 81-88

Barlow PW, Rathfelder ER (1985) Cell division and regeneration in primary root meristems of *Zea mays* recovering from cold treatment. Env Exp Bot 25: 303-314

Barlow PW, Pilet P-E (1984) The effect of abscisic acid on cell growth, cell division and DNA synthesis in maize root meristem. Physiol Plant 62: 125-132

Bensen RJ, Zeevaart JAD (1990) Comparison of *ent*-kaurene synthetase A and B activities in cell-free extracts from young tomato fruits of wild-type and *gib-1*, *gib-2*, and *gib-3* tomato plants. J Plant Growth Regul 9: 237-242

Clowes FAL (1956) Localization of nucleic acid synthesis in root meristems. J Exp Bot 7: 307-312

Clowes FAL (1971) The proportion of cells that divide in root meristems of *Zea mays* L. Ann Bot 35: 249-261

Clowes FAL (1972) The control of cell proliferation within root meristems. In: The dynamics of meristem cell populations, eds MW Miller and CC Kuehnert, pp 133-145 (Adv Exp Med Biol 18). Plenum, New York

Clowes FAL, Wadekar R (1989) Instability in the root meristem of *Zea mays* L. during growth. New Phytol 111: 19-24

Cohen RL, Muirhead KA, Gill JE, Waggoner AS, Horan PK (1981) A cyanine dye distinguishes between cycling and non-cycling fibroblasts. Nature 290: 593-595

Cooper S (1982) The continuum model: application to G1-arrest and G(0). In: Cell Growth, ed. C Nicolini, pp 315-336. Plenum, New York

Demchenko NP, Ivanov VB (1977) Synchronization of passage through the S period by sister cells in wheat root meristems. Soviet J Devel Biol 8: 349-354

Feldman LJ (1975) Cytokinins and quiescent center activity in roots of *Zea*. In: The development and function of roots, eds JG Torrey and DT Clarkson, pp 55-72. Academic Press, London

Hejnowicz Z, Karcewski J (1993) Modelling of meristem growth of root apices in a natural coordinate system. Am J Bot 80: 309-315

Jaunin F, Hofer R-M, Pernet J-J (1985) Transport and radioactivity from tritiated abscisic acid in intact and freeze-decapped maize roots. J Plant Physiol 121: 407-415

McQueen-Mason S, Durachko DM, Cosgrove DJ (1992) Two endogenous proteins that induce cell wall extension in plants. Plant Cell 4: 1425-1433

Malacinski GM, Neff AW (1990) An essay on redundancy within developmental processes. In: Cytoplasmic organization systems, ed. GM Malacinski, pp 123-152. McGraw Hill, New York

Métraux J-P (1987) Gibberellins and plant cell elongation. In: Plant hormones and their role in growth and development, ed PJ Davies, pp 296-317. Nijhof, Dordrecht

Müller M, Pilet P-E, Barlow PW (1993) An excision and squash technique for analysis of the cell cycle in the quiescent centre of maize. Physiol Plant 87: 305-312

Sabelli P, Burgess SR, Carbajosa JV, Parker JS, Halford NG, Shewry PR, Barlow PW (1993) Molecular characterization of cell populations in the maize root apex. In: Molecular and cell biology of the plant cell cycle, eds JC Ormorod and D Francis, pp 97-109. Kluwer, Dordrecht

Smith JA, Martin L (1973) Do cells cycle? Proc Natl Acad Sci USA 70: 1263-1267
Sossountzov L, Maldiney R, Sotta B, Sabbagh I, Habricot Y, Bonnet M, Miginiac E (1985) Immunocytochemical localization of cytokinins in Craigella tomato and a sideshootless mutant. Planta 175: 291-304
Tanimoto E (1988) Gibberellin regulation of root growth with change in galactose content of cell walls in *Pisum sativum*. Plant Cell Physiol 29: 269-280
Tyson JJ (1987) Size control of cell division. J Theor Biol 126: 381-391
Wheales AE (1982) Size control models of *Saccharomyces cereviseae* cell proliferation. Mol Cell Biol 2: 361-368
Zavala ME, Brandon DL (1983) Localization of a phytohormone using immunocytochemistry. J Cell Biol 97: 1235-1239

GENETIC REGULATION OF ROOT EXPANSION IN *ARABIDOPSIS THALIANA*

Marie-Theres Hauser and Philip N. Benfey
Department of Biology
New York University
New York, N.Y. 10003
USA

INTRODUCTION

The development of plant organs depends on the precise regulation of the timing and orientation of the plane of cell division, the orientation and extent of cell expansion and the differentiation of specialized tissues (Steeves and Sussex, 1989). The coordinated expression of genes involved in these processes influences and probably determines morphogenesis.

Arabidopsis thaliana roots provide an ideal model to study organ morphogenesis because their post embryonic development occurs in a continuous manner (Esau, 1977). Thus, along a single root, all developmental stages are represented and apparent at all times (Steeves and Sussex, 1989; Schiefelbein and Benfey, 1991). Furthermore the root is a simple organ with few differentiated cell types and radial symmetry in the external layers of cells (Dolan *et al.*, 1993).

Here we will focus on one aspect of root morphogenesis - cell expansion. The physiology of cell expansion during cell growth has been studied extensively (Taiz, 1984; Cosgrove and Kniefel, 1987). The genetic regulation of cell expansion has only recently begun to be explored (Baskin *et al.*, 1992, Benfey *et al.*, 1993)

Our goal it to use a genetic approach to dissect the processes involved in cell expansion in roots of *Arabidopsis thaliana* and to reveal the mechanistic and genetic basis for specific expansion patterns. This paper will report on the genetic and physiological characterization of a new expansion mutant and will provide a comparison with the expansion mutants described previously (Benfey *et al.*, 1993).

NATO ASI Series, Vol. H 81
Plant Molecular Biology
Edited by G. Coruzzi and P. Puigdomènech

MATERIAL AND METHODS

Plant strains, plant growth conditions and genetic crosses

The wild type *Arabidopsis thaliana* plants used were of the Columbia (Col) and Wassilewskija (Ws) ecotypes. The *sabre* and *pom pom* mutants were isolated on the basis of their root phenotype from individual T-DNA insertion lines kindly provided by K.A. Feldmann, University of Arizona (Feldmann, 1991). The *cobra* and *lion's tail* mutants were isolated on the basis of their root phenotype following EMS mutagenesis of Col seeds (Benfey *et al.*, 1993). *Arabidopsis* seedlings were sterilized, plated and incubated as previously described (Benfey *et al.*, 1993). Crosses were performed as described by Benfey *et al.* (1993).

Growth measurements and cell area calculations

For measurement of the root length, seedlings were photographed at roughly 24 hour intervals with a 35 mm camera under a Nikon stereo microscope, using Kodak Ektachrome 160T film. The root length was measured from the projected slides with a curvimeter (run-mate ™, Holland). The cell area calculations were performed as previously described (Benfey *et al.*, 1993)

Whole mount preparations

Seedlings were fixed in FAA (formaldehyde (3.7%) : ethanol (50%) : acetic acid (5%)) and mounted with glycerol on microscope slides. Phenotypes were analyzed with a Leitz Laborlux S compound microscope. Photographs were taken using Kodak Ektachrome 160T film.

RESULTS

Mutants with abnormal root expansion

The architecture of the post-embryonic root of *Arabidopsis* is remarkably simple (Schiefelbein and Benfey, 1991; Dolan *et al.*, 1993). The number of cells of certain tissues of the primary root and their cross-sectional areas appear to be nearly invariant (Benfey *et al.*, 1993; Dolan *et al.*, 1993). This simple architecture and conserved dimensions of the *Arabidopsis* root facilitates the screening for mutants with altered cell expansion.

As reported previously, the mutants *cobra, lion's tail* and *sabre* have abnormal radial expansion of their root (Benfey *et al*., 1993). In order to characterize the expansion, fresh transverse sections were cut through the specialization zone of the mutants. A visual comparison of these sections indicated that the abnormal expansion was due to expansion of different tissues in the different mutants.

In *cobra*, it appears that the expansion is primarily due to the expansion of epidermis cells (Fig. 1C, D). Whole mount preparations of *cobra* roots revealed that the epidermis is radially expanded with apparently reduced epidermis elongation (Fig. 1C).

In *lion's tail* the stele appears to be more expanded than the other tissues (Fig. 1E, F). The whole mount preparations of *lion's tail* roots revealed that the epidermis, cortex and endodermis cells also showed reduced elongation and therefore were much more cuboidal in shape when compared to wild-type tissues (Fig. 1A, E).

The expansion of the mutant *sabre* appeared to be primarily due to expansion of the cortex layer (Fig. 1G, H).

In contrast to these three mutants which exhibited expansion primarily in one tissue, the new mutant, *pom pom* appeared to have abnormal radial expansion in two root tissues, the epidermis and the cortex (Fig. 1I, J, K, L). The whole mount preparations of one week old root revealed that the epidermis and cortex have a much more cuboidal shape than wild type (Fig. 1A, I). Roots from three week old plants of *pom pom* exhibited a slightly different phenotype. In whole mounts of these roots the files of root cells were difficult to distinguish (Fig. 1K). The cross section of these roots revealed more disorganized epidermis and cortex cells (Fig. 1K).

In *cobra, lion's tail* and *pom pom* the elongation zone, which is characterized by the presence of smaller cells with dense cytoplasm, is shorter than in wild type (Fig. 1C, E, I, K). In contrast, the expansion of *sabre* seems to start later than in *cobra, lion's tail* and *pom pom* (Fig. 1G). Therefore the elongation zone appears to be less affected by the expansion.

These visual observations were confirmed by measurements of the cross-sectional areas of the mutants (Fig. 2). The root of *pom pom* was slightly more expanded than *cobra* and significantly more expanded than *sabre* and *lion's tail* (Fig. 2B). Abnormal cell expansion occurred in *pom pom* in all root tissues as is the case for *cobra, lion's tail* and *sabre* (Benfey *et al*., 1993). But the degree and pattern of expansion of the root tissues was different among the different mutants.

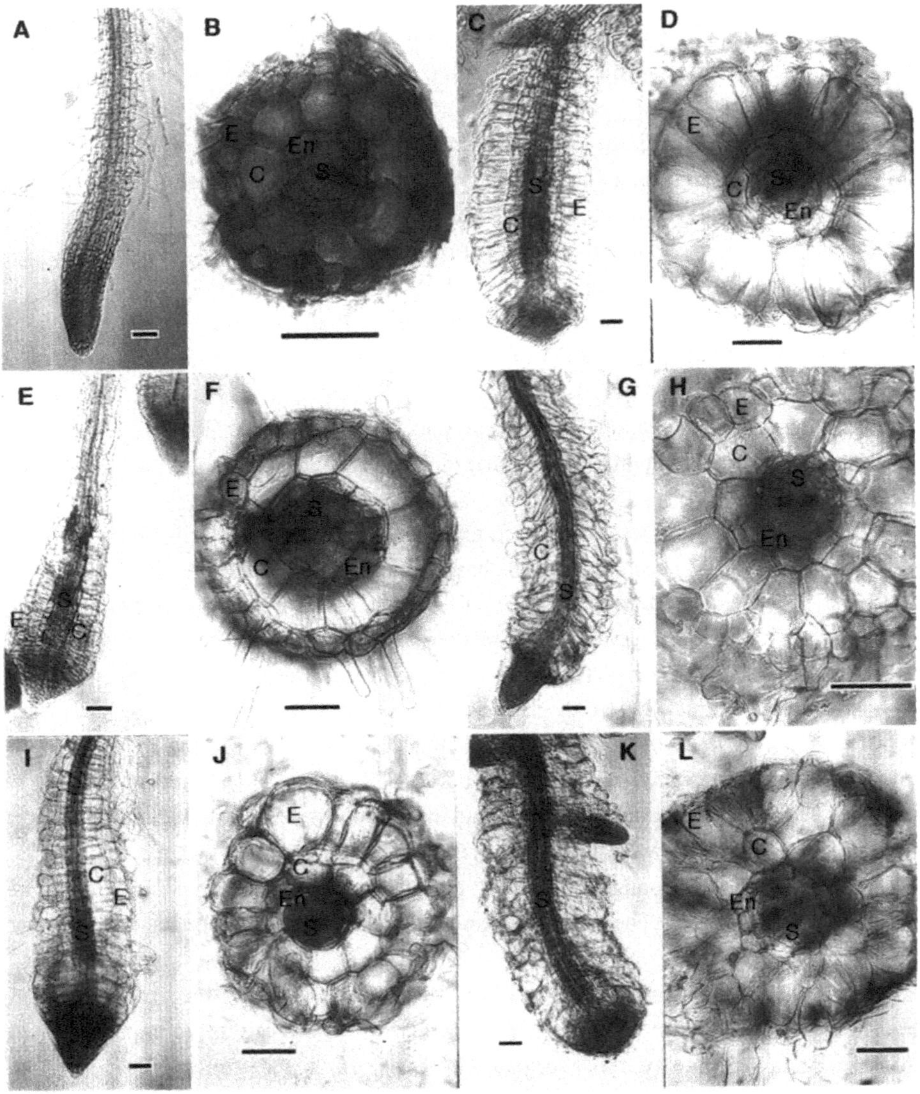

Figure 1: Root morphology of wild type and the expansion mutants.
A.) Whole mount of a wild type (Col) root tip. **B.**) Fresh transverse section of a wild type (Col) root. **C.**) Whole mount of a *cobra* root. **D.**) Fresh transverse section of a *cobra* root. **E.**) Whole mount of a *lion's tail* root. **F.**) Fresh transverse section of a *lion's tail* root. **G.**) Whole mount of a *sabre* root. H.) Fresh transverse section of a *sabre* root. **I.**) Whole mount of a one week old *pom pom* root. **J.**) Fresh transverse section of a one week old *pom pom* root. **K.**) Whole mount of a three week *pom pom* root. **L.**) Fresh transverse section of a three week old *pom pom* root.
Bar = 50 µm. Abbreviations: C, cortex; E, epidermis; En, endodermis; S, stele.

The cross-sectional areas of the epidermal cells in *pom pom* and *cobra* were approximately the same and 15 times larger than wild type (Fig. 2A). Also the cross-sectional areas of the endodermis and the stele tissue were very similar for these two mutants. But in *cobra* the cross-sectional area of the cortex cells was approximately 2.5 times larger than wild type whereas in *pom pom* this area was 5.5 times larger. Therefore the cortex cells of *pom pom* appeared more radially expanded than those of *cobra*.

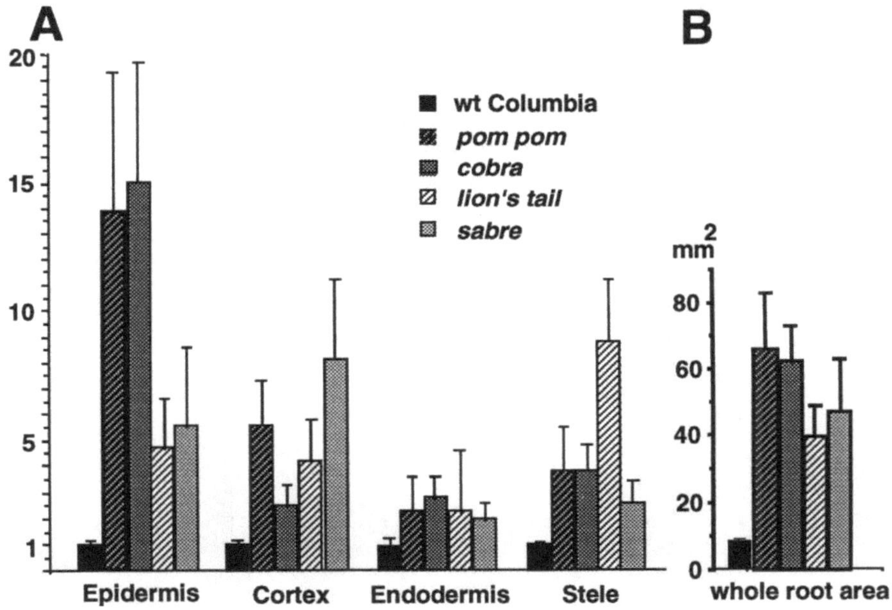

Figure 2: Comparison of cross-sectional root cell areas between *Arabidopsis* wild type (ecotype Columbia) and the expansion mutants.

A.) The cross-sectional areas of individual cells were calculated from digitized images of 5 - 7 fresh sections. The cross-sectional area of each root tissue of wild type (ecotype Columbia) was normalized to one and compared with the cross-sectional area of the same tissue of the four expansion mutants *cobra, lion's tail, pom pom* and *sabre*.
B.) The mean and standard deviation of whole root cross-sectional areas were calculated from digitized images of 5 - 7 fresh sections of wild type (ecotype Columbia) and the expansion mutants *cobra, lion's tail, pom pom, sabre*.

A comparison between *pom pom* and *lion's tail* cross-sectional areas revealed that the epidermis cells, the cortex cells and the stele cells differed significantly between these two mutants. The degree of cell expansion between *pom pom* and *sabre* differed significantly in the epidermis, the cortex and the stele tissue (Fig. 2A). The cell type with the least difference in cross-sectional area for all the mutants was the endodermis (Fig. 2A).

Conditional phenotype of root expansion mutants

Initial observations suggested that the roots of some of the expansion mutants exhibited different responses when grown under varying growth conditions (Benfey *et al.*, 1993). In order to characterize this response, we grew the plants under conditions that changed the root growth rate. In wild-type plants the root growth rate increased with increasing sucrose concentration to a maximum at 4.5% (135 mM) sucrose and then decreased slitghtly as the sucrose concentration was increased to 6% (Fig. 3).

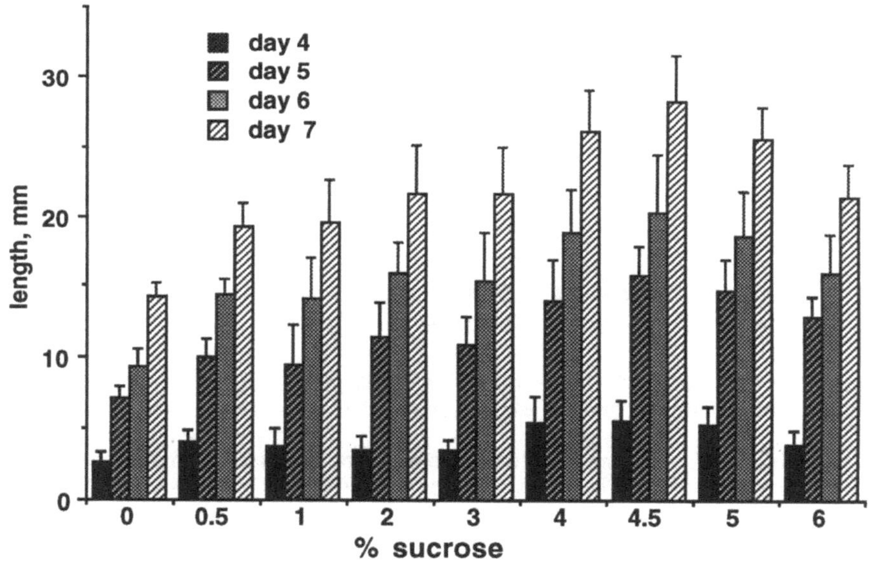

Figure 3: Analysis of root growth of wild type *Arabidopsis* (ecotype Columbia) on nutrient agar medium containing increasing sucrose concentrations.
Mean root length (mm) and standard deviation at different times after germination were plotted as a function of the sucrose concentration in the media.

Interestingly, the response to improved growth conditions differed among the mutants. *Sabre* exhibited no detectable change in root expansion when grown on media with different sucrose concentrations whereas the phenotype of *lion's tail* and *cobra* was observed only at higher sucrose concentrations (Table 1, Benfey *et al.*, 1993). The phenotype of *pom pom* was observed initially at a sucrose concentration of 1% (Table 1). Whereas 100% of *cobra* and *lion's tail* expressed the expansion phenotype on 2% sucrose, the expansion phenotype of *pom pom* was observed in 100% of the plants only on 3% sucrose (Table 1).

Table 1. Response of expansion mutants to sucrose

Percent sucrose in nutrient agar	Percent mutants with expanded roots			
	pom pom	*cobra*	*lion's tail*	*sabre*
0	0	0	0	100
0.5	0	0	11	100
1	11	21	62	100
2	52	100	100	100
3	100	100	100	100
4	100	100	100	100
4.5	100	100	100	100
5	100	100	100	100
6	100	100	100	100

We measured the growth of the primary root of the four expansion mutants. All four mutants showed drastically reduced root length compared to wild type (Fig. 4A, B, C, D). Five days after germination, the wild type root (ecotype Columbia) measured approximately 19 mm whereas the root length of *lion's tail* was only 2.7 mm and that of *cobra* 2.2 mm. In addition, the wild type root (ecotype Wassilewskija) measured approximately 9 mm at five days after germination whereas *sabre* measured only 2 mm and *pom pom* 3.3 mm. Under these conditions the growth rate of the wild type root (ecotype Col) was constant at approximately 6 mm per day and for the ecotype Ws at approximately 3 mm per day. *Cobra* had the lowest root growth rate of 0.2 mm per day followed by *lion's tail* with 0.4 mm per day. *Sabre* had a growth rate of approximately 0.45 mm per day and *pom pom* approximately 1 mm per day.

In all four mutants the extent of lateral root initiation appeared to be altered when compared to wild type. Whereas under standard conditions (see methods) lateral root initiation of wild type roots usually started between the sixth and seventh day after germination the lateral root initiation of *cobra*, *lion's tail* and *pom pom* appeared between the fourth and fifth day after germination. The first lateral root of *sabre* was observed on the eighth day after germination under standard conditions.

When grown in soil, *pom pom* had smaller leaves and was slightly stunted as compared to wild type.

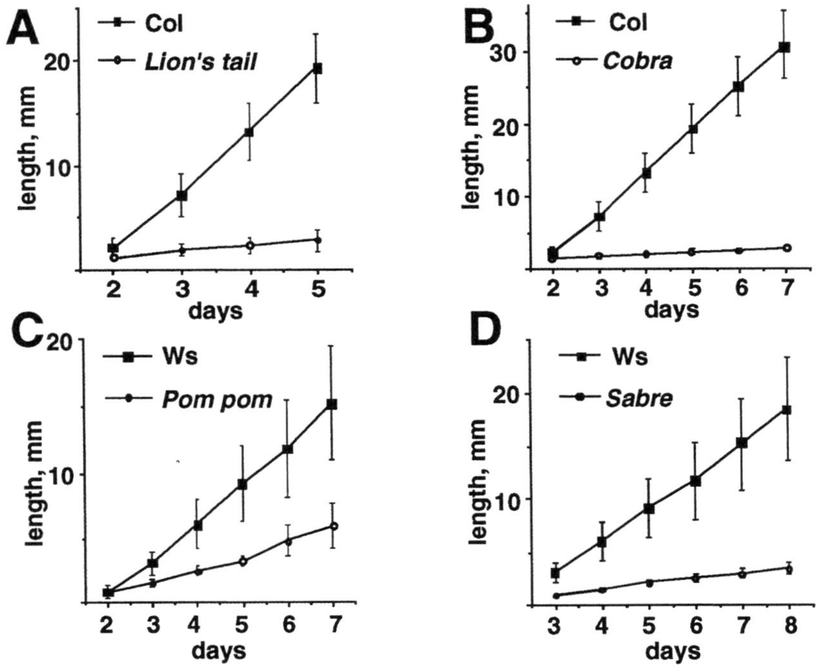

Figure 4: Root growth of wild type ecotype Columbia (Col **A**, **B**) and ecotype Wassilewskija (Ws **C**, **D**) and the expansion mutants *lion's tail* (**A**), *cobra* (**B**), *pom pom* (**C**) and *sabre* (**D**).

In conclusion, the four expansion mutants differ in their response to growth conditions and in the degree and direction of cell expansion of their root tissues.

Genetic Analysis

The genetic characterization of *pom pom* was initiated by crossing it to wild type. All plants of the F1 progeny had a wild type phenotype. This indicated that this mutation was recessive to wild type. The F1 plants were allowed to self-fertilize and segregation of the phenotypes in the F2 generation was scored. *Pom pom* segregated in a ratio that was statistically indistinguishable from 3:1 (Table 2). As a result of these data, we conclude that the phenotype of *pom pom* results from a recessive mutation at a single locus.

Table 2. Segregation analysis in the F2 generation of *pom pom* crossed to wild type.

N= total number of plants observed. Obs.= number of plants with wild type or mutant phenotype. Exp.= number expected given a 3:1 ratio of segregation. The X^2 value indicates no significant deviation form the expected ratios.

Phenotype	N	wild type obs.	wild type exp.	mutant obs.	mutant exp.	X^2
pom pom	785	557	588.75	208	196.2	2.42

SUMMARY

Cell expansion is one of the factors which influences morphogenesis in plants (Steeves and Sussex, 1989). We previously have characterized two recessive and one semidominant nonallelic mutants that exhibited abnormal cell expansion properties (Benfey *et al.*, 1993). In this report we have characterized a new recessive root expansion mutant, *pom pom*. The phenotype of *pom pom* was conditional upon growth conditions that induced higher growth rates of wild type roots. Under conditions that reduced the root growth rate such as cultivation in soil, the root of *cobra*, *lion's tail* and *pom pom* failed to develop the mutant phenotype. Other conditions which reduced the root growth rate, such as cold treatment or exogenous hormone supply also decreased root expansion of *cobra* and *lion's tail* (Benfey *et al.*, 1993). These results suggest that *pom pom* is a new member of the class of conditional root expansion mutants that includes *cobra* and *lion's tail*.
Interestingly, *pom pom* is the first expansion mutant that we have characterized in which radial expansion occurs primarily in two root tissues, the epidermis and the cortex. The radial expansion of the epidermis differed dramatically depending on the age of the root whereas the cortex cells exhibited a more constant radial expansion. As previously described the radial expansion was proportionally greatest in the epidermis of *cobra*, in the stele of *lion's tail* and in the cortex of *sabre* (Benfey *et al.*, 1993).
Also the direction of expansion differed in the different mutants as microscopic analysis of the whole mounts revealed. In *cobra* and *pom pom* the expansion was radial and perpendicular to the root axis whereas the elongation of the cells is drastically reduced. The direction of the expansion in the stele cells of *lion's tail* couldn't be determined with the methods used. But the cells in the surrounding tissues of this mutant appeared cuboidal rather than cylindrical. The expansion of the

cortex of *sabre* is also primarily radial but the cortex cells appeared to lie at an acute angle to the root axis.

One explanation for the shorter root of these mutants could be that there is a switch in the direction of cell elongation from parallel to perpendicular to the root axis.

We have isolated a new mutant which together with the previously described root expansion mutants (Benfey *et al.*, 1993) indicate that there is a degree of tissue specificity in the regulation of cell expansion and that cell expansion can be influenced by conditions which alter the growth rate. These mutants should allow us to identify genes that regulate the tissue specific expansion processes as well as to study the coordination of expansion among the root tissues.

Acknowledgments

We thank L. Di Laurenzio and R. A. Aeschbacher for careful reading of the manuscript. We thank K. Schultheiss for expert technical assistance with many aspects of this work. Special thanks to Carolyn Schultz for her help with the computer programs. M-Th. H was supported by a Schrödinger fellowship from "Fonds zur Förderung der wissenschaftlichen Forschung" (J0676-MOB). The work in PNB's laboratory was supported by a grant (GM43778) from the NIH.

REFERENCES

Baskin TI, Betzner AS, Hoggart R, Cork A and Williamson RE (1992) Root Morphology Mutants in *Arabidopsis thaliana.* Aust J Plant Physiol 19:427-438

Benfey PN, Linstead PJ, Roberts K, Schiefelbein JW, Hauser M-Th and Aeschbacher RA (1993) Root development in *Arabidopsis*: Four Mutants with dramatically altered root morphogensis. Development in press

Cosgrove DJ and Knievel DP (1987) Physiology of cell expansion during plant growth. American Society of Plant Physiologists, Rockville, Maryland

Esau K (1977) Anatomy of Seed Plants. New York: John Wiley & Sons

Dolan L, Janmaat K, Willemsen V, Linstead PJ, Poethig S, Roberts K. and Scheres B. (1993) The cellular and developmental organization of the *Arabidosis* root. Development in press

Feldmann KA (1991) T-DNA insertion mutagenesis of *Arabidopsis*: Mutational spectrum. Plant J 1:71-82

Steeves TA and Sussex IM (1989) Pattern in plant development. Cambridge: Cambridge University Press

Taiz L (1984) Plant cell expansion: regulation of cell wall mechanical properties. Ann Rev Plant Phys 35:585-657

Williamson R.E. (1991) Orientation of Cortical Mircotubules in interphase Plant cells. Int Rev Cytol 129:135-206

ANALYSIS OF ROOT DEVELOPMENT IN ARABIDOPSIS THALIANA

Ben Scheres[1*], Viola Willemsen[1], Kees Janmaat[1], Harald Wolkenfelt[1], Liam Dolan[2] and Peter Weisbeek[1].

1: Department of Molecular Cell Biology, University of Utrecht, Padualaan 8, 3584 CH The Netherlands.
2: Department of Cell Biology, John Innes Institute, Colney Lane, Norwich, United Kingdom.
*: Corresponding author.

Introduction

Mature plants are the result of two formally separable developmental processes. Embryogenesis provides a mature embryo with axis, cotyledons, and the root and shoot apical meristem. The embryo has a structure of limited complexity which is relatively uniform in flowering plants. Upon seed germination post-embryonic development starts. The primary root and shoot meristems, together with newly formed secondary meristems, elaborate the architecture of the mature plant. We are interested in finding out how meristematic cells in the embryo are programmed to form organs with appropriately spaced tissues and cell types.

Meristem formation can be considered as a special case of cell specification according to position. Programming cells for a specific role in accordance with their position is a central issue in development of all multicellular organisms. The spatial organisation of specified cells, organised in tissues, is heritable and thus necessarily under some form of genetic control. The term 'pattern formation' refers to the ontogeny of this spatial arrangement of cells. Identification of genes involved in pattern formation in Drosophila has been an essential step towards our general understanding of regional specification mechanisms (e.g. Ingham, 1988). The applicability of genetic analysis to plant development is demonstrated by the progress in studies on flower development, initiated by the description of floral homeotic mutants (reviewed by Meyerowitz & Coen, 1991). On these grounds we also favor the genetic approach to analyse cellular specification within meristems.

How is pattern formation at the cellular level achieved? The general picture that emerges from studies on Drosophila embryogenesis is one of a coarse regional specification by concentration gradients of gene products, mostly transcription factors. Combinatorial codes of these products define regions in the embryo (St Johnston & Nüsslein-Volhard, 1992; Gaul and Jäckle, 1989; Hülskamp et al, 1990). This definition of spatial regions is subsequently refined to ultimately yield cellular specification at appropriate positions. An important role for cell-cell signalling in the final stages of cellular specification is not only evident from studying Drosophila embryogenesis (Ingham and Martinez Arias, 1992), but also from studies on Drosophila eye development and vulva formation in Caenorhabditis elegans (Rubin, 1991; Horvitz & Sternberg, 1991). The concept of coarse positional information which is gradually refined to the level of cellular interactions may well be of general importance.

NATO ASI Series, Vol. H 81
Plant Molecular Biology
Edited by G. Coruzzi and P. Puigdomènech

In plants, genes involved in regional specification have also been identified. During Arabidopsis embryogenesis a set of zygotically active genes are postulated to define the main regions of the embryo, e.g. shoot meristem, cotyledons, hypocotyl, and root (Mayer et al, 1991). At the post-embryonic stage of plant development, genes involved in primordium specification in flowers have been identified and analyzed at the molecular level (reviewed by Meyerowitz & Coen, 1991). Noteworthy combinatorial codes are proposed to be important for regional specification in both these studies, analogous to the mechanisms unraveled in Drosophila. Genes which regulate pattern formation at the cellular level have not been analysed systematically in plants. It is our aim to identify such genes, involved in programming meristematic cells.

Root meristems produce cell files which can be traced back to meristem initial cells. This simplifies the explanation of an observed pattern defect in terms of altered activity of meristematic cells. On the other hand, clonal analysis of the Arabidopsis shoot meristem indicated that only a loose association between meristematic cells and differentiated cells exists (Irish & Sussex, 1992; Furner & Pumfrey, 1992). Hence pattern defects in tissues derived from the shoot meristem will not always be relatable to defects in specific meristematic cells. In order to be able to take our study to the cellular level we therefore chose to analyse the root meristem of Arabidopsis, a plant well suited for genetic analysis (Meyerowitz, 1989). In this paper we will first describe the anatomy of the Arabidopsis root including the arrangement of the meristematic initial cells. Subsequently the embryonic origin of the root and the proposed meristematic initials will be elaborated on, referring to both anatomical studies and clonal analysis data. Finally we will present the results of a genetic screen designed to pinpoint genes involved in pattern formation within the root meristem. We will discuss these results within the conceptual framework of gradually refined positional information as sketched above.

Descriptive analysis of root development

Exact knowledge on the position of specific cell types and their clonal origin greatly adds to the interpretation of mutant phenotypes, and is an indispensable prerequisite for detailed genetic analysis of pattern formation at the cellular level. This is impressively demonstrated by the follow-up research leaning on the description of the Caenorhabditis elegans cell lineage (Sulston et al, 1983; Sternberg and Horvitz, 1984; Horvitz & Sternberg, 1991).

Meristematic cells in the ARABIDOPSIS root.

Anatomical analysis reveals a simple and constant organization of the mature Arabidopsis root. Only single layers of each of the cell types outside the central cylinder are present, and the numbers of cells in these layers and inside the central cylinder are relatively invariant. On the outside of the root one layer of epidermal cells is present comprising on average 19 cells. Depending on the distance from the tip one or more lateral root cap layers (32 cells on average) can be distinguished, overlying the epidermal cell layer. Epidermal cells can be divided in root-hair forming cells (trichoblasts) and non-root hair forming cells which are arranged in a regular and predictable fashion. To the inside of the epidermis one layer of 8 cortex cells and one layer of 8 endodermis cells is present. Each cortex cell is neighbouring one endodermal cell. Underneath the endodermis the outer layer of the central cylinder, consisting of on average 12 pericycle cells, is situated. The remaining cells form the central cylinder and contain the vascular cells which show a diarch organisation with two protoxylem and two protophloem poles.

The origin of cell files that form the mature root tissue can be inferred from median longitudinal sections of root tips. The Arabidopsis root contains separate tiers of cells from which the different tissue layers originate, to be called 'initials' hereafter. Such an organization of the root meristem is referred to as the 'closed meristem' type (Clowes, 1981). Epidermis and lateral root cap share common initial cells. Cortex and endodermis also share initial cells. A prepatterned set of initial cells is present for cells inside the vascular cylinder. At the distal end of the root each of the twelve cell files of the columella part of the root cap has its own initial. Four 'central' cells are laterally continuous with the cortex and endodermis cells. These cells do not need to divide to preserve the observed structural integrity of the root. In vivo ^{3}H-Thymidine incorporation studies indicate that these cells have a much slower cell cycle than the surrounding cells, and therefore we regard them as 'quiescent centre' cells (Clowes, 1956). Noteworthy, these cells are facing all the cells which we propose to act as initials for the different root tissues.

Anatomical arguments alone can not prove that the cells we propose to be initials truly participate in pattern formation. However, ^{3}H-Thymidine incorporation was clearly observed in all these cells, indicating that they divide actively. In addition, mitotic figures can be found in the cells predicted to be initials. Derivatives of the initial cells in the meristematic zone appear to have a more rapid cell cycle time than the initials themselves. The derivatives of the columella initials are an exception in that they do not divide again.

Transverse sections of several root tip regions containing the initial cells invariantly depict 16 epidermal initials, 8 cortex/endodermal initials, 9-10 pericycle initials, and 12 columella initials. We take this as an indication that the relative invariance in cell numbers of the mature root finds its origin in even more stringently fixed numbers of initials.

In summary, the developmental history of the Arabidopsis root can be traced back to the activity of a fixed number of initials surrounding a small quiescent centre. It must be noted that an eventual ability of the cells in the quiescent centre to participate in pattern formation under certain specific conditions cannot be ruled out by our experiments. Such activity has been proposed for quiescent centre cells and a comparison to animal stem cells has been made (Barlow, 1978). We have observed alterations in cell shape and division planes in the meristematic area in ageing roots. Nevertheless we have not been able so far to obtain clear evidence for division of the central cells.

Embryonic origin of the root meristem initials

The tiered organisation of the Arabidopsis root meristem initials puts forth the question to the origin of these cells. Transverse sections from the root tip of imbibed seeds prior to germination demonstrate that the complete setting of rings and plates of initials, surrounding the four central quiescent cells, is alreadly apparent in the mature embryo. It is therefore not established during post-embryogenic development from a simpler version of the root meristem. The numbers of different initials in the mature embryo equal the numbers found in the active root meristem of the seedling. In order to investigate how these initials, which constitute the root meristematic core, come about during embryogenesis we have performed anatomical studies of embryos at successive developmental stages. The development of the root apex has been described in several Brassicaceae including Arabidopsis (von Guttenberg, 1947; Tyrkaska, 1979; Mansfield & Briarty, 1991). We have extended these studies to understand the exact patterning of root meristem initials during embryogenesis.

Median longitudinal sections at the mature embryo stage reveal an organisation of the root meristem very similar to what is observed in seedlings. However, fewer layers of both columella and lateral root cap are present. The same coordinate divisions as proposed for the post-embryonic activity of the root meristem initials can explain the ontogeny of the embryonic root (radicle). Sections of earlier stages of embryogenesis are consistent with the view that this sequence of divisions first starts at the late heart stage, and is anatomically marked by the appearance of a periclinal division in epidermal cells that give rise to the first layer of lateral root cap in the embryo. In this view the structure of initials, surrounding the quiescent centre cells, is defined in mid-heart stage embryos just before this characteristic division is apparent.

A next question which is important is the clonal origin of the different initials that appear specified at mid-heart stage. Sections of globular stage embryos indicate that the central 'quiescent' cells, and the columella initials are derivatives of the hypophyseal suspensor cell. The hypophysis in turn originates from divisions of the basal cell after the first zygotic cleavage (for a description and definition of embryonic stages see Jürgens and Mayer, 1993). In contrast, all the other initial cells derive from the apical cell that results from the first zygotic cleavage. Henceforth initials of the root meristem appear to have different clonal origin.

Analysis of the numbers of cells in the different tiers of initials shows that the number of epidermal and cortical initials is already established at the globular stage of embryogenesis. The numerical invariance of these sets of initials therefore appears to be a consequence of the rigid control of cell division at that stage. Cell numbers in the vascular cylinder are established much later in embryogenesis. Nevertheless the pattern of vascular initials, like the pattern of epidermis and cortex, is a result of radial pattern formation in the complete embryonic axis which will give rise to both hypocotyl and root.

The embryonic root: not entirely derived from meristem initials?

In the mature embryo the region of the root which is formed by the activity of the meristem initials is marked by the presence of the lateral root cap. On the other hand the hypocotyl region of the mature embryo is marked by the presence of a second cortex cell layer, not present in the root. Furthermore the hypocotyl region contains more epidermal cells in circumference.

Noteworthy, a region in the mature embryo in between the meristematic root and the hypocotyl is characterized by the presence of either one cortex cell layer, or a second, inner, cortex cell layer that consists of wedge-shaped cells. This layer is not derived from the meristematic initials as no lateral root cap is present. At the late heart stage this 'intermediate' layer can be assumed as the region with the height of about three epidermal cell tiers immediately above the initials which already gave rise to the lateral root cap.

At the seedling stage the connection between root and hypocotyl reveals a region which shares root and hypocotyl characteristics. The inner cortex cells in this region are wedge-shaped and the number of epidermal cells in circumference increases more proximal to the root tip. Therefore this region acquires more 'hypocotyl characteristics' when the boundary with the true hypocotyl, which has a uniform radius, is approached. However, the presence of root hairs defines the region as part of the root in terms of cellular differentiation. This zone, which we call 'interzone' hereafter, can be presumed to arise from the previously mentioned intermediate layer in the embryo. In summary, anatomical data suggest that the root meristem initials are not the only progenitors of the root.

Clonal analysis of the ARABIDOPSIS root using transposon excision

In the previous paragraph two embryonic regions involved in root formation were distinguished. One which gives rise to the tier of initials derived from the apical zygotic cell. A second layer of cells, defined at the late heart stage, was proposed to form a region of the root not elaborated by the initials. This assumption cannot be proven with anatomical studies alone. An alternative scenario may be proposed in which cortical cells within the intermediate embryonic layer undergo further periclinal divisions. In this way the complete intermediate region could be incorporated in the 'true' hypocotyl of the seedling. Such a model predicts that the complete root derives from the meristematic initials. In order to test which of these alternative developmental scenarios is happening, we performed clonal analysis.

For this purpose transgenic plants, kindly provided by Emily Lawson and Carolyn Dean (John Innes Inst., U.K.), were used. These plants contain a T-DNA insert containing a 35S CaMV promoter, a maize Ac element, and ß-glucuronidase (GUS) coding sequences. Upon excision of the Ac element in embryonic cells, clones of descendant cells can be formed in which GUS activity can be histologically detected. Embryonic clonal sectors were sampled from appr. 100,000 Xgluc-stained seedlings derived from a transgenic line with moderate excision frequency. This particular line was chosen because excision frequency was high enough to collect embryonic sectors, and low enough to avoid frequent sector overlap.

Sectors spanning the complete root, interzone, hypocotyl, or a combination of these regions were analyzed. All sectors spanning the complete hypocotyl ended at a sharp lower boundary where the hypocotyl was ending and the root hair containing epidermis cells became first evident ('lower' is referring to positions closest to the root apex, and 'upper' to positions closest to the shoot apex). Sectors spanning the complete root were found and they had their upper boundaries at the same position mentioned above where the first root hairs become apparent. We took this as an indication that root and hypocotyl arise from two embryonic tiers in which at least epidermal cells, after the separation of these tiers, have very limited potential to switch root or hypocotyl fate. The next-largest root sectors found with similar boundaries fall in two classes: interzone sectors in which the lower sector boundary coincides approximately with the region in the root where the interzone-specific second cortex layer disappears; and root sectors originating from the meristem and having their upper sector boundary in the same region.
Currently we are saturating this data set. So far, the sector analysis is consistent with the proposition that three distinct embryonic tiers give rise to hypocotyl, interzone, and initials. In this view the interzone derives from the 'intermediate' cell tiers apparent in the embryo. The remaining basal portion of the root in the seedling appears to derive from the meristematic initials.

Identifying 'root meristem formation genes': rationale

From the anatomical and clonal analyses on the Arabidopsis root we draw the following main conclusions. First, root meristem initials appear specified at the late heart stage of embryogenesis. Second, the radial pattern of root meristem initials, and henceforth the radial anatomy of the root, is a reflection of the divisions which establish the radial arrangement of cells in the complete embryonic axis. Third, the Arabidopsis root consists of a region which is derived from the meristematic initials, and an interzone region at the basis of the hypocotyl which exhibitis root-specific differentiation but appears to derive from a seperate embryonic cell layer.

These three features have consequences for mutant phenotypes which are expected to define genes involved in the specification of the root meristematic cells. First, defective specification of the meristematic initials is expected to yield deviations in embryo development from late heart stage onward. Second, radial pattern mutations evident in the root can be expected to have an embryonic origin. Third, mutations affecting root meristem formation may not be completely rootless, since the interzone can differentiate to root specific cell types.

Bearing these consequences in mind we performed a screen for mutants defective in root meristem formation. Since the majority of the mutantions we were interested in would be expected to yield embryonic phenotypes we anticipated that many mutants could be seedling-lethal. Therefore we performed a screen using single siliques from 10,000 separate F1 families, to be tested for root meristem mutations in the F2 generation. In this way heterozygous plants carrying mutant alleles can be rescued (Jürgens et al, 1991). From this screen we collected all families that segregated phenotypes with an affected root. We reasoned that we could not on forehand anticipate the severity of phenotypes associated with meristematic defects. This bears on the simple radial anatomy of the root, which is an advantage in interpreting the cause of a particular phenotype, but a disadvantage in finding phenotypes. Therefore we considered every family showing a root defect for a somewhat more detailed analysis using 'whole mount' preparations of cleared roots.

Mutations perturbing root meristem development

141 F1 lines from our screen segregated mutants in a Mendelian fashion with abnormalities in root growth or development throughout two subsequent generations. These lines were grouped in classes ranging from severe to mild distortions of root development. Below we will briefly describe three classes of mutants which we believe to be significant in the light of the description of wildtype development as discussed above. These classes of mutants are not covering all different root mutants which we have detected and which are currently under analysis. For example, radial pattern mutants and shape mutants have been found but they often exhibit pleiotropic effects which hampers interpretation of the phenotype.

'Early' embryonic pattern mutants

Seedlings with severely affected development of the basal region, and hence the root, have been found in our screen and they fully fit the phenotypic description of plants carrying gnom and monopteros alleles, described by Mayer et al (1991). The majority of gnom seedlings lack differentiation of the root. Mutation in the gnom gene have been shown to already disturb the first assymetric division of the zygote, but may also be involved in assymetric divisions defining the hypophyseal suspensor cell (Mayer et al, 1993). Mutations in the monopteros gene can lead to the complete absence of root and hypocotyl. The phenotype can be traced back to the octant stage of embryogenesis where supernumerary cell tiers are present in the embryo proper (Berleth & Jürgens, in press). The cell in monopteros mutant embryos at the position of the hypophyseal cell in wildtype embryos appears to be only able to perform transverse divisions.
In conclusion, both gnom and monopteros genes are involved in defining the basal region of the embryo at early stages of development. Their function is a prerequisite for setting up the conditions for correct development of the root meristem.

'Heart stage' embryonic pattern mutants

We identified several lines segregating mutant phenotypes that lack the characteristic anatomical features of the root meristematic area. Noteworthy, several tiers of cells at the basal end of the hypocotyl of mutant seedlings bear root hairs and show the decrease in radial diameter characteristic for the interzone. Therefore the embryonic root appears to have developed normally.

An example of this class of mutants is hobbit. Using Nomarsky optics on whole mounts of hobbit embryos deviations from the typical division sequences of the initials can clearly be observed at the heart stage. At that stage no clear alterations in the apical part of the embryo can be observed. At seedling stage, however, it becomes evident that the presence of the mutation leads to smaller seedlings with altered cotyledon shape. The shoot meristem produces leaves but we have not been able so far to induce flowering. In contrast, another member of this class lacking of a patterned root meristem, orc, exhibits no shoot meristem activity.

Meristem activation mutants

Mutants of this class produce very short roots and normal, mostly fertile shoots. This indicates very limited or no activity of root meristematic cells. The root tip contains the typical setting of initials surrounding a quiescent centre. However, division patterns of initial cells which apparently could be performed during embryogenesis appear are not perpetuated at the post-embryonic stage. An example of this class of mutants is elf. The primary root remains inactive while the hypocotyl elongates normally. Normal sized cotyledons are present at seedling stage, and a morphologically normal albeit sterile shoot it made. After several weeks on sucrose containing medium, lateral roots are formed with normal appearance. This indicates that the lack of meristem activation is specific for the primary meristem. In contrast, other members of this class initiate lateral roots which are, like the primary root, unable to perpetuate meristematic activity.

Concluding remarks

Anatomical analysis of the Arabidopsis root reveals an elegantly simple and highly ordered arrangement of different cell types. Meristematic initials can be identified, and their active participation in pattern formation is inferred from in vivo ^{3}H-Thymidine labeling experiments. A 'core meristem' has been pinpointed, containing a small centre of quiescent cells surrounded by initials for all the different root tissues, and the embryonic origin of this core meristem is elucidated. This description of root development at the cellular level seems to make the Arabidopsis root ideally suited for genetic dissection of root meristem formation and function. One can question whether the genetic determinants of root development of the tiered meristem type, as in Arabidopsis, are significant for understanding the activity of untiered, 'open' root meristems (Clowes, 1981). Initials in open meristems face developmental decisions leading to more different cell types. However, epidermal initials in the Arabidopsis root have to produce sister, epidermis and lateral root cap cells. Likewise, cortex initials have to produce sister, endodermal and cortical cells. We can envisage molecular mechanisms underlying these developmental decisions identical to the ones that direct initials of untiered meristems to produce more different cell types.
Furthermore one can wonder whether understanding the formation of a root with a rather strict lineage-dependent meristematic activity has any relevance for understanding shoot development. In this context it is worthwile to mention that the development of the worm C. elegans, a nematode with invariant cell lineages, involves many cellular interactions. Likewise, the development of the Arabidopsis root may be continuously relying on positional cues to a greater extent than expected from the cell lineage relations (Barlow, 1984). If this is the case overlapping mechanisms for root and shoot meristem activity may exist, since for shoot development it is clear that positional information is important (e.g. Sussex, 1989).

The classes of mutations that we have found, clearly show that formation of a functional root meristem can be blocked at several stages. We envisage a gradual process in which first the basal part of the seedling is defined, depending on the products of genes like gnom and monopteros. This specification can then be refined by products of genes like hobbit, ultimately leading to cellular delimitation of the cells of the core meristem. Finally, upon germination, genes defined by the meristem activation mutations enable perpetuation of cell division by the initials and their derivatives.

As in the case of Drosophila, these propositions will have to be substantiated by both genetic and molecular analysis. Extending the analogy to animal systems, it will be interesting to see whether we have to envisage the broad specification genes as transcription factors, and the ones involved in cellular specification as products involved in communication between cells.

During the spatial definition of the root meristem in the apical-basal axis, the radial pattern must also be defined. We have shown that the radial pattern, set up early in embryogenesis, defines the setting and the numbers of root meristem initials, and hence is directly related to the radial anatomy of the root. Therefore we envisage the radial anatomy of the root as an embryonic property. In a collaboration with L. Di Laurenzio and P. Benfey (New York University) we showed that the shortroot mutation, which lacks an endodermis in the root, results from a radial pattern defect during embryogenenis which also penetrates into the hypocotyl. This is in accordance with our view that mutant phenotypes with altered radial pattern in the root may generally be traced back to embryogenesis.

References

Barlow,P.W. (1978). The concept of the stem cell in the context of plant growth and development. In: stem cells and tissue homeostasis, B.I. Lord, C.S. Potten & R.J. Cole, eds. Cambridge University Press, 87-113.

Barlow,P.W. (1984). Positional controls in root development. In: positional controls in plant development. P.W. Barlow & D.J. Carr, eds. Cambridge University Press.

Berleth,T. and Jürgens,G. (1993). The role of the monopteros gene in organising the basal body region of the Arabidopsis embryo. Development 118 (in press).

Clowes,F.A.L. (1956). The cytogenerative centre in roots with broad columellas. New Phytol. 48-57.

Clowes,F.A.L. (1981). The difference between open and closed meristems. Ann. Bot. 48, 761-767.

Furner,I.J., and Pumfrey,J.E. (1992). Cell fate in the shoot apical meristem of Arabidopsis thaliana. Development 115, 755-764.

Gaul,U., and Jäckle,H. (1989). Analysis of maternal effect mutant combinations elucidates regulation and function of the overlap of hunchback and Krüppel gene expresssion in the Drosophila blastoderm embryo. Development 107, 651-662.

Guttenberg, von,H. (1947). Studien über die Entwicklung des Wurzelvegetationspunktes der Dikotyledonen. Planta 35, 360-396.

Horvitz,R.H. and Sternberg,P.W. (1991). Multiple intercellular signalling systems control the development of the Caenorhabditis elegans vulva. Nature 351, 535-541.

Hülskamp,M., Pfeifle,C., and Tautz,D. (1990). A morphogenetic gradient of hunchback protein organises the expression of the gap genes Krüppel and knirps in the early Drosophila embryo. Nature 346, 577-580.

Ingham,P.W. (1988). The molecular genetics of embryonic pattern formation in Drosophila. Nature 335, 25-34.

Ingham,P.W., and Martinez Arias,A. (1992). Boundaries and fields in early embryos. Cell 68, 221-235.

Irish,V.F., and Sussex,I.M. (1992). A fate map of the Arabidopsis embryonic shoot apical meristem. Development 115, 745-753.

Jürgens,G., Mayer,U., Torres Ruiz,R.A., Berleth,T., Miséra,S. (1991). Genetic analysis of pattern formation in the Arabidopsis embryo. Development 91 (Suppl. 1), 27-38.

Jürgens,G., and Mayer,U. (1993). Arabidopsis. In: a colour atlas of developming embryos. J. Bard (ed.), Wolfe Publishing, London.

Mansfield,S.G. and Briarty,L.G. (1991). Early embryogenesis in Arabidopsis thaliana. II. The developing embryo. Can. J. Bot. 69, 461-476.

Mayer,U., Torres Ruiz,R.A., Berleth,T., Misera,S., and Jürgens,G. (1991). Mutations affecting body organization in the Arabidopsis embryo. Nature 353, 402-407.

Mayer,U., Büttner,G. and Jürgens,G. (1993). Apical-basal pattern formation in the Arabidopsis embryo: studies on the role of the gnom gene. Development 117, 149-162.

Meyerowitz,E.M. (1989). Arabidopsis, a useful weed. Cell 56, 263-269.

Meyerowitz,E.M. and Coen,E.S. (1991). The war of the whorls: genetic interactions controlling flower development. Nature 353, 31-37.

Rubin,G.M. (1991). Signal transduction and the fate of the R7 photoreceptor in Drosophila. TIG 7, 372-377.

Sternberg,P.W., and Horvitz,H.R. (1984). The genetic control of cell lineage in nematode development. Ann. Rev. Genet. 18, 489-524.

St. Johnston,D. and Nüsslein-Volhard,C. (1992). The origin of pattern and polarity in the Drosophila embryo. Cell 68, 201-219.

Sulston,J.E., Shierenberg,E., White,J.G., and Thomsom,J.N. (1983). The embryonic cell lineage of the nematode Caenorhabditis elegans. Dev. Biol. 100, 64-119.

Sussex,I.M. (1989). Developmental programming of the shoot meristem. Cell 56, 225-229.

Tyrkarska,T. (1979). Rape embryogenesis. II. Development of the embryo proper. Acta Socoetatis Botanicorum Poloniae 48, 391-421.

GENES SPECIFYING FLORAL MERISTEM IDENTITY IN *ARABIDOPSIS*

Martin F. Yanofsky, Takashi Araki, Cindy Gustafson-Brown,
Sherry A. Kempin, M. Alejandra Mandel, and Beth Savidge

Department of Biology and Center for Molecular Genetics,
University of California at San Diego, La Jolla, CA 92093-0116

Abstract

Genetic studies indicate that the *Arabidopsis* floral homeotic gene *APETALA1* is one of several genes involved in the generation of floral meristems, the first step in flower development. Molecular analyses indicate that *APETALA1* RNA is expressed in young floral primordia and in sepals and petals, and that *APETALA1* encodes a transcription factor with a MADS-domain. These molecular studies, together with the *apetala1* mutant phenotype, suggest that *APETALA1* acts locally to specify the identity of the floral meristem, and to determine sepal and petal development. Further studies demonstrate that the organ identity gene *AGAMOUS* negatively regulates *APETALA1* RNA accumulation in the two inner whorls of wild-type flowers. Here we discuss the molecular characterization of the *APETALA1* gene and its possible interactions with other genes involved in the specification of floral meristem identity in *Arabidopsis*.

Wild-type flower development and specification of organ identity

Flowers originate as small clumps of cells called floral meristems on the flanks of inflorescence meristems. Although inflorescence meristems are capable of producing an indeterminate number of meristems in a spiral arrangement, floral meristems in contrast are determinate and produce organs arranged in four concentric whorls. *Arabidopsis* produces typical cruciferous flowers with four sepals in the first (outermost) whorl, followed by four petals, six stamens, and two fused carpels in the second, third and fourth whorls respectively (Figure 1). Genetic studies in *Arabidopsis thaliana* and the distantly related dicot species *Antirrhinum majus*, have led to the identification of early-acting genes that specify the identity of floral meristems, and later-acting genes that determine organ fate (For reviews, see Schwarz-Sommer et al., 1990; Coen and Meyerowitz, 1991).

Considerable attention has focused on the later-acting genes that specify organ fate. Detailed genetic analyses led to a model in which three distinct activities, designated A, B, and C, are proposed to act alone and in combination to determine the fate of floral organ primordia (Bowman et al., 1992, Coen et al., 1991, Coen and Meyerowitz, 1991). Each activity is proposed to function in two adjacent whorls (A: whorls 1 and 2; B: whorls 2 and 3;

NATO ASI Series, Vol. H 81
Plant Molecular Biology
Edited by G. Coruzzi and P. Puigdomènech

C: whorls 3 and 4) with A alone specifying sepals, C alone specifying carpels, and the combined activities of AB or BC specifying petals or stamens respectively (See figure 6). One additional feature of this model is that the A and C functions are mutually antagonistic, such that A prevents the activity of C in the outer two whorls, and C prevents the activity of A in the two inner whorls. In *Arabidopsis* the genes necessary for the A, B and C activities are *APETALA2* (*AP2*), *APETALA3* (*AP3*) and *PISTILLATA* (*PI*), and *AGAMOUS* (*AG*) respectively. Although it is clear that these genes are necessary for the A, B, and C activities, it is unclear if they are sufficient to fully provide these functions. For example, at least two genes are needed for the B function.

Figure 1. Photographs of wild-type (left), *ap1-1*(middle), and *ap1-1 cal-1* (right) flowers/inflorescences. Arrows indicate axillary flowers developing in the axils of first whorl bracts in *ap1-1* flowers.

The first of the floral homeotic genes to be isolated from *Arabidopsis* was *AGAMOUS*, and consistent with the genetic data, *AG* RNA was shown to accumulate specifically in stamens and carpels (Yanofsky et al., 1990). In *ap2* mutant flowers, *AG* RNA accumulates in all four whorls demonstrating that *AP2* negatively regulates *AG* RNA accumulation in the two outer whorls of wild-type flowers (Drews et al., 1991). These data confirm the predictions based on genetic observations that *AP2* prevents *AG* activity in the outer two whorls. In order to determine if *AG* is sufficient for the C function, transgenic tobacco and *Arabidopsis* plants were generated that express *AG* in all four whorls. Ectopic *AG* expression was indeed sufficient to convert the perianth (sepals and petals) into reproductive organs (carpels and stamens) (Mandel et al., 1992; Mizukami et al., 1992; Kempin et al., 1993).

The *AG* gene encodes a putative transcription factor since a portion of its deduced protein product shares similarity to transcription factors from mammals and yeast. This region of sequence similarity is part of a sequence specific DNA binding domain, referred to as the MADS domain, and is also present in the products of several recently isolated floral homeotic genes from *Antirrhinum* (Schwarz-Sommer et al., 1990). Reduced stringency hybridizations using the *AG* gene as probe allowed the isolation from *Arabidopsis* of a family of genes, *AGL1-AGL6* (AG Like) which also encode this putative DNA binding domain (Ma et al., 1991). We have recently extended these observations by isolating eight additional members of this *Arabidopsis* gene family, designated *AGL7-14* (Mandel et al., 1992b; Rounsley, Kempin, and Yanofsky, unpublished). In addition to *AG* and the *AGL1-14* genes, the homeotic genes *AP3* and *PI* are also members of this gene family (Jack et al., 1992) bringing the total to at least 17 related MADS-box genes in *Arabidopsis*.

Genes specifying floral meristem identity

As mentioned above, a great deal of progress has been made towards identifying and characterizing the genes that determine organ fate. Comparatively little is understood about the earlier-acting genes that determine the identity of the floral meristem itself. These early-acting genes are necessary for the generation of floral meristems from inflorescence meristems , and are thus good candidates for activators of the later-acting organ identity genes (Weigel et al., 1992; Weigel and Meyerowitz, 1993). At least four *Arabidopsis* genes have been shown to play important roles in determining floral meristem identity (Figure 2). One of these is *LEAFY (LFY)*, mutations in which lead to a partial transformation of flowers into inflorescence shoots (Weigel et al., 1992; Huala et al., 1992). This phenotype is

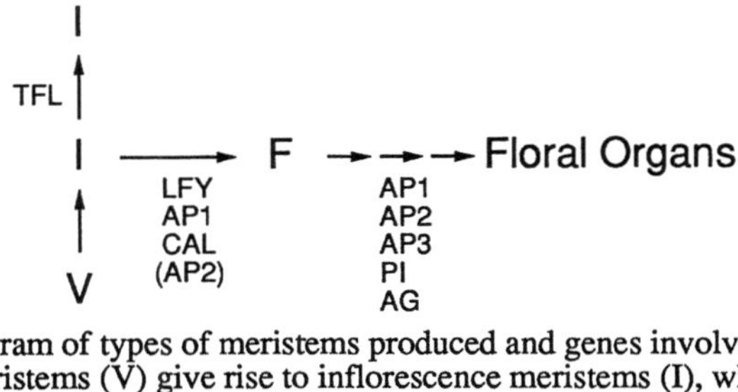

Figure 2. Diagram of types of meristems produced and genes involved.
Vegetative meristems (V) give rise to inflorescence meristems (I), which can continue to give rise to additional inflorescence meristems or instead to floral meristems (F). The activation of the organ identity genes within floral meristems leads to the specification of floral organ fates. The *TERMINAL FLOWER (TFL)* gene has been proposed to promote the maintenance of the indeterminate inflorescence meristem (Shannon et al., 1991; Alvarez et al., 1992).

significantly enhanced when *lfy* mutations are combined with mutations in another homeotic gene, *APETALA1* (*AP1*), suggesting that *LFY* and *AP1* act to determine the identity of floral meristems (Weigel et al., 1992).

Mutations in the *AP1* gene lead to alterations in the first and second whorl organs, whereas the stamens and carpels occupying the two inner whorls develop essentially as in wild-type (Figure 1). The second whorl organs generally fail to develop, and the first whorl organs develop as bracts. These first whorl bracts have axillary meristems which develop into secondary flowers that reiterate the phenotype of the primary *ap1* mutant flower, complete with higher order meristems developing in the axils of the first whorl bracts of internal flowers (Irish and Sussex, 1990; Bowman et al., 1993). The repeated development of axillary meristems has been interpreted as a partial transformation of floral meristems into inflorescence meristems (Weigel et al., 1992) (Figure3). A nearly complete transformation of floral meristems into inflorescence meristems (Figure 3) is observed when the *ap1* mutation is combined with mutations in another gene, *CAULIFLOWER* (*CAL*) (Bowman et al., 1993). *ap1 cal* double mutant plants continue to proliferate inflorescence meristems in place of floral meristems resulting in a structure that is virtually indistinguishable from the common cauliflower of the dinner table (Figure 1). The *AP2* gene, whose primary functions are to determine the identities of the outer two whorls of organs and to negatively regulate *AG*, also appears to play a minor role in determining floral meristem identity (Irish et al., 1990; Bowman et al., 1993).

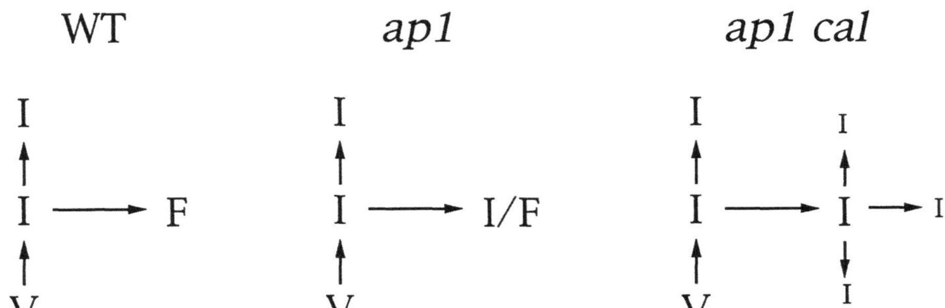

Figure 3. Types of wild-type and mutant meristems produced. The inflorescence meristem gives rise to floral meristems (F) in wild-type, to a partial conversion to floral meristems (I/F) in the *ap1* mutant, and to additional inflorescence meristems in the *ap1 cal* double mutant. These inflorescence meristems in the double mutant continue to give rise to additional inflorescence meristems.

Molecular cloning and characterization of the *AP1* gene

We have isolated the *AP1* gene and shown that it contains a MADS-box and thus likely encodes a transcription factor (Mandel et al., 1992b). RFLP mapping, rescue of the mutant phenotype in transgenic plants, and DNA sequence analysis of mutant alleles all support the conclusion that this is the *AP1* gene. The positions of introns within the *AP1* transcript are virtually identical to the intron positions for *AG*, although these related genes share very little sequence similarity outside their MADS-boxes (Figure 4).

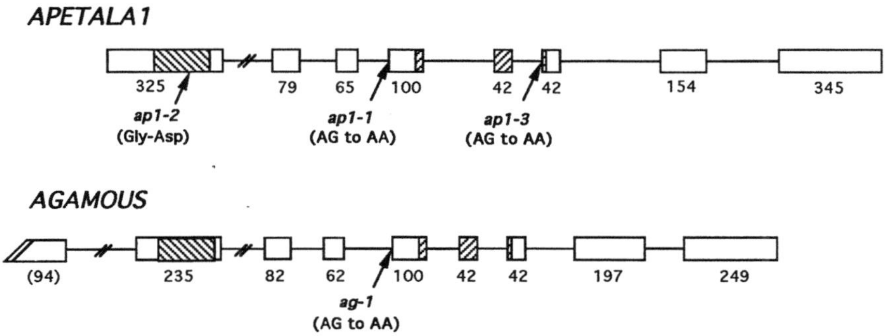

Figure 4. Intron/exon structure for *AP1* and *AG* genes. Boxes indicate exons and thin lines introns. The positions of several characterized mutant alleles are indicated. The nucleotide size of exons is indicated.

A large number of *ap1* mutant alleles have been identified and characterized, and these result in phenotypes ranging from severe to weak (Bowman et al., 1993). We have determined the molecular lesions in thirteen of these alleles in order to begin to define the functional domains of AP1 (Mandel et al., 1993). The most extensively characterized allele is *ap1-1*, which is suggested to be a loss of function allele since it is the most severe allele thus far described (Bowman et al., 1993). The *ap1-1* allele has a single nucleotide substitution at the splice acceptor site of the third intron (Mandel et al., 1992b) which leads to an altered transcript (Mandel et al., 1993) with the potential to encode a truncated product (Figure 4). However, the corresponding region of other MADS-domain factors is sufficient for DNA binding (Mueller and Nordheim, 1991) suggesting that the putative *ap1-1* mutant protein could still retain some DNA binding capability. Interestingly, the *ag-1* mutation, also suggested to represent a loss of function allele, has an analogous splice acceptor site mutation (Yanofsky et al., 1990) (Figure 4). Analysis of the RNA produced by the *ag-1* allele similarly reveals that correct splicing is disrupted, presumably resulting in a truncated putative protein product (Mandel et al., 1993).

We were interested in determining if the *ag-1* and *ap1-1* alleles indeed represent loss of function alleles, or if perhaps they still retain some residual activity. Since *AG* is proposed to be necessary for determining stamen and carpel identity, and since carpelloid features can develop in some double mutant combinations that are homozygous for the *ag-1* allele, it is of interest to determine if these carpelloid features can occur in the absence of *AG* function or simply result from residual *AG* activity. Similarly, petals occasionally develop in plants homozygous for the severe *ap1-1* allele, and it is important to determine if this can occur in the absence of *AP1* activity, or if they develop due to residual *AP1* activity. One of the mutant alleles that we analyzed, *ap1-15*, contains a single nucleotide deletion within the MADS-box near the amino terminus of the putative protein (Mandel et al., 1993). Since the *ap1-15* allele almost certainly represents a loss of function allele, and since petals occasionally develop in *ap1-15* mutant plants, we conclude that *AP1* activity is not essential for petal development. Since the *ap1-15* allele proudces a phenotype that is indistinguishable from *ap1-1* mutant plants, these data support the suggestion that the *ap1-1* allele represents a loss of function allele. Furthermore, since the molecular lesions in the *ap1-1* and *ag-1* alleles occur in similar positions in their respective genes, these data also suport the notion that *ag-1* is a loss of function allele and that carpelloid features can develop in the absence of *AG* activity.

Analysis of *AP1* RNA accumulation

Genetic studies indicate that *AP1* has dual roles in flower development, an early-acting role in the specification of floral meristem identity, and a later-acting role in the initiation and proper development of the first and second whorl organs. To determine if the accumulation of *AP1* RNA correlates with its proposed functions, we analyzed *AP1* expression by RNA blotting and tissue *in situ* hybridizations with an *AP1*-specific probe (Mandel et al., 1992b). *AP1* RNA is preferentially expressed in flowers as no hybridizing RNAs were detected in vegetative tissues. RNA *in situ* hybridization experiments show that *AP1* is expressed as soon as floral primordia arise (Summarized in figure 5). Expression is uniform in stage 1 and 2 flowers, and increases during this period. As soon as sepal primordia are visible (stage 3), *AP1* expression is no longer detected in the central dome, the cells that give rise to the inner whorl organs.. At later stages, *AP1* RNA is localized to sepals and petals, and no expression was detected in stamens and carpels, or in the cells that give rise to these organs from stage 3 and beyond. The expression of *AP1* early in young floral primordia supports the conclusion that *AP1* is involved in determining the identity of the floral meristem. The accumulation of *AP1 RNA* in sepal and petal primordia supports the idea that *AP1* is involved in specifying the identity of these organ types. *AP1* RNA also accumulates throughout flower development in pedicels (Mandel et al., 1992b), consistent with its role in suppressing ectopic flower development that often occurs from pedicels of *ap1* mutant flowers (Bowman et al., 1993).

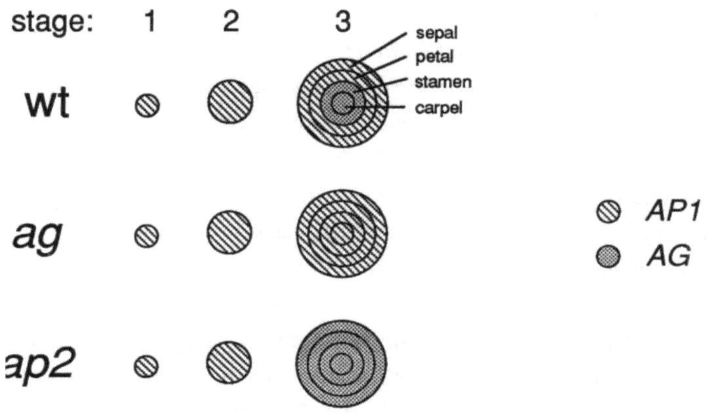

Figure 5. Schematic representation of *AP1* and *AG* RNA accumulation.
The first three stages of flower development are represented as defined by Smyth et al. (1991). Accumulation of *AP1* and *AG* RNAs are shown for wild-type and the indicated mutants. Data are from Gustafson-Brown and Yanofsky (1993). Although not shown in this diagram, we note that *AG* RNA does accumulate in the inner whorls of *ag* mutant flowers (Savidge and Yanofsky, in preparation). We have not indicated this here since the resulting RNA does not produce a functional product.

AP1 RNA accumulates uniformly in young floral meristems, but later is excluded from cells that will give rise to the two inner whorl organs. Interestingly, the loss of *AP1* expression in the two inner whorls coincides with the onset of *AG* expression in this same region (Mandel et al., 1992b; Drews et al., 1991), suggesting the possibility that *AP1* and *AG* interact. To investigate this interaction, we used RNA blotting to compare the levels of *AP1* RNAs in wild-type and several mutant inflorescences. *AP1* RNA levels are elevated in *ag* mutant flowers, consistent with the idea that *AG* negatively regulates *AP1* expression. *AP1* RNA levels are significantly reduced in *ap2* mutant flowers suggesting that *AP2* positively regulates *AP1* RNA accumulation. The opposing effects of *AG* and *AP2* on *AP1* expression are explained in part by the proposed mutual antagonism between *AG* and *AP2*.

To investigate the regulation of *AP1* further, we analyzed the pattern of *AP1* RNA accumulation in mutant inflorescences by RNA *in situ* hybridization analyses (Gustafson-Brown and Yanofsky, 1993), and these data are summarized in figure 5. If *AG* is a negative regulator of *AP1*, then the expression domain of *AP1* should expand to include all of the inner whorls of *ag* mutant flowers. Indeed this is the case, as *AP1* is uniformly expressed throughout *ag* mutant flowers. These results demonstrate that *AP1* is negatively regulated by *AG*, but do not allow us to determine if this interaction is direct. For example, the ectopic expression of *AP1* in the inner whorls *ag* mutant flowers could be due to the loss of *AG*, or instead to the proposed ectopic *AP2* activity. To gain further insight into this question, we analyzed *AP1* expression in *ap2* mutant flowers. *AP1* RNA accumulates at wild-type levels during stages 1 and 2 of *ap2* mutant flowers, but beyond stage 3, *AP1* RNA is no longer detected. The expression of *AP1* early in *ap2* mutant flowers demonstrates that *AP2* is not required for the initial activation of *AP1*. The loss of *AP1* expression in later developing flowers is likely due to the ectopic expression of *AG* that is known to occur in *ap2* mutant flowers (Drews et al., 1991). However, from these data it cannot be ruled out that *AP2* is required for maintenance of *AP1* expression at later stages of flower development. To address this question, we analyzed *AP1* expression in the *ap2 ag* double mutant. As in the *ag* single mutant, *AP1* RNA accumulates in all whorls of the *ap2 ag* double mutant throughout flower development. These observations allow us to conclude that *AG* is a negative spatial regulator of *AP1* RNA accumulation, and that *AP2* is not required for either the initial activation of *AP1* or the maintenance of *AP1* expression. Based on these molecular data, together with the *ap1* mutant phenotype which shows aberrant first and second whorl organ development, we have incorporated *AP1* into the organ identity model originally proposed by Bowman et al. (1991) (Figure 6). These studies suggest that the ectopic petals and sepals that develop in *ag* mutant flowers are in part due to the ectopic expression of *AP1*.

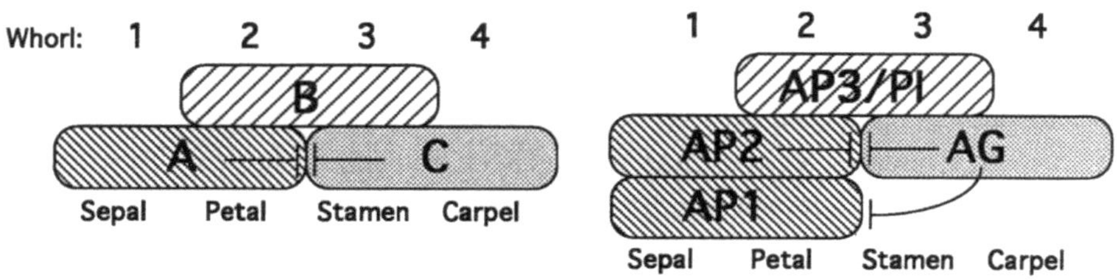

Figure 6. Revised model for genes specifying floral organ identity.
As originally proposed by Bowman et al. (1991), one half of a floral primordium is represented with the positions of the four whorls indicated. Three activities, A, B, and C specify the four organ types. The genes necessary for the A, B, and C activities are indicated in diagram on the right. Lines are drawn to indicate negative interactions between genes, such as the negative regulation of *AP1* by *AG*.

Molecular cloning of *CAULIFLOWER*?

The *ap1* mutant phenotype is dramatically enhanced when the *ap1* mutation is combined with mutations in another gene, *CAL* (Bowman et al., 1993), suggesting that *AP1* and *CAL* interact to promote the generation of floral meristems. The *cal-1* allele, which results in no obvious phenotype on its own, was identified when the *ap1-1* mutation was crossed into the apparently wild-type Wassilewskija (WS) ecotype. This indicated that the WS ecotype is homozygous for the *cal-1* allele. Since *cal-1* has no phenotype as long as a wild-type allele of *AP1* is present, it appears as though *AP1* is able to fully complement the defect in *cal-1*. However, since the *ap1* phenotype is strongly enhanced in plants homozygous for the *cal-1* allele, it seems that the wild-type *CAL* gene is partially able to complement for the loss of *AP1* activity. These intriguing observations suggest the possibility that the *AP1* and *CAL* genes might be functionally related, and could in part substitute for one another. In support of this idea, we have recently identified a new gene, designated *AGL10*, that shares extensive sequence similarity with the *AP1* gene (Kempin, Savidge, and Yanofsky, unpublished). We compared the putative AGL10 protein from the Landsberg *erecta* ecotype with that of the WS ecotype and observed three amino acid differences that might be responsible for the *cal-1* phenotype. To gain further support for the possibility that *AGL10* corresponds to the *CAL* gene, we screened EMS mutagenized *ap1-1* seeds for additional alleles of *CAL*. We obtained a new allele, designated *cal-2*, that contains a single nucleotide alteration in the *AGL10* MADS-box. These data support the notion that *AGL10* corresponds to the *CAL* gene, although further data are needed to confirm this hypothesis. RNA *in situ* hybridization studies indicate that *AGL10*, like *AP1* is expressed in floral meristems (during stage 1), supporting the idea that *AGL10* (*CAL*?) and *AP1* interact to specify the identity of floral meristems.

Negative Regulation: An emerging theme?

The past several years has produced rapid progress in the isolation and characterization of genes that control flower development in two diverse dicot plant species, *Arabidopsis* and *Antirrhinum*. Interestingly, the underlying mechanisms controlling floral morphogenesis in these distantly related plants share remarkable similarities. One intriguing theme that appears to be emerging is that many of the genes controlling flower development interact antagonistically. For example, we have shown that *AP1* is negatively regulated in the inflorescence meristem by the product of the *TERMINAL FLOWER* gene, and *AP1* is negatively regulated in the inner two whorls of wild-type flowers by *AG* (Gustafson-Brown and Yanofsky, 1993).Previous genetic and molecular studies revealed that the *AP2* and *AG* genes are mutually antagonistic, with *AP2* preventing *AG* RNA accumulation in the two outer whorls, and *AG* preventing *AP2* activity in the two inner whorls (Drews et al., 1991;

Bowman et al., 1991). In addition, the *AP3* gene is negatively regulated in the innermost whorl by the product of the *SUPERMAN* gene (Bowman et al., 1992). It should also be noted that *LFY* and *AP1* may act as positive regulators, since the expression of the organ identity genes *AP3* and *AG* is dramatically reduced in *lfy ap1* double mutants (Weigel et al., 1993). However, further studies are needed to determine if *LFY* and *AP1* act directly to activate the organ identity genes, or instead to prevent the activity of a negative regulator of these genes. Whether the negative interaction between the genes controlling flower development is an emerging theme that dictates their spatially restricted expression domains remains to be determined.

References

Alvarez, J., Guli, C.L., Yu, X.-H., and Smyth, D.R. (1992) TERMINAL FLOWER: a gene affecting inflorescence development in *Arabidopsis thaliana*. *Plant J.* **2**:103-116.

Bowman, J.L. , Smyth, D.R. and Meyerowitz, E.M. (1989) Genes Directing Flower Development in *Arabidopsis*. *Plant Cell* **1**:37-52.

Bowman, J.L, Smyth, D.R., and Meyerowitz, E.M. (1991) Genetic interactions among floral homeotic genes of Arabidopsis. *Development* **112**:1-20.

Bowman, J.l., Sakai, H., Jack, T., Weigel, D., Mayer, U., and Meyerowitz, E.M. (1992) *SUPERMAN*, a regulator of floral homeotic genes in *Arabidopsis*. *Development* 114:599-615.

Bowman, J.L., Alvarez, J., Weigel, D., Meyerowitz, E.M., and Smyth (1993) Control of flower development in *Arabidopsis* thaliana by *APETALA1* and interacting genes. Submitted.

Coen, E.S., and Meyerowitz, E.M. (1991) The war of the whorls: genetic interactions controlling flower development. *Nature* **353,** 31-37.

Coen, E.S., Romero, J.M., Doyle, S., Elliott, R., Murphy, G. and Carpenter, R. (1990) *floricaula*: a homeotic gene required for flower development in *Antirrhinum majus*. *Cell* **63,**1311-1322.

Drews, G.N., Bowman, J.L, and Meyerowitz, E.M. (1991) Negative regulation of the Arabidopsis homeotic gene *AGAMOUS* by the *APETALA2* product. *Cell* **65**:991-1002.

Gustafson-Brown, C. and Yanofsky, M.F. (1993) The *Arabidopsis* floral homoeotic gene *APETALA1* is negatively regulated by *AGAMOUS*. In preparation.

Huala, E. and Sussex, I.M. (1992) *LEAFY* interacts with floral homeotic genes to regulate *Arabidopsis* floral development. *Plant Cell* **4**, 901-913.

Irish, V.F. and Sussex, I.M. (1990) Function of the *APETALA1* gene during *Arabidopsis* floral development.*The Plant Cell* **2,** 741-753

Kempin, S.A., Mandel, M.A., and Yanofsky, M.F. (1993) Ecotpic expression of the tobacco floral homeotic gene *NAG1* in transgenic tobacco converts the perianth into reproductive organs. In preparation.

Ma, H, Yanofsky, MF, and Meyerowitz, EM. (1991) A family of genes, *AGL1-6*, similar to the floral homeotic genes *AG* and *defA* and genes encoding transcription factors. *Genes and Devel.* **5**:484-495.

Mandel, M.A., Bowman, J.L., Kempin, S.A., Ma, H., Meyerowitz, E.M., and Yanofsky, M.F. (1992) Manipulation of flower structure in transgenic tobacco. *Cell* **71**:133-143.

Mandel, M.A., Gustafson-Brown, C.., Savidge, B., and Yanofsky, M.F. (1992b) Molecular Characterization of the *Arabidopsis* floral homeotic gene *APETALA1*. *Nature* **360**:273-277.

Mandel, M.A., Kempin, S.A., and Yanofsky, M.F. (1993) Molecular analysis of mutations in the *Arabidopsis* floral homeotic genes *APETALA1* and *AGAMOUS*. In preparation.

Mizukami, Y. and Ma, H. (1992) Ectopic expression of the floral homeotic gene *AGAMOUS* in transgenic *Arabidopsis* plants alters floral organ identity. *Cell* **71**:119-131.
Mueller, C.G.F. and Nordheim, A. (1991) A protein domain conserved between yeast MCM1 and human SRF directs ternary complex formation. *EMBO J.* **10**:4219-4229.
Sadik, S. (1962) Morphology of the curd of cauliflower. *Amer. J. Bot.* **49**:290-297.
Shannon, S., and Meeks-Wagner, D.R. (1991) A mutation in the *Arabidopsis TFL1* gene affects inflorescence meristem development. *Plant Cell* **3**:877-892.
Smyth, D.R., Bowman, J.L., and Meyerowitz, E.M. (1990) Early flower development in *Arabidopsis*. *Plant Cell* **2**:755-767.
Schwarz-Sommer, Z., Huijser, P., Nacken, W., Saedler, H. and Sommer, H. (1990) Genetic control of flower development: homeotic genes in *Antirrhinum majus*. *Science* **250**, 931-936.
Weigel, D., Alvarez, J., Smyth, D.R., Yanofsky, M.F. and Meyerowitz, E.M. (1992) *LEAFY* controls floral meristem identity in *Arabidopsis*. *Cell* **69**:843-859
Weigel, D. and Meyerowitz (1993) Activation of *Arabidopsis* floral homeotic genes. Submitted.
Yanofsky, M.F., Ma, H., Bowman, J.L., Drews, G., Feldmann, K. and Meyerowitz, E.M. (1990) The protein encoded by the Arabidopsis homeotic gene *AGAMOUS* resembles transcription factors. *Nature* **346**:35-39.

Acknowledgments

Special thanks to J. Bowman and D. Weigel for helpful discussions and for communicating results prior to publication, D. Weigel for assistance in the isolation of the *cal-2* allele, B. Shunong and R. Sung for providing the *ap1-15* allele, S. Rounsley for allowing us to cite unpublished work, and Ellen Marsh for technical assistance. We thank R. Schmidt and N. Crawford for helpful discussions. This work was supported by grants from the NSF (DCB 9018749), the David and Lucile Packard Foundation, and the Arnold and Mabel Beckman Foundation. T.A. is supported by an HFSP long term fellowship; B.S. is supported by an NIH training grant.

Dedication

Dedicated to the wonderful memories of my mother, Carol Yanofsky, who died recently of cancer. She inspired all who knew her, and not one day passes without my missing her deeply.

The *S* Gene Family: Genes with Diverse Structures and Functions

June B. Nasrallah and Mikhail E. Nasrallah
Section of Plant Biology
Division of Biological Sciences
Cornell University
Ithaca, New York 14853
U.S.A.

The analysis of the self-incompatibility system of Brassica and the isolation of genes derived from the self-incompatibility or *S* locus has led to the identification of a group of related sequences that is rapidly becoming an extended family. Members of this *S* gene family have now been reported in Brassica, Arabidopsis, and maize. Thus, this family appears to be ubiquitous in plants, and *S* genes may yet be discovered in a large number of plant species in both dicots and monocots.

The two major sub-families of the *S* gene family

Based on the structure of the genes and their predicted products, the *S* gene family can be divided into two major sub-families: one sub-family encodes secreted glycoproteins and another sub-family encodes receptor-like protein kinases. These two sub-families are represented by the highly polymorphic *S*-locus linked *S-Locus Glycoprotein* (*SLG*) and *S-Receptor Kinase* (*SRK*) genes. Genes that are expressed exclusively in reproductive structures, like *SLG* and *SRK*, as well as genes that are primarily expressed in vegetative tissues have been identified. The reported members of each sub-family are listed in Table 1.

In Brassica, the secreted glycoprotein-encoding sub-family includes, in addition to *SLG*, two genes that are unlinked to the *S* locus and that have been designated *S-Locus Related* (*SLR*) genes. *SLR1* (Lalonde et al., 1989; Trick & Flavell, 1989; Trick, 1990; Isogai et al., 1991) and *SLR2* (Boyes et al., 1991)

NATO ASI Series, Vol. H 81
Plant Molecular Biology
Edited by G. Coruzzi and P. Puigdomènech

are expressed in pistils and anthers of self-incompatible and self-compatible strains. In the self-fertile *Arabidopsis thaliana, AtS1* is a gene with floral-specific expression that is also predicted to encode a secreted glycoprotein (Dwyer et al., 1992). Where reported, i.e. in the case of *SLG* and *SLR1*, the glycoproteins encoded by these genes are between 50-60 kDa and migrate as a cluster of two or more molecular weight forms on SDS-PAGE. This molecular weight heterogeneity has been ascribed to differences in the extent of glycosylation of one polypeptide chain (Umbach et al., 1990).

Table 1. The *S* family of plant genes.

	PROTEIN KINASE SUBFAMILY		SECRETED GLYCOPROTEIN SUBFAMILY
	Site of Expression		Site of Expression
	Reproductive Tissues	Vegetative Tissues	Reproductive Tissues
Brassica	*SRK* (*S* locus)		*SLG* (*S* locus) *SLR1* *SLR2*
Arabidopsis		*ARK1* *RLK1*, *RLK4*	*AtS1*
Maize		*ZmPK1*	

Several genes in the receptor-like protein kinase sub-family have been reported. The predicted products of these genes contain an extracellular domain which exhibits sequence similarity to *SLG* and is designated the *S* domain, followed by a membrane-spanning domain and a cytoplasmic domain containing a putative protein kinase catalytic center that shares a high degree of sequence similarity with serine/threonine protein kinases. In fact, the SRK protein was shown to exhibit intrinsic serine/threonine kinase activity (Goring and Rothstein, 1992; Stein and Nasrallah, 1993). Among members of this sub-family, only one gene, *SRK* , has been reported that is expressed exclusively in reproductive structures. The other reported members of this sub-family are all expressed in vegetative tissues. In maize, amplification of root cDNAs with the polymerase chain reaction using primers based on conserved regions of protein kinases led to the isolation of *ZmPK1* (Walker and Zhang, 1990). In Arabidopsis, the screening of genomic libraries with *SRK*-derived probes led to the identification of *ARK1* (*Arabidopsis Receptor Kinase*) (Tobias et al., 1992) and two other receptor-like protein kinase genes (Dwyer, Tobias, and Nasrallah, unpublished observations). More recently, the *Receptor-Like Kinase RLK1* and *RLK4* genes were isolated from Arabidopsis by screening a cDNA library of above-ground tissues with a probe corresponding to the catalytic domain of *ZmPK1* under low-stringency conditions (Walker, 1993).

The *S* domain

The common denominator of members of the family is the 1.3-kb *S* domain. *S* gene family members share varying extents of sequence similarity in this domain. Most similar are the two *S*-locus linked genes, *SLG* and *SRK*. These genes form a physically linked gene pair (Boyes and Nasrallah, 1993) that share over 90% sequence identity in their *S* domains (Stein et al., 1991). These genes are also unique among characterized members of the gene family in being highly polymorphic, as expected from their role in specific self-pollen recognition, with alleles of these genes sharing DNA sequence identity of 67% to 90%. In contrast, the *SLR1* and *SLR2* genes are highly conserved among various *Brassica* strains, and exhibit over 95% sequence identity even when derived from two distinct species, *B. oleracea* and *B. campestris*. On average, the *SLR* genes share approximately 65% DNA sequence identity with the majority of allelic variants of the *S*-locus genes. The Arabidopsis *AtS1* gene is most similar to *SLR1* with which it shares 74% nucleotide sequence identity (Dwyer et al., 1992). The *S* domain of the Arabidopsis *ARK1* gene which shares only 58% DNA sequence identity with *AtS1*, is more similar to the Brassica *SLG* and *SRK* genes exhibiting approximately 60% DNA sequence identity with these genes. More distantly related are the *S* domains of *RLK1* and *RLK4* which exhibit approximately 25% sequence identity with the *S*-locus genes and appear to represent a distinct sub-class of receptor-like genes. Another distinct sub-class is defined by *ZmPK1*, the extracellular domain of which is 26% identical on average to the *S*-locus genes and 22% identical to the *S* domains of *RLK1* and *RLK4* (Walker, 1993).

The *S* domain predicts a polypeptide chain composed of 435 amino acids on average. Common features of the S domains of the various family members is the presence of an N-terminal hydrophobic region consisting of 31 amino acids which has been shown, at least for SLG and SLR1, to function as a signal peptide for transport of the proteins to the cell surface. This region is followed by a 400 amino acid sequence that exhibits several potential N-glycosylation sites and a cysteine-rich region at the C-terminus. In members of the *S* gene family of crucifers, twelve cysteine residues are conserved and are arranged in an invariant pattern (Figure 1). Of these, nine cysteines are also conserved in ZmPK1, RLK1, and RLK4. The analysis of additional *S* genes from diverged species should narrow down the number of conserved cysteines that are necessary for function.

```
AtS1      TE-KCDLYQICGRDSYCDTKTSPTCNCIKGFVPKNVTAWALGDTFE----GCVR
ZmPK1     MTQPCNIHGLCGPNGICHYSPTPTCSCPPGYATRNPGNW-----TE----GCMA
RLK1      GNMACGYNNICSLG----NNKRPKCECPERFVLKDPSNEYGDCLPDFEMQTCRP
RLK4      PEDPCRVYNLCGQLGFCSSELLKPCACIRGFRPRNDAAWRSDDYSD----GCRR
              *     *           * *                           *
              C..X5.C..X5.C...X7..CXC...........Xn............C..

AtS1      KSRLNC-HRDG----FFLLMKRMKLPGTSTAIVDKTIGLNECKERCSKDCNCTG
ZmPK      IVNTTCDRYDKRSMRFVRLPNTDFWGSDQQHLL--SVSLRTCRDICISDCICKG
RLK1      ENQTA--NSDVNLYEFITLEKTNWPFGDYE-SYANYDEER-CKASCLSDCLCAA
RLK4      ENGDSGEKSDT----FEAVGDLRY-DGDVKMSRLQVSKSS-CAKTCLGNSSCVG
                                                   *   *    *
          .X7..C                                   C.X3C.X3CXC..

AtS1      --FANKDIQNGGSGCVIWTG
ZmPK      --FQYQEGT---GSCYPKAY
RLK1      VIFGTNRDLK----CWKKKF
RLK4      --FYHKEKSNL---CKILLE
                        *
          .....Xn.......C
```

Figure 1. The cysteine-rich C-terminal region of the S domain. The consensus spacing of the cysteines is shown below the sequences, with the underlined residues indicating the cysteines conserved in the *S* genes of crucifers, and asterisks indicating the cysteine residues that are invariant in all members of the *S* gene family analyzed to date.

S gene structure

A comparison of cDNA and genomic sequences has revealed a dichotomy with respect to gene structure among members of the *S*-gene family in Brassica and Arabidopsis (Figure 2). The secreted glycoprotein-encoding *SLR1, SLR2,* and *AtS1,* and the majority of *SLG* alleles reported to date, have been shown to produce a 1.6-kb transcript (Nasrallah et al., 1985b, 1988; Trick and Flavell, 1989; Dwyer et al., 1992; Goring et al., 1992a and 1992b). Where genomic clones are available, this transcript was found to be colinear with the genomic sequence, leading to the conclusion that these genes are intronless (Nasrallah et al., 1988; Lalonde et al., 1989; Trick, 1990).

By contrast, the *SRK* gene encodes a 3.0-kb transcript, and consists of six exons separated by seven introns (Stein et al., 1991). Like many other eukaryotic genes, the *SRK* gene appears to have originated by shuffling of exons that represent distinct functional domains. Thus, exon 1 encodes the extracellular *S* domain, exon 2 encodes the transmembrane and juxtamembrane domain, and exons 3 through 7 encode the cytoplasmic kinase domain (Figure 2). The nature of exon 1 as a distinct structural entity is further underscored by its physical separation from the remaining exons by the longest intron, which varies from 896 base pairs to over 5.5 kilobases (Stein et al., 1991; Tantikanjana et al., 1993; J.B. Nasrallah, unpublished data). This gene organization has also been described for the Arabidopsis *ARK1* gene in which the positions of the introns are conserved relative to *SRK* and the longest intron is the 411-bp intron 1 (Tobias et al., 1992). It is interesting to note that in both *SRK* and *ARK1*, intron 1 contains an in-frame stop codon two base pairs downstream of the donor splice site (Stein et al., 1991; Tobias et al., 1992). In addition, alternative transcripts of both genes have been detected by RNA blot analysis. These two observations suggest the possibility that each of the two genes encodes a secreted form of the extracellular domain as well as the full-length transmembrane protein kinase.

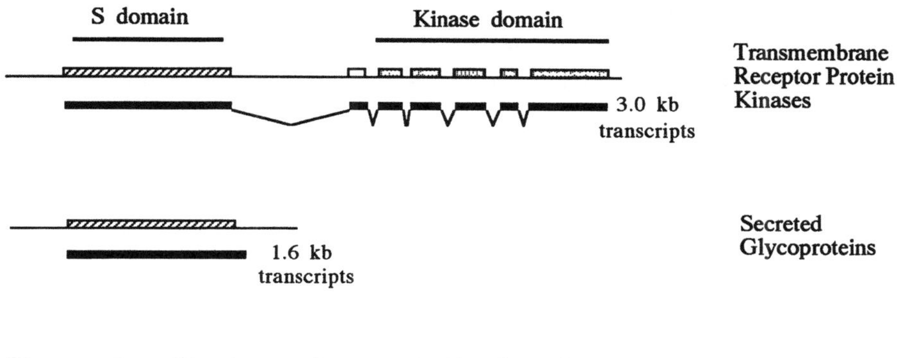

Figure 2. Structures of genes in the Brassica and Arabidopsis *S* gene families. The gene structure is shown with boxes denoting the exons. Hatched boxes: *S* domains; stippled boxes: kinase domain; white box: transmebrane domain. The transcripts are shown as black boxes below the genes.

Because very few sequences have been analyzed at the genomic level, it is not known at this time if all members of the *S* gene family will turn out to have the basic gene structures described for the Arabidopsis and Brassica secreted glycoprotein and transmembrane protein kinase genes. The characterization of genomic clones for the more diverged genes such as *RLK1, RLK4,* and *ZmPK1* will be revealing in this regard. It has become evident however, that even in Brassica, variant gene structures do occur. We have recently characterized an *SLG* allele isolated from a self-incompatibility genotype that is pollen-recessive and exhibits a weak stigma self-incompatibility phenotype (Tantikanjana et al., 1993). This *SLG* allele was found to have an unusual structure consisting of two exons: the first exon encodes the *S* domain of the gene, and the second exon shares a high degree of sequence similarity with the second transmembrane-encoding exon of *SRK*. Two transcripts that differ at their 3' ends are produced by this *SLG* gene. One transcript contains sequences from the first exon, terminates within the first intron, and encodes a secreted glycoprotein. The second, larger transcript contains sequences from the first and second exons and encodes a membrane-anchored form of SLG. The production of this membrane-anchored molecule is likely to have important functional consequences and may explain the pollen-recessiveness and weakness of the genotype under study. In addition, the unusual structure of this *SLG* gene suggests that *SLG* is derived from *SRK* by duplication. It is possible that the ancestral duplication event involved the entire *SRK* gene and was followed by the deletion of a varying number of exons to generate the present-day *SLG/SRK* gene pairs. Alternatively, the duplication event may have occurred more than once in the evolutionary history of the *S* locus. In any event, the duplication of *SRK*-like genes and the generation of genes that encode secreted glycoproteins appears to have occurred with some frequency during the evolution of the *S* gene family. Independent evidence for such duplications has been recently found in the analysis of a secreted glycoprotein-encoding member of the *S*-gene family that also maintains an intron downstream of its *S* region (Tantikanjana, T., Nasrallah, M.E., and Nasrallah, J.B., unpublished observations).

Functional relationships of secreted glycoprotein and receptor protein kinase genes

With the exception of the *S*-locus genes of Brassica, *S* family genes have not been associated with any biological function. However, in the case of the floral-specific *SLR1*, *SLR2,* and *AtS1* genes, the spatial pattern of their expression and the distribution of their protein products are suggestive. For example, the

SLR1 gene, which has been characterized in most detail, encodes a glycoprotein that accumulates in the papillar cell walls of the Brassica stigma (Umbach et al., 1990). Further, a chimeric gene consisting of the *SLR1* promoter fused to the ß-glucuronidase gene was shown, at least in transgenic tobacco, to be expressed in pistils and in pollen (Hackett et al., 1992), in a manner similar to the activity exhibited by the *SLG* promoter in this transgenic host (Thorsness et al., 1991). In transgenic Arabidopsis, an *SLR1*:: ß-glucuronidase gene fusion is predominantly expressed in stigmatoid tissue and most prominently in the papillar cells (Lalonde, Thorsness, and Nasrallah, unpublished observations), again paralleling the activity of the *SLG* promoter in transgenic Arabidopsis (Toriyama et al., 1991). In view of their ubiquitous nature, their conserved sequences, and their expression in self-compatible as well as self-incompatible strains (Lalonde et al., 1989; Umbach et al., 1990), the *SLR* genes are not thought to be direct determinants of self-incompatibility. Similarly, *AtS1* cannot be involved in the discrimination against self-pollen because Arabidopsis is a self-fertile species that does not exhibit a self-incompatibility system. Rather, it is likely that the *SLR1*, *SLR2*, and *AtS1* genes may be involved in some general aspect of pollination responses, perhaps as components of a system that promotes the development of pollen tubes in compatible pollinations.

In any event, the close evolutionary relationships between the secreted glycoprotein and receptor protein kinase genes raise several questions relating to the functional relationships between the two classes of genes. For example, it is of interest to determine if the secreted glycoprotein products have independent functions or if they function exclusively in association with the receptor protein kinases. While this question cannot be fully addressed at this stage, it has become clear from the analysis of Brassica self-incompatibility that both SLG and SRK are required for the operation of the self-incompatibility system (for review, see Nasrallah and Nasrallah, 1993). Based on this requirement, we propose that interactions between the transmembrane receptor and the secreted form of the extracellular "recognition" domain of the receptor are essential for signaling by this class of molecules across the plant cell wall. This hypothesis is supported by the properties of the Arabidopsis *ARK1* gene. Although we have found no evidence that this gene occurs in close physical linkage with a secreted glycoprotein-encoding gene similar to *SLG*, the *ARK1* gene itself, as described above, has the potential to encode a secreted form of the extracellular domain as well as the full-length transmembrane receptor. The observation that, among members of the *S* gene family, two different mechanisms have evolved to generate secreted forms of receptor extracellular domains argues that such secreted forms are important for receptor function. The *SLG/SRK* gene pair would thus represent an extreme case. It is possible that the duplication of the extracellular domain of *SRK* in the form of the *SLG* gene occurred as a result of unique requirements in pollen-stigma signaling. A large excess of the secreted form of the extracellular domain relative to the transmembrane receptor may be required for the efficient functioning of the self-incompatibility system. Such a requirement would have been met by the evolution of the *SLG* gene with its highly active promoter and the consequent "amplification" of the secreted form of the extracellular domain. A functional requirement for high levels of this molecular form could also provide the selective pressure to maintain the tight linkage of *SLG* to *SRK* and the extensive sequence similarity between the two genes.

Because of its complexity, the *S* gene family is well suited for investigating the evolution and functional diversification of gene families in plants. The intrinsic variability of the *S*-locus genes has already provided some clues regarding the relationship of the secreted glycoprotein genes to the extracellular domains of the receptor-like protein kinases. In the future, the isolation of additional *S* genes, the characterization of their structure and patterns of expression, and the identification of potential ligands and substrates should shed light on the evolution and functions of this diverse gene family.

References

Boyes DC, Chen CH, Tantikanjana T, Esch JJ, Nasrallah ME, Nasrallah, JB (1991) Isolation of a second S-locus related cDNA from *Brassica oleracea*: genetic relationships between the S locus and two related loci. Genetics 127: 221-228.

Boyes DC, Nasrallah, JB (1993) Physical linkage of the *SLG* and *SRK* genes at the self-incompatibility locus of *Brassica oleracea*. Mol. Gen. Genet. 236: 369-373.

Dwyer KD, Lalonde BA, Nasrallah JB, Nasrallah ME (1992) Structure and expression of *AtS1*, an *Arabidopsis thaliana* gene homologous to the *S*-locus related genes of *Brassica*. Mol. Gen. Genet. 231: 442-448.

Goring DR, Banks P, Beversdorf WD, Rothstein SJ (1992a) Use of the polymerase chain reaction to isolate an S-locus glycoprotein cDNA introgressed from *B. campestris* into *B. napus* ssp. *oleifera*. Mol. Gen. Genet. 234: 185-192.

Goring DR, Banks P, Fallis L, Baszczynski CL, Beversdorf WD, Rothstein, SJ

(1992b) Identification of an S-locus glycoprotein allele introgressed from *B. napus* ssp. *rapifera* to *B. napus* ssp. *oleifera*. Plant J. 2: 983-989.

Goring DR, Rothstein SJ (1992) The S-locus receptor kinase gene in a self-incompatible *Brassica napus* line encodes a functional serine/threonine kinase. Plant Cell 4: 1273-1281.

Hackett RM, Lawrence MJ, Franklin FCH (1992) A *Brassica S*-locus related gene promoter directs expression in both pollen and pistil of tobacco. *Plant J.* 2: 613-617.

Isogai A, Yamakawa S, Shiozawa H, Takayama S, Tanaka H, Kono T, et al (1991) The cDNA sequence of NS1 glycoprotein of *Brassica campestris* and its homology to S-locus related glycoproteins of *B. oleracea.* Plant Mol. Biol. 17: 269-271.

Lalonde B, Nasrallah ME, Dwyer KD, Chen CH, Barlow B, Nasrallah JB (1989) A highly conserved *Brassica* gene with homology to the S-locus specific glycoprotein structural gene. Plant Cell 1: 249-258.

Nasrallah JB, Kao TH, Goldberg ML, Nasrallah ME (1985) A cDNA clone encoding an *S*-locus specific glycoprotein from *Brassica oleracea.* Nature 318: 617-618.

Nasrallah JB, Yu SM, and Nasrallah ME (1988) Self-incompatibility genes of *Brassica:* expression, isolation and structure. Proc. Natl. Acad. Sci. USA 85: 5551-5555.

Stein JC, Howlett B, Boyes DC, Nasrallah ME, Nasrallah JB (1991) Molecular cloning of a putative receptor protein kinase gene encoded at the self-incomaptibility locus of *Brassica oleracea.* Proc. Natl. Acad. Sci. USA 88: 8816-8820.

Stein JC, Nasrallah JB (1993) A plant receptor-like gene, the *S*-locus receptor kinase of *Brassica oleracea,* encodes a functional serine/threonine kinase. Plant Physiol. 101: 1103-1106.

Tantikanjana T, Nasrallah ME, Stein JC, Chen CH, Nasrallah, JB (1993) An alternative transcript of the *S*-locus glycoprotein gene in a class II pollen-recessive self-incompatibility haplotype of *Brassica oleracea* encodes a membrane-anchored protein. Plant Cell 5 (in press).

Thorsness MK, Kandasamy MK, Nasrallah ME, Nasrallah JB (1991) A Brassica *S*-locus gene promoter targets gene expression and cell death to the pistil and pollen of transgenic Nicotiana. Dev. Biol. 143:173-184.

Tobias CM, Howlett B, Nasrallah JB (1992) An *Arabidopsis thaliana* gene with sequence similarity to the S-locus receptor kinase of *Brassica oleracea:* sequence and expression. Plant Physiol. 99: 284-290.

Toriyama K, Thorsness MK, Nasrallah JB, Nasrallah ME (1991) A Brassica *S*-locus gene promoter directs sporophytic expression in the anther tapetum of transgenic Arabidopsis. Dev. Biol. 143: 427-431.

Trick M, Flavell RB (1989) A homozygous *S* genotype of *Brassica oleracea* expressed two *S*-like genes. Mol. Gen. Genet. 218: 112-117.

Trick M (1990) Genomic sequence of a *Brassica* S locus related gene. Plant Mol. Biol. 15: 203-205.

Umbach AL, Lalonde BA, Kandasamy MK, Nasrallah JB, Nasrallah ME (1990) Immunodetection and post-translational modification of two products encoded by two independent genes of the self-incompatibility multigene family of *Brassica*. Plant Physiol. 93: 739-747.

Walker J, Zhang R (1990) Relationship of a putative receptor kinase from maize to the *S*-locus glycoproteins of *Brassica*. Nature 345: 743-746.

Walker J (1993). Receptor-like protein kinase genes of *Arabidopsis thaliana*. Plant J. 3: 451-456.

The Genetics and Physiology of the Totipotent Cell

Fiorella Lo Schiavo
Department of Biology
University of Padua
Via Trieste 75
35121 Padova
Italy

INTRODUCTION

Plants, as compared to animals, are characterized by a great flexibility in their differentiative programmes. Plants are capable of regenerating part of their organs; somatic cells may become embryogenic and thus reproduce all developmental stages up to adult plants.

These statements of general nature cannot be applied as such to all plant species. Experience tells us that primary explants of various species (hypocotyls of germinating seedlings, cotyledons, leaves, roots etc.) respond to hormonal treatments (in general auxin, most typically 2,4-D) in various ways. Explants of some species give rise to cell lines that proliferate without showing any morphogenetic capacity; some acquire organogenetic capacity (e.g. tobacco, Skoog and Miller 1957) i.e. the capacity to regenerate organs, usually shoots, that, upon transfer to different hormonal conditions, will complete the formation of the radical system. Other plant species (e.g. carrot,Halperin 1966, or alfalfa, Dudits 1991) go through the various steps of embryogenesis in a process, called *somatic* embryogenesis, whose steps are virtually indistinguishable from those of the real zygotic one.

The attempts to increase the number of plant species capable to give morphogenetic responses *in vitro* (particularly in the form of somatic embryogenesis) were done in the past by

NATO ASI Series, Vol. H 81
Plant Molecular Biology
Edited by G. Coruzzi and P. Puigdomènech

using all sort of combinations of culture media and phytohormones, in different concentrations; in more recent times greater success was achieved by using different types of explanted tissues. In fact, in many recalcitrant species, e.g. cereals, morphogenetic capacities are acquired by lines originated from embryonal tissues (Vasil and Vasil 1986) or young leaf tissue (Wernicke and Brettel 1980) which eventually proved capable of forming somatic embryos.But the fact that success depends so much on the use of (slightly) modified protocols tells us how little we know on the cellular and molecular basis of somatic embryogenesis, in spite of the fact that the first description of this process is now 35 years old (Steward et al.1958; Reinert 1958).

Some of the questions that may help us define and analyze in more detail the process, are the following:

1) Are all plant somatic cells totipotent?

2) What are the characteristics of the cells capable to acquire totipotency?

3) What are the events that make a somatic cell into a totipotent one?

THE PROGENITORS OF AN EMBRYOGENIC CELL LINE

A systematic description of the early events that occur when a primary explant gives rise to an embryogenic line is our attempt to give an answer to the preceding questions.
Our experimental system is the formation of an embryogenic cell line in carrot, starting from hypocotyls of 7-days old seedlings cultured, from time zero to the end of the experiment, in a synthetic medium (Gamborg's B5) supplemented with 2.3 µM 2,4-D. We followed with a histo-cytological analysis the reaction of the different tissues present in the explant to high concentrations of auxin, and we identified the cells that respond to the hormonal treatment and those that give rise to the embryogenic cell line (Guzzo et al. *in preparation*) .Two types of cytological response were apparent: i) the epidermal layer and the parenchyma cells enlarge

showing the typical response to auxin; ii) procambium and the non-differentiated cells of the xylem and phloem, proliferate. After 5 to 7 days of exposure to auxin, the cells of the cortical parenchyma, fully enlarged, come off in layers from the central cylinder and, no matter what the concentration of auxin, do not proliferate; this tissue therefore will not contribute to the formation of the embryogenic cell line.
In the procambial cells a series of events take place:
a) increase of the cytoplasmatic mass and reduction of the vacuolar volume; b) divisions along both longitudinal and transverse plane that originate proliferating masses of small isodiametric cells; c) enlargement and detachment of the afore-mentioned cells from the cylinder into the culture medium: these will proliferate and give rise to the embryogenic cell line.

Working on soybean seedlings, Guilfoyle and coworkers identified, by *in situ* hybridization, two classes of auxin-induced genes, differentially expressed and designated SAUR and GH (Gee et al.1991). SAUR transcripts were more abundant in the tissues that responded to auxin with an enlargement: epidermis and cortical parenchyma. GH3 transcripts were present in minimal amounts, if any, in untreated soybean seedlings, but, after incubation with high concentrations of auxin they were expressed particularly in the differentiating vascular tissue.

All these results concur in indicating that different auxin receptors are present in tissues that respond differently to auxin; cell enlargement is accompanied by specific mRNA's (SAURs in soybean) and so is cell proliferation (GH3 mRNA's); the latter type of response in the carrot system correlates with the formation of embryogenic cell lines.

When making primary explants from different parts of the carrot plant (roots, cotyledons) cell proliferation and detachment is induced by auxin on procambium and adjacent cells, as already shown in hypocotyls; in all these primary explants of carrot, embryogenic cell lines are always originated by those pro-cambial cells.Stelar cells are shed from the stele into the medium, keep on proliferating and generate a cell line. When the explanted tissue comes from a plantlet regenerated *in vitro*, the results are markedly different, in that also the epidermis and cortical parenchyma proliferate in response to auxin and then contribute to the formation of a cell line. The fact that somatic embryos originated from the epidermis of regenerated carrot plantletswas established long ago (Haccius and Lakshman 1969).

THE ONSET OF EMBRYOGENESIS

In about 20 days after explantation of carrot hypocotyls a particular cell type appears, very small (10-20μm) which after a series of very regular divisions generates pro-embryogenic masses (PEM)(Halperin 1966) the initial embryonal stage.PEM's constitute always a fraction in an embryogenic cell population and are continuosly produced in the established cell line.

A systematic cytological investigation on the type of division undertaken by the single cells in our embryogenic cell line showed that 50% of those cells underwent an asymmetric division thus generating small cells (10-20μm) and large cells (up to 150μm); the remaining half divided symmetrically, generating equal cells that successively enlarged. The small cells, or type I (corresponding to Komamine's state 0)(Komamine et al.1990) are characterized, besides their diameter and lack of enlargement, by the series of regular division that eventually lead to the formation of PEM's. Unequal divisions were associated with earliest embryogenic events in the observation by Backs-Hüsemann and Reinert (1976). We may add that this type of unequal division that generates type-I cells is only present in embryogenic cell lines and, in those populations, only in the presence of auxin.

In other plant species the same type of primary explant (hypocotyl from germinated seedlings) may generate cell lines capable of indefinite proliferation without morphogenetic capacity (e.g. tomato), or cell lines with organogenetic capacities (among these we are currently studying Arabidopsis and Cichorium).

The cytological analysis of Arabidopsis shows that, in this system too, tissues respond to auxin in different ways: enlargement in epidermis and cortical cells, and proliferation in the vascular tissue of the stele.But the cell line that originates in this species is unable to undergo asymmetric divisions, does not form type-I cells (and their successors:PEM's) and generates exclusively type-II cells without embryogenic capacity.

Working with alfalfa mesophyll protoplasts, Dudits et al.(1991) made the observation that asymmetric divisions occur under high 2,4-D concentrations, whereas, at lower concentrations, the same cell population, did only divide symmetrically. In the former case an embryogenic callus (i.e. generating somatic embryos) is seen, whereas in the latter only unorganized callus growth was observed.

. Fujimura and Komamine (1980) showed emergency of polarity during embryogenesis from carrot single cells. Pattern of cell division was followed by autoradiography using ^{3}H-thymidine(^{3}H-TdR). In the cell clusters consisting of 4 cells,polarised incorporation of ^{3}H-TdR in the cell clusters was observed throughout development up to globular stage embryos.

These examples suggest that on the cells that respond to auxin by proliferating there are two types of receptors: one responsible for cell division as such, and another one responsible for the asymmetric divisions that generate embryogenic cells. On the other hand, if we consider the zygotic development of both dicots (e.g. Arabidopsis, Mansfield and Briarty 1991) and monocots (e.g. maize, Sheridan and Clark 1987) a transverse asymmetric division generates a relatively small apical cell and a larger basal cell; this is the first cytological event we see in the zygote after fertilization and it marks the onset of the embryogenic programme.

In carrot, PEM's -which are generated by type-I cells after a series of regular divisions in the presence of auxin-consist of cells that no longer respond to auxin by enlarging. The cells, in the presence of auxin, do not resemble meristematic cells, having a large vacuole and scarce cytoplasm. If auxin is removed, these cells in few hours fill up with cytoplasm, the vacuole is fragmented, and they acquire the cytoplasmatic aspect of embryonal cells. In 3 days, cell proliferation becomes extremely fast (doubling time=3hrs) and PEM's give rise to globular (in 5 days) hearts(in 6 days) and torpedo-shaped embryos (8-10 days).

We have established what type of cells in the hypocotyl of D.carota gives origin to an embryogenic cell line and the type of cytological events that accompany the formation of PEM's. However important remaining questions are:

1) Why, using the same type of explant and the same cultural conditions, some species generate lines without any morphogenetic capacity, other species generate lines with embryogenic capacity and other species lines with organogenic capacity?

2) What is the mechanism whereby a cell becomes morphogenetically competent?

The approach that we are following is to study the differential responses of cell cultures to auxins. The idea is that only an analysis of the auxin receptors may give us a clue on why the same cell types of different plant species may respond to auxin in different ways. We began to investigate the behaviour and the sensitivity to 2,4-D of embryogenic carrot cell lines; 2,4-D was chosen because it is the most efficient auxin in inducing embryogenic capacity, in carrot and other species(Dudits 1991).The sensitivity to auxin was measured as the auxin-binding capacity (ABC) of a crude membrane preparation of cells exposed to different concentrations of 2,4-D. An increase in the concentration of the hormone in the medium causes an increase in ABC (Lo Schiavo et al.1991); We repeated this type of experiment of auxin-dependent induction of ABC on primary explants. In carrot, the analysis was carried out on cotyledons, hypocotyls

and roots incubated in the presence of hormones. Hypocotyl turned out to be the strongest modulating part and cotyledons the weakest one: they have a constitutive low level of ABC that remains constant in the presence of auxin; roots show an intermediate level between the two other types of organs analyzed.
The limit to those experiments on organs is that one is dealing in each case with several types of cells that respond differentially to auxin.

BIOLOGICAL SIGNIFICANCE OF MODULATION

ABC-modulation is shown by our model system with different characters. Can we find alterations of such a type as to give us indications on the biological significance of the phenomenon? To this end we have used mutant carrot cell lines, altered in the differentiative processes, isolated in our laboratory, previously or during this investigation.
We measured modulation in:

- auxin-resistant lines (incapable to embryogenize),that were obtained by plating cells in the presence of 45µM 2,4-D;
- emb^- lines isolated for their incapacity to embryogenize
- a habituated line (E9) characterized by alterations in the IAA metabolism (Lo Schiavo et al.1989)

The results indicated that every time modulation was reduced or absent, -either because of a high constitutive level or low inducibility- embryogenesis was also altered or totally abolished (Filippini et al.1992). The isolation, from carrot embryogenic cell lines, of cell lines that proliferate normally but without modulating capacity and without differentiative capacity, indicates that the type of hormonal response that leads to cell division is independent of the type of response that leads to acquisition of totipotency and generation of PEM's in the presence of auxin. As a consequence, we should think that a class of auxin-binding proteins (ABP) exists, responsible for cell division, and other ABP's, capable to induce, beside cell division, regeneration. The class of ABP's that are induced in response to auxin, determines the cell morphogenetic fate. Plant species recalcitrant to differentiative events may perhaps generate their lines starting from cells only capable of expressing the class of ABP involved in cell division. To assay modulation on primary explants could tell us if the starting material is good, or if we are making the wrong selection for generating cell lines. If the starting material does not modulate in response to variations of auxin concentration we are either using inappropriate cultural conditions or using the inappropriate tissue.

CELLULAR MECHANISMS THAT GENERATE TOTIPOTENCY

To the first step, in which some cells of the primary explant respond to auxin and give rise, in carrot, to cell lines capable of modulating, a second one follows, morphologically definable with the appearance of PEM's, which characterizes the embryogenic lines. PEM's have been purified from cultures of proliferating cells and incubated with different concentrations of auxin. This sub-population of cells turns out to be "insensitive" to auxin during all embryonal stages up to mature globular embryo. In a proliferating cell culture, thus two cell populations are present: modulating cells and PEM's, the latter generated from modulating cells but insensitive to auxin.At later embryonal stages (from heart-shaped embryos on) some tissues lose the characteristics of the primary meristem and start responding to auxin. In order for the totipotent cell to arise through the asymmetric division previously described, perhaps it is necessary -but also sufficient- reaching a certain configuration of the auxin receptor, which, after complete loading, remains unable to bind further amounts of the hormone. This hypothesis might explain why some plant species are embryogenic whereas others, unable to reach the postulated configuration of the auxin receptor, never gain totipotency.

If this hypothesis is correct, it should become possible, in a more or less distant future, to purify, characterize, clone the single components of the receptorial complex from, say, carrot, and upon gene transfer, express the proper component, capable of giving embryogenic capacity to recalcitrant species

Another suggested mechanism capable of giving totipotency is *somatic meiosis*. Nuti-Ronchi et al.(1992) have described in cell cultures a series of events exactly matching the steps of meiosis as it occurs during gametogenesis and they suggest that this mechanism generates in vitro a cell (the progenitor cell of the somatic embryo) which assumes a role similar to the fertilized ovule at the onset of zygotic embryogenesis (Nuti-Ronchi et al.1992). A consequence of such a mechanism would be haploidization and segregation with generation of a great amount of genetic variability, such as the one that we see, e.g. in the phenomenon of somaclonal variation.To this haploidization, an endoreduplication, giving rise to a completely homozygous diploid, will soon follow. Results of genetic experiments performed with RFLP and RAPD markers support this hypothesis. (Nuti-Ronchi et al. *submitted*).

SUCCESSIVE EMBRYO STAGES

When the cells get firmly committed to embryogenesis they lose auxin sensitivity, which in fact is absent up to the advanced globular stage. With the appearance of the first cells capable to enlarge, in the advanced heart stage and, later, in torpedoes, auxin sensitivity is regained. In practical terms this is demonstrated by the impossibility to obtain callus (in carrot but in Arabidopsis as well) from globular embryos; callus is obtained easily instead from later embryonal stages.

With the formation of advanced globular embryos an important phase in plant embryogenesis is concluded: that is the phase when the genes responsible for the basic body pattern of the plant are activated. Mutations that alter pattern formation in Arabidopsis, have been isolated by Jürgens and coworkers

(Mayer et al.1991) and the deviations from normal development are all apparent before the conclusion of the heart stage. Mutations affecting embryo development have also been isolated in maize (Clark and Sheridan 1991) and, using somatic embryogenesis, temperature-sensitive mutants have been isolated by three groups in carrot (Breton and Sung 1982, Giuliano et al.1984, Schnall et al.1988). The characterization of these mutants has already, and will in the future, contribute to define the details of developmental processes.

We think that the advantages of somatic embryogenesis, viz. great amounts of embryogenic cells, possibility to isolate gram-quantities of synchronized embryo stages (and hence possibility of biochemical investigations), speed of the developmantal process (10-15 days), possibility to manipulate great quantities of embryos chemically or physically, are so great that the system cannot be dispensed with.

Few genes have been described that are expressed during the early events of embryogenesis (Aleith and Richter 1991); some genes, specific to the embryogenic process have been identified (Choi et al.1984, Borkird et al.1988, Thomas and Wilde 1985, 1987) and of these, a few were also found to be expressed during zygotic embryogenesis (Borkird et al.1988). Recently gene products that are essential for somatic embryogenesis were purified from the glycoproteins secreted into the culture medium from carrot cells (de Vries et al.1988, van Engelen and de Vries 1992). The functional importance of at least some of these glycoproteins was recognized by their ability to rescue temperature-sensitive mutant embryos that were developmentally arrested (de Jong et al.1992). More details about the secreted glycoproteins and other embryonic markers will be found in the contributions of de Vries (this volume).

This work has been supported by MURST and MAF (Sviluppo tecnologie avanzate).

REFERENCES

Aleith F., Richter G.(1991) Gene expression during induction of somatic embryogenesis in carrot cell suspensions.Planta 183: 17-24

Backs-Hüsemann H., Reinert J.(1970) Embryobildung durch isolierte Einzellen aus Gewebekulturen von *Daucus carota.* Protoplasms 70:49-60

Borkird C., Choi J.H., Jin Z.H., Franz G., Hatzopoulos P., Chorneau R., Bonas U., Pelegri F., Sung Z.R. (1988) Developmental regulation of embryonic genes in plants. Proc. Nat.Acad.Sci.85:6399-6403

Breton A.M., Sung Z.R. (1982) Temperature-sensitive carrot variants impaired in somatic embryogenesis. Dev. Biol. 90:58-86

Choi J.H., Sung Z.R. (1984) Two-dimensional gel analysis of carrot somatic embryogenesis proteins.Plant Mol.Biol.Rep.2:19-25

Clark J.K., Sheridan W.F.(1991) Isolation and characterization of embryo-specific mutations of maize. Plant Cell 3:935-951.

De Jong A.J., Cordewener J., LoSchiavo F., Terzi M., Vandekerkove J., Van Kammen A., De Vries S.C.(1992) A carrot somatic embryo mutant is rescued by chitinase. Plant Cell 4:425-433

De Vries S.C., Booij H., Janssens R., Vogel R., Saris L., LoSchiavo F., Terzi M., van Kammen A. (1988) Carrot somatic embryogenesis depends on the phytohormone-controlled expression of correctly glycosylated extracellular proteins.Genes Dev.2:462-476

Dudits D., Bogre L, Gyorgyev J. (1991) Molecular and cellular approaches to the analysis of plant embryo development from somatic cells in vitro. J. Cell Sci.99:475-484.

Filippini F., Terzi M., Cozzani F., Vallone D., Lo Schiavo F.(1992) Modulation of auxin binding proteins in cell suspensions.II. Isolation and initial characterization of carrot lines impaired in somatic embryogenesis. Theor.Appl.Genet.84:430-434.

Fujimura T., Komamine A. (1980) Aspects of DNA, RNA and protein synthesis during somatic embryogenesis in a Daucus carota cultivar Kurodagosun cell suspension culture. Physiol.Plant.49:255-260.

Gee M.A., Hagen G., Guilfoyle T.J. (1991) Tissue-specific and organ-specific expression of soybean auxin responsive transcripts GH3 and SAURs. Plant Cell 3:419-430.

Giuliano G., LoSchiavo F., Terzi M.(1984) Isolation and developmental characterization of temperature-sensitive carrot cell variants. Theor.Appl.Genet.67:179-183.

Haccius B., Lakshman K.C. (1969)Adventiv-Embryonen-Embryoide-Adventiv-knospen-Ein Beitrag aus klärung der Begriffe. Österr.bot.Z.116:145-158.

Halperin W.(1966) Alternative morphogenetic events in cell suspensions. Am.J.Bot. 53:443-453.

Komamine A., Matsumoto M., Tsukahara M., Fujiwara A., Kawahara R., Ita M., Smith J., Fujimura T. (1990) Mechanisms of somatic embryogenesis in cell cultures: physiology, biochemistry and molecular biology. In" Progress in Plant Cellular and

Molecular Biology" (J.J. Nijkamp, L.H.van der Plas, J. van Aartrijk eds.) Kluwer Academic Pbshrs, Dordrecht (NL) pp.307-313

Lo Schiavo F., Pitto L., Giuliano G., Torti G., Nuti-Ronchi V., Marazziti D., Vergara R., Orselli S., Terzi M.(1989) DNA methylation of embryogenic carrot cell cultures and its variationa as caused by mutation, differentiation, hormones and hypomethylating drugs. Theor.Appl.Genet.77:325-333

Lo Schiavo F., Giuliano G., De Vries S.C., Genga A., Bollini R., Pitto L., Cozzani F., Nuti-Ronchi V., Terzi M.(1990) A carrot cell variant temperature-sensitive for somatic embryogenesis reveals a defect in the glycosylation of extracellular proteins. Mol.Gen.Genet.233:385-393.

Lo Schiavo F.,Filippini F., Cozzani F., Vallone D., Terzi M.(1991) Modulation of auxin binding proteins in cell suspensions. I. Differential responses of carrot embryo cultures. Plant Physiol.97:60-64.

Mansfield S.G., Briarty L.G.(1991) Early embryogenesis in *Arabidopsis thaliana*. II. The developing embryo. Canad.J.Bot.69:461-476.

Mayer V., Torres Ruiz R.A., Berleth T., Miséra S., Jürgens G.(1991) Mutations affecting body organization in the Arabidopsis embryo . Nature 353:402-407

Nuti-Ronchi V., Giorgetti L., Tonelli M.G., Martini G.(1992) Ploidy reduction and genome segregation in cultured carrot cell lines. II. Somatic meiosis. Plant Cell Tiss.Org.Cult. 30:115-120

Reinert J. (1958) Morphogenese und ihre Kontrolle an Gewebekulturen aus Karotten. Naturwiss.45:344-345

Schnall J.A., Cooke T.J., Cress D.E.(1988) Genetic analysis of somatic embryogenesis in carrot cell cultures:initial characterization of six classes of temperature-sensitive variants. Dev.Genet.9:49-67.

Sheridan W.F., Clark J.K.(1987) Maize embryogeny: a promising experimental system. Trends Genet.3:3-6

Skoog F., Miller C.O.(1957) In: Symp.Soc.Exp.Biol.11:118-131

Steward F.C., Mapes M.O., Hears K.(1958) Growth and organized development of cultured cells.II. Organization in cultures grown from freely suspended cells.Am;J;Bot.45:705-708.

Thomas T.L., Wilde H.D.(1985) Analysis of gene expression in carrot somatic embryos. In "Somatic Embryogenesis" (M.Terzi, L. Pitto, Z.R. Sung eds.) IPRA, Rome, pp.77-85

Thomas T.L., Wilde H.D. (1987) Analysis of carrot somatic embryo gene expression programs. In Proc.VI Int.Congr.Plant Cell Tiss.Cult.(Alan R.Liss, New York) pp 83-93

Vasil I.K., Vasil V. (1986) Regeneration in cereal and other grass species. In: I.K.Vasil (ed.):Cell Culture and somatic Cell Genetics of Plants. Vol.3. Plant Regeneration and genetic Variability. Academic Press, Orlando, pp.121-150.

Wernicke W., Brettel.(1950) Somatic embryogenesis from *Sorghum bicolor* leaves. Nature 287:138-139

The Carrot Extracellular Lipid Transfer Protein EP2: Quantitative Aspects With Respect to its Putative Role in Cutin Synthesis.

Theo Hendriks*, Ellen A. Meijer, Sharon Thoma[a], Jean-Claude Kader[b], and Sacco C. De Vries

Department of Molecular Biology,
Agricultural University of Wageningen
Dreijenlaan 3
6703 HA Wageningen
The Netherlands.

Introduction

Plant lipid transfer proteins (LTPs) have been isolated from different sources in both monocot and dicot plant species (see for review Kader, 1990). The purified proteins were shown to be small basic proteins capable of transferring several types of lipids between various types of membranes *in vitro*. Based on this observation, it was suggested that, analogous to cytosolic LTPs from mammals (Wirtz, 1991), the transfer of phospholipids between organelle and endoplasmatic reticulum membrane systems would be their function *in vivo* (Arondel and Kader, 1990; Kader, 1990). However, the presence of a putative signal peptide sequence in mRNAs of previously identified plant LTPs (Bernhard et al., 1991; Skriver et al., 1992; Tchang et al., 1988), as well as of putative LTPs identified on the basis of cDNA-derived amino acid sequence homology (Bernhard and Sommerville, 1989; Fleming et al., 1992; Foster et al., 1992; Hughes et al., 1992; Sterk et al., 1991; Torres-Schumann et al., 1992; Weig et al., 1992), indicated that they most likely represent secreted proteins. In barley, an LTP previously identified as an α-amylase/protease inhibitor, was found to be present in the medium of an aleurone cell culture (Bernhard and Sommerville, 1989; Mundy and Rogers, 1986). Similarly, immunological studies on a putative LTP in *Arabidopsis* revealed that the protein is present extracellularly in cell walls of epidermal cells (Thoma et al., 1993). These results indicate that plant LTPs are extracellular proteins. Consequently, their function needs to be readdressed.

Sterk et al. (1991) reported the isolation of a cDNA clone corresponding to a 10 kDa protein in the medium of carrot somatic embryo cultures. The cDNA-derived amino acid sequence of this protein, designated extracellular protein 2 (EP2), showed homology to LTPs from several other plant species. Employing *in situ* mRNA localization, the EP2 gene was shown to be expressed in the protoderm of both zygotic and somatic embryos, and in epidermal cells of leaf

* Corresponding author (tel.: +31 8370 84706; fax: +31 8370 84325)

[a] MSU-DOE-Plant Research Laboratory, Michican State University, East Lansing, Michican, USA.

[b] Physiologie Cellulaire et Moléculaire, CNRS (URA 1180), Université Pierre et Marie Curie, 4 place Jussieu, 75252 Paris Cédex 05, France.

NATO ASI Series, Vol. H 81
Plant Molecular Biology
Edited by G. Coruzzi and P. Puigdomènech

primordia, flower organs and maturing seeds (Sterk et al., 1991). The expression of the gene in epidermal cells and the extracellular location of the encoded protein, led us to propose a role for the carrot EP2 protein in the transfer of cutin monomers through the cell wall of epidermal cells to sites of cutin polymerization (Sterk et al., 1991).

More recently, after purification of the EP2 protein from the medium of carrot embryo cultures, it was shown that the protein was able to enhance the transfer of fluorescent phospholipid analogs between artificial membranes, and that it was the only secreted protein with this property (Meijer et al., 1993). In addition, it was found that the EP2 protein was capable of binding palmitic acid, oleic acid and oleoyl-CoA. Since in plants these molecules are considered to be precursors of cutin monomers (Kolattukudy, 1980; Holloway, 1982b), these results suggest that the EP2 protein has the biochemical properties to either transfer precursors of cutin monomers or cutin monomers themselves.

In this paper another important aspect in the envisaged role of the EP2 protein as a carrier of precursors of cutin monomers is addressed, namely whether the amount of EP2 protein produced is sufficient to cope with the demand for cutin monomers. Evidence is presented confirming the indentity of the EP2 protein purified previously (Meijer et al., 1993), by showing that the amino acid composition of the isolated protein was similar to that predicted by the EP2-cDNA. The amino acid composition of the purified EP2 protein provided a more accurate quantitation of the amount of EP2 protein purified and it was found that the protein binds oleoyl-CoA in a near equimolar ratio. Calculations are presented indicating that when the EP2 protein is discarded after delivery of a cutin monomer, the amount of EP2 protein needed to form a cuticle around a globular embryo probably exceeds by far the amount of EP2 protein produced. This indicates that the EP2 protein may be involved in the transport of a particular subclass of cutin monomers only, or that the EP2 protein can shuttle between sites of monomer synthesis and cutin polymerization.

Materials and Methods

Plant Material and Culture Conditions. Suspension, embryo, and proliferating cultures from carrot (*Daucus carota* L. cv. Trophy, obtained from Zaadunie B.V., Enkhuizen, The Netherlands) were maintained as described (De Vries et al., 1988).

Protein Isolation and Purification. The EP2 protein was purified from the secreted proteins in the conditioned medium of 12 d embryo cultures as described previously (Meijer et al., 1993).

Amino Acid Composition Analysis. The amino acid composition of the purified EP2 protein was analysed according to Barkholt and Jensen (1989) in the laboratories of the MARIBO Seed Company, Copenhagen, Denmark.

Fatty Acid Binding Assay. Different amounts of purified EP2 were incubated for 30 min at room temperature with 2 nmol (220,000 cpm) [1-^{14}C]-labeled oleoyl-CoA (Amersham) in 20 mM Tris-HCl, pH 7.4, containing 150 mM NaCl, 1 mM EDTA and 1 mM EGTA, in a total volume of 100 µL. The mixtures were applied to 1 mL Sephadex G50 (Pharmacia LKB Biotechnology Inc.) columns in disposable 1 mL syringes to separate the free from the bound oleoyl-CoA molecules. Fractions of 100 μL were collected and monitored by scintillation counting.

Results

Isolation and Identification of EP2

To identify the EP2 protein during its purification as described previously (Meijer et al., 1993) an antiserum raised against the EP2-β-galactosidase fusion-protein (Sterk et al., 1991) was used. More recently, however, experiments indicated that the specificity of this antiserum was questionable, since the reaction observed appeared mainly due to an interaction between the EP2 protein and the second antibody-alkaline phosphatase conjugate (Gibco-BRL Life Technologies). The reaction was specific in that amongst all other proteins present in the embryo culture medium only the EP2 protein was recognized, while no reaction was observed with proteins isolated from the medium of non-embryogenic cell lines. The latter observation is in accordance with the absence of EP2 gene expression in non-embryogenic cell lines (Sterk et al., 1991). On immunoblots (not shown) the purified EP2 protein showed a strong reaction with antisera raised against LTPs from maize (Grosbois et al., 1987) and spinach (Bernhard et al., 1991) and with an antiserum raised against a fusion protein corresponding to a putative LTP from *Arabidopsis* (Thoma et al., 1993). In these experiments another second antibody-alkaline phosphatase conjugate (BioRad) was used for detection, since this gave only a very weak signal with the EP2 protein. Immunoblots obtained from proteins present in media of carrot cell suspension cultures employing the *Arabidopsis* LTP antiserum are shown in Fig.1.

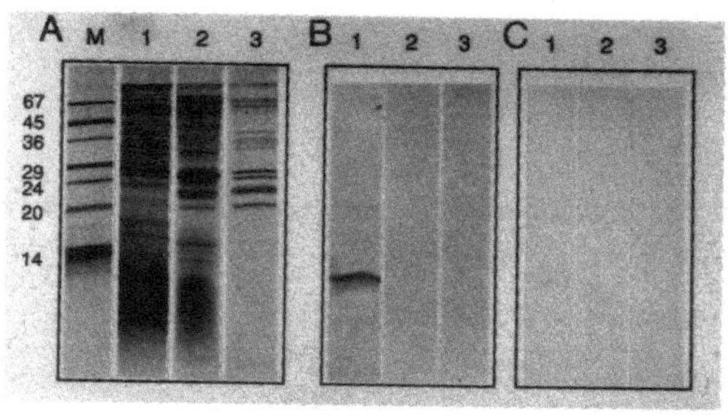

Fig. 1. SDS-PAGE and immunoblots of proteins present in the medium of embryo and suspension cultures of an embryogenic cell line and a suspension culture of a non-embryogenic cell line. A) Silver stained SDS-PAGE gel. B) Immunoblot obtained from a gel as in A) using an antiserum raised against an *Arabidopsis* LTP:MalE fusion protein. C) Immunoblot as in B) using only the alkaline phosphatase conjugated goat-anti-rabbit antiserum. *Lanes 1*: Proteins present in the medium of an embryo culture. *Lanes 2*: Proteins present in the medium of a suspension culture of an embryogenic cell line. *Lanes 3*: Proteins present in a suspension culture of a non-embryogenic cell line. *Lane M* in A): Marker proteins with their molecular mass (kDa) indicated on the left.

Table 1. Amino acid composition of purified carrot EP2 protein. The data are based on two amino acid analyses of the purified EP2 protein (Analysis), and are expressed as residues per 100 amino acid residues. For comparison the amino acid composition based on the amino acid sequence predicted by the EP2-cDNA (Sterk et al., 1991) is given (Calculated).

Amino Acid	Analysis	Calculated
Asx	11.6	11.7
Glx	5.0	3.2
Ser	3.2	2.1
Gly	10.2	8.5
His	0.3	0.0
Arg	5.3	6.4
Thr	10.2	10.6
Ala	11.7	11.7
Pro	6.3	6.2
Tyr	2.1	2.1
Val	9.8	11.7
Ile	2.0	2.1
Leu	11.6	11.7
Phe	0.2	0.0
Lys	3.2	3.2
Trp	- [a]	0.0
Cys	7.3[b]	8.5
Met	0.1	0.0

[a] Not determined. [b] Determined as the sum of 1/2-cys.

Further identification of the purified protein as the EP2 gene encoded protein was confirmed by determining its amino acid composition. As shown in Table 1 the actual amino acid composition of the purified protein was very similar to that predicted by the EP2-cDNA. In view of the observed discrepancies in determining the amount of protein by methods used routinely for protein content determinations (Meijer et al., 1993), the final yield of purified EP2 protein was calculated from its amino acid composition (Table 2).

Table 2. Purification of carrot EP2. The amounts of protein obtained during the purification of the EP2 protein from 25 L medium of 12 d embryo cultures were determined by three methods based on different principles Meyer et al., 1993). In all methods the amounts of protein given are relative to BSA which was used as a standard. In the last column the amount of EP2 protein was calculated from the amino acid composition (AAC; cf. Table 1).

	Amount of protein (mg)			
Fraction	Bradford[a]	Smith et al.[b]	A280nm	AAC
Embryo culture medium proteins	16	-	251	-
S-Sepharose flow through	13	-	211	-
S-Sepharose salt pulse	6.3	38	45	-
Superdex 75 fractions 28-34	0.24	9.6	-	-
Mono S fractions 10-15 (EP2)	0.16	6.5	2.2	2.4

[a] according to Bradford, 1976; [b] according to Smith et al., 1985; - = not determined

Fatty Acid Binding Activity of EP2

Employing a simple gel permeation assay it was shown previously (Meijer et al 1993) that the EP2 protein was capable of binding palmitic acid, oleic acid and oleoyl-CoA. Accurate quantitation of the binding ratio, however, was hampered by difficulties in determining the exact amount of protein applied. With the amino acid composition known, more accurate quantitation of the amount of EP2 protein was possible. By varying the amount of EP2 protein in the gel permeation assay, it could be determined that the EP2 protein binds oleoyl-CoA molecules in a near equimolar ratio (Fig. 2).

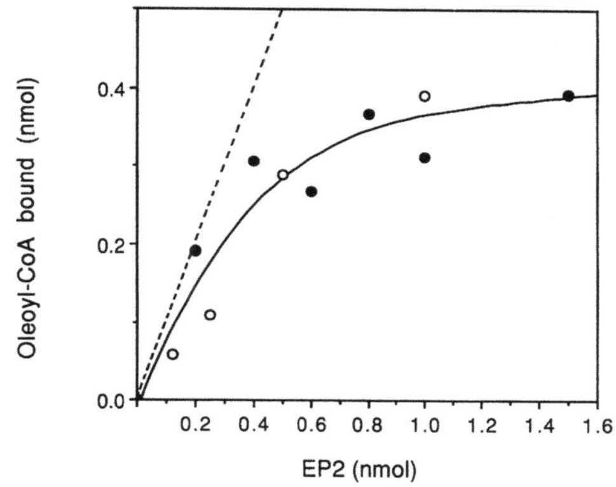

Fig. 2. Binding of oleoyl-CoA by carrot LTP EP2. ^{14}C-Labeled oleoyl-CoA (2 nmol; 220,000 cpm) dissolved in buffer (20 mM Tris-HCl, pH 7.4, containing 100 mM NaCl, 1 mM EDTA and 1 mM EGTA) was incubated with different amounts of purified EP2 protein. After incubation for 30 min, the mixtures were applied to 1 mL Sephadex G50 gel filtration columns and eluted with buffer to separate the free and bound oleoyl-CoA. Different symbols indicate independent experiments. The dashed line indicates an 1:1 binding-ratio.

Quantitative Aspects of the Putative Role of EP2 in Cutin Synthesis

Another important aspect in the envisaged role of the EP2 protein as a carrier of cutin monomers, is whether the amount of EP2 protein produced is sufficient to cope with the demand for cutin monomers. In theory, the synthesis of a 25 nm thick cuticle around a globular embryo, 200 µm in diameter on average, would require the synthesis of ~ 3 ng cutin/globular embryo (according to Schreiber and Schönherr (1990), cutin weighs ~1000 kg/m^3), and thus the transfer of 3 ng monomers/globular embryo. Based on the results presented in Fig. 2, we may assume that the EP2 protein (MW ~10000) binds a single cutin monomer (MW of the acyl group ~300). So, if the EP2 protein is used as a carrier only once, the synthesis of a 25 nm thick cuticle would require the production of (3x10000/300 ng=) 0.1 µg EP2 protein/globular embryo. As can be calculated from the data presented in Fig. 3, about 80 globular embryos/mL are present in our embryo cultures at day 4. When all of them acquired a cuticle as mentioned in the example above, 8 µg EP2 protein/mL should have been produced.

The results presented in Table 2 show that the medium of an embryo culture contained at least 0.1 mg EP2/L after 12 days of culture. As shown in Fig. 3, there is a steady increase in the amount of EP2 protein in the medium during somatic embryogenesis. Assuming that the EP2 protein production is linear, only about 0.033 µg EP2/mL would be present in the medium at day 4. Even if we consider that only half of the amount of EP2 protein is present in the medium, and the remainder in the cell wall (Sterk et al., 1991), this would still be at least 120-fold lower as required for the transfer of the cutin monomers as calculated above. Though largely hypothetical, this calculation shows that a model in which the EP2 protein is discarded after delivery of cutin monomers can not be valid as such.

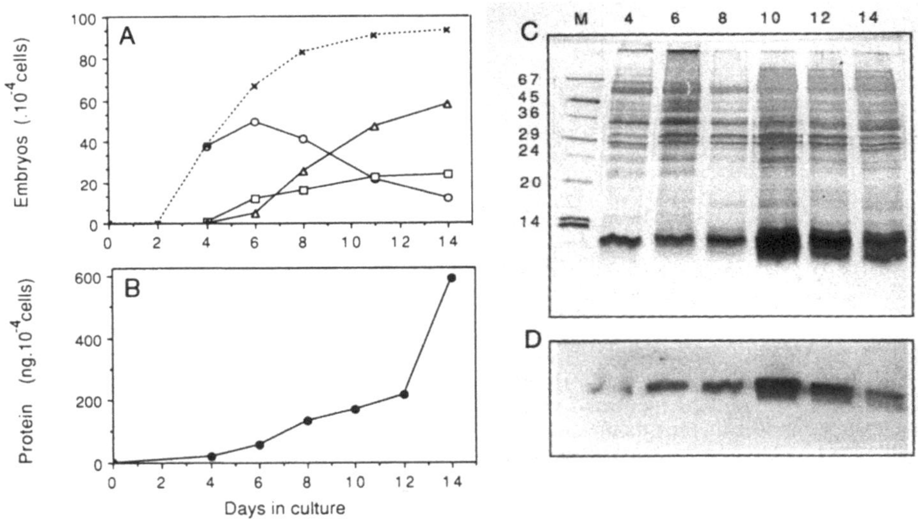

Fig. 3. Production of carrot LTP EP2 during somatic embryogenesis.A) Formation of somatic embryos. Indicated are the number of globular embryos (o), heart-stage embryos ( ) torpedo-stage embryos (Δ) and the total number of embryos (x), expressed per 10000 cells initially present in the culture. At the start of the experiment the culture contained 20000 cells/mL. B) The amount of protein present in the medium during somatic embryogenesis expressed per 10000 cells initially present in the culture. C) SDS-PAGE analysis of the proteins (silver stained) present in the medium. Equal amounts of protein (1.5 µg) were applied in each lane. Numbers above the lanes correspond to the day the proteins were sampled from the culture. *Lane M*: Marker proteins with their molecular mass (kDa) indicated on the left. D) Immunoblot from a gel as in C), using the antiserum against the *Arabidopsis* LTP.

Discussion

We have proposed previously that the carrot EP2 protein transports cutin monomers from their site of synthesis through the cell wall towards the growing cutin polymer (Sterk et al., 1991).

This hypothesis was based on three findings: 1) the cDNA-derived EP2 amino acid sequence shows homology with other plant LTPs, 2) the EP2 protein is secreted, and 3) the EP2 gene is expressed only in (pro)epidermal cells that synthesize a primary cuticle. Subsequently we have shown that the EP2 protein is really an LTP capable of enhancing the transfer of fluorescent phospholipid analogs between membranes *in vitro* (Meijer et al., 1993), and therefore is similar to LTPs from other plants (Kader 1990). In addition it was found that the EP2 protein was able to bind palmitic acid, oleic acid and oleoyl-CoA (Meijer et al., 1993), molecules shown to be incorporated in cutin *in vitro* (Croteau and Kolattukudy, 1974) and *in vivo* (Lendzian and Schönherr, 1983).

Cutin is a polyester of fatty acids, consisting mainly of hydroxy-, epoxy- and carboxy-derivatives of palmitic and/or stearic acid, although unsubstituted palmitic and oleic acid may be present as minor components (Kolattukudy, 1980; Holloway, 1982b). In contrast to its composition, the knowledge about the biosynthesis of cutin is still scanty. Two pathways for the biosynthesis of C_{16} and C_{18} monomers have been proposed, based on *in vitro* studies using labelled precursors, including palmitic and oleic acid (Kolattukudy, 1980). Besides the final products of these pathways, intermediates are also incorporated in cutin (Kolattukudy, 1980), and probably arrive at the site of cutin polymerization as their respective CoA-esters. Therefore, oleoyl-CoA can be considered to be a true cutin monomer, and thus the binding of oleoyl-CoA by the EP2 protein is in agreement with our hypothesis. Studies on the binding of the EP2-protein with other, more commonly found modified cutin monomers, are needed to answer whether the EP2 protein is capable of binding cutin monomers in general. Activation of the carboxyl group by CoA-ester formation is considered an essential step in the synthesis of the cutin monomers as well as in their incorporation into the cutin polymer by a cell wall bound enzyme (Croteau and Kolattukudy, 1974). Therefore, the function of the EP2 protein in cutin synthesis may be two-fold: 1) The targetting of cutin monomers and precursors thereof towards the site of cutin synthesis, and thus preventing them from being trapped in endoplasmic reticulum or plasmalemma membranes or in other cell wall components. 2) Protection of these molecules from enzymatic and/or non-enzymatic loss of the CoA-ester linkage essential for the incorporation into the cutin polymer.

The calculation of the amount of EP2-protein required for the synthesis of a cuticle surrounding globular embryos, though largely hypothetical, was meant as a demonstration that the model in which the EP2 protein is discarded after delivery of cutin monomes may be to simple. Both structure and chemical composition of cuticles change during development (Heide-Jørgensen, 1991; Holloway, 1982a+b; Kolattukudy, 1980; Riederer and Schönherr, 1988), and it could be that the EP2 protein is only involved in the transfer of a particular subset of cutin monomers. In accordance with his hypothesis the EP2 gene was shown to be expressed in young (meristematic) epidermal tissues, but not in the epidermis of mature leaves or stems (Sterk et al., 1991). Although the the presence of the EP2 protein in mature organs has not been investigated yet, this may indicate that the role of the EP2 protein is restricted to an early stage in cuticle formation, for instance in the deposition of the cuticle proper or procuticle. The deposition of a cuticle proper was shown to start very early in the formation of the epidermis and to be completed before elongation starts (Riederer and Schonherr, 1988). The cuticle proper may be derived from a procuticle, an osmiophilic layer found in electron microscopic images of very young epidermal tissues (Heide-Jørgensen, 1991). At present no biochemical data concerning the composition or biosynthesis of a procuticle are available. Alternatively, it may be that the EP2 protein is reused and functions as a shuttle in the transfer of cutin monomers from their site of synthesis to the site of cutin polymerization. In this case other, membrane-bound, proteins would probably be required for receptor-mediated endocytosis of the free EP2 protein, or transfer of cutin monomers to the plasmalemma and loading of the free EP2 protein at the exterior surface of the plasmalemma (Fig. 4). These possibilities are more easy to envisage in the plant, where the EP2 protein is likely to be localized in the cell wall, similar as shown for the putative LTP from *Arabidopsis* (Thoma et al., 1993). In embryo cultures the shuttle will function very inefficiently, since a major portion of the EP2 protein will be lost from the cell wall by diffusion into the medium.

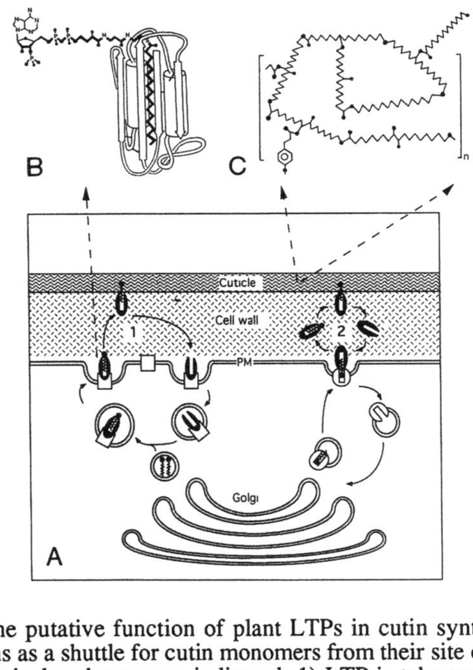

Fig. 4. Model for the putative function of plant LTPs in cutin synthesis. A) Plant LTP is secreted and functions as a shuttle for cutin monomers from their site of synthesis towards the cuticle. Two hypothetical pathways are indicated: 1) LTP is taken up by receptor-mediated endocytosis and loaded intracellularly by fusion of an endocytic vesicle with cutin monomer-containing vesicles derived from Golgi or endoplasmic reticulum. 2) LTP remains extracellular and is loaded by a membrane-bound cutin monomer carrier. B) A hypothetical model of a plant LTP (after Madrid and von Wettstein, 1991) containing an oleoyl-CoA molecule. C) A tentative model of cutin (after Kolattukudy, 1980)

Taken together, the available evidence on most plant LTPs suggest that they are biochemically capable of binding cutin monomers (Arondel et al., 1990; Rickers et al., 1984; Rickers et al., 1985) and are synthesized in the right place and time (Foster et al., 1992; Skriver et al., 1992; Sossountzov et al., 1991; Sterk et al., 1991; Thoma et al., 1993) to have a function in cuticle formation. Definite answers should come from studies where the LTP gene is underexpressed or mutated and from the identification of the natural substrate present in the protein *in vivo.*

Acknowledgements.

We thank the DANISCO Seed Company for amino acid analysis, and Peter van Druten for photography. These investigations were supported (EAM, TH) by the European Commission Biotechnology Program BRIDGE.

Literature

Arondel V, Kader J-C (1990) Lipid transfer in plants. Experientia 46: 579-585

Arondel V, Vergnolle C, Tchang F, Kader J-C (1990) Bifunctional lipid-transfer: fatty acid-binding proteins in plants. Mol Cell Biochem 98, 49-56

Barkholt V, Jensen AL (1989) Determination of cysteine plus half-cystine in proteins after hydrochloric acid hydrolysis with a disulfide compound as additive. Anal Biochem 177, 318-322

Bernhard WR, Somerville CR (1989) Coidentity of putative amylase inhibitors from barley and finger millet with phospholipid transfer proteins inferred from amino acid sequence homology. Arch Biochem Biophys 269, 695-697

Bernhard WR, Thoma S, Botella J, Somerville CR (1991) Isolation of a cDNA clone for spinach lipid transfer protein and evidence that the protein is synthesized by the secretory pathway. Plant Physiol 95, 164-170

Bradford M (1976) A rapid and sensitive method for the quantitation of microgram quantities of protein utilizing the principle of protein-dye binding. Anal Biochem 71, 248-254

Croteau R, Kolattukudy PE (1974) Biosynthesis of hydroxyfatty acid polymers. Enzymatic synthesis of cutin from monomer acids by cell-free preparations from the epidermis of *Vicia faba* leaves. Biochemistry 15: 3193-3202

De Vries SC, Booij H, Janssens R, Vogels R, Saris L, LoSchiavo F, Terzi M, Van Kammen A (1988) Carrot somatic embryogenesis depends on the phytohormone-controlled expression of correctly glycosylated extracellular proteins. Genes Develop 2, 462-476

Fleming AJ, Mandel T, Hofmann S, Sterk P, De Vries SC, Kuhlemeier C (1992) Expression pattern of a tobacco lipid transfer protein gene within the shoot apex. Plant J 2, 855-862

Foster GD, Robinson SW, Blundell RP, Roberts MR, Hodge R, Draper J, Scott RJ (1992) A *Brassica napus* mRNA encoding a protein homologous to phospholipid transfer proteins, is expressed specifically in the tapetum and developing microspores. Plant Science 84, 187-192

Grosbois M, Guerbette F, Douady D, Kader J-C (1987) Enzyme immunoassay of a plant phospholipid transfer protein. Biochim Biophys Acta 917, 162-168

Heide-Jørgensen HS (1991) Cuticle development and ultrastructure: evidence for a procuticle of high osmium affinity. Planta 183, 511-519

Holloway PJ (1982a) Structure and histochemistry of plant cuticular membranes: an overview.*In* DF Cutler, KL Alvin, CE Price, eds, The plant cuticle. Academic Press, Londen New York, pp 1-31

Holloway PJ (1982b) The chemical constitution of plant cutins. *In* DF Cutler, KL Alvin, CE Price, eds, The plant cuticle. Academic Press, Londen New York, pp 45-85

Hughes MA, Dun MA, Pearce RS, White AJ, Zhang L (1992) An abscisic-acid-responsive, low temperature barley gene has homology with a maize phospholipid transfer protein. Plant Cell Environ 15, 861-865

Kader J-C (1990) Intracellular transfer of phospholipids, galactolipids, and fatty acids in plant
Kolattukudy PE (1980) Cutin, suberin, and waxes. *In* PK Stumpf, ed, Biochemistry of Plants, A Comprehensive Treatise. Lipids: Structure and Function, Vol 4. Academic Press, New York, pp 571-645
Lendzian KJ, Schönherr J (1983) In-vivo study of cutin synthesis in leaves of *Clivia miniata* Reg. Planta 158, 70-75
Madrid SM, von Wettstein D (1991) Reconsiling contradictory notions on lipid transfer proteins in higher plants. Plant Physiol Biochem 29,
Meijer EA, De Vries SC, Gadella Jr DWJ, Wirtz KWA, Hendriks T (1993) Characterization of the non-specific lipid transfer protein EP2 from carrot (*Daucus carota* L.). Mol Cell Biochem, in press.
Mundy J, Rogers JC (1986) Selective expression of a probable amylase/protease inhibitor in barley aleurone cells: Comparison to the barley amylase/subtilisin inhibitor. Planta 169, 51-63
Rickers J, Spener F, Kader JC (1985) A phospholipid transfer protein that binds long-chain fatty acids. FEBS Letters 180, 29-32
Rickers J, Tober I, Spener F (1984) Purification andbinding characteristics of a basic fatty acid binding protein from *Avena sativa* seedlings. Biochim Biophys Acta 794, 313-319
Riederer M, Schönherr J (1988) Development of plant cuticles: fine structure and cutin composition of *Clivia miniata* Reg. leaves. Planta 174, 127-138
Schreiber L, Schönherr J (1990) Phase transitions and thermal expansion coefficients of plant cuticles. The effects of temperature on structure and function. Planta 182, 186-193
Skriver K, Leah R, Müller-Uri F, Olsen FL, Mundy J (1992) Structure and expression of the barley lipid transfer protein gene *Ltp1*. Plant Mol Biol 18, 585-589
Smith PK, Krohn RI, Hermanson GT, Mallia AK, Gartner FH, Provenzano MD, Fujimoto EK, Goeke NM, Olson BJ, Klenk DC (1985) Measurement of protein using bicinchoninic acid. Anal Biochem 150, 76-85
Sossountzov L, Ruiz-Avila L, Vignols F, Jolliot A, Arondel V, Tchang F, Grosbois M, Guerbette F, Miginiac E, Delseny M, Puigdomenèch P, Kader J-C (1991) Spatial and temporal expression of a maize lipid transfer protein gene. Plant Cell 3, 923-933
Sterk P, Booij H, Schellekens GA, Van Kammen A, De Vries SC (1991) Cell-specific expression of the carrot EP2 lipid transfer protein gene. Plant Cell 3, 907-921
Tchang F, This P, Stiefel V, Arondel V, Morch M-D, Pages M, Puigdomenech P, Grellet F, Delseny M, Bouillon P, Huet J-C, Guerbette F, Beauvais-Cante F, Duranton H, Pernollet J-C, Kader J-C (1988) Phospholipid transfer protein: Full length cDNA and amino acid sequence in maize. J Biol Chem 263, 16849-16855
Thoma S, Kaneko Y, Somerville CR (1993) The non-specific lipid transfer protein from *Arabidopsis* is a cell wall protein. Plant J, .
Torres-Schumann S, Godoy JA, Pintor-Toro JA (1992) A probable lipid transfer protein gene is induced by NaCl in stems of tomato plants. Plant Mol Biol 18, 749-757
Weig A, Komor E (1992) The lipid-transfer protein C of *Ricinus communis* L.: isolation of two cDNA sequences which are strongly and exclusively expressed in cotelydons after germination. Planta 187, 367-371
Wirtz, KWA (1991) Phosholipid transfer proteins. Ann Rev Biochem 60, 73-99

Early Events in Apical-basal Pattern Formation in *Arabidopsis*

Gerd Jürgens, Ramón A. Torres Ruiz, Thomas Laux, Ulrike Mayer and Thomas Berleth
Institut für Genetik und Mikrobiologie
Lehrstuhl für Genetik
Universität München
Maria-Ward-Str. 1a
D-8000 München 19
Federal Republic of Germany

Plant development is dominated by meristematic growth which continually adds new structures to the pre-existing body (Steeves and Sussex, 1989). A different mode of growth appears to prevail during a brief period of the life cycle: the developing embryo grows by disperse cell divisions, and this growth pattern may thus be termed embryonic. However, the switch from embryonic to meristematic growth is not coupled to physiological changes that signify the transition from the heterotrophic embryo to the autotrophic seedling, such as seed maturation, dormancy or germination. Embryonic growth rather seems related to developmental processes that generate the primary organisation of the plant body, including the meristems of the shoot and the root. In some flowering plant species, such as maize, the shoot meristem becomes active in the embryo, producing several leaf primordia while in others, including *Arabidopsis*, the shoot meristem remains nearly indistinct before seed germination (Medford, 1992). By contrast, the root meristem contributes to the formation of the seedling body in most flowering plant species (Rutishauser, 1969).

The apical-basal axis is the main axis of polarity in flowering plants, and the meristems of the shoot and the root occupy the opposite ends of the axis. In the seedling, the two meristems are separated from each other by the following intervening pattern elements: cotyledons, hypocotyl and embryonic root (radicle). Here, we discuss genetic aspects of processes that generate the apical-basal pattern in the *Arabidopsis* embryo.

NATO ASI Series, Vol. H 81
Plant Molecular Biology
Edited by G. Coruzzi and P. Puigdomènech

Development of the apical-basal pattern in the *Arabidopsis* embryo

The elements of the apical-basal pattern of the seedling have been traced back to cell groups in the early embryo, which was facilitated by the regular patterns of cell division in *Arabidopsis* embryogenesis (Fig. 1; Mansfield and Briarty, 1991; Jürgens and Mayer, 1993).

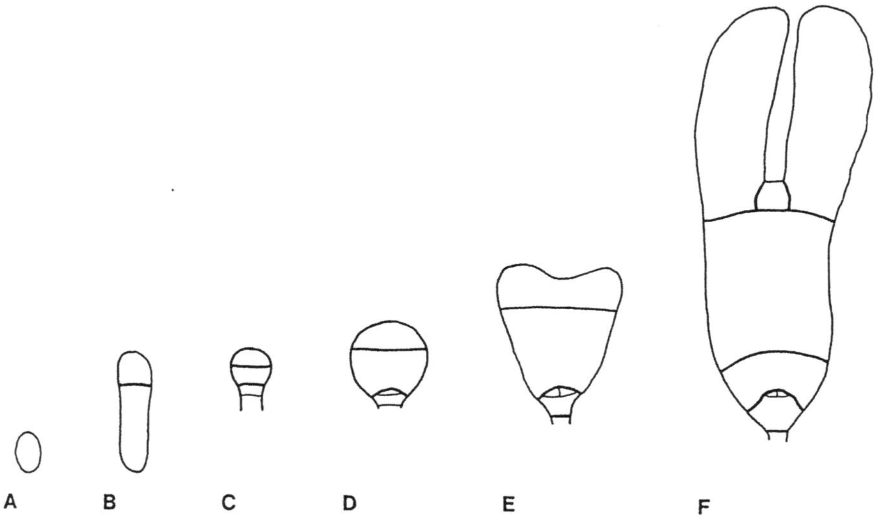

Fig. 1. Development of the apical-basal pattern in the embryo. (A) zygote, (B) 1-cell, (C) octant, (D) globular, (E) early-heart, (F) torpedo stages. For details, see text.

Following fertilisation, the egg cell expands in the future apical-basal axis about three-fold before dividing asymmetrically to give a small apical and a large basal daughter cell. The basal cell produces, by repeated transverse divisions, a file of 6-9 cells, and only the uppermost derivative becomes the hypophysis which joins the embryo proper to give rise to the very basal end of the embryo. The apical daughter cell of the zygote undergoes three rounds of stereotyped pseudo-cleavage divisions, resulting in two tiers each of four cells which make up the octant-stage embryo proper. The upper tier will eventually give rise to the shoot meristem and the cotyledons while the derivatives of the lower tier will form the hypocotyl and the embryonic root (radicle), and also contribute to the root meristem. In the globular-stage embryo, the descendants of the two tiers begin to differ in shape and pattern of cell division, and the hypophysis divides asymmetrically, with the upper lens-shaped daughter cell generating the quiescent centre of the root meristem, while the lower daughter cell will produce the central portion of the root cap. Three major regions of the apical-basal axis can be distinguished in the heart-stage embryo: apical, central and basal. The basal region corresponds to the incipient root primordium derived from the hypophysis. From the late-heart stage, a layer of root precursor cells (initials) above the quiescent centre give off files of cells which constitute the embryonic root, thus subdividing the central region into an upper hypocotyl and a lower radicle. In the apical region, the incipient shoot meristem is marked off against the cotyledonary primordia so that the mature embryo contains all apical-basal pattern elements differentiated in the seedling.

Apical-basal pattern mutants

How the apical-basal pattern is generated in the embryo is not known for any flowering plant species. In order to analyse underlying mechanisms, we have taken a genetic approach by isolating and characterising relevant mutants in *Arabidopsis*. Four genes have thus been identified on the basis of their mutant seedling phenotypes (Mayer et al., 1991), and mutant alleles of another gene were recognised by the failure of mutant seedlings to produce rosette leaves (T. Laux and G. Jürgens, unpubl. data). Each of the five genes seems to make a specific contribution to pattern formation along the apical-basal axis, as suggested by their mutant phenotypes (Fig. 2).

Mutant *gnom* seedlings lack the apical and basal ends: the root is always missing while the shoot end is variably reduced. In the extreme case, the seedling is ball-shaped, showing no morphological signs of apical-basal polarity (Mayer et al., 1993). Mutations in the other genes cause defects that are confined to specific regions along the axis. Mutant *monopteros* seedlings lack structures derived from the central and basal regions of the heart-stage embryo: hypocotyl, radicle and root meristem, which are replaced by a basal peg attached to the cotyledons (Berleth and Jürgens, 1993). The apical end can be normal such that a shoot meristem produces rosette leaves; however, a considerable proportion of mutant seedlings have fused cotyledons. Strong mutations in the *gurke* gene cause complementary defects: mutant seedlings seem to lack cotyledons and shoot meristem which are derived from the apical region of the heart-stage embryo, while the other structures are normal (Mayer et al., 1991; R.A. Torres Ruiz and G. Jürgens, unpubl. data). In mutant *fackel* seedlings, the root seems directly attached to the cotyledons, suggesting that the hypocotyl is missing (Mayer et al., 1991). Finally, mutant *zwille* seedlings look normal upon germination but subsequently fail to produce rosette-leaf primordia, thus revealing a defect of the shoot meristem (T. Laux and G. Jürgens, unpubl. data).

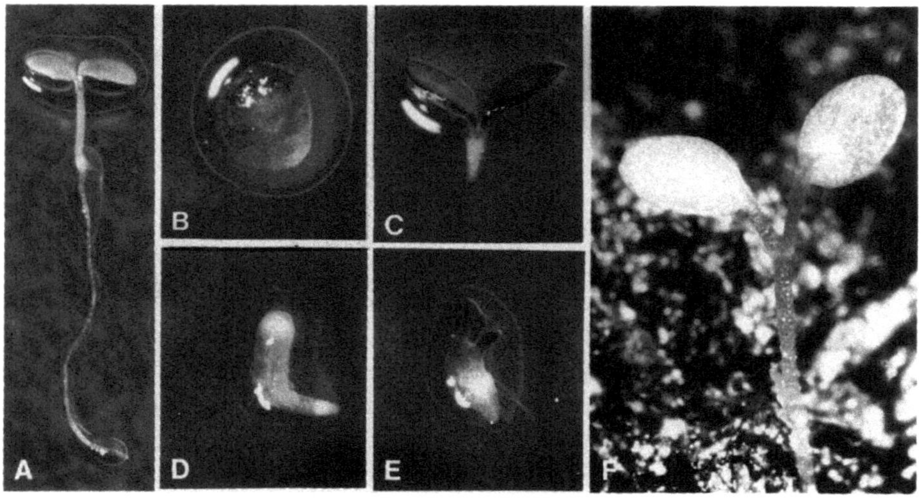

Fig. 2. Mutant seedling phenotypes of genes involved in apical-basal pattern formation. (A) wild-type, (B) *gnom*, (C) *monopteros*, (D) *gurke*, (E) *fackel*, (F) *zwille*.

Development of mutant seedlings

In order to define more precisely how mutations in the genes identified affect development, we performed two sets of experiments on mutant seedlings, depending on their pattern defects (Fig. 3). Mutant seedlings that lack basal structures, such as *monopteros* and *gnom*, were assayed for their ability to form roots under inducing conditions. As a control, wild-type seedlings were bisected below the cotyledons, and the apical pieces were found to regenerate the missing basal structures when cultured on root-inducing medium. The same result was obtained with mutant *monopteros* seedlings that had been wounded or bisected below the cotyledons, which suggests that the *mp* gene is only required for organising basal development in the embryo but not for forming root structures (Berleth and Jürgens, 1993). Under the same experimental conditions, mutant *gnom* seedlings did not produce roots but formed callus instead. Thus, the *gnom* gene appears to be required for root formation, both in the embryo and in tissue culture (Mayer et al., 1993).

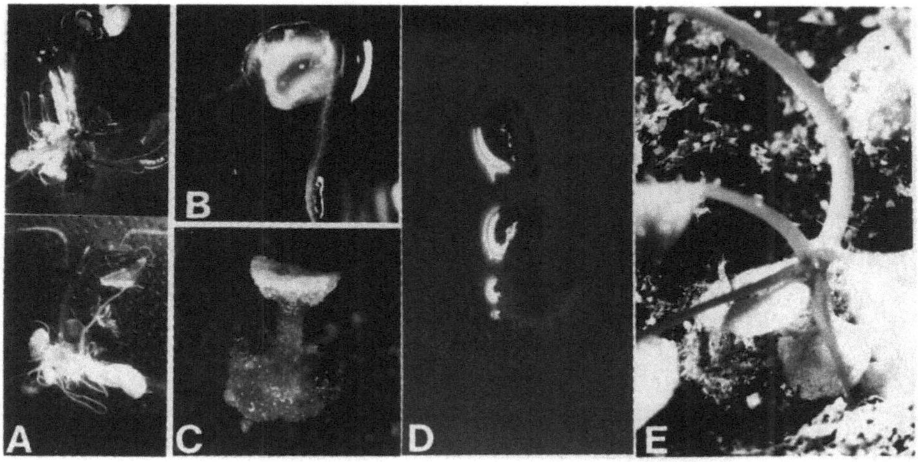

Fig. 3. Development of mutant seedlings. (A-C) Response to root-inducing conditions following wounding: (A) wild-type, (B) *monopteros*, (C) *gnom*; (D) *gurke* seedling cultured in vitro; (E) *zwille* seedling with secondary shoot.

Mutant seedlings with apical pattern defects, such as *gurke* and *zwille*, were left intact. Mutant *gurke* seedlings, which appear to lack the apical region, produced green leaf-like tissue from the apical end when cultured in vitro. However, the growing apical tissue did not acquire the normal organisation of shoot and leaves, which suggests that the *gurke* gene may be required for organising the apical region (R.A. Torres Ruiz and G. Jürgens, unpubl. data). Mutant *zwille* seedlings, when grown on soil, do not form rosette leaves but after about two weeks, they can form secondary shoots. Evidently, the *zwille* gene is required for shoot formation in the embryo but not at later stages (T. Laux and G. Jürgens, unpubl. data).

Early phenotypes of apical-basal pattern mutants

The development of mutant embryos has been studied in order to determine how the seedling phenotypes originate. In each case, embryogenesis was found to deviate from normal in a specific manner, as indicated by the time and position of abnormal cell behaviour (Fig. 4).

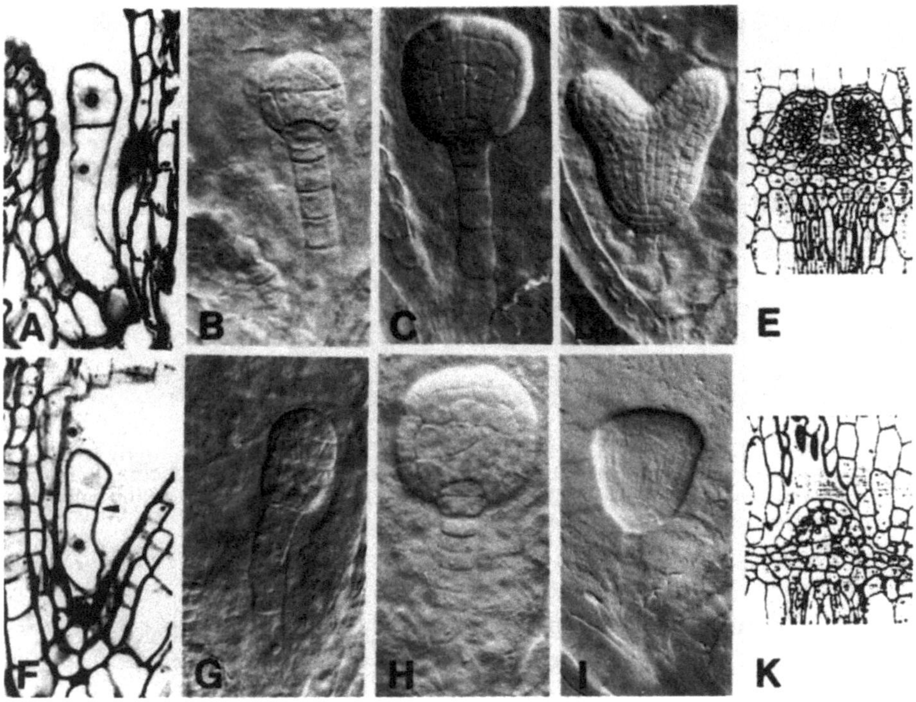

Fig. 4. Early phenotypes of apical-basal pattern mutants. (A-E) wild-type, (F) *gnom*, (G) *monopteros*, (H) *fackel*, (I) *gurke*, (K) *zwille*. Corresponding developmental stages of wild-type and mutant embryos are shown: (A,F) one-cell, (B,G) octant, (C, H) globular, (D,I) early-heart. (E,K) Seedlings, 2 days after germination. (A,E,F,K) Histological sections, (B-D,G-I) whole-mount preparations.

Mutant *gnom* embryos are abnormal from the earliest stage: the zygote tends to divide symmetrically, giving an enlarged apical cell at the expense of the basal cell (Mayer et al., 1993). The apical cell divides abnormally, with the plane of division being oblique or perpendicular, rather than in parallel, to the apical-basal axis. It is conceivable that these very early changes cause the irregularities seen later in *gnom* embryogenesis. Mutant *monopteros* embryos have supernumerary cells at the octant stage, but subsequent development shows that the derivatives of the lower tier(s) lack the organised growth pattern of the basal region (Berleth and Jürgens, 1993). In addition, a central pile of cells appears in the place of the incipient root primordium which is normally produced by the hypophysis. Mutant *fackel* embryos are abnormal at the globular stage: the cells that normally give rise to the vascular tissue in the hypocotyl do not divide unequally (U. Mayer and G. Jürgens, unpubl. data). Mutant *gurke* embryos are distinguishable at the early-heart stage when in wild-type but not in *gurke* embryos, prominent cotyledonary primordia begin to bulge out (R.A. Torres Ruiz and G. Jürgens, unpubl. data). The mutant *zwille* phenotype is expressed at the cellular level immediately after germination when the cells in the shoot-meristem region enlarge while the cells of the wild-type shoot meristem remain small and densely cytoplasmatic (T. Laux and G. Jürgens, unpubl. data).

Gene action in apical-basal pattern formation: a model

How the genes identified might bring about the subdivisions of the apical-basal axis has been studied in double mutants in order to distinguish between combinatorial and hierarchical modes of gene action. In all combinations involving the *gnom* gene, only the two parental phenotypes appeared in the progeny of trans-heterozygotes, suggesting epistatic relations between the genes involved. However, due to the lack of linked markers that would enable independent assessment of the genotypes, identifying the epistatic gene entirely rests on statistical arguments for most combinations tested. Only in one case could we recognise the epistatic gene because the two genes involved are closely linked such that in the cis-double heterozygote the progeny should predominantly show the phenotype of the epistatic gene, and it was thus found that the *gnom* gene is epistatic to the *monopteros* gene (Mayer et al., 1993).

Apical-basal pattern formation proceeds step-wise in the *Arabidopsis* embryo, generating successively smaller regions along the axis. Conceptually, two different processes may be distinguished, partitioning and region-specific development. Our present view of how the genes identified contribute to apical-basal pattern formation is summarised in the following model (Fig. 5).

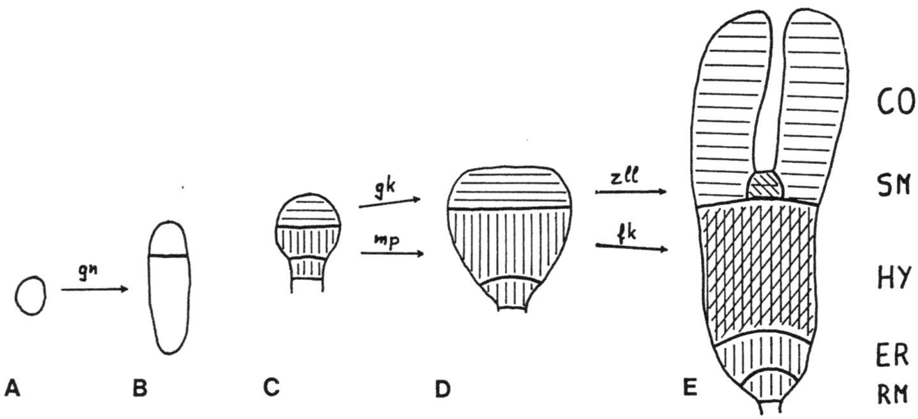

Fig. 5. A model of apical-basal patterning in the *Arabidopsis* embryo.(A-C) Partitioning of the axis: *gn* action results, via asymmetric division of the zygote (A,B), in major subdivisions of the octant-stage embryo (C). Subsequent region-specific development of upper and lower tiers (C-E) establishes, via the triangular stage (D), different primordia at the torpedo stage (E): shoot meristem (SM), cotyledons (CO), hypocotyl (HY), embryonic root (ER) and root meristem (RM). Differently hatched regions correspond to putative domains of action of different genes: *gk* and *zll* in the apical, *mp* and *fk* in the central portions of the embryo. For time of gene action, see text.

The *gnom* gene seems necessary for proper partitioning of the apical-basal axis which is initiated by the asymmetric division of the zygote. The small apical daughter cell undergoes three rounds of stereotyped divisions, resulting in the two-tiered octant-stage embryo. As a consequence of the partitioning, the upper and the lower tiers seem to acquire different developmental fates which are initially expressed as different patterns of cell division. This region-specific development depends on the activity of the *monopteros* gene for the lower-tier derivatives, while the *gurke* gene may play a comparable role for the upper-tier derivatives. The *monopteros* gene is also essential for normal development of the incipient root primordium which originates from the hypophysis. It is not clear whether the *monopteros* gene acts in both regions autonomously or whether only one of the two requires the activity directly. By contrast, the domain *fackel* gene action seems confined to the primordium of the hypocotyl from the globular stage. It is not until the torpedo stage that a comparable subdivision becomes apparent in the apical region derived from the upper tier of the octant-stage embryo: the incipient shoot meristem is marked off against the cotyledonary primordia, and the *zwille* gene seems to direct its development.

Literature references

Berleth T, Jürgens G. (1993) The role of the *monopteros* gene in organising the basal body region of the *Arabidopsis* embryo.Development 118 (in press)

Jürgens G, Mayer U (1993) *Arabidopsis*. In: A Colour Atlas of Developing Embryos. Bard J (ed). Wolfe Publ. London

Mansfield SG, Briarty LG (1991). Early embryogenesis in *Arabidopsis thaliana*. II. The developing embryo. Can J Bot 69: 461-476

Mayer U, Torres Ruiz RA, Berleth T, Miséra S, Jürgens G (1991) Mutations affecting body organization in the *Arabidopsis* embryo. Nature 353: 402-407

Mayer U, Büttner G, Jürgens G (1993) Apical-basal pattern formation in the *Arabidopsis* embryo: studies on the role of the *gnom* gene. Development 117: 149-162

Medford JI (1992) Vegetative apical meristems. The Plant Cell 4:1029-1039

Rutishauser A (1969) Embryologie und Fortpflanzungsbiologie der Angiospermen. Springer-Verlag, New York, Wien

Steeves TA, Sussex IM (1989) Patterns in Plant Development. Cambridge University Press, Cambridge

Diversity of Embryonic Mutants Identified Following *Agrobacterium* Mediated Seed Transformation in *Arabidopsis thaliana*

David W. Meinke
Department of Botany
Oklahoma State University
Stillwater, OK 74078 USA

Introduction

Arabidopsis thaliana has clearly become the model system of choice for a wide range of studies in plant developmental and molecular genetics (Koncz et al. 1992; NSF Publication 92-112; Meinke 1993). My laboratory has for many years pursued the use of *Arabidopsis* as a model system for genetic analysis of plant embryogenesis (Meinke and Sussex 1979). My approach was based initially on the pioneering work of Andreas Müller (1963). The history of this field has been the subject of several extensive reviews (Meinke 1991a,b; Lindsey and Topping 1993). The purpose of this minireview will be to focus primarily on the isolation and characterization of T-DNA insertional mutants of *Arabidopsis* with defects in embryo development. Additional details on these mutants can be obtained from recent publications (Errampalli et al. 1991; Castle and Meinke 1993; Castle et al. 1993).

Although embryonic mutants have been known for many years (Meinke 1986), large numbers of insertional mutants defective in embryogenesis have only recently been identified among transposon lines of maize (Clark and Sheridan 1991; Sheridan and Clark 1993) and T-DNA tagged families of *Arabidopsis* (Errampalli et al. 1991; Castle et al. 1993). The importance of these insertional mutants is that flanking plant sequences can be recovered through various methods including plasmid rescue (Yanofsky et al. 1990; Castle et al. 1993; Kieber et al. 1993), inverse PCR (Deng et al. 1992), and library construction (Herman and Marks 1989). Thus, the identification of tagged embryonic mutants is an important first step in the molecular isolation of genes with essential functions during plant embryogenesis. Although considerable progress has been made towards the isolation of transposon tagged mutants in *Arabidopsis*, currently the most efficient method of generating

NATO ASI Series, Vol. H 81
Plant Molecular Biology
Edited by G. Coruzzi and P. Puigdomènech

tagged mutants is to score transgenic families produced by Agrobacterium-mediated seed transformation for the presence of T-DNA insertional mutants (Feldmann 1991). My laboratory has collaborated for several years with Ken Feldmann (originally at DuPont in Wilmington, Delaware, and presently at the University of Arizona, Tucson) on screening transgenic families for the presence of embryonic mutations. Additional families have been examined by Robert Goldberg and colleagues at the University of California (West et al. 1993).

Embryogenesis presents a challenge to plant developmental geneticists because a large number of genes perform essential functions at this stage of the life cycle. As a result, many recessive mutants defective in embryogenesis have been identified following chemical mutagenesis in both maize (Sheridan 1988) and Arabidopsis (Meinke 1991b). The question then becomes which mutants should be chosen for detailed analysis. Two different strategies have been followed in research with Arabidopsis. The approach taken by Gerd Jürgens and colleagues at the University of Munich has been to focus on a relatively small group of "pattern" mutants that exhibit striking phenotypes consistent with defects in critical regulatory genes (Jürgens et al. 1991; Mayer et al. 1991, 1993). A similar approach has been pursued with considerable success in the genetic dissection of early embryogenesis in Drosophila. There are two major limitations to this approach: (1) many regulatory genes are likely to be missed because the loss-of-function mutant phenotype is not particularly novel; and (2) a large collection of embryonic mutants is needed to place the results of studies with pattern mutants in a broader framework. The approach taken by my laboratory has been to study a wide range of embryonic mutants and attempt to establish the relationship between mutant phenotype and normal gene function. One problem with this approach is that many of these mutants are likely to be defective in general housekeeping functions. Thus, it may be difficult to identify genes with more global regulatory functions among large collections of embryonic mutants. The purpose of my research has therefore not been to define plant embryogenesis in relation to a small number of regulatory genes, but rather to explore what types of embryonic mutations can be identified, and what they reveal about the diversity of cellular and developmental functions occurring during plant embryogenesis. This approach has led to the discovery of a biotin auxotroph that becomes lethal during embryo development (Schneider et al. 1989) as well as a homeotic "leafy cotyledon" mutant that appears to be defective in a regulatory gene with a critical role in embryonic maturation (Meinke 1992). Thus we have examples of mutants with defects in genes at both ends of the housekeeping vs regulatory spectrum. The question that must now be addressed, and for which T-DNA insertional mutants are particularly valuable, is what can be said about the remainder of embryonic mutants defective in genes with unknown functions.

Screening Transgenic Families for Embryonic Mutants

The method of seed transformation was originally developed by Feldmann and Marks (1987) to create transgenic Arabidopsis plants without tissue culture and thus avoid problems associated with somaclonal variation. Over 10,000 transgenic plants have now been produced by this method (Feldmann 1991). Many different laboratories have participated in screening these families for mutant phenotypes that might be associated with T-DNA insertion. Over the past 3 years, my laboratory has screened approximately 5,000 of these families for the presence of embryonic mutations. One hundred and seventy-eight recessive embryonic mutants with a wide range of developmental abnormalities have been identified (Errampalli et al. 1991; Castle et al. 1993). This corresponds to roughly 3% of the transgenic families screened. Mutants were identified by screening immature siliques for the presence of 25% defective seeds following self-pollination. A single silique was examined from 6 to 8 T_3 plants chosen at random from each of 5,000 T_2 families derived from independent transformation events. Seeds from putative mutants were then grown again to confirm the presence of a recessive embryonic mutation. Mutant lines were maintained by planting the phenotypically normal seeds produced by heterozygous plants and then scoring the resulting plants for the presence of defective seeds.

Phenotypic Diversity of Embryonic Mutants

Three broad and overlapping classes of embryonic mutants have been described in Arabidopsis: embryonic lethals, defectives, and pigment mutants (Meinke 1991b). Pattern mutants identified by Jürgens et al. (1991) represent an intriguing type of embryonic defective with particularly dramatic alterations in body organization. The distinction between embryonic lethals and defectives has been difficult to follow in the literature, in part because some of the first mutants isolated in my laboratory were called lethals, following the tradition of Müller (1963), even though mutant embryos were capable of producing defective seedlings in culture. In some cases, a single mutant locus has at different times been referred to as an embryonic lethal, an embryonic defective, and a pattern mutant. For example, one mutant (112A-2A; emb30) originally isolated and characterized as a "lethal" in my laboratory (Meinke 1985) was subsequently found to produce rootless plants in culture (Baus et al. 1986) and has recently been shown to be allelic to the gnom pattern mutant of Jürgens et al. (1991). To avoid further confusion in the future, I have started to refer to all recessive embryonic mutants that consistently produce 25% defective seeds as embryo defectives. Most of these emb mutants can be identified either by screening immature siliques for the presence of defective seeds following self-pollination, or by examining the morphology and germination response of mature seeds in culture.

Table 1 presents an overview of the embryonic defectives identified after screening 5,000 transgenic Arabidopsis families. Mutants have been placed into phenotypic classes based on general morphology and the apparent extent of embryo development prior to desiccation. These classes do not correspond exactly to normal stages of development. For example, arrested embryos from transition mutants were often heart shaped but were usually much larger than wild-type torpedo embryos and may have in some cases corresponded at the cell level to more advanced stages of embryo development. The classes shown in Table 1 are therefore simply an approximation of the size and shape of mutant embryos produced by each family.

Table 1. Embryonic Mutants Identified in Transgenic Families

Mutant Class	Number of Embryonic Mutants Identified			
	Tagged	Not tagged	Unresolved	Total
Preglobular	9	24	16	49
Globular	14	19	30	63
Transition	7	12	7	26
Cotyledon	6	11	7	24
Fusca	4	1	1	6
Other	1	7	2	10
Total	41	74	63	178

One problem with summarizing data in the format shown in Table 1 is that striking examples of abnormal development are ignored. In fact, many interesting developmental abnormalities were observed among mutants in this collection, ranging from disorganized, reduced, and duplicated cotyledons, to distorted embryos with giant cells, enlarged suspensors, and an irregular protoderm layer. A defective endosperm mutant that apparently cannot form cell walls, a twin mutant with a high frequency of polyembryony, and a homeotic mutant with leafy cotyledons have also been identified in this collection. Other mutants that produce aborted seeds blocked at early stages of development may reveal even more interesting defects in morphogenesis once they are examined in more detail with Normarski optics. The spectrum of mutant phenotypes represented in this collection appears to be similar to that obtained following chemical mutagenesis, but some new phenotypes have been identified because we are still far from saturating for embryonic defectives in Arabidopsis. Nevertheless there should be plenty of material available to begin a detailed analysis of gene function during plant embryogenesis.

Genetic Analysis of Embryonic Mutants

The first objective in genetic analysis of these mutants was to confirm the presence of a recessive embryonic mutation by noting the percentage of mutant seeds produced by selfed heterozygotes. The next step was to determine which mutants were indeed tagged. This turned out in many cases to be a difficult task because a significant number of mutant families contained multiple sites of T-DNA insertion. The general approach was to determine whether the mutation could be separated from dominant T-DNA markers for kanamycin resistance and nopaline synthesis. The identification of a nopaline-negative heterozygous (emb/EMB) plant that produced all kanamycin-sensitive progeny was usually considered sufficient evidence that a given embryonic mutation was not tagged. In some cases, Southern blot hybridizations with genomic plant DNA and T-DNA probes were used to demonstrate that heterozygous plants did not contain inactive or severely truncated inserts. This part of the project involved large numbers of nopaline assays, germination tests on kanamycin, and scoring kanamycin-resistant transplants for the presence of the embryonic mutation. Of the 115 mutants resolved with respect to tagging, 41 mutants (36%) appeared from kanamycin studies and nopaline assays to be tagged. Included among these tagged mutants are many different examples of abnormal development. This material represents the starting point for a long-term project in my lab involving the molecular analysis of embryonic mutants and their corresponding wild-type genes, which appear to perform a variety of essential functions during plant embryogenesis.

Many of the untagged mutants have been incorporated into a large-scale mapping project designed to saturate the linkage map of Arabidopsis with embryonic defectives. Our mapping strategy has been to determine recombination percentages between embryonic defectives and linked visible markers using multiple marker lines (DP23, DP24, DP28) with readily scored visible markers obtained from Maarten Koornneef (Patton et al. 1991). To date, we have assigned 135 of the 250 emb mutants in our collection to linkage groups and placed 75 of these genes on the linkage map. Embryo-defective mutants are therefore clearly the most common class of morphological marker on the Arabidopsis genetic map. At least four tagged emb mutants have been assigned to each of the five linkage groups. Seven tagged mutant families appear to contain chromosomal translocations because the mutant genes exhibit linkage to visible markers on two different chromosomes. Thus chromosomal rearrangements may be fairly widespread following seed transformation. Knowing the chromosomal locations of untagged mutant genes should eventually provide an alternative approach to gene isolation through chromosome walking. Identification of new alleles at a locus of interest can also be facilitated by knowing the chromosomal locations of mutant genes. Allelism tests can then be limited to mutants with similar map locations. Several mutants in our collection were found to be allelic after their defective genes were mapped to similar locations. Multiple alleles are particularly valuable in developmental genetics because they help to reveal the null phenotype and confirm the identity of cloned mutant genes. Identification of additional emb alleles has been particularly challenging because so many different genes are involved.

Molecular Analysis of Tagged Embryonic Mutants

The ultimate goal of research with tagged mutants is to isolate the corresponding wild-type genes and determine their functions in plant growth and development. We found that in light of the vast complexity of T-DNA inserts in many transgenic Arabidopsis families, it was helpful to work first on constructing models of T-DNA insert structures at the mutant locus to allow efficient recovery of appropriate T-DNA/plant junction fragments. Genomic Southern blot hybridizations with 34 tagged embryonic mutants and three T-DNA probes were therefore performed in advance of plasmid rescue experiments (Castle et al. 1993). A wide range of T-DNA duplications, rearrangements, and truncations was observed. This detailed analysis of T-DNA insert patterns should benefit many other investigators working with mutants isolated from related populations of transgenic plants. The value of such models was demonstrated in our laboratory by using plasmid rescue to clone flanking plant DNA from four tagged mutants with different insert patterns and mutant phenotypes. Two of these mutants (emb76 and emb177) produce aborted seeds with large suspensors; one mutant (emb88) is characterized by developmental arrest at the globular stage with less enlargement of the suspensor; and the remaining mutant (emb78) is an example of a fusca mutant (Müller 1963) that completes morphogenesis but accumulates anthocyanin in developing cotyledons and then dies shortly after germination. These genes are currently being examined in detail in my laboratory.

We have made the most progress with emb78, which turns out to be allelic to the fus6 complementation group as defined by Miséra, Jürgens, and Müller (personal communication). We have sequenced FUS6 and found that it encodes a novel 50 kDa protein that has consensus phosphorylation sites and may bind metals and ATP or GTP. Searches of existing databases did not reveal any significant homologies with known genes. Fusca mutants are a particularly intriguing class of embryonic mutant because they appear to be unique to Arabidopsis. We have recently shown that several photomorphogenic (cop and det) mutants of Arabidopsis also accumulate anthocyanin during embryogenesis and are thus fusca mutants as well. These results are consistent with our model that fusca genes play important but diverse roles in the transduction of environmental and developmental signals during plant growth and development (Castle and Meinke, manuscript in preparation).

Abnormal development of the suspensor is a characteristic feature of many embryo-defective mutants of Arabidopsis (Yeung and Meinke 1993). One model proposed to explain this phenotype is that continued growth of the suspensor is normally inhibited by the embryo proper, and only when this inhibitory effect is removed by a mutation lethal to the embryo proper is the full developmental potential of the suspensor revealed. Now that several tagged mutants with abnormal suspensors are available, it should be possible to test this model at the molecular level. Cloned genomic fragments flanking the T-DNA inserts in emb76 and emb177 are currently being used to probe Northern blots and cDNA libraries prepared from immature siliques in hopes of identifying complementary sequences. Although it may be difficult to recover the desired cDNAs if these genes are expressed at low levels and only during early embryo development, the availability of tagged alleles should still provide a distinct advantage over chromosome walking. I believe that with coordinated efforts between various laboratories involved in genetic analysis of plant embryogenesis, it should be possible to make significant advances over the next 10 years in our understanding of the cellular and developmental mechanisms involved in this critical stage of the life cycle.

Acknowledgements

This research was supported by grants from NSF (DCB 8905137; DIR 9104215) and USDA (88-37261-3708). Deena Errampalli helped to coordinate the initial characterization of transgenic families; Linda Castle worked extensively on the molecular characterization of T-DNA inserts and molecular isolation of fusca genes; Linda Franzmann and Elizabeth Yoon were in charge of gene mapping; and Brian Schwartz and Daniel Vernon are continuing the analysis of specific tagged mutants. Many undergraduates at Oklahoma State University provided valuable technical asssistance. Ken Feldmann (University of Arizona) produced all of the transgenic families.

References

Baus AD, Franzmann L, Meinke DW (1986) Growth in vitro of arrested embryos from lethal mutants of Arabidopsis thaliana. Theor Appl Genet 72:577-586.

Castle LA, Errampalli D, Atherton TL, Franzmann LH, Yoon ES, Meinke DW (1993) Genetic and molecular characterization of embryonic mutants identified following seed transformation in Arabidopsis. Mol Gen Genet (in press).

Castle LA, Meinke DW (1993) Embryo-defective mutants as tools to study essential functions and regulatory processes in plant embryo development. Semin Devel Biol 4:31-39.

Clark JK, Sheridan WF (1991) Isolation and characterization of 51 embryo-specific mutations of maize. Plant Cell 3:935-951.

Deng X-W, Matsui M, Wei N, Wagner D, Chu AM, Feldmann KA, Quail PH (1992) COP1, an Arabidopsis regulatory gene, encodes a protein with both a zinc-binding motif and a GB homologous domain. Cell 71:791-801.

Errampalli D, Patton D, Castle L, Mickelson L, Hansen K, Schnall J, Feldmann K, Meinke D (1991) Embryonic lethals and T-DNA insertional mutagenesis in Arabidopsis. Plant Cell 3:149-157.

Feldmann KA (1991) T-DNA insertion mutagenesis in Arabidopsis: mutational spectrum. Plant J 1:71-82.

Feldmann KA, Marks MD (1987) Agrobacterium-mediated transformation of germinating seeds of Arabidopsis thaliana: a non-tissue culture approach. Mol Gen Genet 208:1-9.

Herman PL, Marks MD (1989) Trichome development in Arabidopsis thaliana. II. Isolation and complementation of the GLABROUS1

gene. Plant Cell 1:1051-1055.
Jürgens G, Mayer U, Torres Ruiz RA, Berleth T, Miséra S (1991) Genetic analysis of pattern formation in the Arabidopsis embryo. In: Molecular and Cellular Basis of Pattern Formation, K. Roberts et al., eds (Cambridge: Company of Biologists Ltd), pp 27-38.
Kieber JJ, Rothenberg M, Roman G, Feldmann KA, Ecker JR (1993) CTR1, a negative regulator of the ethylene response pathway in Arabidopsis, encodes a member of the Raf family of protein kinases. Cell 72:427-441.
Koncz C, Chua N-H, Schell J (eds) (1992) Methods in Arabidopsis research. World Scientific, New Jersey.
Lindsey K, Topping JF (1993) Embryogenesis: a question of pattern. J Exp Bot 44:359-374.
Mayer U, Büttner G, Jürgens G (1993) Apical basal pattern formation in the Arabidopsis embryo: studies on the role of the gnom gene. Development 117:149-162.
Mayer U, Torres Ruiz RA, Berleth T, Miséra S, Jürgens G (1991) Mutations affecting body organization in the Arabidopsis embryo. Nature 353:402-407.
Meinke DW (1985) Embryo-lethal mutants of Arabidopsis thaliana: Analysis of mutants with a wide range of lethal phases. Theor Appl Genet 69:543-552.
Meinke DW (1986) Embryo-lethal mutants and the study of plant embryo development. Oxford Surv Plant Molec Cell Biol 3:122-165.
Meinke DW (1991a) Perspectives on genetic analysis of plant embryogenesis. Plant Cell 3:857-866.
Meinke DW (1991b) Embryonic mutants of Arabidopsis thaliana. Devel Genet 12:382-392.
Meinke DW (1992) A homoeotic mutant of Arabidopsis thaliana with leafy cotyledons. Science 258:1647-1650.
Meinke DW (ed) (1993) Plant Developmental Genetics. Semin Devel Biol, Vol 4. Saunders Publications, New York.
Meinke DW, Sussex IM (1979) Embryo-lethal mutants of Arabidopsis thaliana: a model system for genetic analysis of plant embryo development. Devel Biol 72:50-61.
Müller AJ (1963) Embryonentest zum nachweis rezessiver letalfaktoren bei Arabidopsis thaliana. Biol Zentralbl 82:133-163.
NSF Publication #92-112 (1992) The multinational coordinated Arabidopsis thaliana genome research project. Progress report: year two.
Patton DA, Franzmann LH, Meinke DW (1991) Mapping genes essential for embryo development in Arabidopsis thaliana. Mol Gen Genet 227:337-347.
Schneider T, Dinkins R, Robinson K, Shellhammer J, Meinke DW (1989) An embryo-lethal mutant of Arabidopsis thaliana is a biotin auxotroph. Devel Biol 131:161-167.
Sheridan WF (1988) Maize developmental genetics: genes of morphogenesis. Annu Rev Genet 22:353-385.
Sheridan WF, Clark JK (1993) Mutational analysis of morphogenesis of the maize embryo. Plant J 3:347-358.
West MAL, Matsudaira KL, Goldberg RB, Fischer RL, Harada JJ (1993) Genes essential for embryogenesis in Arabidopsis thaliana. J Cell Biochem Suppl 17B:D347.

Yanofsky MF, Ma H, Bowman JL, Drews GN, Feldmann KA, Meyerowitz EM (1990) The protein encoded by the Arabidopsis homeotic gene agamous resembles transcription factors. Nature 346:35-39.

Yeung EC, Meinke DW (1993) Embryogenesis in angiosperms: development of the suspensor. Plant Cell (in press).

The Use of *Arabidopsis* Genetics to Analyze Synthesis, Structure, and Function of the Plant Cell Wall

Wolf-Dieter Reiter
MSU-DOE Plant Research Laboratory
Michigan State University
East Lansing, MI 48824-1312
USA

Introduction

The cell walls of higher plants form a contiguous scaffold surrounding the individual protoplasts within the plant body. In addition to providing mechanical support for the cells, plant cell walls have a pivotal role in growth and development, in cell-cell recognition and interaction, and in defense responses (see Carpita and Gibeaut, 1993 for a recent review). In contrast to animal systems, cell migrations do not contribute to plant development; instead, the ordered formation of plant tissues and organs relies solely on the combination of two processes: The division of cells in specific planes, and the subsequent three-dimensional expansion of the cells in specific dimensions. This latter process of extension growth is primarily a function of the mechanical properties of the walls, and the synthesis of new cell wall material. Since all cells within a growing tissue or organ are held together via their middle lamella, extension growth needs to be tightly coordinated between the various cell types in growing parts of the plant body. Since different cell types within a growth zone are not only distinguished by their shape and size,

NATO ASI Series, Vol. H 81
Plant Molecular Biology
Edited by G. Coruzzi and P. Puigdomènech

but also by the thickness of their walls, a very complex network of regulatory mechanisms is required to ensure the ordered formation of plant tissues and organs. Cell wall formation in higher plants essentially takes place in three stages: cell plate formation, extension growth, and secondary growth. The cell plate is a structure formed during the final stages of cell division by the fusion of Golgi-derived vesicles containing cell wall material. Outward growth of the cell plate from the center of the dividing cell connects it to the mother cell wall; the membranes of the cell plate fuse with the plasma membrane of the mother cell completing cell separation. The newly formed daughter cells then undergo extension growth by deposition of new cell wall material throughout the existing wall area. This mode of cell expansion is unique to plant cell walls, and is quite distinct from the tip-growth mechanism found in other wall-forming organisms such as filamentous fungi where new cell wall material is inserted at localized growth zones pushing the existing wall apart. Since the various components of the growing plant cell wall are covalently and non-covalently cross-linked (Fry, 1986), it is generally believed that synthesis of new cell wall material is accompanied by biochemical "loosening" events that permit a continuous reorganization of fibrillar elements within the existing wall structure (Taiz, 1984). Most workers believe that cleavage of cell wall polysaccharides is required to maintain an extensible wall structure; however, this view has recently been challenged by the isolation of cell wall proteins that induce cell wall extension but lack detectable glycanase or transglycosylase activities (McQueen-Mason *et al.*, 1992). The cell wall formed during extension growth is called the primary cell wall; its thickness is usually 100-200 μm although some cell types have much thicker primary walls. Since these walls are actually composed of two parts separated by the middle lamella, the wall synthesized by a meristematic cell may only be about 50 μm thick, permitting the deposition of no more than four layers of cellulose microfibrils (Roberts, 1990). The dimensions of cell wall extension appear to be determined by the orientation of cellulose microfibrils which wrap around cells essentially forming a coccoon restricting cell extension. A role of microfibril orientation in determining cell shape is primarily inferred from electron microscopic observations where the walls of isodiametric cells are characterized by a random orientation of cellulose microfibrils whereas unidimensionally extending cells are characterized by microfibrils oriented essentially transverse to the dimension of extension (McCann *et al.*, 1990). Extension growth appears to be terminated by a reduction in the yielding properties of the primary wall accompanied by a cessation of the synthesis of cell wall material; however, the exact mechanisms governing extension growth and its termination are poorly understood. Some cell types continue the synthesis of cell wall material after reaching their final size; the resulting thickened wall structure is termed a secondary wall which differs from the primary wall in its polysaccharide composition. Pronounced secondary wall structures are formed by cell types that have mechanical support functions within the plant body. In many cases, protoplasts synthesizing secondary walls undergo programmed cell death, leaving the wall as the only functional element (Esau, 1977).

Plant cell walls are primarily composed of polysaccharides which are classified as cellulose, hemicelluloses, and pectins based on their solubility properties and chemical composition. Furthermore, they contain structural proteins classified according to their amino acid composition as hydroxyproline-rich glycoproteins (HPRGs), proline-rich proteins (PRPs), and glycine-rich proteins (GRPs). Many secondary cell walls are reinforced by lignin, a threedimensional network of cross-linked phenylpropanoid residues. Furthermore, walls of specific cell types are impregnated by hydrophobic components such as cutin, a polyester composed of hydroxy-fatty acids, and suberin, a copolymer of phenylpropanoid and fatty acid residues. Reviews on the biochemistry and function of cutin and suberin are available (Kolattukudy, 1980, 1984).

In addition to the structural wall components listed above, plant cell walls contain arabinose- and galactose-rich proteoglycans (so-called arabinogalactan proteins; AGPs), and a large number of non-structural proteins such as glycanases and peroxidases, many of which appear to be involved in cell wall modification events. These proteins, glycoproteins, and proteoglycans are considered apoplastic macromolecules but not integral parts of the cell wall. Arabinogalactan-proteins have been reviewed by Fincher and Stone (1983), and Showalter and Varner (1989); non-structural cell wall proteins are included in a recent review by Cassab and Varner (1988).

The literature on the structure, composition, mechanical properties, and digestibility of plant cell walls is immense, partly due to the generally recognized importance of cell walls for various aspects of plant development, and partly because of the importance of plant cell walls for applied fields. The vast amount of descriptive literature on plant cell walls contrasts sharply with the current knowledge about the biosynthesis of cell wall material and its regulation with regard to cell cycle, extension growth, secondary growth, and environmental parameters. This lack of knowledge is particularly evident in case of cell wall polysaccharides, where not a single biosynthetic enzyme has been purified to homogeneity, and not a single gene for any such enzyme has been cloned. The primary reason for this gap in our understanding of cell wall synthesis is an almost complete failure to address this issue by conventional biochemical approaches. To understand the function of individual cell wall components and to dissect the mechanisms involved in regulating cell wall synthesis, a genetic approach is clearly required. Although a few mutants with changes in cellulose content and lignin composition have been identified in grasses, genetic work in these plants is hampered by long generation times and complex genomes. Recently, several groups have begun to use the powerful tools of *Arabidopsis* genetics to address cell wall-related issues. The purpose of this chapter is to outline the ways in which *Arabidopsis* genetics can be used to get a better understanding of the role of cell walls for various aspects of plant biology.

Mutants of *Arabidopsis* with Changes in Cell Wall Polysaccharide Composition

To isolate *Arabidopsis* mutants affected in the synthesis of cell wall polysaccharides, four different approaches have been taken over the last few years: (1) The isolation of mutants resistant to inhibitors of cell wall synthesis, (2) the isolation of mutants with visible phenotypes predicted to arise from perturbations in cell wall synthesis, (3) the isolation of mutants non-reactive with antibodies against carbohydrate epitopes, and (4) the isolation of mutants altered in the monosaccharide composition of cell wall material.

Heim *et al.* (1989, 1990a, 1990b) isolated *Arabidopsis* plants resistant to the herbicide isoxaben which has a mode of action very similar to 2,6-dichlorobenzonitrile (DCB), an inhibitor of cellulose synthesis (Delmer, 1987). Like DCB, isoxaben does not completely eliminate cellulose synthesis but reduces it by about 80%. Approximately 50 mutant lines with various degrees of isoxaben resistance were recovered from a population of about one million M2 plants. Genetic analyses of three highly resistant lines revealed two loci involved in isoxaben resistance, *ixr*A (two known alleles: *ixr*A1 and *ixr*A2), and *ixr*B1 (one known allele). Although both loci map to chromosome 5, they are sufficiently separated to appear unlinked in genetic crosses (Heim *et al.*, 1990b). It is not known whether isoxaben interacts directly with cellulose synthase or some other component connected with cellulose synthesis, making it difficult to arrive at conclusions concerning the gene products encoded by the *IXR*A and *IXR*B loci. Although the essentially recessive nature of the *ixr* mutations speaks against the idea that they represent "gain of function" mutations in cellulose synthase components, it is conceivable that multi-subunit synthase molecules containing both isoxaben-sensitive and isoxaben-resistant components are non-functional leading to dominance of the wild type allele. Based on this possibility, the *ixr* mutants may be very useful to clone genes encoding cellulose synthase components, a task that has frustrated workers using conventional biochemical techniques.

Since the precise function of most cell wall polysaccharides is not known, it is difficult to make predictions about visible phenotypes associated with alterations in cell wall composition and structure. One exception to this rule are mutants deficient in cellulose synthesis which are expected to be lethal in a homozygous state but could be recovered as heterozygotes segregating for lethals, or by screening for conditional lethals such as temperature-sensitive mutants. Since the orientation of cellulose microfibrils is believed to determine the dimensions of extension growth, perturbations in cellulose synthesis or in its ordered deposition should cause all cells in a growth zone to assume an isodiametric shape and possibly become enlarged due to a weakened wall structure. In agreement with this prediction, plants exposed to the cellulose synthesis inhibitors DCB and isoxaben exhibit a lateral swelling of their root tips (Potikha and Delmer, 1993). In an effort to isolate *Arabidopsis* mutants with altered cell walls, Baskin *et al.* (1992) screened chemically mutagenized plants for temperature-sensitive radial swelling of their roots, and obtained several plants whose root tips were almost indistinguishable from wild type at ≤21°C but doubled their diameter within 48 h of transfer to 31°C. Biochemical and cell biological studies on these mutants will reveal whether they are affected in some aspect of cytoskeleton formation, or whether they have lesions in genes involved in the synthesis of structural cell wall components.

A different approach to isolate mutants in cellulose synthesis or deposition was taken by Potikha and Delmer (1993). They screened mutagenized M2 seedlings for alterations in shape and birefringence of leaf trichomes and xylem elements. Birefringence of cell walls is thought to be caused by the ordered deposition of cellulose microfibrils, and is reduced in plants grown in the presence of the cellulose synthesis inhibitor DCB (Potikha and Delmer, 1993). One of the most interesting mutants obtained in this screen appears to be impaired in the synthesis of cellulose [ß-(1→4)-D-glucan], and the structurally related callose [ß-(1→3)-D-glucan] which is found in the walls of some specialized cell types and is rapidly synthesized in response to wounding (Delmer, 1987). This ß-D-glucan mutant is characterized by a total lack of birefringence in the trichomes, reduced birefringence in most other cell walls including xylem elements, and an abnormal pattern of callose synthesis in response to wounding. The mutation is dominant, leads to an approximately 40% reduction in the amount of cellulose in leaf material, but does not seriously affect plant viability and seed-set. The alterations in both cellulose and callose synthesis in this mutant provides genetic evidence for some link in the synthesis of the two ß-D-glucans confirming biochemical data on cellulose and callose synthesis (Delmer, 1987).

The use of antibodies directed against cell wall epitopes should be a promising approach to identify cell wall mutants, especially when used in combination with the tissue printing technique that allows the detection of cell wall epitopes on the level of individual cells (Cassab and Varner, 1987). The usefulness of immunological screening procedures is currently limited by the number of antibodies directed against specific polysaccharide epitopes, and the small stem diameter of *Arabidopsis* plants which complicates the use of tissue printing procedures; however, the first limitation is likely to be overcome as more experience is being gained with the production of polysaccharide-specific antibodies. The recent identification of an *Arabidopsis* mutant non-reactive with an antibody directed against a carbohydrate epitope (von Schaewen *et al.*, 1992) indicates the general usefulness of immunological procedures for mutant isolation.

Since cell wall polysaccharides vary widely in their monosaccharide composition (e.g. rhamnose is almost only found in pectins, and xylose is primarily found in hemicelluloses), many changes in the composition of cell wall polysaccharides will be reflected by changes in the relative amounts of monosaccharides in hydrolyzed cell wall material. Quantitation procedures for these monosaccharides should therefore be useful to screen for mutants which are partially or completely deficient in cell wall polysaccharides, or which synthesize cell wall components in abnormal ratios due to regulatory defects. Since the monosaccharide composition of cell wall material shows a considerable plant to plant variation in wild type *Arabidopsis*, such a screening protocol would primarily identify mutants with relatively major changes in monosaccharide composition. A gas-chromatographic quantitation procedure for neutral monosaccharides released from leaf-derived cell wall material has recently been used to screen more than 5000 chemically mutagenized *Arabidopsis* plants for unusual monosaccharide composition values leading to the identification of several mutant lines representing more than ten complementation groups (Reiter *et al.*, 1993). Based on the observed alterations in monosaccharide composition, these mutant lines can essentially be placed in one of the following categories: (1) lines with complex changes in the relative amounts of several monosaccharides, (2) lines with a substantial reduction in the amount of a single monosaccharide, and (3) lines completely deficient in a specific monosaccharide. Most of the mutants in category 1 show alterations in growth habit, morphology, and physiology that were genetically inseparable from the changes in cell wall composition. Visible phenotypes associated with altered cell wall compositions include extremely stunted morphologies, distortions in leaf shape and size, strongly reduced fertility, and changes in the control of flowering. Most of the mutants in category 2 show up to 50% reductions in the amounts of cell wall-derived arabinose or fucose but are morphologically indistinguishable from wild type plants. The complete absence of a monosaccharide was only observed in the case of fucose where several allelic lines at a locus designated *mur1* were obtained. These lines appear to be defective in the *de novo* synthesis of fucose, and are characterized by a slightly dwarfed growth habit, and a reduced mechanical stability of their primary walls, possibly due to disturbances in the structure of the xyloglucan-cellulose network.

The Use of *Arabidopsis* Genetics to Study Lignin Synthesis and Structure

Recently, a mutant of *Arabidopsis* has been described that lacks syringyl units in its lignin due to an inability to convert ferulate into sinapate

(Chapple *et al.*, 1992). This change in lignin composition does not lead to readily observable alterations in the growth habit of the plants suggesting that the presence of syringyl units in angiosperm lignin is not essential for the structural integrity of the plant but plays a more subtle role that remains to be elucidated. One T-DNA-tagged allele of this mutation is available, leading to the cloning of the affected gene. Preliminary sequence data indicate a high degree of sequence similarity with P-450-dependent hydroxylases (C. Chapple, pers. comm.), suggesting that the mutated gene encodes ferulate-5-hydroxylase which is in agreement with biochemical data on the mutant (Chapple *et al.*, 1992). This work demonstrates that *Arabidopsis* genetics can successfully be use to address questions related to cell wall components that are often believed to be confined to woody species. The possibility to modify lignin structure in a variety of plants using genes cloned from *Arabidopsis* has a considerable potential for applied fields since lignin structure and lignification processes have major implications for the processing of wood and the digestibility of forage.

Conclusions and Perspectives

Most of the literature on plant cell walls is largely descriptive, dealing with issues like cell wall composition, ultrastructure, and enzymic digestibility. In contrast, very little information is available on the enzymology and molecular biology of cell wall synthesis, primarily due to difficulties with conventional biochemical approaches. A considerable amount of information is available on the molecular biology of structural cell wall proteins, but the currently available data on gene expression patterns and protein localization do not directly address questions concerning their function within the wall.

To obtain direct evidence on the roles of the various cell wall components, it is necessary to use genetic approaches, i.e. to eliminate specific cell wall components in intact plants, and to evaluate these plants for changes in development and physiology. The availability of genes for structural cell wall proteins allows the use of reverse genetic approaches such as the introduction of antisense constructs. Mutants in cell wall polysaccharides and lignin have already been obtained by conventional genetic approaches, and will be of great utility to study the function of individual cell wall components. The isolation of polysaccharide and lignin mutants in *Arabidopsis* has the additional advantage that these lines can be used to clone genes involved in the synthesis of cell wall polysaccharides and lignin components that are difficult to obtain by other means. In summary, it can be concluded that the use of *Arabidopsis* genetics will be of crucial importance to address complex biological questions concerning the mechanisms of cell wall assembly, and how the synthesis and the properties of cell walls are being regulated.

References

Baskin TI, Betzner AS, Hoggart R, Cork A, Williamson RE (1992) Root morphology mutants in *Arabidopsis thaliana. Aust. J. Plant Physiol.* **19**:427-437.

Carpita NC, Gibeaut DM (1993) Structural models of primary cell walls in flowering plants: consistency of molecular structure with the physical properties of the walls during growth. *Plant J.* **3**:1-30.

Cassab GI, Varner JE (1987) Immunocytolocalization of extensin in developing soybean seed coats by immunogold-silver staining and by tissue printing on nitrocellulose paper. *J. Cell Biol.* **105**:2581-2588.

Cassab GI, Varner JE (1988) Cell wall proteins. *Annu. Rev. Plant Physiol. Plant Mol. Biol.* **39**:321-353.

Chapple CCS, Vogt T, Ellis BE, Somerville CR (1992) An *Arabidopsis* mutant defective in the general phenylpropanoid pathway. *Plant Cell* **4**:1413-1424.

Delmer DP (1987) Cellulose biosynthesis. *Annu. Rev. Plant Physiol.* **38**:259-290.

Esau K (1977) Anatomy of seed plants. 2nd edition. Wiley, New York.

Fincher GB, Stone BA (1983) Arabinogalactan-proteins: structure, biosynthesis, and function. *Annu. Rev. Plant Physiol.* **34**:47-70.

Fry SC (1986) Cross-linking of matrix polymers in the growing cell walls of angiosperms. *Annu. Rev. Plant Physiol.* **37**:165-186.

Heim DR, Roberts JL, Pike PD, Larrinua IM (1989) Mutation of a locus of *Arabidopsis thaliana* confers resistance to the herbicide isoxaben. *Plant Physiol.* **90**:146-150.

Heim DR, Skomp JR, Tschabold EE, Larrinua IM (1990a) Isoxaben inhibits the synthesis of acid insoluble cell wall materials in *Arabidopsis thaliana. Plant Physiol.* **93**:695-700.

Heim DR, Roberts JL, Pike PD, Larrinua IM (1990b) A second locus, *ixr* B1 in *Arabidopsis thaliana*, that confers resistance to the herbicide isoxaben. *Plant Physiol.* **92**:858-861.

Kolattukudy PE (1980) Cutin, suberin, and waxes. In *The biochemistry of plants. A comprehensive treatise* (Stumpf PK, Conn EE eds). Vol. **4**, pp. 571-645. Academic Press, New York.

Kolattukudy PE (1984) Biochemistry and function of cutin and suberin. *Can. J. Bot.* **62**:2918-2933.

McCann MC, Wells B, Roberts K (1990) Direct visualization of cross-links in the primary plant cell wall. *J. Cell Sci.* **96**:323-334.

McQueen-Mason S, Durachko DM, Cosgrove DJ (1992) Two endogenous proteins that induce cell wall extension in plants. *Plant Cell* **4**:1425-1433.

Potikha T, Delmer DP (1993) A search for *Arabidopsis* mutants impaired in cell wall biosynthesis. *J. Cell. Biochem.* Suppl. **17A**:16.

Reiter W-D, Chapple CCS, Somerville CR (1993) Cell wall mutants of *Arabidopsis. J. Cell. Biochem.* Suppl. **17A**:34.

Roberts K (1990) Structures at the plant cell surface. *Curr. Opin. Cell Biol.* **2**:920-928.

von Schaewen A, Sturm A, O'Neill J, Chrispeels MJ (1992) Conversion of high-mannose asparagine-linked glycans to complex glycans is blocked in a mutant of *Arabidopsis thaliana. Plant Physiol.* **99S**, 44.

Showalter AM, Varner JE (1989) Plant hydroxyproline-rich glycoproteins. In *The biochemistry of plants. A comprehensive treatise* (Stumpf PK, Conn EE eds.), Vol. **15**, pp. 485-520. Academic Press, New York.

Taiz L (1984) Plant cell expansion: Regulation of cell wall mechanical properties. *Ann. Rev. Plant Physiol.* **35**:585-657.

MOLECULAR GENETICS OF NITRATE ASSIMILATION IN SOLANACEOUS SPECIES

Françoise DANIEL-VEDELE, Marie-France DORBE, Christian GODON, Hoai-Nam TRUONG and Michel CABOCHE
Laboratoire de Biologie Cellulaire, INRA route de Saint Cyr, 78026 VERSAILLES, FRANCE

INTRODUCTION

Nitrate can be used as the sole nitrogen source to sustain growth, in both microorganisms and higher plants. In plants, two successive enzymatic steps reduce nitrate to ammonium, generally in the leaves. First, nitrate is converted into nitrite in a two electron transfer reaction catalysed by nitrate reductase (NR, EC1. 6. 6. 1), a cytoplasmic enzyme. Nitrite is then translocated to the chloroplast, where it is reduced to ammonium by nitrite reductase (NiR, EC1. 7. 7. 1) (for review see Wray 1988, Solomonson and Barber 1990). Ammonium is subsequently incorporated into the amino acid pool through glutamine and glutamate biosynthesis (Guerrero et al 1981).
NADH:NR is the most common form of nitrate reductase in higher plants, and appears to be a homodimer (monomer size between 100 and 120 kDa), each monomer containing three redox prosthetic groups: FAD, heme and a molybdenum cofactor (MoCo). Many mutants defective in nitrate assimilation have been isolated in different species (reviewed in Kleinhofs et al. 1985, Wray 1988, Wray and Kinghorn 1989, Caboche and Rouzé 1990, Crawford and Campbell 1990). These mutants have been found to be defective in nitrate reduction due to disruptions in either the nitrate reductase structural gene (*nia* mutants) or one of the six or seven genes required for synthesis and assembly of the MoCo (*cnx* or MoCo mutants).
Nitrate assimilation is a highly regulated process. In higher plants, factors such as light, cytokinins and circadian rhythm affect NR expression (Crawford and Campbell 1990, Caboche and Rouzé 1990). In addition, nitrate induces, and ammonium and/or glutamine down-regulate, NR expression (Caboche and Rouzé 1990, Solomonson and Barber 1990). Using cDNA as probes, we have studied the influence of light, nitrogen source and circadian rhythm on *nia* gene transcription.

REGULATIONS OF *NIA* GENE TRANSCRIPTION

Light and carbohydrates

As for light-regulated genes involved in photosynthesis such as ribulose biphosphate carboxylase (rbcS) (Silverthorne and Tobin 1987), NR and NiR transcript levels in leaves of mature green plants decrease in darkness and increase again in the light (Cheng et al 1991, Deng et al 1990). Because nitrate assimilation into glutamine and glutamate is largely dependent on the availability of carbons skeletons, we asked to what extent regulation of NR and NiR gene expression by light could be related to carbohydrate supply, which would be limiting after an extended period of darkness. Therefore, we analysed the consequences of providing carbohydrate to detached leaves of dark-

NATO ASI Series, Vol. H 81
Plant Molecular Biology
Edited by G. Coruzzi and P. Puigdomènech

adapted *Nicotiana plumbaginifolia* plants. Our results (Vincentz et al 1993) suggest that light regulation of *nia* gene expression is, at least in part, mediated by carbohydrates but that light is most likely the signal missing for a full induction.

Nitrate inducibility and N-metabolite repression

NR activity is substrate-inducible (Guerrero et al 1981), and NR mRNA accumulates in leaves of plants grown on nitrate-supplemented medium (Cheng et al 1986, Crawford et al 1988). Like NR, NiR activity is inducible by nitrate (Guerrero et al 1981), and NiR mRNA accumulates after nitrate addition to greening rice seedlings (Ogawa and Ida 1987) or to maize plantlets (Back et al 1988). We studied the expression of NiR mRNA in four *nia* mutants of *N.plumbaginifolia* which differ in the expression level of the NR mRNA (Faure et al 1990). Our results indicate that NR and NiR are co-regulated at the mRNA level in wild-type plants and are still nitrate inducible in NR-deficient mutants. In addition , the lack of NR activity leads to an overproduction of NiR and, in most cases, of NR mRNAs. This observation leads us to propose that N-metabolites derived from nitrate assimilation or the NR protein itself repress the expression of NiR and NR genes. In order to discriminate between these two hypothesis, one experimental approach would consist in the isoaltion of a NiR mutant where the NR protein would be unmodified and still functional. Such a mutant was never identified in a dicotyledon, probably because several genes may encode NiR proteins, as it is the case in *Nicotiana tabacum* and *N.plumbaginifolia* (Kronenberger et al 1993). In order to bypass this major drawback, we analysed the effect of expressing an antisense NiR gene in transgenic tobacco , inhibiting nitrate assimilation and N-metabolite production without modifying the NR gene product (Vaucheret et al 1992). The overexpression of the NR mRNA in transgenic plant demonstrate that the NR protein is not involved in its own metabolic repression, in contrast to what has been postulated in the autocatalytic model of NR expression in fungi (Scazzochio et al 1989).

Circadian rhythm and N-metabolite repression

When tobacco or tomato plants are grown under a dark-light regime, the concentrations of NR transcripts are high at the beginning of the day period but almost undetectable at the end of this day period (Galangau et al 1988). This fluctuation of NR mRNA expression appears to be under the control of a circadian rhythm that can be abolished by biochemical or genetic impairement of NR catalytic activity. Mutations in *nia* or *cnx* (Pouteau et al 1989), as well as tungstate (Deng et al 1989), an inhibitor of NR activity, lead to high and stable expression of NR mRNA. Amino acids contents, specially glutamine, show also marked 24h rhythmic variations, inversely correlated to NRmRNA fluctuations (Deng et al 1991). Using detached leaves from *N.plumbaginifolia* plants grown under low light conditions (limited photosynthesis), we tested the effect of glutamine, glutamate and asparagine on NR and NiR gene expression. Our results (Vincentz et al 1993) indicate that gutamine and glutamate down-regulate NR expression in green leaves whereas asparagine has a weaker inhibitory effect.

Model of NR gene regulations

Taken together, our results suggest the following model of transcriptional regulation of *nia* gene expression (Figure 1):

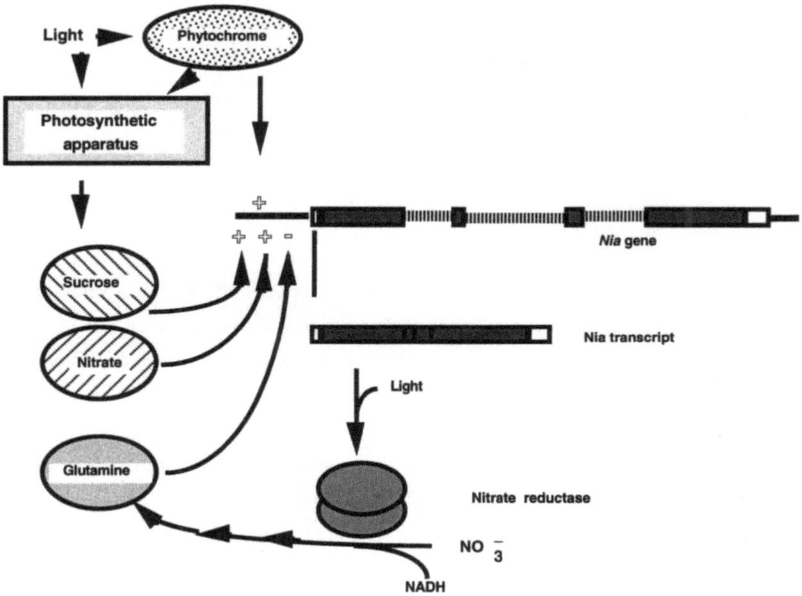

Figure 1: Model of *nia* gene regulation
+ : transcriptional inducer
- : transcriptional repressor

Nitrate and light are strong inducers of *nia* gene transcription. The physiological significance of up-regulation by sugars and down-regulation by N-metabolites may be to regulate the reduction of nitrate in response to changes in the cellular ratio of N to C metabolites.

INVOLVEMENT OF *NIA* GENE PROMOTER IN TRANSCRIPTIONAL REGULATIONS

Three *nia* genes have been isolated from tobacco and tomato (Vaucheret et al 1989, Daniel-Vedele et al 1988). The general organization of the genes is very similar and an overall homology of 81% is found in the exon sequences. Comparison of the three promoters reveals only one conserved region, containing the TATA and CAAT boxes. Two complementary experiments were performed in order to localize the cis-acting sequences involved in *nia* gene regulations.

Complementation of a NR-deficient mutant with the tomato nia gene

The E23 *N.plumbaginifolia* mutant is deficient in NR activity and cannot reduce nitrate into ammonium. As a consequence, it cannot grow on its roots in the green house. We have shown (Dorbe et al 1992) that this mutant can be restored for NR activity by transformation with the cloned tomato *nia* gene, surrounded by 3kb upstream and downstream non-coding regulatory sequences. The transgenic plants expressed from undetectable to 22% of the wild type activity and a good correlation was observed between growth rates and levels of NR activities in transgenic plants. The analysis of the transgene expression showed that the tomato *nia* gene transcription was regulated by light, nitrate and circadian rhythm as in tomato plants. These results suggest that all the cis-acting sequences involved in these regulations are contained in our tomato gene.

Complementation of a NR-deficient mutant with a chimaeric NR gene

The same E23 *N.plumbaginifolia* mutant was restored for NR activity by transformation with a full-length tobacco NR cDNA, fused to the CAMV 35S promoter and to termination signals from the tobacco NR gene (Vincentz and Caboche 1991). The transgenic plants expressed from one-fifth to three times the wild-type NR activity in their leaves. The analysis of chimeric NR gene expression showed that the regulation of NR expression by light, nitrate or circadian rhythm was no longer maintained in transgenic plants. In the same manner, expression of chimeric NR gene was not influenced by darkness and carbohydrate supply (Vincentz et al 1993).

These results demonstrate that the *nia* gene expression is mediated by transcriptional regulations. Work is in progress to further delineate precisely the localisation of the involved cis-acting sequences.

ISOLATION OF A TOBACCO ZINC FINGER PROTEIN HOMOLOGOUS TO REGULATORS OF NITROGEN METABOLISM IN FUNGI

Very little is known about the trans-acting factors that are required for the regulation of NR in higher plants. There are significant similarities between the NR regulation found in plants and fungi. In *Neurospora crassa* and *Aspergillus nidulans,* the NR expression is highly regulated at the mRNA level and requires nitrogen de-repression and simultaneous induction by nitrate. The regulation of the NR and NiR genes is governed on one hand by the pathway-specific positive regulatory genes, respectively *nit-4* and *nirA*, involved in the nitrate inducibility. On the other hand, *nit-2* and *areA*, the major positive regulatory genes are involved in N-metabolites repression. The nucleotide sequence of the *nit-2* and *areA* genes revealed that they encode transcription factors of the GATA-1 family, characterized by a DNA-binding element consisting of a single Cys2/Cys2 "zinc-finger" and an adjacent basic region (Fu and Marzluf 1990a, Kudla et al. 1990). In vitro gel-band mobility-shift and DNA footprinting studies showed that the NIT2 protein recognizes specifically sequences in the 5′ region of N-repressible genes of *N.crassa* and of the *A. nidulans* nitrate and nitrite

reductase genes (Fu and Marzluf 1990b). A core consensus sequence, TATCT (or on the complementary strand AGATA), was identified in all of these binding sites.
Recent studies have demonstrated that two fragments of the tomato *nia* gene promoter region were specifically recognized and bound by the NIT2 protein (Jarai et al. 1992). Both fragments contain the putative GATA recognition sequence. These data suggest that a homolog of NIT2, belonging to the GATA-binding protein family, may exist in tomato and other higher plants and play a central role in controlling NR expression.

Isolation of a partial cDNA from N. plumbaginifolia

The PCR method (Polymerase Chain Reaction) has been successfully used to clone plant transcription factors by homology (reviewed in Katagiri and Chua 1992). There is 98% amino acid identity between the DNA-binding domains of the NIT2 and AREA proteins. Two degenerate oligonucleotides (ol1 and ol2) were derived from the C-X_2-C amino acids of the NIT2 zinc finger and used in PCR experiments on *N.plumbaginifolia* cDNA (Figure 2).

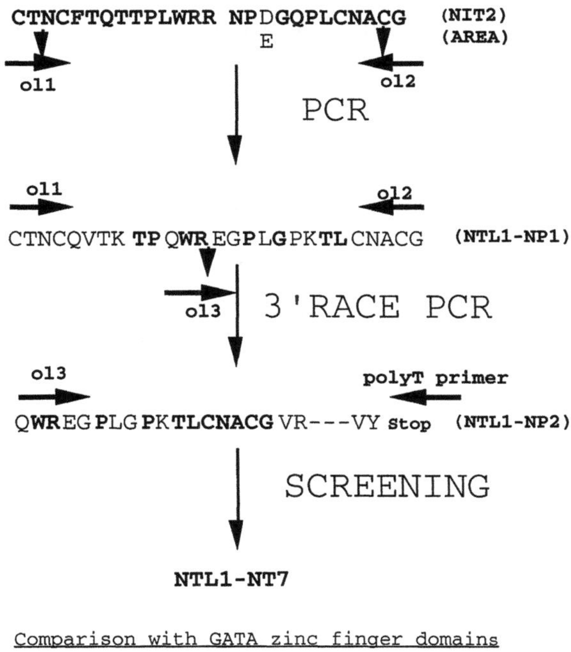

Figure 2: PCR approach

Conserved amino acids between the NIT2, AREA and NTL1 sequences are in **bold face**

The first cDNA strand was synthesized using oligo-d(T) as primer. A faint band of 100 bp, corresponding in size to the amplification product of the *nit*-2 gene, was observed for the *N.plumbaginifolia* sample and not in the corresponding control (cDNA synthesis minus reverse transcriptase). The deduced amino-acid sequence of this Ntl1-Np1 (Ntl for nit-2 like) partial cDNA is shown in figure 2. A main open reading frame, covering the entire length of the fragment was found and the comparison of the corresponding amino acid sequence with the sequence of the NIT-2 zinc finger loop (shown in bold face in Fig. 2) revealed a conservation of 7 out of 17 amino acids.
As a second step, we performed a 3′ RACE PCR experiment (Rapid Amplification of CDNA Ends). A 550 bp cDNA fragment was obtained and cloned as Ntl1-Np2. A part of the deduced amino acid sequences are shown in Fig. 1. Only one large open reading frame of 100 amino acids is found in the entire sequence. The homology with the NIT2 zinc finger region extends to the succession C-N-A-C-G, which may represent the end of a hypothetic zinc finger in the *N.plumbaginifolia* protein. We further used this partial cDNA as a probe to isolate a full length cDNA .

Isolation of a tobacco cDNA encoding a NIT2 like protein
Attemps to isolate a full length *N.plumbaginifolia* Ntl1 cDNA being unsuccessful so far, a full length tobacco homolog of the partial cDNA was therefore screened for. Using *N.plumbaginifolia* cDNA as probe, we screened 4 X 10^5 phages of a cDNA library made from young tobacco leaf, constructed in λgt10. From this screening, we have obtained two recombinant phages that showed either strong or faint hybridization signals (respectively Ntl1-Nt7 and Ntl1-Nt2).
Analysis of the predicted protein sequence reveals the presence of different regions (Figure 3)

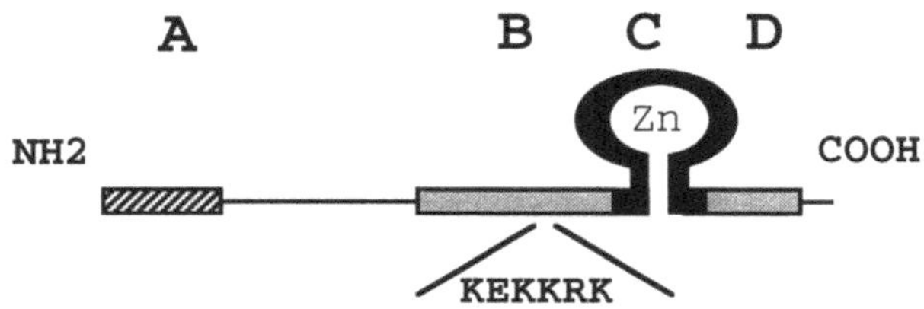

Figure 3: Structural organisation of the NTL1-NT7 amino acid sequence

A: 12-48, acidic region (net charge= -13)
B : 132-200, basic region (net charge= +17)
C : 201-227, zinc finger
D : 230-266, basic region (net charge= +9)

First, a single putative zinc finger DNA-binding domain, covering 26 residues is localised in the C-terminus of the protein. This domain is composed of two pairs of cysteine residues, separated by a loop of 18 amino acids (instead of 17 amino acids as in the NIT2 protein). Two highly basic regions occur immediately upstream and downstream of this putative zinc finger: respectively, from amino acid 132 to 201 with a net charge of +17 and from amino acid 230 to 266 with a net charge of +9 (basic amino acids are underlined in Fig. 4). This basic region, on the carboxy side of the zinc finger, is also found in the NIT2 protein and was shown to represent part of the DNA-binding domain (FU and Marzluf 1990c). The other basic region, upsteam of the zinc finger, contains the amino acid sequence KEKKRK (169 to 174). This sequence is very similar to the well characterized SV40 nuclear localisation signal, KKKRK (Kalderon et al. 1984). The presence of a putative nuclear localisation signal in the NRG1-NT7 protein suggest the import of the protein from the cytoplasm into the nucleus. An acidic region is localised in the N-terminus of the protein, from amino acid 12 to 48, with a net charge of -13. Such acidic regions have been shown to be responsible for transcriptional activation by the yeast GAL4 and GCN4 proteins (Hope and Struhl, 1986 and Ma and Ptashne, 1987).

Computer analysis of the NTL1-NT7 amino acid sequence

A comparison of the predicted primary amino acid sequence of the large open reading frame of the Ntl1-Nt7 cDNA with the GenEMBL data base (GCG; TFASTA seach) revealed that the zinc finger region is related to previously reported protein sequences. Best scores are obtained with the following proteins: AREA (Kudla et àl. 1990), human GATA-1 (Trainor et al. 1990), mouse GATA-1 (Tsai et al. 1989), NIT2 (Fu and Marzluf 1990a), yeast GLN3 (Minehart and Magazanik 1991), and chicken GATA-1 (Evans and Felsenfeld 1989).

We have made a multiple sequence alignment (GCG; PILEUP program) between the amino acid sequences of the NIT2, AREA, yeast GLN3, chicken GATA-1 and the NTL1-NT7 proteins. Except for the zinc finger region, only two amino acids, the S_{141} and the K_{259} are conserved in the five proteins. The alignment which is obtained in the zinc finger region is shown in figure 4. In addition to the four cysteine residues, which are possibly involved in the coordination of a zinc atom and conserved in the five proteins, eight identical amino acids are found at the same positions in these proteins. In the consensus sequence, derived from this alignment, these conserved amino acids are written in capital letter (fig. 4).

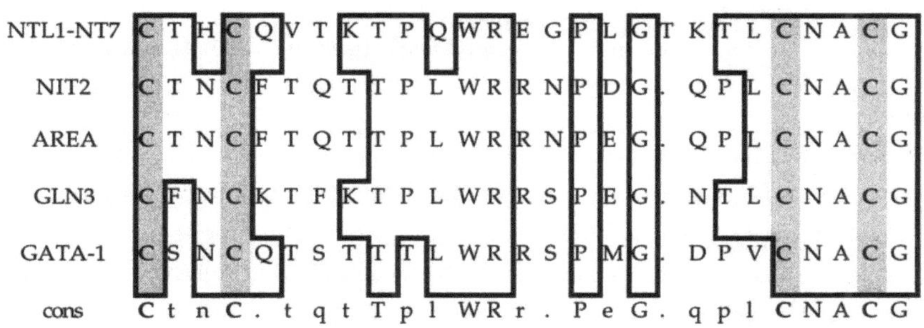

Figure 4: Multiple alignment of zinc finger domains

The four cysteines residues are shadowed and conserved amino acid between the NTL1-NT7 and at least one of the other sequences are boxed. The consensus sequence (cons) shows in capital lettersthe amino acids which are conserved in the five proteins

Interaction of the NTL1-NT7zinc finger domain with a NIT2 target sequence

As a first step of characterisation, we have analysed the interaction of the putative DNA-binding region of NTL1-NT7 with an oligonucleotide containing the target sequence of NIT2 protein. The zinc finger and adjacent basic domains of these two proteins were subcloned in the pGEX expression vector. The fusion proteins were purified on glutation sepharose beads and eluated with glutation according to the recommandations of the manufacturer. Gel-schift analysis (Figure 5) reveals that the two fusion proteins interact in the same manner with the target sequence. The specificity of the binding was provided by competition assays: the retarded bands progressively desappear after addition of the same cold oligonucleotide but were only slightly affected by the addition of an irrelevant oligonucleotide.

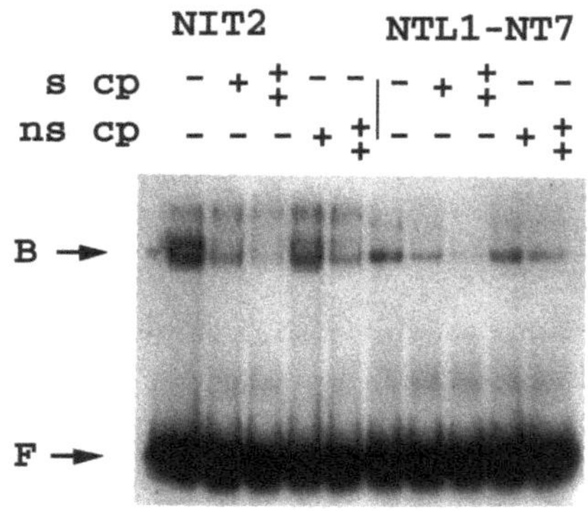

Figure 5: Interactions between NIT2 and NTL1-NT7 fusion proteins with a NIT2 target sequence

F : free probe
B : retarded band
s cp : specific competitor
ns cp: non specific competitor
+,‡ : competitors are added respectively in 50 or 100 molar excess

Perspectives

Whether or not the NTL1 protein is involved in nitrate reductase expression remains to be demonstrated. Some arguments are in favor of a possible role of NRG1 in the nitrate assimilation pathway. Despite the low temperature of the first three PCR cycles and the use of degenerate oligonucleotides, we have isolated only one class of cDNA from total RNAs. If there is a gene homologous to *nit-2* in higher plants, then *Ntl1* may be the only one. In order to verify this hypothesis, we will study, by gel retardation assays, the interactions between the NTL1-NT7 zinc finger domain with the tomato *nia* gene promoter and compare with the NIT2/*nia* promoter results (Jarai et al. 1992). An antisense strategy will be developed against the expression of the *Ntl1* gene in *N.plumbaginifolia* and the resulting transgenic plants will be characterized with respect to the expression of the nitrate assimilatory pathway.

References

Back E, Burkhart W, Meyer M, Privalle L, Rhohstein S (1988) Isolation of cDNA coding for spinach nitrite reductase: complete sequence and nitrate induction. Mol Gen Genet 212: 20-26

Caboche C , Rouzé P (1990) Nitrate reductase: a target for molecular and cellular studies in higher plants. Trends in Genetics 6: 187-192

Cheng CL, Dewdney J, Kleinofs A, Goodman H (1986) Cloning and nitrate induction of nitrate reductase mRNA. Proc Natl Acad Sci USA 83: 6825-6828

Cheng CL, Acedo GN, Dewdney J, Goodman HM, Conkling MA (1991) Differential expression of the two *Arabidopsis* nitrate reductase genes. Plant Physiol 96: 275-279

Crawford NM, Campbell HW, Davis RW (1986) Nitrate reductase from squash: cDNA cloning and nitrate regulation. Proc Natl Acad Sci USA 83: 8073-8076

Crawford NM, Campbell WH (1990) Fertile fields. Plant Cell 2: 829-83

Daniel-Vedele F, Dorbe M-F, Caboche M, Rouzé P (1989) Cloning and analysis of the tomato nitrate reductase-encoding gene: protein structure and amino acid homologies in higher plants. Gene 85: 371-380

Deng MD, Moureaux T, Leydecker MT, Caboche M (1990) Nitrate reductase expression is under the control of a circadian rhythm and is light inducible in *Nicotiana tabacum* leaves. Planta 180: 257-261

Deng MD, Moureaux T, Lamaze T (1991) Diurnal and circadian fluctuation of malate levels and its close relationship to nitrate reductase activity in tobacco leaves. Plant Science 65: 191-197

Dorbe MF, Caboche M, Daniel-Vedele F (1992) The tomato *nia* gene complements a *Nicotiana plumbaginifolia* nitrate reductase-deficient mutant and is properly regulated. Plant Molec Biol 18: 363-375

Evans T, Fesenfeld G (1989) The erythroid-specific transcription factor Eryf1: a new finger protein. Cell 58: 877-885

Faure J-D, Vincentz M, Kronenberger J, Caboche M (1991) Co-regulated expression of nitrate and nitrite reductases. The Plant J 1: 107-113

Fu YH, Marzluf GA (1990a) *nit-2*, the major nitrogen regulatory gene of *Neurospora crassa*, encodes a protein with a putative zinc finger DNA-binding domain. Mol Cell Biol 10: 1056-1065

Fu YH, Marzluf GA (1990b) *nit-2*, the major positive-acting nitrogen regulatory gene of Neurospora crassa, encodes a sequence-specific DNA-binding protein. Proc Natl Acad Sci USA 87: 5331-5335

FuYH, Marzluf GA (1990c) Site-directed mutagenesis of the 'zinc finger' DNA-binding domain of the nitrogen-regulatory protein NIT2 of *Neurospora*. Mol Microbiol 4: 1847-1852

Galangau F, Daniel-Vedele F, Moureaux T, Dorbe MF, Leydecker MT, Caboche M (1988) Expression of leaf nitrate reductase gene from tomato and tobacco in relation to light dark regimes and nitrate supply. Plant Physiol 88: 383-388

Guerrero MG, Vega JM, Losada M (1981) The assimilatory nitrate-reducing system and its regulation. Annu Rev Plant Physiol 44:1197-1199

Hope LA, Struhl K (1986) Functional dissection of a eukaryotic transcriptional activator protein, GCN4 of yeast. Cell 46: 885-894

Jarai G, Truong HN, Daniel-Vedele F, Marzluf G (1992) NIT2, the nitrogen regulatory protein of *Neurospora crassa,* binds upstream of *nia,* the tomato nitrate reductase gene, in vitro. Curr Genet 21: 37-41

Kalderon D, Roberts BL, Richardson WD, Smith AE (1984) A short amino acid sequence able to specify nuclear location. Cell 37: 801-813

Katagiri F, Chua NH (1992) Plant transcription factors: present knowledge and future challenges. Trends in Genet 8: 22-271010

Kleinhofs A, Warner RL, Narayanam KR (1985) Current progress towards understanding of the genetics and molecular biology of nitrate reductase in higher plants. In: MiflinB (ed) Oxford surveys of plant molecular biology, vol 2. Oxford University Press, New-York, pp 91-121

Kronenberger J, Lepingle A, Caboche M, Vaucheret H (1993) Cloning and expression of distinct nitrite reductases in tobacco leaves and roots. Mol Gen Genet 236: 203-208

Kudla B, Caddick MX, Langdon T, Martinez-Rossi NM, Bennett CF, SibleyS, Davies RW, Arst HNJr (1990) The regulatory gene *are*A mediating nitrogen metabolic repression in Aspergillus nidulans. Mutations affecting specificity of gene activation alter a loop residue of a putative zinc finger. EMBO J9: 1355-1364

Ma J, Ptashne M (1987) Deletion analysis of GAL4 defines two transcriptional activating segments. Cell 48: 847-853

Minehart PL, Magazanik B (1991) Sequence and expression of *GLN3,* a positive nitrogen regulatory gene of *Saccharomyces cerevisiae* encoding a protein with a putative zinc finger DNA-binding domain. Mol Cell Biol 11: 6216-6228

Ogawa M, Ida S (1987) Biosynthesis of ferredoxin nitrite reductase in rice seedlings. Plant Cell Physiol 28: 1501-1508

Pouteau S, Chérel I, Vaucheret H, Caboche M (1989) Nitrate reductase mRNA regulation in *Nicotiana plumbaginifolia* nitrate reductase deficient mutants. Plant Cell 1: 1111-1120

Silverthorne J, Tobin EM (1987) Phytochrome regulation of nuclear gene expression. Bioassays 7: 18-23

Scazzochio C, Arst HN Jr (1989) Regulation of nitrate assimilation in *Aspergillus nidulans.* In *Molecular and Genetic Aspects of Nitrate Assimilation* (Wray J and Kinghorn JR es) New York: Oxford University Press, pp299-313

Solomonson LP, Barber MJ (1990) Assimilatory nitrate reductase: functional properties and regulation. Annu Rev Plant Physiol Plant Mol Biol 41: 225-253

Trainor CD, Evans T, Felsenfeld G, Boguski MS (1990) Structure and evolution of a human erythroid transcription factor. Nature 343: 92-96

Tsai SF, Martin DIK, Zorn LI, D'Andrea AD, Wong GG, Orkin SH (1989) Cloning of cDNA for the major DNA-binding protein of the erythroid lineage through expression in mammalian cells.Nature 339: 446-451

Vaucheret H, Vincentz M, Kronenberger J, Caboche M, Rouzé P (1989) Molecular cloning and characterization of the two homeologous genes coding for nitrate reductase in tobacco. Mol Gen Genet 216: 10-15

Vaucheret H, Kroneneberger J, Lepingle A, Vilaine F, Boutin JP, Caboche M (1992) Inhibition of tobacco nitrite reductase activity by expression of antisense RNA. The Plant J 2: 559-569
Vincentz M, Caboche M (1991) Constitutive expression of nitrate reductase allows normal growth and developement of *Nicotiana plumbaginifolia* plants. EMBO J 10: 1027-1035
Vincentz M, Moureaux T, Leydecker M-T, Vaucheret H, Caboche M (1993) Regulation of nitrate and nitrite reductases expression in *Nicotiana plumbaginifolia* leaves by nitrogen and carbon metabolites. The Plant J 3: 315-324
Wray JL (1988) Molecular approaches to the analysis of nitrate assimilation. Plant Cell Environ 11: 369-382
Wray JL, Kinghorn JR (1989) Molecular and genetics aspects of nitrate assimilation. Oxford Science Publications, Oxford

Molecular-Genetics of Nitrogen Assimilation into Amino Acids in *Arabidopsis thaliana*.

Gloria Coruzzi, Karen Coschigano, Hon-Ming Lam,
Rosana Oliveira, Sheila Peng and Carolyn Schultz

New York University,
Department of Biology,
Washington Square,
New York, N.Y. 10003

INTRODUCTION

In all higher plants, nitrogen is first assimilated into the amino acids glutamine and glutamate and is then subsequently channeled into the amino acids aspartate, and asparagine. These amino acids serve as nitrogen donors and/or nitrogen transport compounds in plants. The enzymes involved in the biosynthesis of these amino acids often exist as multiple isoenzymes located in distinct subcellular compartments. Using conventional biochemical approaches it has been difficult, if not impossible, to sort out the function of each distinct isoenzyme in plant metabolism.

In the past, many different species of plants were used to study amino acid biosynthetic enzymes. This was largely due to the fact that researchers purified a particular enzyme from the species in which it was most abundant or from which it was most stable when isolated *in vitro*. Due to this heterogeneous sampling, it has been impossible to determine whether differences in isoenzymes represent species-specific differences or true differences related to their function. By using a molecular-genetic approach to study all of the enzymes involved in the nitrogen assimilation pathway in Arabidopsis, we hope to construct a molecular-genetic blueprint of amino acid biosynthesis and metabolism in a single model plant.

We have begun to study the Arabidopsis genes encoding enzymes involved in the metabolism of glutamine (glutamine synthetase, GS); glutamate (glutamate synthase, GOGAT; glutamate dehydrogenase, GDH); aspartate (aspartate aminotransferase, AspAT); and asparagine (asparagine synthetase, AS) (see Fig. 1). By studying the genes for the above enzymes and their patterns of regulation, we hope to uncover the global factors which

NATO ASI Series, Vol. H 81
Plant Molecular Biology
Edited by G. Coruzzi and P. Puigdomènech

coordinate the regulation of genes in a metabolic pathway. Mutants in each of these enzymes may be used to identify the physiological function of a particular isoenzyme *in vivo.* We have been successful in isolating mutants which are lacking a particular isoenzyme activity (e.g. for AspAT and GDH). In addition, we are also using the cloned genes to characterize previously isolated mutants lacking a particular enzyme activity (Fd-GOGAT). Since several of the enzymes in this pathway are encoded by multigene families even in Arabidopsis, we are currently identifying which structural gene is mutated in each of the isoenzyme mutants. By studying the phenotype of mutations in a particular gene, we will be able to assign a physiological role to each encoded isoenzyme.

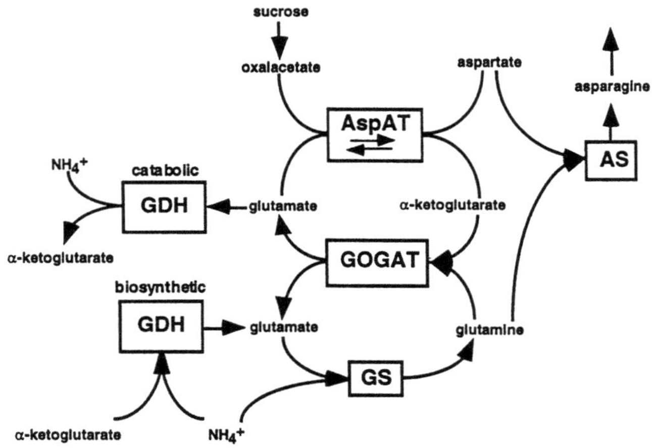

Figure 1. An overview of the enzymes involved in nitrogen assimilation and transport in higher plants. Glutamine synthetase (GS); Glutamate synthase (GOGAT); Glutamate dehydrogenase (GDH); Aspartate aminotransferase (AspAT); Asparagine synthetase (AS)

RESULTS AND DISCUSSION

Glutamine synthetase (GS). Previous studies on GS genes in *Pisum sativum* (Tingey et al., 1987; Tingey et al., 1988) and *Phaseolus vulgaris* (Cullimore and Bennett, 1988) have shown that these species contain a single nuclear gene for chloroplast GS2 and multiple genes for cytosolic GS. It has been suggested that legumes possess multiple genes for GS because special additional cytosolic GS genes were required to assimilate nitrogen fixed by Rhizobia in nodules. It was therefore surprising to find that there was the same array of GS genes in Arabidopsis, i.e. a single nuclear gene for chloroplast GS2 (Atgsl1), and at least three genes for cytosolic GS (Atgsr1, Atgsr2, and Atkb6) (Peterman and Goodman, 1991). Thus, multiple genes for cytosolic GS do not occur only in nitrogen-fixing legumes, but also in a non-legume such as Arabidopsis. The existence of multiple GS genes in the simple genome of Arabidopsis supports the notion that the encoded GS isoenzymes perform distinct functions.

Studies on the regulation of genes for chloroplast GS2 demonstrated light-regulation of GS2 gene expression in pea (Edwards and Coruzzi, 1989), Phaseolus (Bennett et al., 1989) and also in Arabidopsis (Peterman and Goodman, 1991). Furthermore, it has been shown that the GS2 gene is also regulated by the photorespiratory state of the plant in pea (Edwards and Coruzzi, 1989) and Phaseolus (Cock et al., 1991). This latter issue remains to be addressed for Arabidopsis GS2. For cytosolic GS, it appears that each gene is also expressed in a distinct fashion with regard to organ-type in pea (Walker and Coruzzi, 1989), Phaseolus (Gebhardt et al., 1986), and Arabidopsis (Peterman and Goodman, 1991). The differences in expression patterns of the GS genes lends further support to the hypothesis that the different GS genes have different functions.

The suggestion that individual GS genes perform distinct physiological functions was confirmed when it was shown (using GS promoter-GUS constructs) that the pea genes for chloroplast GS2 and cytosolic GS3A were expressed in distinct cell-types in transgenic tobacco (Edwards et al., 1990). The finding that chloroplast GS2 and cytosolic GS3A are expressed in distinct cell-types indicates that they serve non-overlapping functions in plant nitrogen metabolism. The mesophyll-specific expression pattern of chloroplast GS2 implicates this gene in the reassimilation of photorespiratory ammonia (Edwards et al., 1990; Edwards and Coruzzi, 1989). This notion was already established by the analysis of barley photorespiratory mutants in GS2 (Wallsgrove et al., 1987). The phloem-specific expression of the pea cytosolic GS3A gene indicates that this GS gene functions to synthesize glutamine for intercellular transport (Edwards et al., 1990). This suggestion is supported by the finding that the GS3A gene is expressed most highly in two nitrogen-rich organs (nodules and cotyledons) (Walker and Coruzzi, 1989).

The phloem-specific expression of pea GS3A in transgenic tobacco (Edwards et al., 1990) was a surprising finding which leads to two major conclusions. Firstly, while it was long known that amino acids are transported in the phloem, this was the first evidence that at least some glutamine is synthesized in the phloem, *in situ*. Secondly, the fact that cytosolic GS is not expressed in leaf mesophyll cells could explain why the barley mutants of chloroplast GS2 die in air, even though they have normal levels of cytosolic GS (Wallsgrove et al., 1987).

Cytosolic GS may not be able to compensate for the loss of chloroplast GS2 in the barley GS2 mutants if it is not expressed in the same cell-type. With regard to the photorespiratory mutants in GS2, it is curious that mutants in chloroplast GS2 were not isolated in a screen for photorespiratory mutants of Arabidopsis (Somerville, 1984; Lea et al., 1988). There are several possible explanations for the lack of GS2 mutants in Arabidopsis. There may be a second yet undiscovered gene for chloroplast GS2 in Arabidopsis which would preclude the isolation of a GS2 mutant. Alternately, it is possible that chloroplast GS2 and cytosolic GS are both expressed in mesophyll cells in Arabidopsis and that cytosolic GS levels are sufficient to compensate for the loss of chloroplast GS2 in mesophyll cells. Another possibility is that a mutation in chloroplast GS2 in Arabidopsis is lethal (e.g. even in high levels of CO_2 where photorespiration is suppressed). Finally, perhaps the screen for Arabidopsis photorepiratory mutants was not saturating, although this is unlikely given the number of alleles isolated at the other loci (Nancy Artus, Ph.D. thesis). These alternatives need to be further investigated.

Glutamate synthase (GOGAT). Biochemical studies of GOGAT enzymes in a variety of species has indicated that there are two distinct forms of the GOGAT enzyme, a ferredoxin-dependent form (Fd-GOGAT) and an NADH-dependent form (NADH-GOGAT) (Suzuki and Gadal, 1984). Fd-GOGAT is more abundant in mature, green tissue whereas NADH-GOGAT is more abundant in immature, nongreen tissue (Matoh and Takahashi, 1982). The biochemical analysis of GOGAT enzymes is further complicated by the fact that each form of (Fd-GOGAT or NADH-GOGAT) exists as multiple isoenzymes (Suzuki et al., 1982; Chen and Cullimore, 1988). It has been difficult to assign roles to these isoenzymes based solely on the biochemical studies conducted in a number of different species. The molecular-genetic analyses of GOGAT genes and mutants in a single species as outlined below, has begun to clarify the roles of these different GOGAT isoenzymes.

cDNAs for Fd-GOGAT have been cloned from maize (Sakakibara et al., 1991), tobacco (Zehnacker et al., 1992), and barley (Avila et al., 1993). Only a single Fd-GOGAT gene was isolated in each case. Recently, we have cloned two distinct Fd-GOGAT cDNAs from Arabidopsis, indicating the existence of two functional Fd-GOGAT genes (Coschigano & Coruzzi, unpublished results). It is curious that there are two expressed Fd-GOGAT genes in Arabidopsis, yet phenotypic mutants lacking Fd-GOGAT activity (gluS) were isolated in the screen for photorespiratory mutants conducted by Somerville and Ogren (Somerville and Ogren, 1980). The isolation of a gluS mutant phenotype suggested that there was a single Fd-GOGAT gene in Arabidopsis. Several possible explanations for this discrepancy exist. Since the Fd-GOGAT mutants were the largest class of photorespiratory mutants obtained in the Arabidopsis screen, it is possible that a mutation in either Fd-GOGAT gene results in a phenotype. Alternately if one Fd-GOGAT gene is expressed at high levels, and the other at low levels, a mutation in the high-level expressed gene could give rise to a phenotype. This latter explanation has been shown to be the case for the duplicated tryptophan synthase genes of Arabidopsis (Last et al., 1991) and may also be true for the Fd-GOGAT genes since one of the genes appears to be more highly expressed than the other (Coschigano & Coruzzi, unpublished results). Finally, since two Fd-GOGAT genes exist, a mutation in one gene may lead to a leaky phenotype. This latter interpretation is supported by the fact that the gluS mutants were by far the leakiest phenotype of all the photorespiratory mutants isolated by Somerville and Ogren (personal communication).

In wild-type Arabidopsis leaves, Fd-GOGAT contributes approximately 95% of the total GOGAT activity, while the remaining 5% activity can be attributed to NADH-GOGAT (Somerville and Ogren, 1980). In the Fd-GOGAT photorespiratory mutants, while Fd-GOGAT activity was nearly undetectable, the NADH-GOGAT activity remained unchanged. The fact that these gluS mutants grew normally in high CO_2 but died in air indicates that the major role of Fd-GOGAT in Arabidopsis is to reassimilate photorespiratory ammonia. In this way, the Fd-GOGAT enzyme and the chloroplast GS2 enzyme would be linked in a cycle to reassimilate photorespiratory ammonia (Fig. 2). This coupling is borne out by initial experiments at the molecular level, as the genes for Arabidopsis Fd-GOGAT (GGT1) and chloroplast GS2 (Atgsl1) are coordinately regulated by light ((Peterman and Goodman, 1991), Coschigano & Coruzzi, unpublished data).

With regard to NADH-GOGAT, while its activity is low in mature leaves as compared to Fd-GOGAT, biochemical data suggests that NADH-GOGAT activity is greater in immature tissue of roots (Matoh and Takahashi, 1982) and cotyledons (Hecht et al., 1988). This organ-specific pattern of enzyme activity indicates that NADH-GOGAT most likely functions in a cycle with cytosolic GS to generate glutamine for intracellular transport (Fig. 2). In *E. coli*, NADH-GOGAT is the sole form of GOGAT and is composed of two subunits, a large and a small subunit (Miller and Stadtman, 1972). In plants, however, it was recently shown that the two subunits of the NADH-GOGAT enzyme are contiguous in one polypeptide (Gregerson et al., 1993). Polypeptide sequence comparsions showed that there is a conserved NADH-binding consensus sequence in both the bacterial and the plant NADH-GOGAT small subunit/domain (Oliver et al., 1987; Gregerson et al., 1993). Homology to the *E. coli* small subunit is missing in the Fd-GOGAT enzyme (Sakakibara et al., 1991). We have recently cloned an NADH-GOGAT cDNA from Arabidopsis which appears to be a single gene with strong homology to the large and small domains of alfalfa NADH-GOGAT (Lam & Coruzzi, unpublished results). Expression studies have shown that NADH-GOGAT and cytosolic GS are coordinately regulated in Arabidopsis roots (Lam & Coruzzi, unpublished results), again consistent with their apparent physiological coupling. The coordinated functions of NADH-

GOGAT and cytosolic GS may act primarily on the initial incorporation of ammonia into root cells and also in the remobilization of ammonia released through processes such as the breakdown of seed storage proteins in cotyledons during germination (Fig. 2).

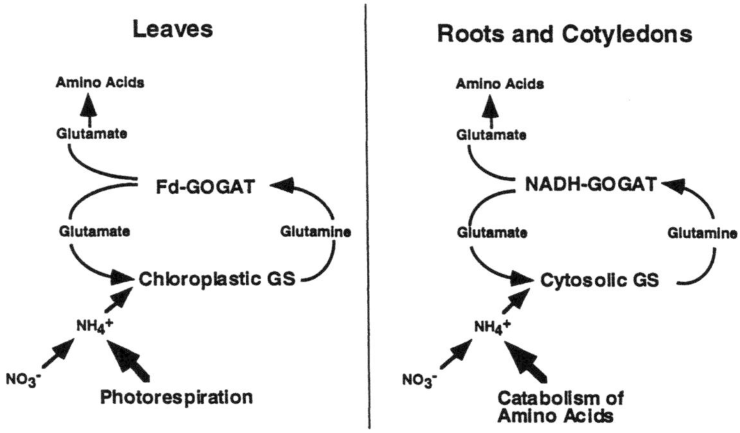

Figure 2. Fd-GOGAT and NADH-GOGAT serve distinct roles in plant nitrogen metabolism.

Aspartate aminotransferase (AspAT). Biochemical studies have shown that there are at least three different forms of AspAT which are localized to the chloroplast, mitochondria or cytosol (Weeden and Marx, 1987). While AspAT plays a role in transamination reactions in each of these compartments, the question remains as to the exact physiological role of each isoenzyme. One reason it has been difficult to assign definitive roles to each AspAT is that the reaction catalyzed by AspAT is reversible, thus increasing the possibility that each isoenzyme has a distinct role. AspAT genes have been cloned from several plant species e.g. alfalfa, *Panicum,* soybean, carrot and lupin ((Gantt et al., 1992; Taniguchi et al., 1992; Wadsworth, 1993; Turano et al., 1992)). While only one or two cDNAs have been isolated from each of these other plant species, our lab has recently cloned four different AspAT cDNAs from a single species, Arabidopsis (Schultz & Coruzzi, unpublished results). These four different AspAT cDNAs most likely encode a single mitochondrial, a single chloroplast and two cytosolic isoenzymes.

Activity staining of non-denaturing gels indicates that Arabidopsis leaves and roots contain two major AspAT isoenzymes (Schultz & Coruzzi, unpublished results). To determine the role of each AspAT isoenzyme, we have begun a search for Arabidopsis mutants which are lacking either of these isoenzymes by directly assaying leaves of EMS mutagenized seedlings for AspAT isoenzyme activity. To date, three different classes of AspAT mutants have been isolated which each have specific, distinct defects in AspAT isoenzymes. We are in the process of correlating the mutants with the cloned AspAT genes and determining the phenotypic effects of the mutation. These studies should enable us to assign a role to each AspAT gene/isoenzyme and to determine whether the multiple forms of AspAT play distinct or overlapping roles in aspartate metabolism in Arabidopsis.

Asparagine synthetase (AS). While glutamine is the primary nitrogen transport compound in most higher plants, asparagine also serves to transport nitrogen in several legumes and non-legume species. Glutamine and asparagine serve different roles in nitrogen metabolism. Since glutamine is metabolically active and asparagine relatively inert, asparagine serves as a nitrogen storage compound in addition to transport (Sieciechowicz et al., 1988). Thus, the ratio of glutamine:asparagine transported in plants is affected by growth conditions. In light-grown plants, glutamine is the predominant nitrogen transport compound whereas in dark-grown or dark-adapted plants, asparagine levels predominate (Urquhart and Joy, 1982). The increased levels of asparagine in plants grown in the dark correlates with the fact that asparagine is a more economical compound with which to transport nitrogen (e.g. has a higher N:C ratio than glutamine).

In pea, we have previously shown that while chloroplast GS2 gene expression is induced by light, the expression of the two pea AS genes (AS1 and AS2) are repressed by light (Tsai and Coruzzi, 1990; Tsai and Coruzzi, 1991). These results are in keeping with the physiological findings of glutamine and asparagine levels in these plants (Sieciechowicz et al., 1988). More recently, we have shown that Arabidopsis has a single gene for glutamine-dependent AS (AS1) which is also regulated by light in a negative fashion (Lam, Peng & Coruzzi, unpublished results). We are interested in determining how light activates the gene for chloroplast GS2 while concomitantly repressing the gene for AS1 in Arabidopsis.

Glutamate dehydrogenase (GDH). Prior to the discovery of glutamine synthetase, GDH was thought to be the primary route of ammonia assimilation in plants, as it was in bacteria and yeast. Subsequent to the discovery of GS (which has a low K_m for ammonia), the biosynthetic role of GDH has been questioned primarily because this enzyme has a high K_m for ammonia (see Fig. 1). At best, GDH may function in a biosynthetic role to provide catalytic amounts of glutamate for the GS/GOGAT cycle. More recently, it has been suggested that the major role for GDH in plants is to catabolize glutamate (Lea et al., 1990) (see Fig. 1). To understand the role of GDH in plants, we have begun to study the genes and mutants for GDH in Arabidopsis.

Activity gel stains for GDH reveal seven activity bands on a non-denaturing gel of wild-type Arabidopsis leaf extracts (Cammaerts and Jacobs, 1985). The seven activity bands of GDH represent two homohexamers and five heterohexamers as a result of the random association of two differently charged peptides encoded by two GDH genes. Arabidopsis mutants which have a single GDH activity band (i.e. a single homohexamer) have been isolated in our lab (Oliveira & Coruzzi, unpublished results), suggesting that a mutation has occurred at one of the two GDH loci. A similar GDH mutant has been described in *Zea mays* but it was not characterized further with regard to the affected gene (Pryor, 1974). We are currently characterizing Arabidopsis GDH mutants phenotypically and will correlate the mutants with the cloned Arabidopsis GDH genes. These studies should help to clarify the roles of GDH in Arabidopsis nitrogen metabolism.

CONCLUSIONS

Molecular and genetic studies of amino acid biosynthesis genes presented here have begun to provide new insights into the possible roles of particular isoenzymes. While biochemical studies first demonstrated the presence of multiple isoenzymes, the more recent molecular studies have shown that genes for individual isoenzymes are expressed in distinct cell types. These cell-specific expression patterns of genes for each isoform suggests that the isoenzymes play non-overlapping roles in metabolism. The gene expression studies have also shown that individual members of each gene family for a particular amino acid biosynthesis enzyme are differentially regulated with regard to environment (e.g. light) and development. This regulation stems from the fact that amino acids in plants are not merely the building blocks of proteins, but are also the precursors to many intermediate products whose synthesis is regulated (e.g. hormones). Generally the pattern that is emerging is that there is more than one gene for each amino acid biosynthesis enzyme and that while one gene may function to make constitutive levels of an amino acid for protein synthesis, another gene(s) of the family function to provide inducible amounts of the product, and are tightly regulated. Thus, the finding that amino acid biosynthesis genes are expressed in a cell-specific fashion along with the data which demonstrates that the genes are tightly regulated (e.g. by light, development, organ-type) indicates that these genes should not be considered "housekeeping genes". Alternatively, "housekeeping" is more complex than was formerly thought. Continuing studies on the amino acid biosynthesis genes in Arabidopsis should uncover the roles of each individual gene and should provide a molecular-genetic blueprint of nitrogen metabolism in the model plant, Arabidopsis.

ACKNOWLEDGMENTS

This work was supported by NIH Grant GM32877, DOE Grant DEFG02-92-ER20071, and NSF Grant MCB 9304913.

REFERENCES

Avila, C., Márquez, A.J., Pajuelo, P., Cannell, M.E., Wallsgrove, R.M. and Forde, B.G. Cloning and sequence analysis of a cDNA for barley ferredoxin-dependent glutamate synthase and molecular analysis of photorespiratory mutants deficient in the enzyme. *Planta* 189:475-483, 1993.

Bennett, M.J., Lightfoot, D.A. and Cullimore, J.V. cDNA sequence and differential expression of the gene encoding the glutamine synthetase polypeptide of Phaseolus vulgaris L. *Plant Mol.Biol.* 12:553-565, 1989.

Cammaerts, D. and Jacobs, M. A study of the role of glutamate dehydrogenase in the nitrogen metabolism of *Arabidopsis thaliana. J.Hered.* 163:517-526, 1985.

Chen, F.L. and Cullimore, J.V. Two isozymes of NADH-dependent glutamate synthase in root nodules of Phaseolus vulgaris L.: Purification, properties and activity changes during nodule development. *Plant Physiol.* 88:1411-1417, 1988.

Cock, J.M., Brock, I.W., Watson, A.T., Swarup, R., Morby, A.P. and Cullimore, J.V. Regulation of glutamine synthetase genes in leaves of Phaseolus vulgaris. *Plant Mol.Biol.* 17:761-771, 1991.

Cullimore, J.V. and Bennett, M.J. The molecular biology and biochemistry of plant glutamine synthetase from root nodules of *Phaseolus vulgaris* L. and other legumes. *Journal of Plant Physiology* 132:387-393, 1988.

Edwards, J.W. and Coruzzi, G.M. Photorespiration and light act in concert to regulate the expression of the nuclear gene for chloroplast glutamine synthetase. *Plant Cell* 1:241-248, 1989.

Edwards, J.W., Walker, E.L. and Coruzzi, G.M. Cell-specific expression in transgenic plants reveals nonoverlapping roles for chloroplast and cytosolic glutamine synthetase. *Proc.Natl.Acad.Sci.USA.* 87:3459-3463, 1990.

Gantt, J.S., Larson, R.J., Farnham, M.W., Pathirana, S.M., Miller, S.S. and Vance, C.P. Aspartate aminotransferase in effective and ineffective alfalfa nodules. *Plant Physiol.* 98:868-878, 1992.

Gebhardt, C., Oliver, J.E., Forde, B.G., Saarelainen, R. and Miflin, B.J. Primary structure and differential expression of glutamine synthetase genes in nodules, roots and leaves of Phaseolus vulgaris. *EMBO J.* 5:1429-1435, 1986.

Gregerson, R.G., Miller, S.S., Twary, S.N., Gantt, J.S. and Vance, C.P. Molecular characterization of NADH-dependent glutamate synthase from alfalfa nodules. *Plant Cell* 5:1993.

Hecht, U., Oelmuller, R., Schmidt, S. and Mohr, H. Action of light, nitrate and ammonium on the levels of NADH- and ferredoxin-dependent glutamate synthases in the cotyledons of mustard seedlings. *Planta* 175:130-138, 1988.

Last, R.L., Bissinger, P.H., Mahoney, D.J., Radwanski, E.R. and Fink, G.R. Tryptophan mutants in *Arabidopsis*: The consequences of duplicated tryptophan synthase beta genes. *Plant Cell* 3:345-358, 1991.

Lea, P.J., Blackwell, R.D., Murray, A.J.S. and Joy, K.W. The use of mutants lacking glutamine synthetase and glutamate synthase to study their role in plant nitrogen metabolism. In: *Recent advances in phytochemistry*, edited by Poulton, J.E., Romeo, J.T. and Conn, E.E. New York and London: Plenum Press, 1988, p. 157-189.

Lea, P.J., Robinson, S.A. and Stewart, G.R. The enzymology and metabolism of glutamine, glutamate, and asparagine. In: *The Biochemistry of Plants*, edited by Miflin, B.J. and Lea,

P.J. New York: Academic Press, 1990, p. 121-159.
Matoh, T. and Takahashi, E. Changes in the activites of ferredoxin and NADH-glutamate synthase during seedling development of peas. *Planta* 154:289-294, 1982.
Miller, R.E. and Stadtman, E.R. Glutamate Synthase from Escherichia coli. *J.Biol.Chem.* 247:7407-7419, 1972.
Oliver, G., Gosset, G., Snachez-Pescador, R., et al. Determination of the nucleotide sequence of the glutamate synthase structural genes of Escherichia coli K-12. *Gene* 60:1-11, 1987.
Peterman, T.K. and Goodman, H.M. The glutamine synthetase gene family of *Arabidopsis thaliana*: Light-regulation and differential expression in leaves, roots, and seeds.*Mol.Gen.Genet.* 230:145-154, 1991.
Pryor, A.J. Allelic glutamic dehydrogenase isozymes in maize - a single hybrid isozyme in heterozygotes? *Heredity* 32:397-419, 1974.
Sakakibara, H., Matanabe, M., Hase, T. and Sugiyama, T. Molecular cloning and characterization of complementary DNA encoding for ferredoxin dependent glutamate synthase in maize leaf. *J.Biol.Chem.* 266:2028-2035, 1991.
Sieciechowicz, K.A., Joy, K.W. and Ireland, R.J. The metabolism of asparagine in plants. *Phytochemistry* 27:663-671, 1988.
Somerville, C.R. The analysis of photosynthetic carbon dioxide fixation and photorespiration by mutant selection. In: *Oxford Surveys of Plant Molecular and Cell Biology, vol.1*, edited by Miflin, B.J. Oxford: Oxford University Press, 1984, p. 103-131.
Somerville, C.R. and Ogren, W.L. Inhibition of photosynthesis in *Arabidopsis* mutants lacking leaf glutamate synthase activity. *Nature* 286:257-259, 1980.
Suzuki, A. and Gadal, P. Glutamate synthase: physicochemical and functional properties of different forms in higher plants and other organisms. *Physiol.Veg.* 22:471-486, 1984.
Suzuki, A., Vidal, J. and Gadal, P. Glutamate synthase isoforms in rice. Immunological studies of enzymes in green leaf, etiolated leaf, and root tissues. *Plant Physiol.* 70:827-832, 1982.
Taniguchi, M., Sawaki, H., Sasakawa, H., Hase, T. and Sugiyama, T. Cloning and sequence analysis of cDNA encoding aspartate aminotransferase isozymes from *Panicum miliaceum* L., a C_4 plant. *Eur.J.Biochem.* 204:611-620, 1992.
Tingey, S.V., Tsai, F.-Y., Edwards, J.W., Walker, E.L. and Coruzzi, G.M. Chloroplast and cytosolic glutamine synthetase are encoded by homologous nuclear genes which are differentially expressed in vivo. *J.Biol.Chem.* 263:9651-9657, 1988.
Tingey, S.V., Walker, E.L. and Coruzzi, G.M. Glutamine synthetase genes of pea encode distinct polypeptides which are differentially expressed in leaves, roots and nodules. *EMBO J.* 6:1-9, 1987.
Tsai, F.-Y. and Coruzzi, G.M. Dark-induced and organ-specific expression of two asparagine synthetase genes in Pisum sativum. *EMBO J.* 9:323-332, 1990.
Tsai, F.-Y. and Coruzzi, G.M. Light represses the transcription of asparagine synthetase genes in photosynthetic and non-photosynthetic organs of plants. *Mol.Cell.Biol.* 11:4966-4972, 1991.
Turano, F.J., Weisemann, J.M. and Matthews, B.F. Identification and expression of a cDNA clone encoding aspartate aminotransferase in carrot. *Plant Physiol.* 100:374-381, 1992.
Urquhart, A.A. and Joy, K.W. Transport, metabolism, and redistribution of Xylem-borne amino acids in developing pea shoots. *Plant Physiol.* 69:1226-1232, 1982.
Wadsworth, G.W.,Marmaras,S.M.,Matthews,B.F. Isolation and characterization of a soybean cDNA clone encoding the plastid form of aspartate aminotransferase. *Plant Mol.Biol.* 21:993-1009, 1993.
Walker, E.L. and Coruzzi, G.M. Developmentally regulated expression of the gene family for cytosolic glutamine synthetase in Pisum sativum. *Plant Physiol.* 91:702-708, 1989.
Wallsgrove, R.M., Turner, J.C., Hall, N.P., Kendally, A.C. and Bright, S.W.J. Barley mutants lacking chloroplast glutamine synthetase-biochemical and genetic analysis. *Plant Physiol.* 83:155-158, 1987.
Weeden, N.F. and Marx, G.A. Further genetic analysis and linkage relationships of isozyme loci in pea. *J.Hered.* 78:153-159, 1987.
Zehnacker, C., Becker, T.W., Suzuki, A., Carrayol, E., Caboche, M. and Hirel, B. Purification and properties of tobacco ferredoxin-dependent glutamate synthase, and isolation of corresponding cDNA clones. *Planta* 187:266-274, 1992.

Genetic Dissection of Tryptophan Biosynthesis in *Arabidopsis thaliana*

Robert L. Last, Andrea J. Barczak, Melissa Ho, Jiayang Li, Kim D. Pruitt, Elaine R. Radwanski and Alan B. Rose
Boyce Thompson Institute for Plant Research
and Section of Genetics and Development
Cornell University
Tower Road
Ithaca, NY 14853-1801

Introduction

A long term goal for this laboratory is to obtain a detailed understanding of the biochemical and genetic mechanisms that plants use to regulate the synthesis of the amino acid tryptophan (Last et al., 1992; Last, 1993). In addition to its role in the production of tryptophan for protein synthesis, this pathway gives rise to many indolic secondary metabolites, including the auxin, indole-3-acetic acid (Wright et al., 1991) and the Arabidopsis phytoallexin called camalexin (Tsuji et al., 1992). Thus, we anticipate that plants regulate the enzymes of tryptophan biosynthesis in response to requirements for both protein and secondary product biosynthesis.

Cloning *Arabidopsis thaliana* tryptophan pathway biosynthetic genes was facilitated by information available from microorganisms. For example, the genes for tryptophan synthase ß subunit (TSB; Berlyn et al., 1989; Last et al., 1991) and anthranilate synthase α subunit (Niyogi and Fink, 1992) were isolated by reduced stringency hybridization screening of cDNA libraries with probes from the bakers yeast, *Saccharomyces cerevisiae*. Another powerful method for isolation of tryptophan pathway enzyme cDNAs is suppression of *Escherechia coli* mutations. This method was used to clone phosphoribosylanthranilate transferase (PAT; Elledge et al., 1991; Rose et al., 1992), PR-anthranilate isomerase (PAI; J. Li and R. Last, manuscript in preparation), indole-3-glycerol phosphate synthase (J. Li and R. Last, unpublished) and tryptophan synthase α subunit (TSA; E. Radwanski and R. Last, manuscript in preparation).

In order to define the pathway(s) operating *in vivo*, we obtained a variety of mutants altered in tryptophan biosynthesis. A number of attributes make the

NATO ASI Series, Vol. H 81
Plant Molecular Biology
Edited by G. Coruzzi and P. Puigdomènech

tryptophan pathway in *A. thaliana* amenable to such genetic dissection. This is the only plant amino acid biosynthetic pathway for which auxotrophs were isolated in whole plants (Last and Fink, 1988; Last et al., 1991). Analysis of these mutants, in addition to the cloned genes described above, suggested that the plant pathway is identical to that in bacteria and fungi.

In a thorough genetic analysis of a biochemical pathway it is desirable to obtain a variety of mutations affecting diverse functions. One goal is to identify mutations in as many steps of the pathway as possible. In this way it is possible to define the steps that operate *in vivo* and to distinguish the rate-limiting regulatory step. In the case of the tryptophan pathway, which potentially contains branches toward various secondary metabolites from various pathway intermediates, biochemical analysis of mutants should help to define where these branches occur. Many secondary metabolites are hypothesized to be important in specific biological processes (for example, phytoallexins in resistance to microbial pathogens). If these mutants are deficient in interesting secondary metabolites it would be possible to directly test the significance of the missing compounds (Li et al., 1993). We developed a variety of screens and selections that allowed the isolation of mutants at several steps in the pathway. Two of these strategies led to the identification of a variety of mutations affecting individual enzymes. Biochemical and DNA sequence analysis of these mutants should lead to insights about the structure and function of these enzymes. We anticipate that these genetic screens and selections will also identify mutations that affect the biochemical and genetic regulation of the pathway.

Identification of mutations affecting a biosynthetic pathway has several advantages over the reverse genetic approaches such as antisense and *in vitro* mutagenesis, which rely upon transgenic plants. One benefit is that both recessive and dominant mutant phenotypes tend to be stable throughout many generations of selfcrosses and outcrosses to plants of similar genetic background. A recessive mutation with a defined biochemical lesion can serve as the starting point for protein engineering by plant transformation. For example, introduction of a specifically altered gene construct into a true null mutant plant allows study of the activity of the protein from the novel allele without interference from the wild-type protein. Ectopic expression of the wild-type protein would give information about the biological significance of cell-type specific expression and transport of pathway intermediates and products.

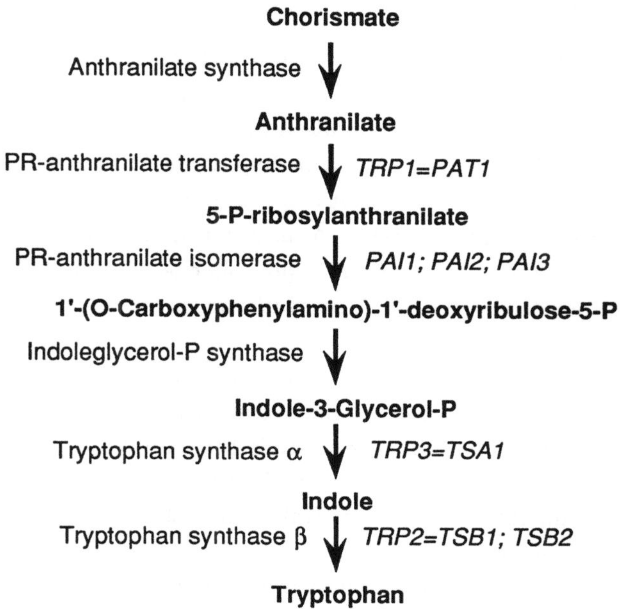

Figure 1. The tryptophan pathway in plants. Enzyme names are on the left of each arrow, and the genes that we are intensively studying are listed on the right.

PR-anthranilate Transferase Mutants

The molecular genetics of the PAT enzyme continues to receive extensive attention because *trp1* mutants defective in this activity have interesting and useful characteristics (Last and Fink, 1988; Rose et al., 1992). These phenotypes include blue fluorescence under UV light caused by accumulation of anthranilate (and its derivatives), tryptophan auxotrophy, slow growth and bushy habit and methylanthranilate resistance. The blue fluorescence phenotype, which can be scored as early as five days post-imbibition, provides a reliable method for obtaining mutants with reduced PR-anthranilate activity. Thus far, eleven blue fluorescent *trp1* mutants were obtained, all of which appear to have *trp1* defects. These mutants define an allelic series that range in severity from an auxotroph that is quite sickly and sterile and must be propagated as a heterozygote, to plants that do not require tryptophan and have a normal growth habit. Preliminary enzyme assay results suggest that the severity of the whole plant phenotype parallels differences in anthranilate substrate affinity of the mutant enzymes (A. Rose, J. Li and R. Last, manuscript in preparation).

A prototrophic *trp1* mutant, which is not blue fluorescent, was identified in a selection for resistance to the analogue 6-methylanthranilate (6MA). This was a surprising result because we expected anthranilate accumulation in any mutant that cannot efficiently convert anthranilate to PR-anthranilate. The evidence that this is a *trp1* mutant is that this mutation fails to complement, and is genetically linked to the *trp1-100* mutation (i.e. the F_1 and F_2 progeny are all resistant to 6MA). We are currently testing hypotheses that may help to explain this mutant phenotype. For example, it is possible that this mutation is so leaky that anthranilate does not accumulate. Alternatively, the mutant enzyme might be unable to convert the analogue 6MA to phosphoribosyl-6-methylanthranilate, but is competent for conversion of the biosynthetic substrate anthranilate to PR-anthranilate.

We isolated cDNA and genomic clones for the wild-type enzyme and demonstrated that the *PAT1* gene is defective in *trp1* mutant plants (Rose et al., 1992). A surprising result from this study is that, unlike the aromatic amino acid biosynthetic enzyme genes previously identified (Klee et al., 1987; Keith et al., 1991; Last et al., 1991; Niyogi and Fink, 1992), there appears to be a single *PAT* gene in Arabidopsis.

The *trp1-1* and *trp1-100* mutations were demonstrated to be alleles of the single copy Arabidopsis PAT structural gene, *PAT1* (Rose et al., 1992). Complementation of the *trp1* mutations by the cloned *PAT1* structural gene in transgenic plants confirmed genetic mapping studies indicating that the *PAT1* gene maps to the same locus as the *trp1-100* mutation. Stable introduction of the wild-type *PAT1* gene into either *trp1* mutant by Agrobacterium-mediated transformation reverses the mutant phenotypes, including blue-fluorescence, tryptophan auxotrophy, methylanthranilate resistance and the morphological defects.

These results indicate that the *PAT1* gene might be useful as a reporter of gene expression or a selectable marker. Plants that express functional PAT protein can be selected using tryptophan prototrophy or screened as plants that are no longer blue fluorescent. It should also be possible to identify loss of function by selection for resistance to methylanthranilate or screening for blue fluorescence, tryptophan requirement and bushy phenotype. Such a reporter gene should be a useful tool in plant molecular biology.

Tryptophan Synthase Beta Subunit

Tryptophan synthase catalyzes the conversion of indole-3-glycerol phosphate to tryptophan (Miles 1991). The structure, function and genetic regulation of this enzyme is particularly well understood in microorganisms (Yanofsky and Crawford, 1987; Miles 1991), making it an attractive target for molecular genetic analysis in plants. In bacteria, and probably in plants, the enzyme is a heterotetramer of structure αßßα, and each subunit can catalyze a half reaction (see Figure 1), albeit less efficiently than in the active complex (Anderson et al., 1991).

There are two unlinked genes (*TSB1* and *TSB2*) that encode highly conserved ß subunit isoenzymes in Arabidopsis (Berlyn et al., 1989; Last et al., 1991). While *TSB1* produces the majority of mRNA in leaf and other tissues, TSB2 mRNA is present at a constant low level in all tissues examined (Last et al., 1991; Pruitt and Last, 1993). *TSB* promoter-GUS fusion studies demonstrated higher GUS activity in plants transformed with the *TSB1*-GUS fusion constructs (Pruitt and Last, 1993). The *TSB1* promoter specifies developmentally regulated GUS reporter gene expression in stem, vascular tissues, root tips, mesophyll cells, and in cells located at the base and tip of the anther filament. *TSB2*-GUS activity is seen later in development in a less complex pattern, which overlaps that of *TSB1*.

The recessive *trp2-1* auxotrophic mutation is in the more highly expressed *TSB1* gene (Last et al., 1991). Apparently, the *TSB2* gene makes a modest contribution to total tryptophan biosynthesis. This is an example where genetic redundancy does not prevent isolation of amino acid requiring mutants. A very different situation is found in maize, where two different TSß subunit genes (*ORP1* and *ORP2*) must be defective for a tryptophan requirement (Wright et al., 1992).

Selection for 5-fluoroindole resistance (FIR) in Arabidopsis yields mutants that are deficient in TSß subunit activity (Last et al., 1991). This selection works because TSß converts indole to tryptophan and presumably can convert exogenously-supplied 5-fluoroindole to the toxic amino acid analogue 5-fluorotryptophan. This method yielded a variety of tryptophan-requiring (Trp^-) *trp2* alleles and prototrophic (Trp^+) mutants (K. Pruitt, A. Barczak and R. Last, manuscript in preparation). A subset of the FIR mutants accumulate reduced TSB mRNA and are candidates for altered regulation of *TSB* gene transcription or mRNA stability. As mentioned above for the *PAT1* gene, the variety of selectable and screenable phenotypes associated with TSß-deficient Arabidopsis mutants make this an attractive reporter gene and selectable marker.

6-Methylanthranilate Resistance

The original Arabidopsis tryptophan pathway mutants were obtained using 5-methylanthranilic acid (5MA), which is presumed to be toxic due to conversion into 5-methyltryptophan (Last and Fink, 1988; Last et al., 1991). However, selection for 5MA resistant mutants is not absolute, and often yields wild-type 'escapers'. We find that the related compound 6-methylanthranilate (6MA) is especially effective for selection of Arabidopsis tryptophan pathway mutants (J. Li and R. Last, unpublished results). Inhibition of Arabidopsis growth by this anthranilate analogue is consistent with a previous report of herbicidal activity on a variety of plant species (Thomas, 1984). We isolated approximately one dozen $6MA^R$ mutants, and demonstrated that several were *trp1*, *trp2*, or *trp3* mutants. We are characterizing the remaining $6MA^R$ plants, to see whether any are altered in the enzymes for which mutants were not yet identified, or are regulatory mutants.

References Cited

Anderson, K. S., Miles, E. W., and Johnson, K. A. (1991) Serine modulates substrate channeling in tryptophan synthase. J. Biol. Chem. 266: 8020-8030

Berlyn, M. B., Last, R. L., and Fink, G. R. (1989) A gene encoding the tryptophan synthase ß subunit of *Arabidopsis thaliana*. Proc. Natl. Acad. Sci. U.S.A. 86: 4604-4608

Elledge, S. J., Mulligan, J. T., Ramer, S. W., Spottswood, M., and Davis, R. W. (1991) λYES: A multifunctional cDNA expression vector for the isolation of genes by complementation of yeast and *Escherechia coli* mutations. Proc. Nat. Acad. of Sci., U.S.A. 88: 1731-1735

Keith, B., Dong, X., Ausubel, F. M., and Fink, G. R. (1991) Differential induction of 3-deoxy-D-*arabino*-heptulosonate 7-phosphate synthase genes in *Arabidopsis thaliana* by wounding and pathogenic attack. Proc. Natl. Acad. Sci. U.S.A. 88: 8821-8825

Klee, H. J., Muskopf, Y. M., and Gasser, C. S. (1987) Cloning of an *Arabidopsis thaliana* gene encoding 5-enolpyruvylshikimate-3-phosphate synthase: sequence analysis and manipulation to obtain glyphosate-tolerant plants. Mol. Gen. Genet. 210: 437-442

Last, R. L. (1993) The genetics of nitrogen assimilation and amino acid biosynthesis in plants: progress and prospects. In International Review of Cytology, Vol. 143 K. W. Jeon, J. Jarvik and M. Friedlander, eds. Academic Press San Diego pp. 297-330

Last, R. L., Barczak, A. J., Casselman, A. L., Li, J., Pruitt, K. D., Radwanski, E. R., and Rose, A. B., 1992. The molecular genetics of tryptophan biosynthesis in *Arabidopsis thaliana*. In Biosynthesis and Molecular Regulation of Amino Acids in Plants. Current Topics in Plant Physiology, Vol. 7 B. K. Singh, H. E. Flores and J. C. Shannon, eds. American Society of Plant Physiology Rockville pp. 37-51

Last, R. L., Bissinger, P. H., Mahoney, D. J., Radwanski, E. R., and Fink, G. R. (1991) Tryptophan mutants in *Arabidopsis*: the consequences of duplicated tryptophan synthase ß genes. Plant Cell 3: 345-358.

Last, R. L., and Fink, G. R. (1988) Tryptophan-requiring mutants of the plant *Arabidopsis thaliana*. Science 240: 305-310

Li, J., Ou-Lee, T.-M., Raba, R., Amundson, R. G., and Last, R. L. (1993) Arabidopsis flavonoid mutants are hypersensitive to UV-B irradiation. Plant Cell 5: 171-179

Miles, E. W. (1991) Structural basis for catalysis by tryptophan synthase. *In* Advances in Enzymology and Related Areas of Molecular Biology., 64, A. Meister, ed. John Wiley and Sons New York, pp. 93-172

Niyogi, K. K., and Fink, G. R. (1992) Two anthranilate synthase genes in Arabidopsis: defense-related regulation of the tryptophan pathway. Plant Cell 4: 721-733

Pruitt, K. P., and Last, R. L. (1993) Expression patterns of duplicate tryptophan synthase ß genes in *Arabidopsis thaliana*. Plant Physiol. In Press.

Rose, A. B., Casselman, A. L., and Last, R. L. (1992) A phosphoribosyl-anthranilate transferase gene is defective in blue fluorescent *Arabidopsis thaliana* tryptophan mutants. Plant Physiol. 100: 582-592

Thomas, G. J. (1984) Herbicidal activity of 6-methylanthranilic acid and analogues. J. Agric. Food Chem. 32: 747-749

Tsuji, J., Jackson, E. P., Gage, D. A., Hammerschmidt, R., and Somerville, S. C. (1992) Phytoalexin accumulation in *Arabidopsis thaliana* during the hypersensitive reaction to *Pseudomonas syringae* pv. *syringae*. Plant Physiol. 98: 1304-1309

Wright, A. D., Moehlenkamp, C. A., Perrot, G. H., Neuffer, M. G., and Cone, K. C. (1992) The maize auxotrophic mutant *orange pericarp* is defective in duplicate genes for tryptophan synthase ß. Plant Cell 4: 711-719

Wright, A. D., Sampson, M. B., Neuffer, M. G., Michalczuk, L., Slovin, J. P., and Cohen, J. D. (1991) Indole-3-acetic acid biosynthesis in the mutant maize *orange pericarp*, a tryptophan auxotroph. Science 254: 998-1000

Yanofsky, C., and Crawford, I. P. (1987) The tryptophan operon. *In Escherechia coli* and *Salmonella typhimurium*: Cellular and Molecular Biology, vol. 2, F. C. Neidhardt, J. L. Ingraham, K. B. Low, B. Magasanik, M. Schaechter and H. E. Umbarger, eds. American Society for Microbiology Washington, pp. 1453-1472

STARCH BIOSYNTHESIS IN STORAGE ORGANS

Alison M. Smith, Kay Denyer and Cathie Martin
John Innes Institue, Colney Lane, Norwich NR4 7UH, UK

In this chapter we shall describe our current understanding of the regulation of starch synthesis in non-photosynthetic storage organs. The regulation of flux through the pathway is the subject of a subsequent chapter, and we shall discuss only briefly the pathway via which ADPglucose, the substrate for the synthesis of the starch polymers, is generated. Most of the chapter will be devoted to the question of how starch structure is determined. In spite of the importance of starch as a carbon reserve in plants, in human nutrition and in industry, this question remains largely unanswered.

THE SYNTHESIS OF ADPGLUCOSE

Storage organs receive their carbon as sucrose, which is metabolised via glycolysis in the cytosol. A glycolytic intermediate enters the plastid as the substrate for starch synthesis. Starch synthesis represents the major fate for carbon in many storage organs - starch is 50-80% of the final dry weight of pea seeds, cereal endosperms and tubers such as potatoes - yet the nature of the intermediate which enters the plastid is still a matter of controversy. Elegant experiments in which specifically-labelled sugars were supplied to intact organs showed that the metabolite must be a six-carbon compound; a hexose or a hexose phosphate (Keeling et al. 1988, Hatzfeld and Stitt 1990, Viola et al. 1991). However, the difficulty of isolation of intact, metabolically-active plastids from most storage organs has thwarted attempts to identify precisely which of these compounds can enter the plastid. We have succeeded in preparing suitable plastids from developing pea embryos, and have shown that glucose 6-phosphate is taken up and made into starch at a rate comparable with that of intact embryos (Hill and Smith 1990). The fact that this synthesis is dependent upon the intactness of the plastids and a supply of ATP, and that other hexoses and hexose phosphates are not taken up and made into starch at significant rates, indicates that glucose 6-phosphate is the metabolite which enters the plastid *in vivo* (Fig. 1).

It is not yet clear whether glucose 6-phosphate is of widespread importance as the metabolite which enters plastids as the substrate for starch synthesis. Plastids from young embryos of oilseed rape - which undergo a transient phase of starch accumulation - can

NATO ASI Series, Vol. H 81
Plant Molecular Biology
Edited by G. Coruzzi and P. Puigdomènech

synthesise starch from glucose 6-phosphate (Kang Fan and S. Rawsthorne, John Innes Centre, Norwich: unpublished data). In contrast, plastids isolated from the endosperm of developing wheat grains synthesise starch from glucose 1-phosphate, and not glucose 6-phosphate (Tyson and ap Rees 1988, I. Tetlow and M. Emes, University of Manchester: unpublished data). Plastids from sycamore suspension cultures can take up ADPglucose and convert it to starch (Potzueta-Romero et al. 1991). On the basis of this result it has been suggested that ADPglucose generated directly from sucrose in the cytosol via sucrose synthase may be the metabolite which enters plastids as the substrate for starch synthesis in vivo. However, this is unlikely to be the case in most storage organs (Okita 1992). First, mutations and antisense transformations which reduce or eliminate the activity of ADPglucose pyrophosphorylase - the plastidial enzyme which catalyses the synthesis of ADPglucose from glucose 1-phosphate - also profoundly reduce the rate of starch synthesis (Lin et al. 1988, Smith et al. 1989, Preiss et al. 1989, Muller-Rober et al. 1992). Second, sucrose synthase is likely to generate UDPglucose rather than ADPglucose in vivo (ap Rees 1988).

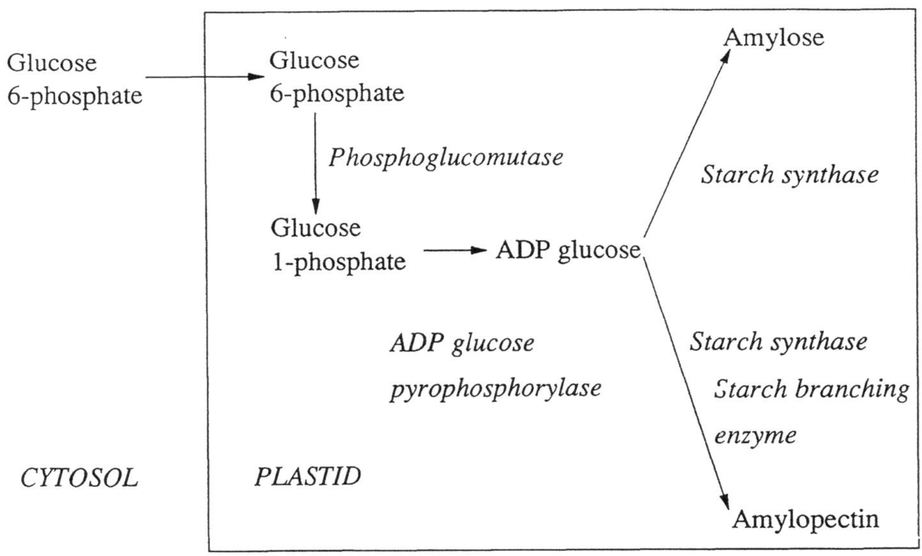

Figure 1

Pathway of starch synthesis inside the plastid in developing pea embryos

Inside the plastid, glucose 1-phosphate is converted to ADPglucose via the enzyme ADPglucose pyrophosphorylase (Fig. 1). The enzyme from most plant sources is a heterotetramer of two different sorts of subunit, usually termed large and small (Smith-White and Preiss 1992). Both are believed to be required for activity . There is increasing evidence that plants may have more than one form of each sort of subunit, and that different forms may be expressed in different parts of the plant, in response to different signals and at different times during development (Müller-Röber et al 1990, Villand et al. 1992). The significance of the great plasticity of subunit composition displayed by this enzyme is unclear. It is, however, tempting to speculate that variation in subunit composition may result in variation in kinetic properties, and that this may reflect the different requirements for regulation of flux through the pathway in different parts of the plant. The role of the enzyme in controlling flux through the pathway is discussed in a subsequent chapter.

THE STRUCTURE OF STARCH

Starch granules are complex and highly-ordered structures, consisting of two sorts of glucose polymer. Amylose, which makes up about 30% of most storage starches, is an essentially linear molecule of 1000-6000 α-1,4-linked glucose residues. Amylopectin is a much larger, highly branched molecule (M_r up to 10^8). It consists of α-1,4-linked chains of about 15-40 residues joined together by α-1,6 linkages. The branches are clustered at regular intervals of 9 nm along the axis of the molecule (French 1984, Kainuma 1988).
Within the granule, amylopectin molecules lie with the non-reducing, free ends of the branches pointing outwards. Adjacent chains within a cluster may form double helices, and these helices pack together in regular arrays which may run across the adjacent clusters of many amylopectin molecules. These crystalline regions may occupy up to half of the volume of the granule. Amylose molecules are believed to exist in single-helical form in the amorphous regions of the granule (French 1984, Gidley and Bociek 1985, Kainuma 1988).

There is great genetic and developmental variation in the structure of starch. The sizes and branching pattern of the molecules, the ratio of amylose to amylopectin, the nature of the packing of helices in the crystalline regions and the size and shape of the granules vary between different organs within a single plant, and between species (Shannon and Garwood 1984, Banks and Muir 1980).

THE SYNTHESIS OF THE STARCH POLYMERS

Two enzymes are required for the synthesis of the starch polymers: starch synthase and starch-branching enzyme (Fig. 1). The former catalyses the addition of the glucosyl moiety from ADPglucose onto the non-reducing end of a chain via an α-1,4 linkage, while the latter cleaves a piece from the non-reducing end of an α-1,4-linked chain, typically of about 15 glucose residues, and rejoins it to the side of the same or an adjacent chain via an α-1,6 linkage to create a branch.

These two enzymes must cooperate in vivo to create, simultaneously, amylose and amylopectin molecules which then pack together to form a granule. The mechanism by which this occurs is not known, and it has not proved possible to create anything resembling an amylopectin molecule, let alone a starch granule, in mixtures of the enzymes *in vitro*. However, two pieces of information give indications about the mechanism of polymer and granule formation *in vivo*. First, multiple isoforms of both enzymes have been reported to occur in many plant organs. These often differ from each other in kinetic properties, and have in some cases been shown to be different gene products. In the case of starch synthase, the isoforms also differ in location. Some are found in a soluble form, in the stroma of the plastid, while others are tightly attached to the granule (Smith and Martin 1993). Starch synthesis *in vivo* may thus involve spatially-separate interactions of specific isoforms of starch synthase and starch-branching enzyme.

Second, mutations which affect a granule-bound isoform of starch synthase provide a great deal of information about the synthesis of the two sorts of polymer. These mutations were first recognised because they give a waxy appearance to the endosperms of cereals, and have subsequently been found in the tuber of potato and the perisperm of *Amaranthus* seeds. They lie in genes (called *waxy* in cereals and amylose-free or *amf* in potatoes) encoding exclusively granule-bound proteins of about 60 kD . These proteins are similar in amino-acid sequence to the glycogen synthase of bacteria, which catalyse the same reaction as starch synthase (Preiss 1991). The mutations cause the loss of granule-bound starch synthase activity, but have no effect on soluble activity and little effect on the total amount of starch in the endosperm. The loss of granule-bound activity results in the elimination of amylose from the granule: the starch of waxy mutant endosperm consists almost entirely of amylopectin (Tsai 1974, Echt and Schwartz 1981, Shure et al. 1983, Hseih 1988, Rohde et al. 1988, Wang et al. 1990, Hovenkamp-Hermelink et al. 1987, Jacobsen et al. 1989, Visser et al. 1989).

The effect of the mutation indicates strongly that the granule-bound isoform of starch synthase is exclusively responsible for amylose synthesis. This has allowed a model for the synthesis of amylose and amylopectin to be proposed (Robyt 1984). It is envisaged that soluble starch synthase and starch-branching enzyme act together at the periphery of the growing granule to form amylopectin. Amylopectin molecules crystallise together to form the granule matrix, to which the granule-bound starch synthase binds. The product of the granule-bound starch synthase is protected from the action of the branching enzyme by its location at or within the surface of the matrix and therefore remains unbranched, forming the amylose component of the granule.

Further progress in understanding the mechanism of polymer and granule synthesis has been hampered by a lack of information about starch synthases and branching enzymes other than the products of the *waxy* and *amf* loci. Until recently, genes encoding these proteins had not been cloned, and the number of different proteins contributing to synthase and branching activities was not known for any storage organ. We have now identified and cloned the genes encoding all of the isoforms of starch synthase and branching enzyme in the developing pea embryo (Smith and Denyer 1992, and unpublished data). In the remainder of this chapter I shall discuss the picture that emerges, and the information it provides about the synthesis of starch.

POLYMER SYNTHESIS IN THE DEVELOPING PEA EMBRYO

Nature and location of the enzymes

As in other storage organs, starch synthase activity in the pea embryo is found both in the soluble fraction and tightly bound to starch granules. The major protein bound to pea starch granules is 59 kD, and is extremely similar in amino-acid sequence to the waxy and amf proteins of cereals and potato respectively. The sequence predicted from a cDNA clone of the pea protein is 59 % identical to that of the waxy protein of maize (Dry et al. 1992). However, this protein does not appear to be responsible for most of the granule-bound starch synthase activity of the pea embryo. Although it has starch synthase activity when purified, most of the activity associated with isolated granules is attributable to a more minor protein of 77 kD (Smith 1990, Denyer et al. 1993). The C-terminal 60 % of the 77-kD protein is similar in amino acid sequence to the waxy class of proteins (it is 31 % identical to the waxy protein of maize), but the N-terminal 203 amino acids consitute a unique, serine-rich domain (Dry et al. 1992). While the possibility that the 59-kD protein is inactivated during granule isolation cannot be ruled out, the 77-kD protein is undoubtedly a distinct and quantitatively-important isoform of granule-bound starch synthase. Its estimated activity in the embryo is more than sufficient to account for the rate of starch synthesis (Smith 1990).

Most of the soluble starch synthase activty of the pea embryo is accounted for by a protein of 77 kD (Denyer and Smith 1992). Comparison of amino-acid sequence, antigenic properties and the sizes of proteolytic products of this soluble protein and the granule-bound 77-kD starch synthase indicates strongly that the two are identical. The possibility that the presence of the 77-kD protein on the granule simply reflects contamination of the granules with soluble starch synthase during their isolation has been ruled out by immunogold labelling experiments. Antibodies to the 77-kD protein label the whole of the interior of granule in sections of developing embryos, and not simply the stroma of the plastid (Denyer et al. 1993).

The branching enzyme activity of developing pea embryos is accounted for by two distinct isoforms, one (SBE I) of 112 kD and the other (SBE II) of 100 kD (Smith 1988). They are similar in amino-acid sequence each other (approximately 50 % identical), and to branching enzymes of rice and maize endosperm and glycogen-branching enzymes of bacteria (Bhattacharyya et al. 1990, Baba et al. 1991, Kawasaki et al. 1993). Both isoforms are present in the soluble fraction of the amyloplast and bound to the starch granule. Towards the end of development about half of the total branching enzyme protein of the embryo is present in the starch. Enzymic digestion of the starch releases both isoforms in an active state (Denyer et al. 1993).

Regulation of polymer synthesis

The fact that all of the isoforms of starch synthase and starch branching-enzyme in the pea embryo are, at least in part, granule bound, seems to contradict the model for the synthesis of amylose and amylopectin discussed above. The model assumes that there is no branching enzyme bound to the granule, hence the products of granule-bound synthase remain unbranched. If synthases and branching enzymes are not spatially separated, how can both branched and unbranched polymers be formed? We are using several approaches to elucidate the roles in polymer synthesis of each of the isoforms we have identified. First, we have examined the expression of the isoforms during the development of the embryo, and the effects of a mutation which dramatically reduces the amylopectin content of the starch in the embryo. Second, we have expressed the 77-kD starch synthase in wildtype and *amf* mutant potatoes. We wish to know whether the location of the protein is the same in potato as in pea, and whether it can complement the *amf* mutation. Third, we have investigated whether species other than pea, including those with waxy mutations, have more than one isoform of granule-bound starch synthase.

1. Expression of starch synthases and branching enzymes during development

Isoforms of starch synthase and branching enzyme are expressed at different times during the development of the embryo. The level of transcript for the 77-kD starch synthase peaks at a relatively early stage and then falls, whereas that for the 59-kD starch synthase peaks much later (Dry et al. 1992). This pattern is consistent with a role for the former protein in amylopectin synthesis, and the latter in amylose synthesis, since the amylose to amylopectin ratio of embryos starch rises from about 15 % in very young embryos to 30 % in mature embryos (Greenwood and Thompson 1962).

Isoforms of starch-branching enzyme are also maximally expressed at different times during development. The level of transcript for SBE I peaks early, and that for SBE II peaks later. Preliminary experiments (Smith 1988) suggested that the two forms make branches of different lengths. Recent, detailed experiments have confirmed that this is the case for the two isoforms of branching enzyme of the developing maize endosperm (Takeda et al. 1993). Thus amylopectin synthesised early in development, when SBE I is the dominant isoform, is likely to be different from that synthesised later when SBE II is also present.

The effects of a mutation at the *r* locus potentially provide further information about the roles of the isoforms of branching enzyme. The mutation reduces the starch content of the embryo by about 50 %, and decreases the amylopectin content of the starch from 70 % to 30 %. The average size of the polymers is reduced, and the packing of helices to form ordered arrays is altered (Wang and Hedley 1991, Colonna et al. 1982, Colonna and Mercier 1984, 1985, We have shown that the mutation consists of a transposon-like insertion in the gene encoding SBE I, which lies at the *r* locus (Bhattacharyya et al. 1990). During the first part of development, when SBE I accounts for most of the activity of wildtype embryos, mutant embryos have almost no detectable branching enzyme activity (Smith 1988).

It is tempting to suggest that a simple comparison of the structure of starch in mature wildtype and *r* mutant embryos will reveal the roles of SBE I and II. However, starch structure may also be altered as a secondary effect of the mutation, and this approach must be treated with caution. For example, the higher ADPglucose content of mutant embryos (Edwards et al. 1988) could affect the nature or rate of production of substrate for branching enzyme through effects on isoforms of starch synthase. The mutation also affects the protein composition of the starch. At least two proteins are present in significant amounts in the mutant starch but barely detectable in wildtype starch. One of these is starch phosphorylase. The reasons for and consequences of these alterations are not yet clear, but the possibility that these proteins themselves influence the structure of starch in the mutant cannot be ruled out.

2. Expression of the 77-kD starch synthase in potato

To discover whether the 77-kD starch synthase is capable of amylose synthesis when granule-bound, we transformed wildtype and *amf* mutant potatoes with a cDNA for this protein. In both cases, a substantial fraction of the 77-kD protein expressed in the tubers is tightly associated with starch granules. Expression of the 77-kD protein in the *amf* mutant results in a very substantial increase in the granule-bound starch synthase activity of minitubers, from virtually undetectable levels prior to transformation to as much as 25 % of wildtype activities in the transformants. However, the starch of both transformed and untransformed minitubers stains red with iodine, indicating that it contains only very low levels of amylose (wildtype starches stain blue with iodine, whereas low-amylose starches stain red). This suggests that the granule-bound 77-kD starch synthase does not restore the ability to make amylose. This view is strengthened by the results of transformation experiments in which tubers with a range of different activities of granule-bound starch synthase were generated by introduction of an antisense construct for the *amf* gene into wildtype potatoes (Visser et al. 1991). Activities as low as 10 % of wildtype still gave rise to blue-staining starch, and had 50 % or more of the wildtype amylose content.

Analysis of the effects of transformation of potatoes with the 77-kD protein is still in progress.

3. Comparisons between pea embryo and other storage organs

Our results suggested that peas might be exceptional among storage organs in having a quantitatively-important, granule-bound starch synthase which is not the waxy gene product. The reported elimination of granule-bound starch synthase activity by *waxy* mutations in a wide range of storage organs could be taken as evidence that in all of these cases there are no other important granule-bound starch synthases. However, in almost all cases this has not been investigated in detail. To provide a comparison with pea, we have examined the granule-bound starch synthase activity in one such organ, the barley endosperm. We wished to establish whether the endosperm has multiple forms of granule-bound starch synthase, and what the fate of these proteins may be in the *waxy* mutant.

We have discovered two isoforms of starch synthase in addition to the waxy gene product in starch from developing barley endosperm. These were identified in two ways. First, proteins were solubilised from the starch by incubation with amylase and fractionated by anion exchange chromatography (Smith 1990). Second, granule-bound proteins were solubilised in SDS, separated by SDS-PAGE, individually reconstituted and tested for starch synthase activity (Denyer et al. 1993). The isoforms are proteins of 77 and 91 kD. The former is strongly antigenically related to the 77-kD starch synthase of pea, the latter more weakly so.

In agreement with other studies, we find that intact starch granules of waxy barley have very little starch synthase activity in relation to wildtype granules. When granules are mechanically disrupted, however, the activity of both wildtype and waxy starch is considerably increased, such that waxy starch has about half the activity of wildtype starch. The simplest interpretation of this result is that the 77 and 91-kD starch synthases are present within both types of starch, but are normally buried within the granule. Disruption of the granule allows ADPglucose to reach them. Examination of the proteins of waxy starch reinforces this view. The waxy protein is absent, but the 77 and 91 kD proteins are not affected by the mutation. Solubilisation and ion exchange chromatography of the granule-bound proteins of waxy starch confirms that both can be recovered from this starch as active starch synthases.

The fact that the non-waxy starch synthases are present in the starch of *waxy* mutant barley, which contains only a very low percentage of amylose, indicates very strongly that these proteins do not make amylose. The ability to synthesise amylose is thus not simply a function of a granule-bound location, as the model presented above suggests.

We have recently shown that wheat, like barley, has two non-waxy isoforms of granule-bound starch synthase. One is a protein of 77 kD and is antigenically related to the 77-kD starch synthase of pea. The other is a protein of about 90 kD. Wheat endosperm has two soluble starch synthases, one of which is a 77-kD protein antigenically related to the 77-kD starch synthase of pea. It seems very likely that the soluble and granule bound 77-kD starch synthases of wheat are the same protein. The other soluble isoform has not yet been identified.

Although the picture for barley and wheat differs in detail from that for pea, important similarities emerge. Both cereals have granule-bound starch synthases other than than the waxy protein. These may be a quantitatively-important component of the total starch synthase activity of the granule. One of these proteins is the same size as, and closely related to, the 77-kD starch synthase of pea. At least in the case of wheat, this isoform is also a major soluble starch synthase. The fact that the nonwaxy starch synthases of barley do not appear to make amylose is consistent with the observation that the 77-kD starch synthase of pea does not restore the ability to make amylose to amf mutant potatoes.

Taken as a whole, our work indicates that the model for amylose and amylopectin synthesis presented above is too simplistic. Organs which synthesise amylose may have granule-bound branching enzyme, and the occurrence of granule-bound starch synthases which do not appear to synthesise amylose is probably very widespread. The synthesis of amylose thus appears to be an exclusive property of the waxy starch synthase, rather than an inevitable consequence of its granule-bound location.

The fact that - in pea at least - all of the granule-bound starch synthases and branching enzymes except the waxy protein are also soluble allows us to speculate about their origins and roles (Denyer et al. 1993). We suggest that these proteins may be active *in vivo* in a soluble zone at the periphery of the granule, where they catalyse the synthesis of amylopectin. As the concentration of polymer increases, it crystallises to form the ordered arrays which constitute the granule matrix. The soluble proteins may simply become trapped within the matrix as their product crystallises around them. Inside the matrix, these proteins may be inactive because they are held with a crystalline structure which is not readily penetrated by ADPglucose.

The waxy starch synthase, in contrast, is almost certainly not active as a soluble starch synthase. The protein appears to have a very high affinity for the insoluble, crystalline matrix of the granule, and we imagine that it may be active at the surface of the matrix *in vivo*. It too becomes trapped within the matrix as polymer forms around it. Thus although all of the isoforms of starch synthase and starch-branching enzyme become trapped within the granule as it grows, we suggest that active fraction of the waxy starch synthase may be in a different location from the active fraction of those isoforms which are both bound and soluble. The product of the waxy starch synthase may thus be unavailable to the active fraction of the branching enzyme, and remain unbranched.

Although our results indicate that the broad picture of enzymes of polymer synthesis may be similar in peas, wheat and barley, there are clear differences in detail. This is expected since the fine structure of the polymers of cereal and embryo starches is different. More extensive studies of the differences between storage organs in their complements of starch synthases and branching enzymes may reveal general correlations between particular classes of polymer synthesising enzymes and the structure of the starch polymers. Such studies will be important in establishing a rational basis for the manipulation of starch structure in the future.

It would, however, be naive to believe that a complete description of the starch synthases and starch-branching enzymes of a starch-storing organ is sufficient to explain the structure of its starch. Many other factors may be involved. These include the timing of expression of isoforms and changes in the concentration of ADPglucose during development, and the possibility that other activities such as priming, debranching and disproportionating enzymes are involved in starch synthesis. Although great progress has been made in the last few years in understanding the biochemistry and molecular biology of starch synthesis, we are still a very long way from a model which explains adequately how the complex structure of the starch granule is determined.

ACKNOWLEDGEMENTS

We are very grateful to our colleagues Rachel Burton, Anne Edwards, Lionel Hill, Chris Hylton, and Jacqui Marshall, whose unpublished work is referred to in this chapter. We also thank Dr Richard Visser for his help and collaboration with the transformation of *amf* potato tubers, and Drs Ian Tetlow and Mike Emes for making unpublished data available to us.

REFERENCES

ap Rees T (1988) Hexose phosphate metabolism by nonphotosynthetic tissues of higher plants. In: Preiss J (ed.) The biochemistry of plants, vol 14. Academic Press, San Diego: 1-33

Baba T, Kimura K, Mizuno K, Etoh H, Ishida Y, Shida O and Arai Y (1991) Sequence conservation of the catalytic regions of amylolytic enzymes in maize branching enzyme-I. Biochem. Biophys. Res. Commun. *181*: 87-94

Banks W and Muir DD (1980) Structure and chemistry of the starch granule. In: Preiss J (ed.) The biochemistry of plants, Vol. 3. Academic Press, New York: 321-269

Bhattacharyya MK, Smith AM, Ellis THN, Hedley C and Martin C (1990) The wrinkled-seed character of pea described by Mendel is caused by a transposon-like insertion in a gene encoding starch-branching enzyme. Cell *60*: 115-122

Colonna P, Buleon A, Lemaguer M and Mercier C (1982) *Pisum sativum* and *Vicia faba* carbohydrates. IV. Granular structure of wrinkled pea starch. Carbohydr. Polymers *2*: 43-59

Colonna P and Mercier C (1984) Macromolecular structure of wrinkled- and smooth-pea starch components. Carbohydr. Res. *126*: 233-247

Colonna P and Mercier C (1985) Gelatinization and melting of maize and pea starches with normal and high-amylose genotypes. Phytochem. *24*: 1667-1674

Denyer K and Smith AM (1992). The purification and characterisation of two isoforms of soluble starch synthase from developing pea embryos. Planta *186*: 609-617

Denyer K, Sidebottom C, Hylton C and Smith AM (1993) Soluble isoforms of starch synthase and starch-branching enzyme are also bound to starch granules in developing pea embryos. Plant J., in press.
Dry I, Smith A, Edwards EA, Bhattacharyya M, Dunn P and Martin C (1992) Characterisation of cDNAs encoding two isoforms of granule-bound starch synthase which show differential expression in developing storage organs of pea and potato. Plant J. *2*: 193-202
Echt CS and Schwartz D (1981) Evidence for the inclusion of controlling elements within the structural gene at the *waxy* locus in maize. Genetics *99*: 275-284
Edwards J, Green JH and ap Rees T (1988) Activity of branching enzyme as a cardinal feature of the Ra locus in *Pisum sativum*. Phytochem. *27*: 1615-1620
French D (1984) Organisation of starch granules. In: Whistler RL, BeMiller JN and Paschall EF (eds) Starch: Chemistry and Technology. Academic Press, Orlando: 183-247
Gidley MJ and Bociek SM (1988) ^{13}C CP/MAS NMR studies of amylose inclusion complexes, cyclodextrins, and the amorphous phase of starch granules. J. Amer. Chem. Soc. *110*: 3820-3829
Greenwood CT and Thompson J (1962) Studies on the biosynthesis of starch granules. 2. The properties of components of starches from smooth and wrinkled seeded peas during growth. Biochem. J. *82*: 156-164
Hatzfeld W-D and Stitt M (1990) A study of the rate of recycling of triose phosphates in heterotrophic *Chenopodium rubrum* cells, potato tubers, and maize endosperm. Planta *180*: 198-204
Hill LM and Smith AM (1991). Evidence that glucose 6-phosphate is imported as the substrate for starch synthesis by the plastids of developing pea embryos. Planta *185*: 91-96
Hovenkamp-Hermelink JHM, Jacobsen E, Ponstein AS, Visser, RGF, Vos-Scheperkeuter GH, Bijmolt EW, de Vries JN, Witholt B and Feenstra WJ (1987) Isolation of an amylose-free starch mutant of the potato (*Solanum tuberosum* L.). Theor. Appl. Genet. *57*: 217-221
Jacobsen E, Hovenkamp-Hermelink JHM, Krigsheld HT, Nijdam H, Pijnacker LP, Witholt B and Feenstra WJ (1989) Phenotypic and geneotypic characterisation of an amylose-free starch mutant of the potato. Euphytica *44*: 43-48
Kainuma K (1988) Structure and chemistry of the starch granule. In: Preiss J (ed.) The biochemistry of plants, vol 14. Academic Press, San Diego: 141-180
Kawasaki T, Mizuno K, Baba T and Shimada H (1993) Molecular analysis of the gene encoding a rice starch branching enzyme. Mol. Gen. Genet. *237*: 10-16
Keeling PL, Wood JR, Tyson RH and Bridges IG (1988) Starch biosynthesis in developing wheat grain. Evidence against the direct involvement of triose phosphates in the metabolic pathway. Plant Physiol. *78*: 311-319
Lin T-P, Caspar T, Somerville C and Preiss J (1988) A starch-deficient mutant of *Arabidopsis* with low ADPglucose pyrophosphorylase activity lacks one of the two subunits of the enzyme. Plant Physiol. *88*: 1175-1181
Müller-Röber B, Kossmann J, Hannah LC, Willmitzer L and Sonnenwald U (1990) One of two different ADP-glucose pyrophosphorylase genes from potato responds strongly to elevated levels of sucrose. Mol. Gen. Genet. *224*: 136-146
Müller-Röber B, Sonnenwald U and Willmitzer L (1992) Inhibition of ADP-glucose pyrophosphorylase in transgenic poataoes leads to sugar-storing tubers and influences tuber formation and expression of tuber storage protein genes. EMBO J. *11*: 1229-1238
Okita TW (1992) Is there an alternative pathway for starch synthesis? Plant Physiol. *100*: 560-564
Potzueta-Romero J, Frehner M, Viale AM and Akazawa T (1991) Direct transport of ADPglucose by an adenylate translocator is linked to starch biosynthesis in amyloplasts.

Proc. Natl Acad. Sci. *88*: 5769-5773
Preiss J, Cress D, Hutny J, Morell M, Bloom M, Okita T and Anderson J (1989) Regulation of starch synthesis. Biochemical and genetic studies. In: Whitaker JR and Sonnet PE (eds) Biocatalysis in agricultural biotechnology. American Chemical Society, Washington: 84-92
Preiss J (1991) Biology and molecular biology of starch synthesis and its regulation. In: Miflin BJ (ed.) Oxford surveys of plant molecular and cell biology, Vol. 7. Oxford University Press: 59-114
Robyt J (1984). Enzymes in the synthesis and hydrolysis of starch. In: Whistler RL, BeMiller JN and Paschall EF (eds) Starch: Chemistry and Technology. Academic Press, Orlando: 87-123
Rohde W, Becker D and Salamini F (1988) Structural analysis of the *waxy* locus from *Hordeum vulgare*. Nucleic Acids Res. *16:* 1785-1786
Shannon, J.C. and Garwood, D. L. (1984) Genetics and physiology of starch development. In: Whistler RL, BeMiller JN and Paschall EF (eds) Starch: Chemistry and Technology. Academic Press, Orlando:
Shure M, Wessler S, Fedoroff N (1983) Molecular identification and isolation of the *waxy* locus in maize. Cell *35*: 225-233
Smith AM (1988). Major differences in isoforms of starch-branching enzyme between developing embryos of round- and wrinkled-seeded peas. Planta *175*: 170-179
Smith AM (1990). Evidence that the "waxy" protein of pea is not the major, starch-granule-bound starch synthase. Planta *182*: 599-604
Smith AM, Bettey M and Bedford ID (1989) Evidence that the *rb* locus alters the starch content of developing pea embryos through an effect on ADPglucose pyrophosphorylase. Plant Physiol. *89*: 1279-1284
Smith AM and Denyer K (1992). Starch synthesis in developing pea embryos. New Phytol. *122*: 21-33
Smith AM and Martin C (1992). Starch biosynthesis and the potential for its manipulation. In: Grierson D (ed) Biosynthesis and manipulation of plant products. Plant Biotechnology, Vol. 3. Blackie and Son, Glasgow: 1-54
Smith-White BJ and Preiss J (1992) Comparison of proteins of ADPglucose pyrophosphorylase from diverse sources. J. Mol. Evol. *34*: 449-464
Takeda Y, Guan H-P and Preiss J (1993) Branching of amylose by the branching enzyme isoforms of maize endosperm. Carbohydrate Res. *240*: 253-263.
Tsai C-Y (1974) The function of the *waxy* locus in starch synthesis in maize endosperm. Biochem. Genet. *11*: 83-96
Tyson RH and ap Rees T (1988) Starch synthesis by isolated amyloplasts from wheat endosperm. Planta *175*: 33-38
Villand P, Aalen R, Olsen O-A, Luthi E, Lonnenborg A and Kleczkowski L A (1992) PCR amplification and sequences of cDNA clones for the small and large subunits of ADP-glucose pyrophosphorylase from barley tissues. Plant Molec. Biol. *19*: 381-389
Viola R, Davies HV and Chudek AR (1991) Pathways of starch and sucrose biosynthesis in developing tubers of potato (*Solanum tuberosum* L.) and seeds of Faba bean (*Vicia faba* L.). Planta *183*: 202-208
Visser RGF, Hergersberg M, Van der Leij FR, Jacobsen E, Witholt B and Feenstra WJ (1989) Molecular cloning and partial characterisation of the gene for granule-bound starch synthase from a wildtype and an amylose-free potato (*Solanum tuberosum* L.). Plant Sci. *64*: 185-192
Visser RGF, Somhorst I, Kuipers, GJ, Ruys NJ, Feenstra, WJ and Jacobsen E (1991) Inhibition of the expression of the gene for granule-bound starch synthase in potato by

antisense constructs. Mol. Gen. Genet. *225*: 289-296
Wang TL and Hedley C (1991) Seed development in peas: knowing your three "r"s (or four, or five). Seed Sci. Res. *1*: 3-14
Wang Z, Wu Z, Xing Y, Zheng F, Guo X, Zang W and Hong M (1990) Nucleotide sequence of the rice *waxy* gene. Nucleic Acids Res. *18*: 5898

Auxin-Regulated Gene Expression

Tom J. Guilfoyle, Gretchen Hagen, Yi Li, Zhanbin Liu, Tim Ulmasov, Tim Strabala, and Xiang Yang Shi
Department of Biochemistry
117 Schweitzer Hall
University of Missouri
Columbia, MO 65211

INTRODUCTION

A variety of auxin-responsive mRNAs have been identified and characterized (reviewed by Hagen, 1989 and Key, 1989). In a number of cases, the genes that encode these mRNAs have also been identified and sequenced (reviewed by Guilfoyle et al., 1993). The promoter elements that confer auxin-responsiveness are under investigation in several laboratories. Here, we summarize some of our recent results on auxin-responsive mRNAs in soybean and *Arabidopsis*, the genes that encode these mRNAs, and the auxin-inducible promoters.

SOYBEAN cDNA AND GENE SEQUENCES AND HOMOLOGS IN OTHER PLANTS

GH1

Soybean GH1 mRNA is moderately abundant in freshly excised elongating hypocotyl sections (EHS). The abundance of this mRNA declines if EHS are incubated in the absence of auxin (Hagen et al., 1984), and subsequent addition of auxin to the incubation medium results in a several-fold increase in abundance of GH1 mRNA (i.e., 3-20-fold depending on the prior incubation period). The transcription of this gene is specifically induced by auxins (Hagen and Guilfoyle, 1985). Both a cDNA and gene have been sequenced (G. Hagen, unpublished results). The soybean gene contains at least five introns; the transcription start site has not been identified. A gene homolog has been isolated from *Arabidopsis* and has been partially sequenced. While the GH1 cDNA does not cross-hybridize with Aux22 [JCW1] and Aux28 [JCW2] (Walker and Key, 1982; Ainley et al., 1988), the deduced amino acid sequence of GH1 indicates that it is related to Aux22 and Aux28. Five islands of conserved amino acid sequence are present in each of these auxin-responsive gene products. Homologs to Aux22 and/or Aux28 from *Arabidopsis* and mung bean have been identified and sequenced (Conner et al., 1990; Yamamoto et al., 1992). Thus, it is likely that these gene products are functionally related, but this function has yet to be identified.

NATO ASI Series, Vol. H 81
Plant Molecular Biology
Edited by G. Coruzzi and P. Puigdomènech

GH2/4

The GH2/4 mRNA is present at a low level in freshly exised EHS and basal hypocotyl sections (BHS), but in contrast to GH1, the abundance of this mRNA increases upon incubation of sections in a medium not containing auxin (Hagen et al., 1984). Addition of auxin to the medium results in a further increase in abundance of GH2/4 mRNA (i.e., 3-20-fold depending on the organ tested and length of incubation). While auxin causes increased transcription rates on the GH2/4 gene (Hagen and Guilfoyle, 1985), this is not the only inducer. A variety of auxin analogs, heavy metals, salicylic acid, wounding, and stresses also induce the accumulation of GH2/4 mRNA (Hagen et al., 1988; T. Guilfoyle, unpublished results). The heavy metal cadmium induces the accumulation of GH2/4 mRNA as well as a transcript that contains an unprocessed intron (Hagen et al., 1988; Czarnecka et al., 1988). A soybean cDNA and gene have been sequenced (Hagen et al., 1988; T. Ulmasov, unpublished results) and have been shown to be identical to the heat shock sequence Gmhsp26-A (Czarnecka et al., 1988). The promoter contains two and the intron contains one TCA-1-like elements that are conserved in a variety of stress-inducible genes (Czarnecka et al., 1988; T. Ulmasov, unpublished results), and this element has recently been reported to bind to a salicylic acid-inducible binding protein (Goldsbrough et al., 1993). A family of related genes exist in soybean (G. Hagen, unpublished results). A variety of auxin-inducible cDNA clones and/or genes that are related to GH2/4 have been isolated from *Nicotiana sp.* (Dominov et al., 1992; Takahashi et al., 1989 and 1991; van der Zaal et al., 1991). The deduced amino acid sequences reveal that these auxin-induced gene products are related to the *E. coli* stringent starvation protein (Serizawa and Fukuda, 1987) and glutathione S-transferase or GST (Guilfoyle et al., 1993). Expression of a soybean GH2/4 in *E. coli* results in a large increase in GST activity in bacterial extracts, and the GH2/4 protein expressed in *E. coli* binds to a glutathione affinity resin (T. Ulmasov, unpublished results). While these auxin-induced GSTs have been proposed to play a role in cell division (Takahashi et al., 1989), we think that these GSTs are expressed in response to variety of stresses, including high exogenous auxin concentrations. The GSTs may function in detoxifying high amounts exogenous auxin.

GH3

Because of its low abundance, the GH3 mRNA is difficult to detect in soybean seedlings that have not been exposed to external auxin (Hagen et al., 1984). When intact seedlings, EHS, or BHS are treated with exogenous auxins, the GH3 mRNA increases some 25-100-fold. Increased transcription rates on the GH3 gene can be detected within five minutes after auxin application (Hagen et al., 1985), and the gene is specifically induced by auxins. The GH3 cDNA and gene have been sequenced (Hagen et al., 1991), and at least one other related gene is found in the soybean genome. The soybean gene contains two introns. Two gene homologs have been isolated from *Arabidopsis* and sequenced (T. Ulmasov and G. Hagen, unpublished results), and the exon/intron borders are conserved among the *Arabidopsis* and soybean GH3 genes.

SAURs

The SAURs (Small Auxin Up RNAs) are moderately abundant in freshly excised EHS, rapidly fall to undectable levels upon incubation of EHS in medium not containing auxin, and increase rapidly in abundance in EHS upon addition of auxin to the incubation medium (McClure and Guilfoyle, 1987). Increases in SAUR abundance are detected within 2.5 minutes after auxin application, and half maximal steady-state induced mRNA levels are observed by 15 minutes after adding auxin. Thus, the SAURs display appropriate kinetics for mRNAs that might play a role in auxin-induced cell elongation (reviewed by Guilfoyle et al., 1990). Increased transcription rates on SAUR genes can be detected within 5-10 minutes after auxin addition to EHS (McClure et al., 1989). The SAURs are asymmetrically expressed in hypocotyls undergoing a gravitropic response [i.e., greater expression on the lower, more rapidly elongating side] (McClure and Guilfoyle, 1989). The SAURs are highly unstable mRNAs which are posttranscriptionally induced by protein synthesis inhibitors such as cycloheximide, anisomycin, and emetine (Franco et al., 1991). Superinduction of SAURs occurs when EHS are treated with cycloheximide plus auxin. The cycloheximide-inducible element within these mRNAs has been localized to the open reading frame or ORF (Y. Li, T. Strabala, G. Hagen, and T. Guilfoyle, manuscript submitted for publication). Five SAUR genes have been found to be clustered within a 7 kb fragment of soybean genomic DNA (McClure et al., 1989). This gene cluster has been sequenced along with three cDNAs that correspond to the three internal genes within this cluster (McClure et al., 1989). None of the genes contain introns, and adjacent genes are oriented and transcribed in opposite directions. Four SAUR gene homologs have been sequenced from *Arabidopsis* (G. Hagen, unpublished results), and SAUR cDNA homologs have been identified in pea (G. Hagen and B. McClure, unpublished results) and mung bean (Yamamoto et al., 1992). The deduced amino acid sequences for these genes are shown in Figure 1.

TISSUE-SPECIFIC AND ORGAN-SPECIFIC GENE EXPRESSION

Two approaches have been taken to identify the tissues and organs that express the GH and SAUR genes either in the absence or presence of exogenous auxin. The first approach relies on *in situ* hybridization with tissue prints or with fixed thin sections. The second approach utilizes the GH and SAUR promoters fused to the reporter gene that encodes β-glucuronidase or GUS, and the transfer of these chimeric genes into tobacco or *Arabidopsis*.

GH1

In situ hybridization with fixed thin sections reveals that GH1 mRNA is most strongly expressed in vascular tissues of soybean hypocotyls and roots; however, lower levels of expression are also detected in the epidermis, cortex, and pith (Gee, 1992). GH1 mRNAs are also expressed in developing flowers, and the strongest expression is again localized to vascular tissues. With excised EHS and roots, auxin-treatment simply enhances the level of expression of GH1 mRNA in the same tissues that express this mRNA in the absence of external auxin.

```
MGFRLPGI------------RKAS--------KAaDAPKGYLAVYVGE-----KlKRFVIPV          15A
*GFRLPGI------------RKAS--------nAVDAP**YLA**V*E-----KMK*FVI*V          X15
*GFRLPGI------------RKtSiSANQASSKsVeVP**YLv**V*d-----KMr*FiI*V          X10A
*GFRiaGIv-----------RrtSFyttQAaSKrVDVP**YaA**V*d-----KMr*FtI*V          10A5
*GFRLPGI------------RKASFSANQASSKAVDVe**YLA**V*E-----KMr*FVI*V          6B
*GFRfsGIm-----------RKASFaANrsaSKtADmP**YLA**V*E-----KqK*cVI*i          PSAUR1
*aiRLsrvin-SKQsqKQqsR---------------VP**hvA**V*EeMEs-K-K*FVv*i          ASAUR1
*GlsrfaIsnatKQILKl--nslanrnrtsSSssdhVP**hvA**V*EQiEmeK-K*FVv*i          ASAUR2
*aiRisrvLQSSKQILKs---------lshSSnnVaiP**hLA**V*EmM----qK*FVv*V          ASAUR3
*aiRiPrvLQSSKQILrQ----AkllssssSSsslDVP**YLA**V*EQn----MK*FVv*V          ASAUR4
*GFRLPGI------------RKtlSArNeASSKvlDAP**YLA**g*En-----MK*FVI*V          ARG7

SYLNQPSFQDLLSQAEEEFGYDHPMGGLTIPCSEDVFqcITScLN                           15A
SYmNQ*S*QDL*tQAE****YD**M******CS*eV*qrITccLN                           X15
SYLNQ*S*QDL*nQAE****YD**M******Ck*De*LtvTShLNdL                         X10A
SYLNe*S*QeL*SQAE****YD**M******Ck*ee*LNvTahLNeL                         10A5
SYLNk*S*QDL*SQAE****YD**n******CS*DV*qhITSfLN                           6B
SYLNQ*S*QDL*ceAk****Yn**M******Ct*DV*qNITSRfNgL                         PSAUR1
SYLNh*S*QeL*SrAE****fn**i******Cr*et*vglsSeifiyslidelyeesiihpglnlt      ASAUR1
SflNh*S*kef*SrAE****fn**M******Cr*eV*Ldliasrlq                          ASAUR2
tYLsh*c*QkL*rkAE****fD**M******Ct*ei*LdlaSRLstss                        ASAUR3
SYLdQ*S*QDL*rkAE****fD**M******CS*ei*LdlaSRfN                           ASAUR4
ShLNQ*l*QDL*SQAE****YD**M******CS*Dl*qhITScLsaq                         ARG7
```

Figure 1. Alignment of amino acid sequences of SAUR cDNAs and genes from soybean [15A, X15, X10A, 10A5, and 6B] (McClure et al., 1989), pea [PSAUR1] (B. McClure and G. Hagen, unpublished results), *Arabidopsis* [ASAUR1, 2, 3, and 4] (G. Hagen, unpublished results), and mung bean [ARG7] (Yamamoto et al., 1992). The most conserved amino acid is indicated by capital letters unless there is complete identity (*).

GH2/4

A 853 bp soybean GH2/4 promoter (777 bp) and 5' leader (76 bp) were fused at the translational start site to the *E. coli uidA* ORF and the 3' untranslated region of nopaline synthase (T. Ulmasov, unpublished results). This chimeric gene was used to tranform tobacco, and R0 and R1 plants and seedlings were examined for GUS activity by histochemical staining. These studies have revealed that the GH2/4 promoter is highly active in root tips. The promoter is expressed to lower levels in a variety of other organs and tissues. Vascular tissues show higher expression levels than surrounding tissues. Mature pollen shows strong expression of the GH2/4 promoter. Wounding results in strong activation of this promoter. Exogenous auxin induces expression of the GH2/4 promoter in most, if not all, organs and tissues. While auxins appear to be the strongest inducers, salicyclic acid, heavy metals, and a variety of other agents induce expression from this promoter. Thus the GH3 promoter is developmentally regulated as well as responsive to a variety of stress agents.

GH3

In situ hybridization of fixed thin sections showed that the GH3 mRNA was not expressed at detectable levels in most organs and tissues of soybean seedlings and plants. Examination of developing flowers revealed, however, that GH3 mRNA could be detected in developing floral organs, especially in developing ovules and in ovary walls (Gee et al., 1991). Exogenous auxin application to EHS, BHS, epicotyls, and roots resulted in strong induction of GH3 mRNA within one hour after hormone application. While expression was strongest in vascular tissues, most, if not all, organs and tissues showed auxin-induced increases in GH3 mRNA. These results indicate that essentially all organs and tissues have the receptor and signal/transduction machinery required for auxin-induced gene expression.

A 753 bp soybean GH3 promoter (592 bp) and 5' leader (161 bp) were fused to the GUS reporter gene as described above, and this chimeric gene was used to transform tobacco and *Arabidopsis* (Hagen et al., 1991; Z. Liu, unpublished results). The expression pattern for the GH3 promoter observed in the transgenic plants mirrored that observed in soybean by *in situ* hybridization. In addition, however, the transgenic studies revealed that the GH3 promoter is active in developing pollen, developing seeds, lateral root meristems, and apical meristems (Y. Li and T. Guilfoyle, unpublished results). In addition, the promoter is asymmetrically activated on the bottom side of gravitropic tobacco stems (Y. Li, unpublished results).

SAURs

Both tissue print hybridization and *in situ* hybridization with fixed thin sections were used to determine the pattern of SAUR gene expression in soybean (McClure and Guilfoyle, 1989; Gee et al., 1991). These studies showed that SAUR gene expression is largely restricted to elongation regions of soybean hypocotyls and epicotyls. The strongest expression of SAURs was observed in epidermal and cortical cells, with little expression in vascular tissues. Expression could be detected in young developing vascular tissues, however, and specifically in developing xylem elements. The abundance of SAURS rapidly declines in excised EHS incubated in the absence of auxin. Addition of auxin, caused a rapid increase in SAURS in the epidermis and cortex of excised, incubated EHS. Expression of SAURs in BHS was not detected prior to or after auxin treatement. Unlike GH3 expression which shows a wide range of tissue and organ expression after auxin treatment of soybean seedlings, SAUR gene expression shows a highly specific and restricted expression pattern in the presence and absence of external auxin.

Each of the SAUR 6B, 10A5, and 15A promoters (about 800 bp) and 5-leaders have been fused to the GUS reporter gene as described for GH1 (Li et al., 1991; Y. Li, unpublished results). These chimeric genes have been used to transform tobacco. Like GH3, the organ and tissue expression patterns in tobacco generally parallel those observed in soybean observed by *in situ* hybridization. However, at least one of the SAUR promoters is expressed in elongation zones of roots and anther filaments in addition to elongation zones of hypocotyls and epicotyls. In older plants, the SAURs are expressed in stems and petioles; most strongly in epidermal and cortical cells, but also in developing vascular tissues. The SAUR promoters are expressed asymmetrically during gravitropism of tobacco stems, and this asymmetric expression is blocked by auxin transport inhibitors (Li et al., 1991). Unlike the GH1 and GH3 promoters, the SAUR promoter is not expressed in mature pollen.

PROMOTER ANALYSES

GH3

The soybean GH3 promoter contains three TGA-elements within 450 bp from the start site of transcription [Fig. 2] (Hagen et al., 1991). The two most proximal elements are the major binding sites observed with proteins in nuclear extracts from soybean, wheat, cauliflower, and pea (T. Ulmasov and T. Guilfoyle, unpublished results). Because related elements containing an ACGT core (a) bind b-ZIP type transcription factors (Katagari and Chua, 1992), (b) are important regulatory elements in ABA, light, and wound-responsive genes, (c) are major binding sites for nuclear proteins, and (d) are also present in the GH3 homolog promoters in *Arabidopsis*, the TGA-elements were prime candidates for Auxin-inducible Response Elements or AuxREs. Deletion analysis of the soybean GH3 promoter in transgenic tobacco plants has indicated that a 182 bp sequence from the start site of transcription is sufficient for auxin-inducibliity (Z. Liu, unpublished results). This sequence contains only the most proximal TGA-element. Site directed mutagenesis of the two most proximal TGA-elements in transgenic tobacco plants has revealed that neither of these elements is required for auxin-inducibility of the GH3 gene (T. Ulmasov, X. Shi, and Z. Liu, unpublished results). More recent transient expression experiments with carrot protoplasts have indicated that the auxin-inducible element or modules are localized between the two most proximal TGA-elements in the GH3 promoter (Z. Liu and T. Ulmasov, unpublished results).

SAURs

The SAUR promoters contain a conserved sequence which has been called the DUE/NDE element [Fig. 4] (McClure et al., 1989). Deletion analysis of the 15A, 10A5, and 6B SAUR promoters in transgenic tobacco plants have indicated that the DUE/NDE element is involved in auxin-inducible expression (Y. Li, unpublished results). Multimerization of the DUE/NDE element supports this element's involvement in auxin-inducible expression.

Since there is no apparent sequence relationship between the AuxREs in the GH3 and SAUR promoters and since the expression patterns of these genes is entirely different, it is likely that more than one type of AuxRE is involved in auxin-regulated gene expression or that a variety of different modules comprise different types of auxin-inducible promoters. This might not be surprising since a variety of different types of auxin-inducible genes have been identified, and while some of these may be involved in cell expansion, others may be involved in cell division or cell differentiation.

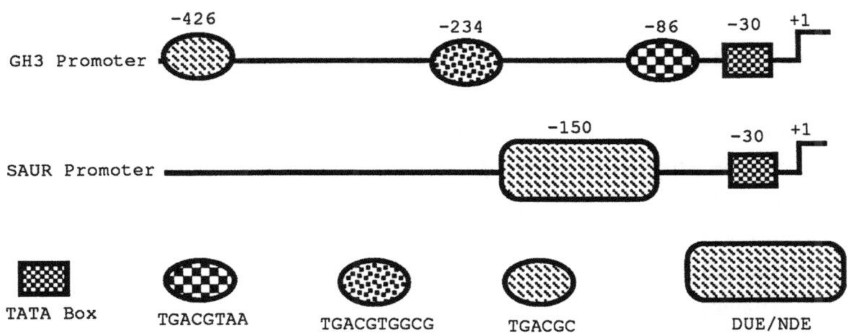

Figure 2. Diagrams of GH3 and SAUR promoters showing the relative positions of the TATA box, TGA-elements, and DUE/NDE elements to the transcription start site (+1).

AUXIN-INDUCIBLE PROMOTERS AS PROBES TO STUDY THE MECHANISM OF AUXIN ACTION

The GH3 and SAUR promoter/GUS fusions have been used to add support to the Cholodney/Went theory on gravitropism (Li et al., 1991; Y. Li, unpublished results). More recently, the GH3 promoter/GUS fusion has been observed to be expressed at early stages of lateral root formation, in apical and lateral shoot meristems, and in early stages of ovule development and seed formation (Y. Li, unpublished results). Because the GH3 and SAUR promoters show restricted patterns of gene expression, we have designed chimeric genes to express auxin and cytokinin biosynthetic genes and auxin conjugating genes for the overexpression or modification of these hormones (Li et al., 1992; Y. Li, unpublished results). For example the asymmetric expression of the GH3 and SAUR promoters during gravitropism should result in the asymmetric production of auxins or cytokinins or synthesis of inactive conjugated auxins with the appropriate chimeric gene constructs. These constructs should be useful in verifying the roles of auxin in processes such as gravitropism, phototropism, apical dominance, rooting, and vascular differentiation.

ACKNOWLEDGMENTS

This research was support by a grant DCB 8904493 from the National Science Foundation and by the Food for the 21st Century Program at the University of Missouri. T.S. is support by a National Science Foundation Postdoctoral Fellowship. This is journal report number 11,918 of the Missouri Agricultural Experiment Station.

REFERENCES

Ainley WM, Walker JC, Nagao RT, Key JL (1988) Sequence and characterization of two auxin-regulated genes from soybean. J Biol Chem 263: 10658-10666

Conner TW, Goekjian VH, Lafayette PR, Key JL (1990) Structure and expression of two auxin-inducible genes from *Arabidopsis*. Plant Mol Biol 15: 623-632

Czarnecka E, Nagao RT, Key JL, Gurley WB (1988) Characterization of Gmhsp26-A, a stress gene encoding a divergent heat shock protein of soybean: Heavy-metal-induced inhibition of intron processing. Mol. Cell. Biol. 8: 1113-1122

Dominov JA, Stenzler L, Lee S, Schwarz JJ, Leisner S, Howell SH (1992) Cytokinins and auxins control the expression of a gene in *Nicotiana plumbaginifolia* cells by feedback regulation. Plant Cell 4: 451-461

Franco AR, Gee MA, Guilfoyle TJ (1990) Induction and superinduction of auxin-responsive mRNAs with auxin and protein synthesis inhibitors. J Biol Chem 265: 15845-15849

Gee MA (1992) Tissue-specific expression of auxin-regulated genes. PhD Thesis, University of Missouri

Gee MA, Hagen G, Guilfoyle TJ (1991) Tissue-specific and organ-specific expression of soybean auxin-responsive transcripts GH3 and SAURs. Plant Cell 3: 419-430

Goldsbrough AP, Albrecht H, Stratford R (1993) Salicylic acid-inducible binding of a tobacco nuclear protein to a 10 bp seqeunce which is highly conserved amongst stress-inducible genes. Plant J 3: 563-571

Guilfoyle TJ, Hagen G, Li Y, Ulmasov T, Liu Z, Strabala T, Gee M (1993) Auxin-regulated transcription. Aust J Plant Physiol 20: in press

Guilfoyle TJ, McClure BA, Hagen G, Brown C, Gee M, Franco A (1990) Regulation of plant gene expression by auxins. In: Gene Manipulation in Plant Improvement. (JP Gustafson ed) Plenum Press, New York. pp 401-418

Hagen G. (1989) Molecular approaches to understanding auxin action. New Biol 1: 19-23

Hagen G, Guilfoyle TJ (1985) Rapid induction of selective transcription by auxins. Mol Cell Biol 5: 1197-1203

Hagen G, Kleinschmidt A, Guilfoyle T (1984) Auxin regulated gene expression in intact soybean hypocotyl and excised hypocotyl sections. Planta 162: 147-153

Hagen G, Martin G, Li Y, Guilfoyle, TJ (1991) Auxin-induced expression of the soybean GH3 promoter in transgenic tobacco plants. Plant Mol Biol 17: 567-579

Hagen G, Uhrhammer N, Guilfoyle TJ (1988) Regulation of an auxin-induced soybean sequence by cadmium. J Biol Chem 263: 6442-6446

Katagiri F, Chua N-H (1992) Plant transcription factors: present knowledge and future challenges. Trends Genet 8: 22-27

Key J (1989) Modulation of gene expression by auxin. BioEssays 11: 52-57

Li Y, Hagen G, Guilfoyle TJ (1991) An auxin-responsive promoter is differentially induced by auxin gradients during tropisms. Plant Cell 3: 1167-1175

Li Y, Hagen G, Guilfoyle TJ (1992) Altered morphology in transgenic tobacco plants that overproduce cytokinins in specific tissues and organs. Develop Biol 153: 386-395

McClure BA, Guilfoyle TJ (1987) Characterization of a class of small auxin-inducible soybean polyadenylated RNAs. Plant Mol Biol 9: 611-623

McClure BA, Guilfoyle TJ (1989) Rapid redistribution of auxin-regulated RNAs during gravitropism. Science 243: 91-93

McClure BA, Hagen G, Brown CS, Gee MA, Guilfoyle TJ (1989) Transcription, organization, and sequence of an auxin-regulated gene cluster in soybean. Plant Cell 1: 229-239

Takahashi Y, Kuroda H, Tanaka T, Machida Y, Takebe I, Nagata T (1989) Isolation of an auxin-regulated gene cDNA expressed during the transition from Go to S phase in tobacco mesophyll protoplasts. Proc Natl Acad Sci USA 86: 9279-9283

Takahashi Y, Kusaba M, Hiraoka Y, Nagata T (1991) Characterization of the auxin-regulated par gene from tobacco mesophyll protoplasts. Plant J 1: 327-332

van der Zaal EJ, Droog FNJ, Boot CJM, Hensgens LAM, Hoge JHC, Schilperoort RA, Libbenga KR (1991) Promoters of auxin-induced genes from tobacco can lead to auxin-inducible and root tip-specific expression. Plant Mol Biol 16: 983-998

Walker JC, Key JL (1982) Isolation of cloned cDNAs to auxin-responsive poly(A^+) RNAs of elongating soybean hypocotyl. Proc Natl Acad Sci USA 79: 7185-7189

Yamamoto KT, Mori H, Imaseki H (1992) cDNA cloning of indole-3-acetic acid-regulated genes: Aux22 and SAUR from mung bean (*Vigna radiata*) hypocotyl tissue. Plant Cell Physiol 33: 93-97

EARLY AUXIN REGULATED GENE EXPRESSION IN *ARABIDOPSIS*

Steffen Abel and Athanasios Theologis
Plant Gene Expression Center
800 Buchanan St
Albany, CA 94710
USA

INTRODUCTION

Extensive physiological studies have implicated the plant growth hormone auxin, typified by indole-3-acetic acid (IAA), in the regulation of cell elongation, cell division and cell differentiation, the driving and shaping forces in plant growth and development (Estelle, 1992). Auxin mediated cell elongation, one of the fastest hormonal responses in plants (Evans, 1974), is associated with rapid changes in gene expression (Guilfoyle, 1986; Theologis, 1986). The signal transduction pathway(s) responsible for auxin-regulated transcriptional activation are unknown.

Primary response genes provide a paradigm to unravel hormonal signal transduction pathways (Herschman, 1991). Members of several auxin-regulated gene families have the characteristic features of primary auxin-responsive genes, i.e. their transcriptional activation is rapid, specific for auxin and independent of *de novo* protein synthesis (Theologis *et al.*, 1985; Hagen and Guilfoyle, 1985; Van der Zaal *et al.*, 1987; McClure *et al.*, 1989; Franco *et al.*, 1990; Takahashi *et al.*, 1991; Oeller *et al.*, 1993). Their auxin-responsive elements (AuxRE) and their encoded proteins whose cellular function(s) remain elusive, can subsequently be used as molecular probes to identify components of the auxin transduction pathway using biochemical and reverse genetic approaches.

We have been focusing our efforts to elucidate the mechanisms of auxin action by studying primary auxin-responsive genes (*PS-IAA4/5* and *PS-IAA6*) in *Pisum sativum* (Theologis *et al.*, 1985; Theologis, 1986; Oeller *et al.*, 1993; Ballas *et al.*, 1993; Oeller and Theologis, 1993; Abel and Theologis, 1993). We have recently extended our studies to auxin-regulated genes in *Arabidopsis thaliana* in order to

NATO ASI Series, Vol. H 81
Plant Molecular Biology
Edited by G. Coruzzi and P. Puigdomènech

be able to introduce molecular genetic approaches for achieving our long term objective. Here we report the cloning and expression characteristics of *IAA1*, a member of a divergent multigene family of auxin-responsive genes in *Arabidopsis thaliana*.

MATERIALS AND METHODS

Plant Material and Tissue Treatment. Six-day-old etiolated seedlings (*Arabidopsis thaliana*, ecotype Columbia) were grown as described by Guzman and Ecker (1990). The seedlings were incubated in buffered (pH 5.5) liquid Murashige and Skoog media (GIBCO) in the absence or presence of various inducers (in the dark at 25 °C). Different organs from light-grown plants (3 weeks) were harvested and stored at -70 °C before RNA extraction.

PCR Cloning. The degenerate oligonucleotides DC2 (512-fold degeneracy) and DC4 (64-fold degeneracy) corresponding to the conserved amino acid sequences I/VVGWPP in domain II and DKDGDWM in domain IV of auxin-regulated polypeptides (Oeller *et al.*, 1993 and Fig. 1A) were used to amplify related genomic sequences of *Arabidopsis*. The reaction contained 100 ng of *Arabidopsis* genomic DNA, 1μM of each primer, 0.5 units of *Taq* polymerase (Perkin-Elmer/Cetus) and was performed for 30 thermal cycles, each consisting of 1 min at 94 °C, 1 min at 46 °C, and 1 min at 72 °C. The PCR products were separated by gel electrophoresis and isolated by adsorption to Schleicher&Schuell NA-45 paper. The amplified genomic DNA fragments were subcloned into the *Eco*RI and *Hind*III sites of pUC19, characterized by cross-hybridization and sequenced. An inverse PCR product was synthesized to determine the open reading frame of IAA1 cDNA, using circular cDNA synthesized with poly(A)$^+$ RNA (Huang *et al.*, 1990) from auxin-treated *Arabidopsis* seedlings as template and amplimers 12/1A (5'TTCACAGTAGGTGAATATTCC3') and 12/1B (5'ACATAACTCACGTT TTTGTTG-3'). These amplimers correspond to amino acids 111-117 and 77-70 of IAA1 (Fig. 1B), respectively, and were designed using the sequence information of the specific DC2-DC4 PCR product p12/1. The inverse PCR product, p12/1i, was subcloned into the *Eco*RV site of pIC20R and sequenced. The coding region of IAA1 cDNA was generated by PCR with the same circular cDNA template and amplimers CD1-1 (5'-GAGAGAATATGGAAGTCACC-3') and CD1-2 (5'-CAAAA GGTGTTTTGAGACAAT-3') complementary to 5'- and 3'-untranslated regions of IAA1 cDNA, as deduced from the sequence of the inverse PCR product p12/1i.

The IAA1 cDNA was subcloned as a *Eco*RI-*Hind*III fragment into pUC19 and the sequence of one strand determined.

Nucleic Acid Techniques. The isolation, fractionation, and analysis of *Arabidopsis* genomic DNA, total and poly(A)+ RNA were performed essentially as described (Sato and Theologis, 1989; Rottmann *et al.*, 1991; Davis *et al.*, 1980). The DNA and RNA hybridization analyses shown in Fig. 2 and 3 were carried out with ^{32}P-labeled *IAA1* gene-specific PCR product p12/1. p12/1 corresponds to amino acids 56-140 of the IAA1 protein (see Fig.1B). Equal loading of RNA was assessed by RNA hybridization with an *Arabidopsis* actin DNA probe.

RESULTS

Cloning *Arabidopsis* Genomic Sequences Encoding Auxin-regulated Polypeptides. A family of auxin-regulated polypeptides from various plant species (Ainley *et al.*, 1988; Conner *et al.*, 1990; Yamamoto *et al.*, 1992; Oeller *et al.*, 1993) share similar physical properties and extensive amino acid sequence identity in four highly conserved regions (Conner *et al.*, 1990 and Fig.1A). Taking advantage of these conserved domains, we used a PCR-based procedure to clone partial genomic sequences of early auxin-inducible genes in *Arabidopsis thaliana*. Partially degenerate oligonucleotides were synthesized corresponding to amino acid sequences I/VVGWPP and DKDGDWM, which represent consensus sequences for discrete and highly conserved regions in domain II and domain IV, respectively (Fig.1A). By using these degenerate amplimers and *Arabidopsis* genomic DNA, twelve distinct PCR fragments were reproducibly obtained (data not shown). The degenerate primers contained linker sequences for the *Eco*RI or the *Hind*III recognition site, allowing directional cloning of the amplified genomic DNA fragments into pUC19. Individual recombinants were classified by cross-hybridization at high stringency. Representative inserts from each class were sequenced. This approach allowed the isolation of fourteen distinct genomic DNA fragments. All of them contain open reading frames coding for peptides similar to the domain II-IV region of auxin-regulated proteins in Fig. 1A (Abel and Theologis, unpublished results).

Cloning the IAA1 cDNA. We have choosen one of the fourteen subcloned genomic fragments to reconstruct the complete coding region of its corresponding cDNA by PCR. The subcloned product p12/1 contains a single,

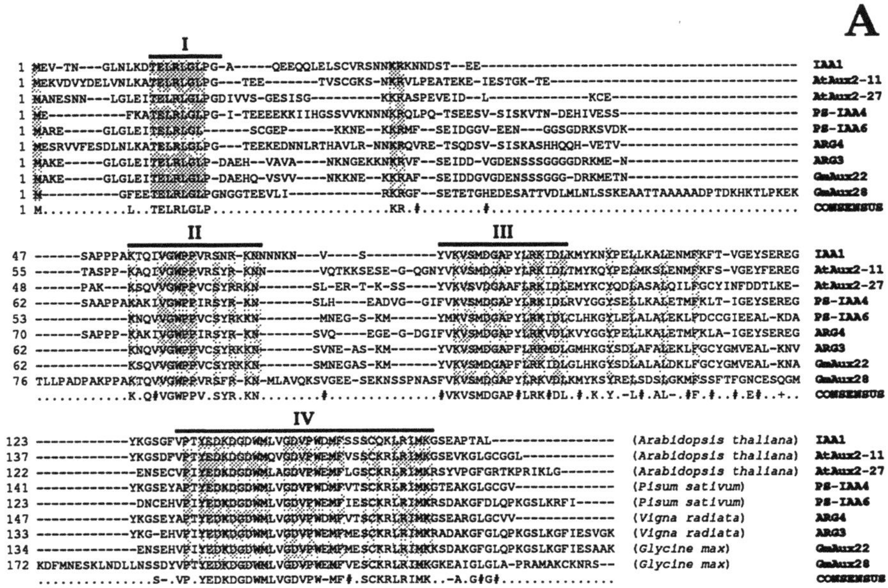
A
I
II
III
IV
IAA1
AtAux2-11
AtAux2-27
PS-IAA4
PS-IAA6
ARG4
ARG3
GmAux22
GmAux28
CONSENSUS
(Arabidopsis thaliana)
(Pisum sativum)
(Vigna radiata)
(Glycine max)

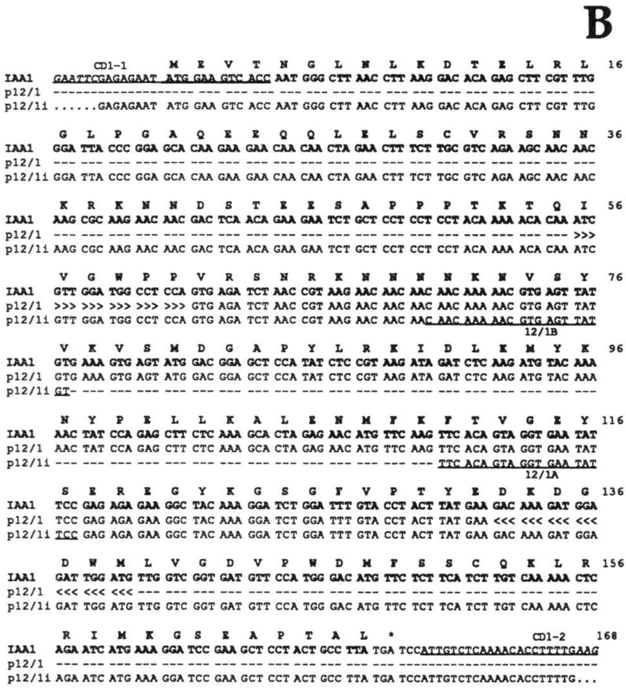
B
CD1-1
IAA1
p12/1
p12/1i
12/1B
12/1A
CD1-2

uninterrupted open reading frame coding for amino acids of domain II to domain IV of the putative auxin-regulated protein IAA1 (Fig.1B). To obtain sequence information for the 5'- and 3'-untranslated regions of the corresponding cDNA, we designed the gene-specific amplimers 12/1A and 12/1B, that allow the amplification of the missing cDNA sequences from a circularized cDNA library template. The template was constructed with poly(A)$^+$ RNA from auxin-treated etiolated *Arabidopsis* seedlings. Inverse PCR yielded a 570-bp fragment, p12/1i, which was subcloned into pIC20R and sequenced. p12/1i partially overlapps with p12/1, indicating the specificity of the reaction (Fig.1B). Primers CD1-1 and CD1-2 were synthesized using the sequence information of p12/1i to amplify the complete coding region (IAA1 cDNA) from the same circularized cDNA template. The IAA1 cDNA fragment was directionally subcloned into pUC19(*Eco*RI/*Hind*III) and sequenced. The sequence of IAA1 cDNA was identical to the composite of p12/1 and p12/1i sequences (Fig.1B). Translation of its open reading frame predicts a polypeptide of 168 amino acids and of 19.1 kDa. Most importantly, the IAA1 protein contains the four conserved domains and invariant amino acids shared among the auxin-regulated polypeptides shown in Fig. 1A. Labeled p12/1 fragment was used as a gene-specific probe for Southern analysis. p12/1 hybridizes to single DNA fragments of the *Arabidopsis* genome, demonstrating the authenticity of the PCR-generated IAA1 cDNA (Fig.2A). The same probe hybridizes to one major transcript of approximately 1,000 nts long (Fig.2B).

Expression Characteristics of *IAA1*. The specificity of the hormonal response is shown in Fig.3A. Two classes of chemicals are able to significantly increase the *IAA1* mRNA level: natural and synthetic auxins (lanes 3, 6, 7, 8, 12, 16, 17) and

Fig. 1. Primary structure of early auxin-responsive gene products. (**A**) Sequence alignment. Identical (shaded) and generally conserved amino acid residues appear in the consensus (at least seven out of nine matches (acidic [D, E], -; basic [R, K], +; or hydrophobic [A, C, V, I, L, M, F, Y, W], #). Conserved domains are overlined and indicated by Roman numerals. Sequences were visually aligned and dashes inserted for alignment improvement. Sources of the sequences are: IAA1 (this study); AtAux2-11 and AtAux2-27 (Conner *et al.*, 1990); PS-IAA4 and PS-IAA6 (Oeller *et al.*, 1993); ARG3 and ARG 4 (Yamamoto *et al.*, 1992); GmAux22 and GmAux28 (Ainley *et al.*, 1988). (**B**) Cloning and coding region of IAA1 cDNA. Nucleotide sequences of PCR-generated IAA1 cDNA and of PCR products p12/1 and p12/1i (inverse PCR) are aligned. Amplimers used to synthesize these fragments are underlined (specific primers) or indicated by arrows heads (DC2 and DC4). The Gene Accession number of IAA1 is L15448.

the protein synthesis inhibitor cycloheximide (lanes 4 and 5). The rapidity of the induction by auxins is shown in Fig.3B. Elevated mRNA concentration is detectable as early as 6 min after IAA application and steadily increases over 30 min (Fig.3B). Prolonged IAA treatment (up to 4 hrs) does not further increase the *IAA1* mRNA level (data not shown). Optimal rate of IAA-regulated *IAA1* mRNA accumulation is found at 10 μM IAA. However, concentrations as low as 10 nM IAA are effective in eliciting the hormonal response (Fig.3C).The *IAA1* mRNA is differentially regulated in untreated etiolated seedlings and in various parts of light-grown plants (Fig.3D). It is highly expressed in etiolated seedlings (lane 1), stems, leaves and flowers (lanes 3-5), at moderate levels in roots (lane 2), and is almost undetectable in siliques (lane 6).

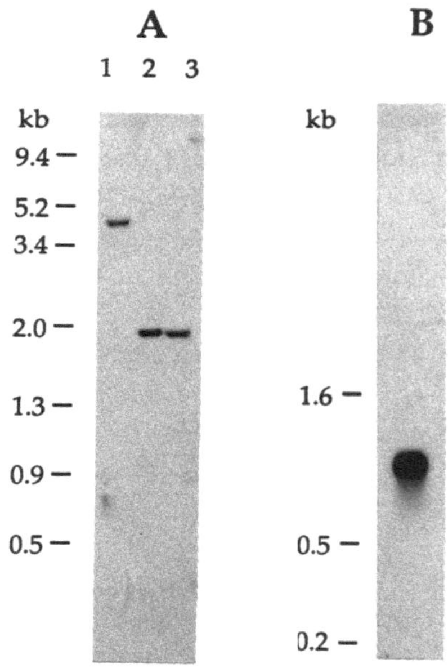

Fig. 2. Southern analysis and estimation of transcript size. (**A**) Southern analysis of *Arabidopsis* genomic DNA (10 μg per lane) digested with *Eco*RI (lane 1), *Hind*III (lane 2) and *Eco*RI/*Hind*III (lane 3). (**B**) Northern analysis of total RNA (25 μg) from auxin-treated etiolated *Arabidopsis* seedlings. Both blots were hybridized with ^{32}P-labeled *IAA1*-specific probe.

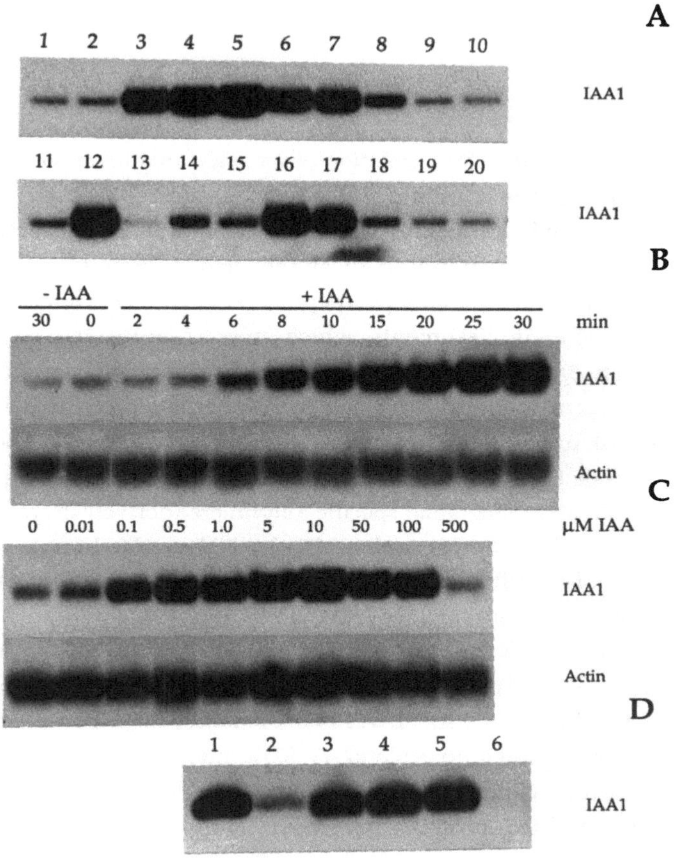

Fig. 3. Expression characteristics of *IAA1*. (**A**) Specificity of the hormonal response. Total RNA (25 μg) from 6-d-old etiolated seedlings treated with various inducers for 60 min were hybridized with ^{32}P-labeled *IAA1*-specific probe. Lanes: 1, untreated control; 2 and 11, mock control; 3 and 12, IAA (20 μM); 4, cycloheximide (50 μM); 5, IAA (20 μM)/cycloheximide (50 μM); 6, 2,4-D (20 μM); 7, NAA (20 μM); 8, PAA (20 μM); 9, L-tryptophane (20 μM); 10, wounding; 13, abcisic acid (20 μM); 14 gibberellic acid (20 μM); 15, benzyl adenine (20 μM); 16, IAA/benzyl adenine (20 μM each); 17, as 16/LiCl (50 mM); 18, LiCl (50 mM); 19, ethylene (10 ppm); 20, anaerobiosis (nitrogene atmospere). (**B**) Rapidity of the hormonal response. Total RNA (25 μg) from 6-d-old etiolated seedlings treated with or without 20 μM IAA for various time intervals indicated were hybridized with ^{32}P-labeled *IAA1*-specific probe or with ^{32}P-labeled actin-specific probe. (**C**) Sensitivity of the hormonal response. Total RNA (25 μg) from 6-d-old etiolated seedlings treated with various concentrations of IAA for 2 hrs were hybridized with ^{32}P-labeled *IAA1*-specific probe or with ^{32}P-labeled actin-specific probe. (**D**) Expression of *IAA1* in *Arabidopsis* plants. Poly(A)$^+$ RNA (2.5 μg) isolated from etiolated seedlings (6-d-old) and from various parts of light-grown plants (25-days-old) were hybridized with ^{32}P-labeled *IAA1*-specific probe. Lanes: 1, etiolated seedlings; 2, roots; 3, stems; 4, leaves; 5, flowers; and 6, siliques.

DISCUSSION

We have decribed the rapid cDNA isolation of an early auxin-responsive gene in *Arabidopsis thaliana*. The cloning procedure is entirely based on PCR, circumventing the need of library screening. The first PCR reaction uses degenerate amplimers corresponding to amino acid sequences in conserved domains II and IV of auxin-regulated polypeptides (Fig.1A). This approach yields fourteen distinct genomic fragments which may indicate the size of the multigene family in *Arabidopsis* (Abel and Theologis, unpublished results). Two of these fragments are identical with the corresponding sequences in auxin-regulated genes AtAux2-11 and AtAux2-27 (Conner *et al.*, 1990). The second (inverse) PCR reveals the sequence information of the untranslated regions of a specific cDNA (IAA1) by using gene specific amplimers (derived from one of the fourteen primary PCR products) and a circularized cDNA library template. Finally, a third set of amplimers specific for the untranslated regions is used to generate the complete coding region of IAA1 cDNA (Fig.1B).

The predicted IAA1 protein contains all the conserved amino acids (mostly clustered in four conserved regions) shared among auxin-regulated polypeptides of this class (Fig.1A). IAA1 is the smallest described member (168 residues) and is most similar to AtAux2-11 (63% nucleotide identity) and PS-IAA4 (62% nucleotide identity). The four conserved domains in the coding region of auxin-regulated proteins are most likely indispensible for their function(s).

We studied the hormonal response of *IAA1* in greater detail. The expression characteristics of this gene meet the criteria of primary response genes (Ringold, 1979; Herschman, 1991): Its mRNA induction is very rapid (well within 6 min; Fig.3B), specific for auxins and independent of *de novo* protein synthesis (Fig.3A). The regulation of *IAA1* gene expression is as complex as that of *PS-IAA4/5* . Experimental evidence indicates that the *PS-IAA4/5* gene induction is due to transcriptional activation as well as to mRNA stabilization (T. Koshiba, N. Ballas, and A. Theologis, unpublished). Transcriptional activation by inhibitors of protein synthesis has been shown for several mammalian genes (Herschman, 1991). The cycloheximide inducibility of auxin-regulated genes suggests that their expression is under the control of short-lived repressors. The rapidity of the *IAA1* mRNA accumulation by IAA is reminiscent of that found for *PS-IAA4/5* and *PS-IAA6* (Theologis *et al.*, 1985) and the SAUR genes (McClure et al., 1989). In a similar study of AtAux2-11 and AtAux2-27 which are members of the same multigene family in *Arabidopsis* (Fig.1A), enhanced mRNA levels are detected 30 min and 90 min after 2,4-D application (100 μM), respectively (Conner *et al.*, 1990). This suggests that this family of auxin-responsive genes is differentially regulated. Characterization of the other members of the family will provide valuable information for understanding auxin transduction in *Arabidopsis thaliana*.

ACKNOWLEDGEMENTS

This work was supported by a postdoctoral fellowship from Boehringer-Ingelheim-Fonds, Germany, to S.A. and by a grant from the National Institute of Health (GM-35447) to A.T.

REFERENCES

Abel, S. and Theologis, A. (1993) *Proc. Natl. Acad. Sci. USA* submitted 5/10/93.

Ainley, W.M., Walker, J.C., Nagao, R.T. and Key, J.L. (1988) *J. Biol. Chem.* **263**: 10658-10666.

Ballas, N., Wong. L.-M. and Theologis, A. (1993) *J. Mol. Biol.* In press.

Conner, T.W., Goekjian, LaFayette, P.R. and Key, J.L. (1990) *Plant Mol. Biol.* **15**: 623-632.

Davis, R.W., Botstein D. and Roth, J.R. (1980) *Advanced Bacterial Genetics* (Cold Spring Harbor Lab., Cold Spring Harbor, NY).

Estelle, M. (1992) *BioEssays* **14**:439-444.

Evans, M.L. (1974) *Annu. Rev. Plant Physiol.* **25**:195-223.

Franco, A.R., Gee, M.A. and Guilfoyle, T.J. (1990). *J. Biol. Chem.* **265**: 15845-15849.

Guilfoyle, T.J. (1986) *CRC Crit. Rev. Plant Sci.* **4**:247-276.

Guzman, P. and Ecker, J.R. (1990) *Plant Cell* **2**: 513-523.

Hagen, G. and Guilfoyle, T.J. (1985) *Mol. Cell. Biol.* **5**:1197-1203.

Herschman, H.R. (1991) *Annu. Rev. Biochem.* **60**:281-319.

Huang, S.H., Hu, Y.Y., Wu, C.H. and Holcenberg, J. (1990) *Nucleic Acids Res.* **18**: 1992.

McClure, B.A., Hagen, G., Brown, C.S., Gee, M.A. and Guilfoyle, T.J. (1989) *Plant Cell* **1**:229-239.

Oeller, P.W., Keller, J.A., Parks, J.E., Silbert, J.E. and Theologis, A. (1993) *J. Mol. Biol.* In press.

Oeller, P.W. and Theologis, A. (1993) *Proc. Natl. Acad. Sci. USA* submitted 5/10/93.

Rottmann, W.H., Peter, G.F., Oeller, P.W., Keller, J.A., Shen, N.F., Nagy, B.P., Taylor, L.P., Campbell, A.D. and Theologis, A. (1991) *J. Mol. Biol.* **222**: 937-961.
Sato, T and Theologis, A. (1989) *Proc. Natl. Acad. Sci. USA* **86**: 6621-6625.
Ringold, G.M. (1979) *Biochem. Biophys. Acta* **560**:487-508.
Takahashi, Y., Kusaba, M., Hiraoka, Y., and Nagata, T. (1991) *Plant J.* **1**:327-333.
Theologis, A., Huynh, T.V. and Davis, R.W. (1985) *J. Mol. Biol.* **183**:53-68.
Theologis, A (1986) *Annu. Rev. Plant Physiol.* **37**:407-438.
Van der Zaal, E.J., Memelink, J., Mennes, A.M., Quint, A. and Libbenga, K.R. (1987) *Plant Mol. Biol.* **10**:145-157.

MOLECULAR AND GENETIC ANALYSIS OF THE CONSTITUTIVE ETHYLENE RESPONSE MUTATION *ctr1*

Joseph J. Kieber and Joseph R. Ecker
Department of Biology
Plant Science Institute
University of Pennsylvania
Philadelphia, PA 19104-6018

INTRODUCTION

The simple olefin ethylene has been known to affect plant growth and development since the early part of this century when the various effects of illuminating gas on pea seedling development were first described (Neljubow, 1901). Since then, it has been shown to influence a diverse array of plant growth and developmental processes, including germination, senescence, epinasty, leaf abscission, stress responses, sex determination in monoecious species and fruit ripening (Abeles *et al.*, 1992). The biosynthetic pathway for ethylene has been elucidated and genes encoding the two key enzymes, ACC synthase and ACC oxidase, have been cloned from a number of plant species (Theologis, 1992).

The "triple response" is a response of dark-grown seedlings to ethylene and in *Arabidopsis* consists of a shortening and radial swelling of the hypocotyl, inhibition of root elongation and exaggeration of the curvature of the apical hook (Fig. 1). This response is extremely amenable to genetic analysis in *Arabidopsis* due to its high reproducibility, early stage of growth at which the screen is done and the ease of screening very large numbers of seedlings ($>10^6$) for mutations. We and others have used this response to identify mutants that are insensitive to ethylene (Bleecker *et al.*, 1988; Guzman and Ecker, 1990) and mutants that constitutively display this response (Kieber *et al.*, 1993). The recessive mutation *ctr1* results in constitutive activation of seedling and adult ethylene responses. The epistatic relationship of *ctr1* with several ethylene-insensitive mutants was used to position this gene in the ethylene signal pathway. The *CTR1* gene was cloned and found to encode a protein that resembles the Raf family of serine/threonine protein kinases (Kieber *et al.*, 1993). The putative CTR1 kinase is postulated to act as a negative regulator in the ethylene signal transduction chain. In this report we further characterize this mutation and its gene.

NATO ASI Series, Vol. H 81
Plant Molecular Biology
Edited by G. Coruzzi and P. Puigdomènech

ISOLATION OF CONSTITUTIVE TRIPLE RESPONSE MUTATIONS

We screened independent lots of ethyl methanesulfonate, diepoxybutane and X-ray mutagenized seeds for mutants that displayed the triple response in the dark in the absence of exogenous ethylene (Fig. 1). More than one million seedlings were screened in this manner, yielding four hundred putative mutants, of which eighteen survived, produced seeds and re-tested for this phenotype. These mutants fell into two classes: those that were revertable by inhibitors of ethylene biosynthesis and binding and those that were not (Kieber *et al.*, 1993). The latter class were affected in ethylene perception or signal transduction. All five were recessive and fell into the same complementation group called *ctr1*. *ctr1* seedlings were indistinguishable from ethylene-treated wild-type seedlings (Fig. 2). The *ctr1* mutation was pleiotropic. In addition to the constitutive triple response phenotype, mutant rosette leaves were greatly reduced in size, the roots were much less extensive, the plants bolted later and the inflorescence was much less extensive than wild-type plants (Kieber *et al.*, 1993). These phenotypes could be

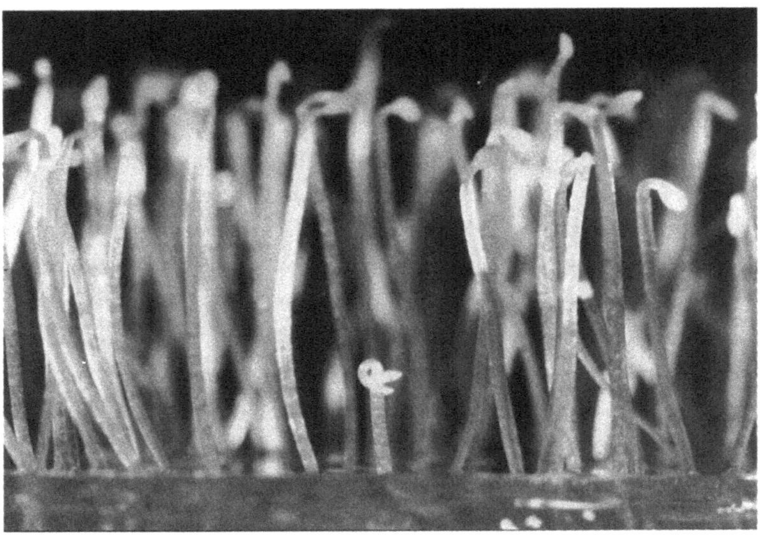

Fig. 1. Screen for constitutive triple response mutants. Seeds were surface sterilized with bleach, plated on MS agar (Gibco) containing 5% sucrose, placed at 4°C for four days and then grown in the dark in air for three days at 23°C. The single *ctr1* mutant is shown amid wild-type seedlings.

phenocopied by growth of wild-type plants in ethylene and result, at least partially, from a defect in cell elongation.

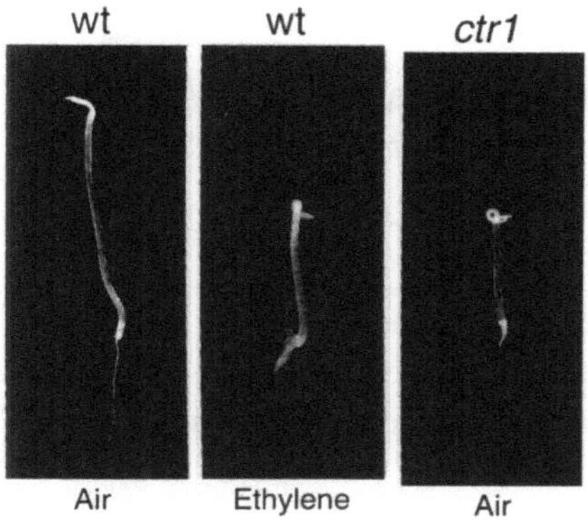

Fig. 2. Three day old wild-type seedlings grown in air or 10 nl ethylene/liter and *ctr1* mutant seedlings grown in air were picked and photographed.

REDUCED TRANSMISSION OF THE *ctr1* ALLELE IS DUE TO A DEFECT IN FEMALE GAMETES

The *ctr1* allele is transmitted at a reduced frequency relative to the wild-type allele (Kieber *et al.*, 1993). This reduced transmission of the mutant allele could be due to a defect in either the male, female or both gametes. To further analyze this, we performed reciprocal crosses between homozygous mutant plants (*ctr1/ctr1*) and plants heterozygous at the *ctr1* locus (Table 1). If wild-type and mutant male and female gametes are equally robust, then equal numbers of wild-type and mutant progeny are expected in the F1. This was the case when the male parent was heterozygous, indicating that mutant pollen competes equally well with wild-type pollen for fertilization. However, when the female parent in these crosses was the heterozygote, there were significantly more wild-type seedlings in the F1, indicating that mutant female gametes were less competitive than their wild-type counterparts. The allele frequency for the female gametes derived from this data (0.3 for the mutant verses 0.7 for wild-type) predicts that 15% of the F2 progeny from a cross of a homozygous *ctr1* mutant to a wild-type plant should be mutant (with the male gametes showing an allele frequency of 0.5%). This is in good agreement with the 17% mutant progeny observed in such a cross (Kieber *et al.*, 1993).

Table 1 Reciprocal Crosses with Heterozygous *ctr1*

Female parent	Male parent	F1 WT	F1 *ctr*	X^2
ctr1/+	*ctr1/ctr1*[a]	130	55	30.4; $p < 0.0005$[b]
ctr1/ctr1	*ctr1/+*[c]	99	93	0.19; $p > 0.5$

[a] summed from six independent crosses
[b] calculated for an expected 1:1 ratio
[c] summed from five independent crosses

CTR1 GENOMIC REGION

A line from a T-DNA tagged population (Feldmann 1991) was identified that segregated a constitutive triple response mutant that failed to complement *ctr1* (Kieber *et al.*, 1993). Plant DNA flanking the site of insertion was rescued by plasmid rescue in *E. coli*, and was used to isolate genomic λ clones. A restriction map of these clones is presented in Figure 3. These clones span a region of approximately 22 kb, 5 kb of which encompasses the *CTR1* coding region. Just upstream of the longest cDNA is 32-bp region of alternating AG residues. This motif has been found upstream of the *Drosophila* heat shock genes *hsp70* and *hsp26*, the *Ultrabithorax* gene and the histone genes *his3* and *his4*, and a factor

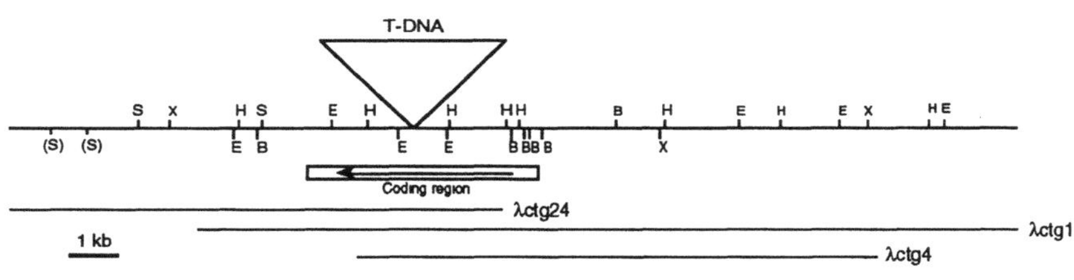

Fig. 3. Restriction map of *CTR1* genomic region. The site of insertion of the DNA in *ctr1*-5 is indicated above the map, the *CTR1* coding region and three genomic clones spanning this region are shown below the map. S: Sal1; H: HindIII; B: BamH1; X: Xho1

that binds specifically to this sequence has been purified. (Biggin and Tjian, 1988; Gilmour *et al.*, 1989). This element may play some role in controlling the expression of *CTR1*.

To prove that these clones are the *CTR1* gene, we examined five independent *ctr1* alleles for restriction fragment length polymorphisms. Southern blot analysis suggested that the X-ray allele *ctr1*-2 contained a small deletion in the *CTR1* coding region. We amplified this region using the polymerase chain reaction with two flanking oligonucleotide primers (Fig. 4). This data confirmed that this allele did indeed contain a small deletion, which subsequent sequence analysis revealed to span 17-bp (Kieber *et al.*, 1993). Further analysis of four other *ctr1* alleles further confirmed that these clones corresponded to the *CTR1* gene.

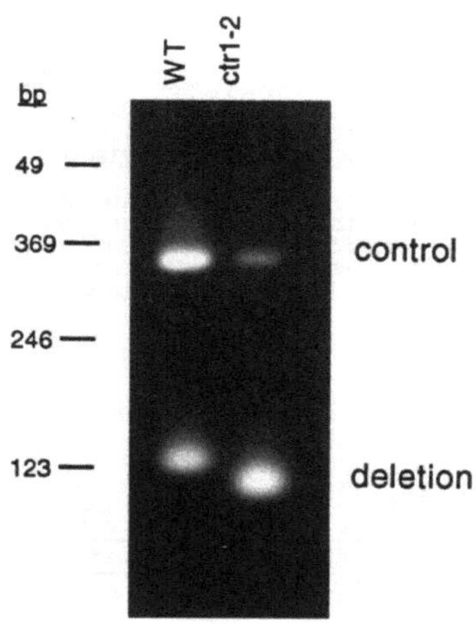

Fig. 4. Polymerase chain reaction (PCR) products using primers flanking the deletion in *ctr1*-2. Two primers flanking the putative deletion in *ctr1*-2 and two control primers from a separate region of *CTR1* were used to amplify DNA from wild-type and mutant genomic DNA. Sizes of size standards from the 123 bp ladder (Bethesda Research Laboratories) are indicated on the right.

CTR1 ENCODES A PROTEIN KINASE

Analysis of the amino acid sequence of the C-terminal half of CTR1 indicated that it contained the eleven subdomains common to all known protein kinases (Hanks et al., 1988; Hanks and Quinn, 1991). The highest degree of homology seen in searches of the various protein databases is found with members of the Raf family of serine/threonine protein kinases (41% identity in the kinase domain). An alignment of all known members of this family is presented in Figure 5. The sequence of the CTR1 protein suggests that it phosphorylates serine/threonine residues (Hanks and Quinn, 1991). Weak, but significant homology to the Raf family also extends through the N-terminal portion of CTR1.

Two of the *ctr1* alleles that were sequenced were single amino acid

A

```
Raf-1   (551) VMlStRIGSG SFGTVYkGkW HG.DVAVKIL kVVDPTPEQF QAFrNEVAVL rKTRHVNILL FMGYmTK.dN
XRaf    (340) VMlSsRIGSG SFGTVYkGkW HG.DVAVKIL kVtDPTPEQL QAFrNEVAVL rKTRHVNILL FMGYmTK.dN
Draf    (355) ILIgpRIGSG SFGTVYRahW HG.pVAVKtL nVktPsPaQL QAFkNEVAmL KKTRHcNILL FMGCVsK.Ps
lin-45  (481) fiIqykVGSG SFGTVYRGeF fG.tVAIKkL nVVDPTPsQM aAFkNEVAVL KKTRHlNVLL FMGWVre.Pe
CTR1    (551) lnIkekIGaG SFGTVhRaeW HGsDVAVKIL meqDfhaErv neFlrEVAIM KrlRHpNIvL FMGaVTqpPN

      Raf-1   LAIVTQWCEG SSLYkHLHVq ETK..FqMFQ LIDIARQtAQ GMDYLHAKN. .IIHRDMKSN NIFLHEgL.T
      XRaf    LAIVTQWCEG SSLYyHLHVl DTK..FqMFQ LIDIARQtAQ GMDYLHAKN. .IIHRDMKSN NIFLHEgL.T
      Draf    LAIVTQWCEG SSLYkHvHVs ETK..FkLnt LIDIgRQVAQ GMDYLHAKN. .IIHRDLKSN NIFLHEdL.s
      lin-45  iAIITQWCEG SSLYrHiHVq EprveFeMga iIDIlkQVsl GMnYLHsKN. .IIHRDLKtN NIFLmDdMsT
      CTR1    LsIVTeYlsr gSLYrlLHks gareqLderr rlsmAydVAk GMnYLHnrNp pIVHRDLKSp NlLv.DkkyT

      Raf-1   VKIGDFGLAT VKsRW..SGS QQvEQPTGSV LWMAPEVIRM QDNNPFSFQS DVYSYGIVLY ELMTGELPYS
      XRaf    VKIGDFGLAT VKTRW..SGS QQvEQlTGSI LWMAPEVIRM QDNNPFSFQS DVYSYGIVLY ELMTGELPYS
      Draf    VKIGDFGLAT aKTRW..SGe kQanQPTGSI LWMAPEVIRM QElNPYSFQS DVYaFGIVMY ELLaecLPYg
      lin-45  VKIGDFGLAT VKTkWtvnGg QQqqQPTGSI LWMAPEVIRM QDdNPYtpQS DVYSFGIcMY EiLsshLPYs
      CTR1    VKVcDFGLsr lKastflSsk saagtPe... .WMAPEVlRd epsNe...kS DVYSFGVILW ELaTlqqPWg

      Raf-1   HINNRDQIIF MVGRGYasPD LSKlyKNCPK AMKRLVaDCV KkvkEERPLF PQILssi
      XRaf    HIrdRDQIIF LVGRGgvvPD LSKlyKNCPK AMKRLVaDsI KklrDERPLF PQILssi
      Draf    HIsNkDQIlF MVGRGlLRPD MSqVrsdarr hsKRLaEDCI KytpkDRPLF rplLnmL
      lin-45  nINNRDQIlF MVGRGYLRPD rSKIrhdtPK sMlkLyDnCI mfDrnERPvF geVLerL
      CTR1    nl.NpaQVVa aVG...Fkck rleIprNlnp qvaaiIEgCw tnEpwkRPsF atIMdlL
```

B

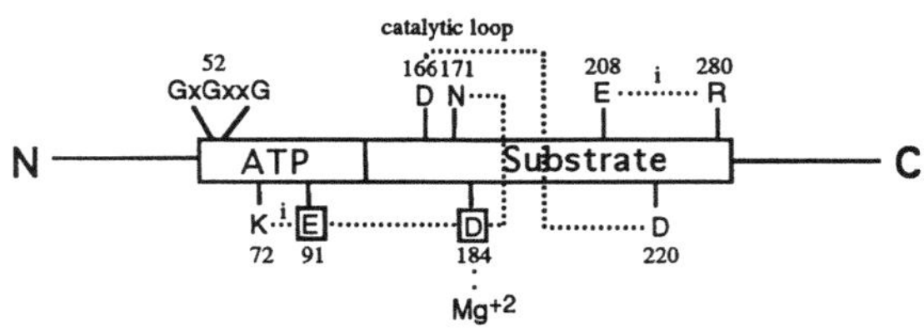

Fig. 5. A) Homology of the catalytic domains from CTR1 and other Raf family members. Lineup was generated using the GCG Pileup program with the default setting. Capital letters indicated 2 or more residues identical or highly conserved (score $\geq$ 1.0 on GCG rescaled PAM 250 matrix). Raf-1, human Raf gene; XRaf, *Xenopus* Raf; Draf, *Drosophila* Raf (also called *polehole*); lin-45, *C. elegans* Raf. B) Schematic of the catalytic domain derived from the crystal structure of mouse recombinant cAMP-dependent protein kinase (cAPK) showing the nearly invariant amino acids and some critical bonds among them (Taylor *et al.*, 1992). The regions of the molecule that bind ATP and substrate are indicated, as is the catalytic loop. Dotted lines with an "i" indicate ionic bonds; dotted lined without an "i" indicate hydrogen bonds. N and C refer to the amino and carboxy termini of the protein respectively. The numbers below the residues indicate the position in the cAPK protein. The two amino acids mutated in *ctr1* allels are boxed.

changes. Both changes were in residues that are invariant or nearly invariant in all known protein kinases (Hanks and Quinn, 1991). Figure 5 shows the location of these residue and some important interactions among them. *ctr1*-4 results in a Glu→Lys change at the position corresponding to Glu-91 in Figure 5. This residue is involved in an ionic interaction with Lys-72, and thus alteration to an oppositely charged amino acid would be expected to disrupt the structure of the protein. The change in *ctr1*-1 is a Asp→Glu change at the position corresponding to Asp-72 in Figure 5. This residue is involved in chelating Mg^{+2} as well as hydrogen binding with both Glu-91 and Asn-171. This hydrogen bonding is important for keeping the two lobe of the catalytic domain together (see Taylor *et al.* 1992). Therefore, it is not surprising that even a conservative change to a Glu residue would disrupt the kinase activity of CTR1. A systematic mutagenesis of yeast cAPK (Gibbs and Zoller, 1991) has shown that when either of these residues is changed to an uncharged alanine, catalytic activity is either very strongly reduced (Glu-91) or essentially abolished (Asp-184).

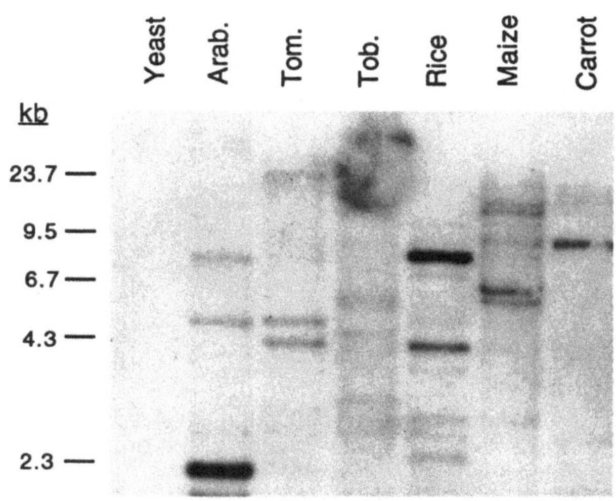

Fig. 6. Southern blot of genomic DNA from various plants. Genomic DNA was isolated from various plants, digested with EcoR1 and electrophoresed on an 0.8% agarose gel. The DNA was transferred to Hybond N (Amersham) and hybrydized with a radiolabled probe containing the *CTR1* catalytic domain. Hybridization was performed in 5X SSPE, 30% formamide, 0.5% SDS, 5X Denharts and 100 ug/ml sonicated salmon sperm DNA. The filters were washed in 2X SSPE, 0.5% SDS at 46°C for 30 min. and autoradiographed. Lambda DNA digested with HindIII was used as a molecular weight standard. Arab. - Arabidopsis; Tom. - Tomato; Tob. - Tobacco

OTHER HIGHER PLANTS CONTAIN SEQUENCES SIMILAR TO *CTR1*

We used a *CTR1* cDNA clone as a probe in a reduced stringency Southern blot hybrydization against genomic DNA from a number of different plant species (Fig. 7). Similar DNA sequences were detected in all the plant species tested, including tomato, tobacco, carrot, beet and monocots such as corn and rice. This suggests that the *CTR1* gene may be a widespread component of ethylene signal transduction.

MOLECULAR MODEL FOR ETHYLENE SIGNAL TRANSDUCTION

By analogy to the protein kinase cascades that the various Raf proteins participate in, ethylene signal transduction in plants may also involve a cascade of kinases. One model that incorporates this idea as well as other ethylene response genes and what has been learned about their genetics is presented in Figure 7. This model proposes that in the absence of ethylene the kinase activity of CTR1 is

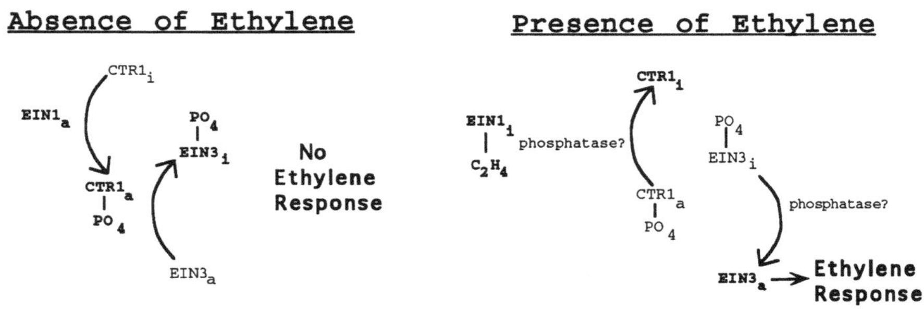

Fig. 7. Model for ethylene signal transduction in *Arabidopsis*. The a and i subscripts refer to the active and inactive states of the enzymes. -PO4 indicates that the protein is phosphorylated.

active and it phosphorylates the genetically downstream *EIN3* gene product. EIN3 then activates a number of terminal ethylene regulated genes. When ethylene is present, it binds to a receptor, possibly the EIN1 gene product. This ethylene-receptor complex then inactivates CTR1, either directly or indirectly and possibly via phosphorylation (as in the case with several Raf proteins). The inactive CTR1 no longer phosphorylates the *EIN3* gene product, which is then activated, possibly via dephosphorylation by a protein phosphatase. This model is consistent with what is known about these genes, but is at this point purely speculative. Cloning and analysis of other genes involved in this pathway should increase our understanding of the molecular basis for ethylene signal transduction.

ACKNOWLEDGMENTS

The authors thank A. Golden, M. Han and P. Sternberg for sharing the unpublished sequence of the *C. elegans lin-45* gene. This work was supported in part by grants (GM R29 38820) and (NSF IBN-92-05342) to J. R. E. and by an NSF Postdoctoral Fellowship in Plant Biology (DCB-9008323) to J. J. K.

REFERENCES

Abeles FB, Morgan PW, Saltveit Jr. ME (1992) Ethylene in Plant Biology, 2nd edition Academic Press, Inc San Diego, CA

Biggin MD, Tjian R (1988) Transcription factors that activate the *Ultrabithorax* promotor in developmentally staged extracts. Cell 53:699-711

Bleecker AB, Estelle MA, Somerville C, Kende H (1988) Insensitivity to ethylene conferred by a dominant mutation in *Arabidopsis thaliana*. Science 241:1086-1089

Feldmann KA (1991) T-DNA insertion mutagenesis in *Arabidopsis*: mutational spectrum. Plant Journal 1:71-82

Gibbs CS, Zoller MJ (1991) Rational scanning mutagenesis of a protein kinase identifies functional regions involved in catalysis and substrate interactions. J Biol Chem 266: 8923-8931

Gilmour DS, Thomas GH, Elgin SCR (1989) *Drosophila* nuclear proteins bind to regions of alternating C and T residues in gene promotors. Science 245:1487-1490

Guzman P, Ecker JR (1990) Exploiting the triple response of *Arabidopsis* to identify ethylene-related mutants. Plant Cell 2: 513-523

Hanks SK, Quinn AM (1991) Protein kinase catalytic domain sequence database: Identification of conserved features of primary structure and classification of family members. Methods in Enzymol 200: 38-62

Hanks SK, Quinn AM, Hunter T (1988) The protein kinase family: Conserved features and deduced phylogeny of the catalytic domains. Science 241: 42-52

Kieber JJ, Rothenberg M, Roman G, Feldmann KA, Ecker JR (1993) *CTR1*, a negative regulator of the ethylene response pathway in *Arabidopsis*, encodes a member of the Raf family of protein kinases. Cell 72:427-441

Neljubow D (1901). Ueber die horizontale nutation der stengel von *Pisum sativum* und einiger anderer. Pflanzen Beih Bot Zentralb 10: 128-139

Rapp UR (1991) Role of *Raf*-1 serine/threonine protein kinase in growth factor signal transduction. Oncogene 6: 495-500

Taylor SS, Knighton DR, Zheng J, Ten Eyck LF, Sowadski JM (1992) Structural framework for the protein kinase family. Annu Rev Cell Biol 8: 429-462

Theologis A (1992) One rotten apple spoils the whole bushel: the role of ethylene in fruit ripening. Cell 70: 181-184

Yang SF, Hoffman NE (1984). Ethylene biosynthesis and its regulation in higher plants. Ann Rev Plant Physiol 35: 155-189

The vetch (*Vicia*) and *Rhizobium leguminosarum* bv. *viciae* symbiosis: A system to study the activity of *Rhizobium* Nod factors.

Irma Vijn[1], Ton van Brussel[2], Albert van Kammen[1] and Ton Bisseling[1,3]

1. Department of Molecular Biology, Agricultural University, Dreijenlaan 3, 6703 HA Wageningen, The Netherlands
2. Institute of Molecular plant Sciences, Leiden University, Nonnensteeg 3, 2311 VJ Leiden, The Netherlands
3. Corresponding author

Introduction

The interactions of leguminous plants and bacteria of the genera *Rhizobium* , *Bradyrhizobium* and *Azorhizobium* -here collectively called rhizobia- result in the formation of root nodules, new organs in which the bacteria are able to fix atmospheric nitrogen into ammonia. The formation of root nodules involves a number of developmental steps (Newcomb, 1976; Newcomb, 1981). First the bacteria attach to the root hairs of the host plant and cause deformation and curling of root hairs. Then the bacteria enter the plant by newly formed tubular structures, the infection threads, starting from the curls of the root hairs. At the same time, cells in the cortex of the root become mitotically active and form a nodule primordium. The infection threads grow towards the nodule primordia, and after penetration of individual primordium cells the bacteria are released in the cytoplasm, by endocytosis. Then the nodule primordium differentiates into a nodule.

During nodule formation several plant genes, called nodulin genes (Van Kammen, 1984), are specifically expressed. The genes of which the expression is induced before nitrogen fixation starts are called early nodulin genes (ENOD genes) (Nap et al., 1989). Different early nodulin genes have been isolated, mainly from pea and soybean, and *in situ* hybridization experiments to study their expression pattern, showed that the expression of individual nodulin genes marks specific stages of nodule development (Franssen et al., 1992).

The first identified early nodulin gene was ENOD2. ENOD2 gene expression is found in the nodule parenchyma (Van de Wiel et al., 1990) and starts at the nodule primordium stage when the nodule peripheral tissues begin to differentiate.

Genes specifically of interest to study initial steps of nodule formation are ENOD5 and ENOD12, because they are already expressed during the first steps of nodule

NATO ASI Series, Vol. H 81
Plant Molecular Biology
Edited by G. Coruzzi and P. Puigdomènech

formation. These genes might be infection related, because ENOD5 gene expression is found in root cells containing the infection thread tip and the ENOD12 gene is transcribed in cells containing the infection thread, in cells preparing for infection thread penetration and in the dividing root cortical cells (Scheres et al., 1990a; Scheres et al., 1990b). In mature indeterminate nodules, which have a persistent meristem, the ENOD5 and ENOD12 genes are expressed in partly overlapping zones of the central tissue. The ENOD5 gene is expressed in infected cells of the whole infection zone II, whereas ENOD12 expression is only found in the distal part of the infection zone II (Franssen et al., 1992). The ENOD5 and ENOD12 genes were initially identified in pea, but homologous genes have been identified in alfalfa (Hirsch et al., 1991; Pichon et al., 1992; Allison et al., 1993) (Barker, pers. comm) and vetch (Vijn, unpublished).

The expression of the ENOD40 gene is induced at another step of nodule formation than the ENOD5 and ENOD12 genes. Its expression is found in cells of the nodule primordia as well as in the pericycle of the root facing the nodule primordia (Yang et al., 1993). Since the expression of these nodulin genes marks specific early steps or stages of nodule development they lend themselves for studying mechanisms by which *Rhizobium* triggers nodule formation.

Besides specific gene expression in the host plant, several genes are specifically expressed in rhizobia during nodule formation. A set of genes that plays a pivotal role in different steps of nodulation are the *nod* genes. In fast growing *Rhizobium* species the *nod* genes are located on a large plasmid (pSym), whereas in the slow growing *Bradyrhizobium* species these genes are present on the chromosome (Long et al., 1989). In free-living bacteria the *nod* genes are not expressed, except the *nodD* gene. Flavonoids secreted by the root of the plants most likely bind to the NodD protein, and the complex acts as a transcriptional activator (Long et al., 1989; Goethals et al., 1992), which induces the transcription of the other *nod* genes (Peters et al., 1986; Redmond et al., 1986). These *nod* genes encode for proteins which are essential for the biosynthesis of the so-called Nod factors. The Nod factors are secreted by the bacteria and are able to achieve in the plant, even in the absence of the bacteria, the first steps of nodule formation, like root hair deformation, cytoplasmic rearrangements in the root outer cortex and cortical cell divisions in the inner cortex (Lerouge et al., 1990; Spaink et al., 1991; Fisher et al., 1992; Van Brussel et al., 1992).

The legume-*Rhizobium* symbiosis has a host specific nature, which implies that a distinct bacterial strain can only nodulate a particular set of host plants. Some bacterial strains have a narrow host range of one or a few plant species, others have a broad host range including many different plant species. The bacterial *nod* genes involved in determining the bacterial host range are the so-called host specific *nod* genes. The class of *nod* genes which are functionally exchangeable between *Rhizobium* species are called the common *nod* genes. The common Nod proteins accomplish the formation of the core molecule of the Nod factor, while the products of the host specific *nod* genes decorate this core molecule and make them most likely host plant specific (Lerouge et al., 1990; Spaink et al., 1991). The Nod factors of several *Rhizobium* species have been purified and their structure has been elucidated. The core molecule of all Nod factors are chitin oligomers, containing an acyl moiety at the non-reducing terminal sugar, but the decoration of these core molecules varies in the different species (Dénarié et al., 1992). The Nod factors produced by *R. leguminosarum* bv. *viciae* consist of a tetra- or pentamer of N-acetyl D-glucosamines with an unsaturated C18 fatty acid at the non-reducing terminal sugar. The acyl chain may either have one or four unsaturated bonds (Spaink et al., 1991). Furthermore the molecules contain an O-acetyl group at the non-reducing terminal sugar (Fig. 1).

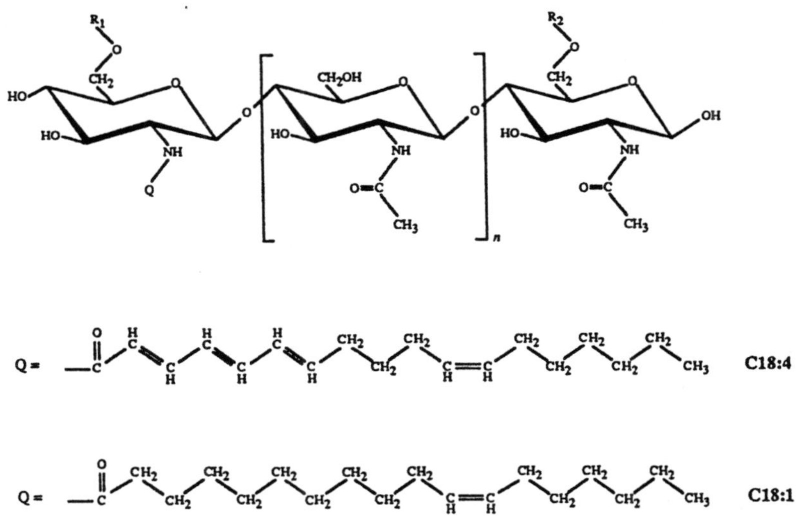

Fig. 1: Structure of the Nod-factors produced by *R. leguminosarum* bv. *viciae*.
n = 2 or 3; Q = C18:1 or C18:4; **R_1** = H or CH_3CO; **R_2** = H. When n = 3, Q = C18:4 and R_1= CH_3CO the nomenclature for the factor is: NodRlv-V(Ac, C18:4).

Thus different *Rhizobium* species produce structurally different Nod factors, but the process of nodule formation induced by the various Nod factors on their host plants is quite similar. The availability of purified Nod factors makes it possible to study the action of each of the Nod factors in different early steps of nodulation. Such a study requires a plant system in which the different steps of nodule formation can be readily induced with purified Nod factors and easily observed. We find that plants of the genus *Vicia* are most suitable host plants to study the effect of the Nod factors produced by *R. leguminosarum* bv. *viciae* as will be described below.

Vetch (*Vicia*): general aspects

Since the symbiosis between legume plants and rhizobia is a host specific interaction, as outlined in the introduction, the choice of the test plant is limited with each *Rhizobium* species. Working with *R. leguminosarum* bv. *viciae* gives the possibility to use plants of the genera *Pisum*, *Vicia*, *Lathyrus* and *Lens* as host plant. The interaction between pea (*Pisum sativum*) and *R. leguminosarum* bv. *viciae* has been studied in considerable detail in view of the agricultural and economic importance of this crop plant. Furthermore, the genetics of pea has been studied for a long time (Mendel, 1866) and pea has been often used as a model plant in plant physiological and biochemical research. For studying the different developmental steps of nodule formation on the other hand, pea is not always such a suitable host plant. It is for example quite a large plant and therefore not easily usable for nodulation tests with large numbers of different *Rhizobium* strains. Furthermore, the phenotypic changes characteristic for the first steps of the interaction, like root hair deformation and nodule primordium induction or the formation of short thick roots can not be easily observed in pea after application of Nod factors produced by *R. leguminosarum* bv. *viciae*.

Plants from the genus *Vicia* turned out not to have such disadvantages. The genus *Vicia* contains a number of plant species varying in size from very small (*V. tetrasperma*) to large (*V. narbonensis*). Especially the small species *V. hirsuta*, *V. tetrasperma*, *V. sativa* and *V. lathyroides* were found to be very suitable for nodulation tests with many different *Rhizobium* strains (Van Brussel et al., 1982). Such tests can easily be performed in test tubes. Nodules are formed four to seven days after inoculation with *R. leguminosarum* bv. *viciae*. Furthermore we have found that *V. sativa* is very useful in testing different biological activities of Nod factors, like root hair deformation, the formation of short thick roots or the induction of nodule primordia. With these assays it is possible to check separate steps in nodule formation and examine how each step depends on specific structural features of Nod factors.

Root hair deformation

A bioasssay for one of the first steps of the interaction between plant and rhizobia is the root hair deformation assay. A very useful method to study deformation of root hairs was developed by Bhuvaneswari (Bhuvaneswari et al., 1985). In her procedure white clover (*Trifolium repens* L.) seeds were germinated and seedlings were transferred to Fåhraeus slides. These are microscope slides covered with a coverglass by applying drops of silicon glue at the corners, which served as spacers as well. The slides can be filled with liquid plant culture medium and are incubated upright in a staining dish for 24 hours in a growth cabinet. Then the medium is decanted and the chamber filled with the solution to be tested for deformation activity. The slides are incubated in a moist chamber at room temperature and root hair deformation can be observed microscopically.

We have used this system with *V. sativa* and examined root hair deformation induced by Nod factors produced by *R. leguminosarum* bv. *viciae*. Four different Nod factors are produced by *R. leguminosarum* bv. *viciae* (NodRlv factors)(see Fig. 1). The factors have four or five glucosamine residues and can contain then either the highly unsaturated C18:4 fatty acid chain or the C18:1 chain. Root hair deforming capacity of these Nod factors was tested by adding only one of the purified Nod factors to the medium within the slides. All Nod factors were added in a concentration range of 10^{-8} - 10^{-13} M to the culture medium and root hair deformation could be observed within three hours with all four factors at concentrations of 10^{-12} or higher (Heidstra, pers. comm.). Not all the root hairs on the root deformed after adding the Nod factors. Only the hairs that had just obtained their mature size showed deformation.

Using this bioassay, the relation between the structure of the Nod factors and the induction of root hair deformation can be studied, particularly now Nod factors and their derivatives can be chemically synthesized (Nicolaou et al., 1992). Since all four NodRlv factors induced root hair deformation on *V. sativa* the total number of glucosamine residues (4 or 5) makes no difference for the induction of root hair deformation nor does the saturation level of the fatty acid chain. However, the presence of an acyl group is a necessity since chitin oligomers are unable to induce root hair deformation (Spaink et al., 1991)(Heidstra, unpublished). This suggests that also Nod factors produced by other *Rhizobium* strains, consisting of a glucosamine backbone and a fatty acid chain would be able to induce root hair deformation on vetch. The major Nod factor produced by *R. meliloti* (NodRm factors) consists of four glucosamine residues, containing an C16:2 fatty acid chain (Q) on the non-reducing terminal sugar and the reducing sugar is substituted with a sulphate group (R_2 = SO_3H Fig.1). The NodRm factors induce root hair deformation on its own host plant alfalfa, but these factors are not able to induce root hair deformation on vetch. After removal of the sulphate group the NodRm factors loose the ability to induce root hair deformation on alfalfa but they gain the ability to induce root hair deformation on vetch (Roche et al., 1991). This shows that the induction of root hair deformation is a host specific reaction, in which the additional groups added to the core molecule, in this case the sulphate group, are important.

Thick Short Root formation (Tsr)

A phenomenon that can be observed with *V. sativa* roots after inoculation with *R. leguminosarum* bv. *viciae* is the formation of Thick short roots (Tsr) (Van Brussel et al., 1982). This Tsr phenotype is not observed in *V. hirsuta* roots after inoculation with *R. leguminosarum* bv. *viciae.* After inoculation of *V. sativa* roots with a very low inoculum concentration of *R. leguminosarum* bv. *viciae* (0.5-5 bacteria / ml) the roots do not develop the Tsr phenotype and the nodules are formed on the primary root. Inoculation with a high inoculum concentration (4.9 x 10^5 / ml) causes the formation of abberant roots and nodulation is delayed for approximately three days, because the nodules are not formed on the primary root but on the sites of lateral-root emergence (Zaat et al., 1989). The formation of Tsr roots on *V. sativa* is not found when the plants are inoculated with strains mutated in the *nodD* gene or in any of the *nodA, B* or *C* genes. Apparently the proteins encoded by these genes are essential for the formation of Nod factors, showing that these compounds are involved in the induction of Tsr formation (Zaat et al., 1987).

Adding purified Nod factors produced by *R. leguminosarum* bv. *viciae* to *V. sativa* roots induced indeed the formation of Tsr (Spaink et al., 1991). Like root hair deformation, Tsr formation is induced by each of the four NodRlv factors at concentrations as low as 10^{-10} to 10^{-11} M, but not by the sulphated Nod factors of *R. meliloti.* On the other hand the non-sulphated form of the NodRm factor is capable to induce the Tsr phenotype (Roche et al., 1991), showing that for the induction of Tsr formation a specific Nod factor structure is required.

The induction of the Tsr phenotype is probably a side effect of the Nod factors, because Tsr formation is not a precondition for nodule formation. Most leguminous plants do not form Tsr roots after inoculation with rhizobia. The formation of Tsr on *V. sativa* can be inhibited with aminoethoxyvinylglycine (AVG), an ethylene synthesis inhibitor. Inhibition of Tsr results in earlier nodulation and the nodules are formed on the primary root (Zaat et al., 1989). Nevertheless, Tsr formation can be used in elucidating structure-function relations of Nod factors, because Tsr is a host specific reaction and a specific Nod factor structure is required.

Nodulin gene expression used as a molecular marker

The interaction between *Rhizobium* bacteria and their host plants involves the specific expression of nodulin genes already in an early stage of nodule formation. As foresaid these ENOD genes are expressed in a developmentally and spatially controlled manner.

It has been found that the expression of the ENOD5 and ENOD12 genes can be induced after inoculation of the plants with Nod factors (Horvath et al., 1993). In this study the expression of the ENOD5 and ENOD12 genes in root hairs of pea (*Pisum sativum*) plants was followed in time, after adding Nod factors produced by *R. leguminosarum* bv. *viciae* to the roots of the plants. Because the presence of ENOD5 and ENOD12 mRNA in root hairs could not be detected on Northern blots after inoculation of the plants with the NodRlv factors, PCR based experiments were used. Both factors NodRlv-V(Ac, C18:1) and NodRlv-V(Ac, C18:4) induce at a concentration of 10^{-8} and 10^{-9} M the expression of the ENOD5 and ENOD12 genes in a transient manner between 12 to 48 hours after addition of the factor. Upon adding each factor alone, NodRlv-V(Ac, C18:1) caused a delay of about 12 hours in the induction of gene expression compared to factor NodRlv-V(Ac, C18:4). If an equimolar mixture of all four NodRlv factors at a concentration of 10^{-8} M was added to the roots of the plants, both genes were expressed during the whole period that the plants were assayed. How the mixture of Nod factors can change the period of expression of the ENOD5 and ENOD12 genes is not understood.

In contrast to root hair deformation and Tsr formation which were not induced by the sulphated Nod factor of *R. meliloti*, the expression of both the ENOD5 and ENOD12 genes was induced after adding this factor. The maximal expression level of each of the two genes was found 48 hours after addition of the NodRm factors at a concentration of 10^{-8} to 10^{-10}, whereas the NodRlv-V(Ac, C18:4) factor induced maximal expression of ENOD12 at 12 hours and ENOD5 at 24 hours after inoculation. Thus NodRm factors which do not accomplish root hair deformation on pea, are able to induce early nodulin gene expression in these plants. The effect in pea is observed at the same concentration of Nod factors required for root hair deformation on alfalfa (Schultze et al., 1992). Apparently the induction of early nodulin gene expression is not a host specific reaction, like root hair deformation and Tsr formation, suggesting that the signal transduction pathway, starting with Nod factor recognition and ending in nodulin gene expression, is probably not the same as the pathway leading from Nod factor recognition to root hair deformation. On the other hand, it can not yet be excluded that the sulphate group is removed before the NodRm factor is recognized by the plant, causing the delay in induction of gene expression.

Induction of expression of the ENOD12 gene in transgenic alfalfa plants (*Medicago varia* A2), carrying a chimaeric *Medicago trunculata* ENOD12 promoter-ß-glucuronidase (GUS) gene revealed that already within 3 to 6 hours after inoculation of the plants with *R. meliloti* the ENOD12 gene is induced (Pichon et al., 1992). GUS expression was found in all epidermal cells in a region of the root that started just above the root tip, continued throughout the zone containing the emerging root hairs and terminated at the mature root hair zone. Studies in which Nod factors produced by *R. meliloti* are added to these transgenic alfalfa have not yet been reported.

Recently also vetch ENOD5, ENOD12 and ENOD40 cDNA clones have been isolated from a nodule cDNA library (Vijn, unpublished). So it will now be possible to examine nodulin gene expression in combination with bioassays like root hair deformation and Tsr formation.

Promoterstudies of early nodulin genes

Now it has been shown that Nod factors can induce the expression of the early nodulin genes, ENOD5 and ENOD12, the next question is how Nod factors induce this gene expression. What are the different steps involved in the signal transduction pathway, starting with the Nod factor recognition and ending in nodulin gene expression? Studies on specific expression of early nodulin genes encompass promoter analyses, which will result in the identification of cis -and trans-acting elements involved in the induction of expression of these genes. Until now, detailed promoter studies have only been performed with late nodulin genes like the leghaemoglobin genes of soybean (*Lbc3*) (Stougaard et al., 1987) and *Sesbania rostrata* (Szabados et al., 1990) and the soybean late nodulin gene N23 (Jørgensen et al., 1988). In these studies chimaeric genes consisting of the promoter of the nodulin gene and a reporter gene were transformed to *Lotus corniculatus* plants with *Agrobacterium rhizogenes* (Petit et al., 1987). These promoter analyses showed that for an efficient expression in nodules at least three different promoter regions are required; a "minimal promoter" containing the CAT and TATAA boxes and, further upstream, an organ specific element (OSE) and a strong positive element (SPE). The OSE of all three genes studied, contained two highly conserved DNA sequences, AAAGAT and CTCTT. Despite the identification of these cis-elements, no trans-acting factors have yet been isolated which interact with cis-elements and regulate nodulin gene expression.

Lotus corniculatus, the plant used in all these promoter studies, forms determinate root nodules after inoculation with *R. loti*. We are studying the interaction of *R. leguminosarum* bv. *viciae* with its host plants, vetch and pea, which both form indeterminate nodules. Determinate nodules are morphologically different from indeterminate nodules, because they lack a persistent meristem. Especially for promoter studies of genes involved in the early steps of nodulation, like ENOD5 and ENOD12, indeterminate nodules on the transgenic roots will be favourable, since the expression of these genes in such nodules can be studied over a much longer period of time than in determinate nodules. Even in mature indeterminate nodules ENOD5 and ENOD12 gene expression can be observed (Scheres et al., 1990a; Scheres et al., 1990b; Franssen et al., 1992). Recently a *A. rhizogenes* based transformation system has been developed for *V. hirsuta* by the group of Pühler (pers. comm.). In this system *V. hirsuta* plants are stem-inoculated with *A. rhizogenes* resulting in the formation of hairy roots. After inoculation of these hairy roots with *R. leguminosarum* bv. *viciae* transgenic indeterminate nodules were formed and transgenic nodules were obtained within four to six weeks. Therefore this system appears to be suitable for promoter studies of early nodulin genes. Working backwards in the signal transduction pathway, the identification of cis-elements might help in finding trans-acting factors involved in the induction of this specific gene expression and in that way contribute to the elucidation of a step in the signal transduction pathway elicited by the Nod factors.

Induction of nodule primordia in vetch roots

After inoculation of vetch plants with *R. leguminosarum* bv. *viciae* the bacteria induce cortical cell divisions in the root inner cortex. These dividing cells form the nodule primordium. The Nod factors produced by *R. leguminosarum* bv. *viciae* have also the ability to induce cortical cell divisions in vetch roots (Spaink et al., 1991). To study the mitotis inducing activity of the NodRlv factors a special assay has been developed (Spaink et al., 1991). In this system the *V. sativa* roots are grown in amber vials, to shield the roots from the light, because otherwise Tsr formation is induced. If an equimolar mixture of the four NodRlv factors at a concentration of 10^{-7}-10^{-9} M is applied to *V. sativa* plants, cortical cell divisions can be observed after 4 to 6 days (Spaink et al., 1991) (Heidstra, unpublished). When purified NodRlv-V(Ac, C18:4) at a concentration of 10^{-7}-10^{-9} M is added to *V. sativa* plants these structures are formed as well, whereas NodRlv-V(Ac, C18:1) does not induce cortical cell divisions (Spaink et al., 1991) (Moerman, pers. comm.). This suggests that the structure of the acyl chain is important for the induction of cell divisions.

An important question to answer is whether these Nod factor induced clusters of dividing cortical cells have the same characteristics as the primordia induced by the bacteria? The nodule primordia induced by the NodRlv-V(Ac, C18:4) factor or the NodRlv mixture are induced in the inner cortical cell layers preferentially opposite a proto-xylem pole, just as *R. leguminosarum* bv. *viciae* does. Furthermore *in situ* hybridization showed the expression of the early nodulin genes ENOD12 and ENOD40. ENOD12 gene expression was found in the dividing cortical cells, whereas ENOD40 gene expression was found in these dividing cortical cells, as well as in the pericycle facing the nodule primordium (Vijn et al., 1993). These specific expression patterns are identical to those in nodule primordia induced by the *Rhizobium* bacteria. This shows that in these early stages the nodule primordia induced by the bacteria are not distinguishable from the nodule primordia induced by the purified Nod factors. However, the nodule primordia induced by the NodRlv factors in vetch roots never developed further into a mature nodule, in contrast to alfalfa, where complete nodule structures are induced by the Nod factors of *R. meliloti* (Truchet et al., 1991). In what stage of development these NodRlv factor induced structures in *V. sativa* are arrested is not known at the moment. To find an answer to this question it will be interesting to study if the ENOD2 gene, which marks the differentiation of the nodule peripheral tissues, is expressed in these NodRlv factors induced primordia.

Increase of *nod* gene inducing activity (Ini)

When *V. sativa* roots are inoculated with *R. leguminosarum* bv. *viciae* an increase of *nod* gene inducing activity, called Ini, of the root exudate can be observed after four days of cocultivation (Van Brussel et al., 1990). This is caused by eight additional flavonoids secreted by the plant. All eight flavonoids are able to stimulate the expression of *nod* genes of *R. leguminosarum* bv. *viciae* (Recourt et al., 1991). The Ini effect is not only observed in the symbiosis between *V. sativa* and *R. leguminosarum* bv. *viciae*, but also in *Trifolium repens* (white clover) inoculated with *R. leguminosarum* bv. *trifolii*. But this increase was also seen when white clover seedlings were inoculated with *R. leguminosarum* bv. *viciae* (Recourt et al., 1991). In *V. sativa* Ini seems a host specific reaction, since after four days of cocultivation of *V. sativa* with *R. leguminosarum* bv. *trifolii* this respons could not be observed (Van Brussel et al., 1990).

Ini can also be observed after the addition of purified NodRlv factors to *V. sativa* roots, but only factors containing both the O-acetyl group and the highly unsaturated acyl moiety (NodRlv-IV or V(Ac, C18:4)) at the non-reducing sugar were able to produce Ini on *V. sativa* roots (Spaink et al., 1991). The lowest concentration of NodRlv factor that could induce Ini in *V. sativa* roots is 10^{-8} M. This concentration is higher than the concentration at which the Nod factors are able to induce root hair deformation, Tsr formation, cortical cell divisions and the expression of the early nodulin genes. In *V. sativa* roots Ini is only induced when the roots are grown in the light, and Ini induction can be inhibited, just like Tsr formation, with aminoethoxyvinylglycine (AVG) (Van Brussel, unpublished). This suggests that the increase of *nod* gene inducing activity of the root exudate induced by *R.leguminosarum* bv. *viciae* or the NodRlv factors is a side effect as well, like Tsr formation.

Concluding remarks

The *V. sativa* and *R. leguminosarum* bv. *viciae* system has several advantages for studying the role of Nod factors in the *Rhizobium*-legume symbiosis. Nod factors induce in *V. sativa* roots various effects which can be easily observed, like root hair deformation, the formation of thick short roots, the formation of nodule primordia and an increase of *nod* gene inducing activity. Convenient assays are available for each of these activities. Some of the effects like root hair deformation and formation of nodule primordia correspond to steps in the development of root nodules as achieved by *R. leguminosarum* bv. *viciae* bacteria. Other effects like the formation of thick short roots or the increase of *nod* gene inducing activity seem to be rather side effects. Also *R. leguminosarum* bv. *viciae* is able to induce these effects, but the formation of thick short roots is only induced at a high inoculum concentration and the increase of *nod* gene inducing activity is only induced in the light and both effects can be inhibited with aminoethoxyvinylglycine (AVG). But in all cases the effects are important in studying the structural requirements of Nod factors for the induction of each of the activities.

Results so far showed that for the induction of root hair deformation and Tsr formation the structural requirements are identical. An additional sulphate group at the reducing terminal sugar completely abolishes the induction of both events. However, the structure of the acyl group is not important, but the presence of an acyl chain is a necessity. The most striking results from the studies on the induction of early nodulin gene expression are that the sulphated factors produced by *R. meliloti* are able to induce the expression of the ENOD5 and ENOD12 genes, suggesting that the signal transduction pathway from Nod factor to root hair deformation is a different pathway than from Nod factor to early nodulin gene expression, but it can not yet be excluded that the sulphate group is removed before induction. Studies on the mitotic activity of the NodRlv factors showed the importance of the acyl chain structure. The NodRlv factors containing the highly unsaturated fatty acid chain (C18:4) were able to induce cortical cell divisions, in contrast to the NodRlv factors with the C18:1 acyl chain.

To obtain more insight in the complex signal transduction pathways involved in nodule formation, more detailed structure-function analyses of Nod factors have to be performed. The isolation of vetch ENOD5, ENOD12 and ENOD40 cDNA clones provides the opportunity to combine studies on early nodulin gene expression with bioassays for the biological activity of the Nod factors, like root hair deformation or the induction of nodule primordia. This will make it possible to obtain an accurate correlation between Nod factor induced phenomena and early nodulin gene expression and might provide additional information on the functions of the proteins encoded by these genes.

The recently developed *A. rhizogenes* transformation system for *V. hirsuta* provides the possibility to work backwards in the signal transduction pathways from Nod factors to the induction of certain early nodulin genes since it will be possible to identify transcription factors that are involved in the expression of these nodulin genes.

In the main, vetch can be used in many more bioassays to study Nod factor activity than pea. Pea however, is probably more suitable for studies aimed at the identification of the function of the plant genes involved in nodule formation, because the pea genetics has been studied extensively and various pea mutants disturbed in the interactions with *Rhizobium* are available. Of the vetch genetics is not that much known. Therefore it will be necessary to use both host plants in unravelling the mechanisms involved in NodRlv factor induced nodule formation.

References

Allison, L.A., Kiss, G.B., Bauer, P., Poiret, M., Pierre, M., Savouré, A., Kondorosi, E., Kondorosi, A. (1993). Identification of two early nodulin genes with homology to members of the pea ENOD12 gene family. Plant Molecular Biology **21**: 375-380.

Bhuvaneswari, T.V., Solheim, B. (1985). Root hair deformation in the white clover/*Rhizobium trifolii* symbiosis. Physiol. Plant. **63**: 25-34.

Dénarié, J., Debellé, F., Rosenberg, C. (1992). Signaling and host range variation in nodulation. Annu. Rev. Microbiology **46**: 497-531.

Fisher, R.F., Long, S.R. (1992). *Rhizobium*-plant signal exchange. Nature **357**: 655-660.

Franssen, H.J., Vijn, I., Yang, W.C., Bisseling, T. (1992). Developmental aspects of the *Rhizobium*-legume symbiosis. Plant Molecular Biology **19**: 89-107.

Goethals, K., Van Montagu, M., Holsters, M. (1992). Conserved motifs in a divergent *nod* box of *Azorhizobium caulinodans* ORS571 reveal a common structure in promoters regulated by LysR-type proteins. Proc. Natl. Acad. Sci. USA **89**: 1646-1650.

Hirsch, A.M., Bochenek, B., Löbler, M., Mckhann, H.I., Reddy, A., Li, H.H., Ong, M., Wong, J. (1991). Patterns of nodule development and nodulin gene expression in alfalfa and Afghanistan pea. In: Advanches in molecular genetics of plant-microbe interactions. Hennecke, H., Verma, D.P.S. (Eds.). Kluwer Academic Publishers., Dordrecht, 317-324.

Horvath, B., Heidstra, R., Lados, M., Moerman, M., Spaink, H.P., Promé, J.-C, Van Kammen, A., Bisseling, T. (1993). Lipo-oligosaccharides of *Rhizobium* induce infection related early nodulin gene expression in pea root hairs. The Plant Journal. In press.

Jørgensen, J.-E., Stougaard, J., Marcker, A., Marcker, K.A. (1988). Root nodule specific gene regulation: Analysis of the soybean nodulin N23 gene promoter in heterologous symbiotic systems. Nucl. Acids Res. **16**: 39-50.

Lerouge, P., Roche, P., Faucher, C., Maillet, F., Truchet, G., Promé, J.-C., Dénarié, J. (1990). Symbiotic host-specificity of *Rhizobium meliloti* is determined by a sulphated and acetylated glucosamine oligosaccharide signal. Nature **344**: 781-784.

Long, S.R. (1989). *Rhizobium*-legume nodulation: life together in the underground. Cell **56**: 203-214.

Mendel, G. (1866). Versuche über Pflanzen-Hybriden. Verh. Naturf. Vereins Brünn **IV**: 3-47. (Reprinted 1951 in J. Hered. **42**, 3-47).

Nap, J.P., Bisseling, T. (1989). Nodulin function and nodulin gene regulation in root nodule development. In: The molecular biology of symbiotic nitrogen fixation. Gresshoff, P.M. (Eds.). CRC Press, Florida, 181-229.

Newcomb, W. (1976). A correlated light and electron microscopic study of symbiotic growth and differentiation of pea root nodules. Canadian Journal of Botany **54**: 2163-2186.

Newcomb, W. (1981). Nodule morphogenesis and differentiation. In: Biology of Rhizobiaceae. Giles, K.L., Atherley, A.G. (Eds.). Academic Press, New York, 247-298.

Nicolaou, K.C., Bockovich, N.J., Carnague, D.R., Hummel, C.W., Even, L.F. (1992). Total synthesis of the NodRm-IV factors, the *Rhizobium* nodulation signals. J. Am. Chem. Soc. **114**: 8701-8702.

Peters, N.K., Frost, J.W., Long, S.R. (1986). A plant flavone, luteolin, induces expression of *Rhizobium meliloti* nodulation genes. Science **233**: 977.

Petit, A., Stougaard, J., Kühle, A., Marcker, K.A., Tempé, J. (1987). Transformation and regeneration of the legume *Lotus corniculatus*: A system for molecular studies of symbiotic nitrogen fixation. Molecular and General Genetics **207**: 245-250.

Pichon, M., Journet, E-P., Dedieu, A., de Billy, F., Truchet, G., Barker, D. (1992). *Rhizobium meliloti* elicits transient expression of the early nodulin gene ENOD12 in the differentiating root epidermis of transgenic alfalfa. The Plant Cell **4**: 1199-1211.

Recourt, K., Schripsema, J., Kijne, J.W., Van Brussel, A.A.N., Lugtenberg, B.J.J. (1991). Inoculation of *Vicia sativa* subsp. *nigra* roots with *Rhizobium leguminosarum* biovar *viciae* results in release of *nod* gene activating flavanones and chalchones. Plant Molecular Biology **16**: 841-852.

Redmond, J.W., Batley, M., Djordjevic, M.A., Innes, R.W., Huempel, P.L., Rolfe, B.G. (1986). Flavones induce expression of nodulation genes in *Rhizobium*. Nature **323**: 632-633.

Roche, P., Debellé, F., Maillet, F., Lerouge, P., Faucher, C., Truchet, G., Dénarié, J., Promé, J.-C. (1991). Molecular basis of symbiotic host specificity in *Rhizobium meliloti*: *nodH* and *nodPQ* genes encode the sulfation of lipo-oligosaccharide signals. Cell **67**: 1131-1143.

Scheres, B., Van de Wiel, C., Zalensky, A., Horvath, B., Spaink, H., Van Eck, H., Zwartkruis, F., Wolters, A.M., Gloudemans, T., Van Kammen, A., Bisseling, T. (1990a). The ENOD12 gene product is involved in the infection process during pea-*Rhizobium* interaction. Cell **60**: 281-294.

Scheres, B., Van Engelen, F., Van der Knaap, E., Van de Wiel, C., Van Kammen, A., Bisseling, T. (1990b). Sequential induction of nodulin gene expression in the developing pea nodule. Plant Cell **8**: 687-700.

Schultze, M., Quiclet-Sire, B., Kondorosi, E., Virelizier, H., Glushka, J.N., Endre, G., Gero, D.S., Kondorosi, A. (1992). *Rhizobium meliloti* produces a family of sulfated lipo-oligosaccharides exhibiting different degrees of plant host specificity. Proc. Natl. Acad. Sci. USA **89**: 192-196.

Spaink, H.P., Sheeley, D.M., Van Brussel, A.A.N., Glushka, J., York, W.S., Tak, T., Geiger, O., Kennedy, E.P., Reinhold, V.N., Lugtenberg, B.J.J. (1991). A novel highly unsaturated fatty acid moiety of lipo-oligosaccharide signals determines host specificity of *Rhizobium*. Nature **354**: 125-130.

Stougaard, J., Sandal, N.N., Grøn, A., Köhle, A., Marcker, K.A. (1987). 5' analysis of the soybean leghemoglobin gene in root nodules of transgenic *Lotus corniculatus*. EMBO Journal **6**: 3565-3569.

Szabados, L., Ratet, P., Grunenberg, B., De Bruijn, F.J. (1990). Functional analysis of the *Sesbania rostrata* leghemoglobin *glb* gene 5' upstream region in transgenic *Lotus corniculatus* and *Nicotiana tabacum* plants. Plant Cell **2**: 973-986.

Truchet, G., Roche, P., Lerouge, P., Vasse, J., Camut, S., de Billy, F., Promé, J.-C., Dénarié, J. (1991). Sulphated lipo-oligosaccharide signals of *Rhizobium meliloti* elicit root nodule organogenesis in alfalfa. Nature **351**: 670-673.

Van Brussel, A.A.N., Tak, T., Wetselaar, A., Pees, E., Wijffelman, C.A. (1982). Small leguminosae as test plants for nodulation of *Rhizobium leguminosarum* and other rhizobia and agrobacteria harbouring a leguminosarum Sym-plasmid. Plant Science Letters **27**: 317-325.

Van Brussel, A.A.N., Recourt, K., Pees, E., Spaink, H.P., Tak, T., Kijne, J.W., Lugtenberg, B.J.J. (1990). A biovar-specific signal of *Rhizobium leguminosarum* biovar *viciae* induces increased nodulation gene-inducing activity in root exudate of *Vicia sativa* subsp. *nigra*. Journal of Bacteriology **172**: 5394-5401.

Van Brussel, A.A.N., Bakhuizen, R., Van Spronsen, P.C., Spaink, H.P., Tak, T., Lugtenberg, B.J.J., Kijne, W.J. (1992). Induction of pre-infection thread structures in the leguminous host plant by mitogenic lipo-oligosaccharides of *Rhizobium*. Science **257**: 70-72.

Van de Wiel, C., Scheres, B., Franssen, H., Van Lierop, M.J., Van Lammeren, A., Van Kammen, A., Bisseling, T. (1990). The early nodulin transcript ENOD2 is located in the nodule parenchyma (inner cortex) of pea and soybean root nodules. EMBO Journal **9**: 1-7.

Van Kammen, A. (1984). Suggested nomenclature for plant genes involved in nodulation and symbiosis. Plant Molecular Biology Reporter **2**: 43-45.

Vijn, I., das Neves, L., Van Kammen, A., Franssen, H., Bisseling, T. (1993). Nod factors and nodulation in plants. Science **260**: In press.

Yang, W.C., Katinakis, P., Hendriks, P., Smolders, A., De Vries, F., Spee, J., Van Kammen, A., Bisseling, T., Franssen, H. (1993). Characterization of GmENOD40, a gene showing novel patterns of cell-specific expression during soybean nodule development. The Plant Journal **3**: 573-585.

Zaat, S.A.J., Van Brussel, A.A.N., Tak, T., Pees, E., Lugtenberg, B.J.J. (1987). Flavonoids induce *Rhizobium leguminosarum* to produce *nodDABC* gene-related factors that cause thick, short roots and root hair responses on common vetch. Journal of Bacteriology **169**: 3388-3391.

Zaat, S.A.J., Van Brussel, A.A.N., Tak, T., Lugtenberg, B.J.J., Kijne, J.W. (1989). The ethylene-inhibitor aminoethoxyvinylglycine restores normal nodulation by *Rhizobium leguminosarum* biovar. *viciae* on *Vicia sativa* subsp. *nigra* by suppressing the 'Thick and short roots' phenotype. Planta **177**: 141-150.

Migratory Determinants of Parasitism

Steven P. Briggs & Gurmukh S. Johal*
Pioneer Hi-Bred International, Inc.
7250 NW 62nd Ave.
Johnston, IA 50131
USA

Molecularly-induced changes in both the host and the parasite determine the outcome of their interaction. What are the host-produced molecules that cause determinative changes in the parasite? What are the parasite-produced molecules that cause determinative changes in the host? The answers to these questions form the foundation for understanding host-parasite interactions. The migration of a factor from one organism to the other is an attribute that distinguishes events occurring between organisms from those occurring within an organism. We will restrict our discussion to the migratory determinants of parasitism.

The diversity of molecular and cellular factors that are changed by the interaction between host and parasite appear to be impenetrably complex. Fortunately, this diversity can be reduced to a few functional categories. The first criterion that permits simplification is the requirement that a factor play a determinative role in the interaction. It is no surprise that many things change when a parasite invades it's host so change alone is not a reliable indicator of significance. Conversely, there are no set rules for establishing significance and most cases are based upon a set of corroborating observations. We will describe one plant disease to illustrate how conclusive evidence for a role in the interaction can be obtained. We also present a definition of five categories into which all diffusible determinants can be placed, so that molecules can be compared by the role which they play rather than by structure or mechanism of action. This categorization simplifies and unifies descriptions of host-parasite interactions. We hope that this focus will help to establish principles without being procrustean. The outcome of the interaction is also dependent upon the targets of

---

* University of Missouri, 205 Curtis Hall, Columbia, MO 65211 USA

NATO ASI Series, Vol. H 81
Plant Molecular Biology
Edited by G. Coruzzi and P. Puigdomènech

migratory factors or by molecules that interfere with the action of migratory factors. Since these agents can be defined in terms of their interactions with migratory determinants, they will not be discussed explicitly.

Cochliobolus carbonum race 1 causes a leaf blight and ear mold of maize. This ascomycete (imperfect stage: *Bipolaris zeicola*, formerly *Helminthosporium carbonum*) can infect all parts of the plant but under field conditions is a foliar and ear molding pathogen. Typical symptoms include expanding lesions on the leaf blade and sheath, containing concentric ovals of dark pigment. The carbonaceous appearance of the ear mold is caused by a black mycelium and conidia. Two other races of *C. carbonum* have been documented. Race 2 is thought to be the progenitor of races 1 and 3, occupying a niche formed when leaves have begun to senesce but are not yet substrates for saprophytes (Scheffer 1989). Race 3 infects non-senescent tissues but has much less virulence than race 1. Single genes for resistance to races 2 and 3 have not been described.

R. P. Scheffer and co-workers discovered a factor in culture filtrates of race 1 that affects only genotypes of maize that are susceptible to race 1 infection (Scheffer and Livingston 1984). Their thinking was influenced by the concepts of toxicology (Scheffer 1989) so their assays were designed to detect factors that inhibit the root growth of seedlings. They named this factor HC-toxin, a designation that reflects the concept of the assay more than the role played by the factor in parasitism. The role of HC-toxin was established by two experiments. In the first, HC-toxin production and the ability to infect were found to co-segregate as a single locus (Scheffer, Nelson et al. 1967). In the second experiment, avirulent, non-producers whose growth on susceptible tissue had stopped were observed to resume colonization following application of HC-toxin (Comstock and Scheffer 1973). These experiments showed that HC-toxin is the determinative factor that permits race 1 to infect maize. Additionally, HC-toxin determines the genotype host-specificity of the pathogen, based upon sensitivity to HC-toxin.

Resistance to race 1 and insensitivity to HC-toxin co-segregate with dominant alleles of the *hm1* locus on chromosome 1 in maize (Nelson and Ullstrup 1964; Scheffer and Ullstrup 1965). The difference in sensitivity between resistant and susceptible seedlings is about 100-fold. A second locus, *hm2*, on chromosome 9 also provides resistance to race 1 but the effect of *hm2* on sensitivity to HC-toxin is not known. The most common allele of *Hm1* is fully dominant and provides constitutive resistance. In contrast, both *Hm2* and *Hm1-A* are co-dominant and their pattern of expression is unusual: throughout the plant, full resistance develops rapidly from nearly full susceptibility as the tassels emerge from the whorl. The expression of the genes appears to be controlled by the developmental state of the organism rather than by the state of an organ or tissue.

Histological studies indicate that HC-toxin does not kill in advance of the growing hyphae (Comstock and Scheffer 1973) and there is no evidence that HC-toxin contributes to symptom development. The target for HC-toxin is unknown. Stimulation of root growth (Kuo, Yoder et al. 1970) and enhanced protoplast viability (Wolf and Earle 1991) are two of the effects of HC-toxin on susceptible maize.

Meeley and Walton discovered an enzyme, HC-toxin reductase (HCTR), that converts HC-toxin to an inactive derivative (Meeley and Walton 1991). Genetic studies revealed that HCTR activity co-segregates with *Hm1*. Several *hm1* mutants were tested and all were found to have lost HCTR activity (Meeley, Johal et al. 1992). This proved that one gene (*Hm1*) controls both resistance and HCTR activity and established a causal relationship between the two, defining the mechanism of disease resistance. The *Hm1* gene was cloned and found to encode HCTR (Johal and Briggs 1992).

Because the role of HC-toxin is to permit infection, we call it a *compatibility factor*. This is one of five functional categories of factors that mediate host-parasite intercourse. Parasitism has not been studied in sufficient detail to reveal whether or not every type of factor is involved in every host-parasite combination. However, there is conclusive evidence for factors from each of the following categories playing a determinative role in at least some interactions. Pathogen-produced **compatibility factors** alter host physiology so that colonization is possible; **virulence factors** alter host physiology so that disease develops; and **avirulence factors** alter host physiology so that colonization is restricted. Host-produced **host-factors** alter parasite physiology so that compatibility factors or virulence factors are produced; **defense factors** alter parasite physiology so that colonization is restricted. These are functional, rather than mechanistic, definitions that describe key roles played by molecular determinants of host-parasite interactions. Some determinants may play more than one role in a given interaction or different roles in different interactions. Selected examples of these factors are listed in Table 1.

We believe that all parasites must produce compatibility factors to colonize their host(s). An alternative is that parasites avoid detection by failing to produce substances that can be recognized by the host. In our view, the likelihood of a microorganism growing within it's host and not producing substances that can trigger defense seems remote. Furthermore, the list of known compatibility factors is growing rapidly and now includes examples from viruses, bacteria, and fungi. Some of the best known examples are the virus p30 movement proteins, the T-DNA of *Agrobacterium*, the Nod factors of *Rhizobium*, and the host-selective toxins produced by *Cochliobolus victoriae* and *Cochliobolus carbonum* race 1. Evidence for a

Table 1. Selected examples of migratory factors in parasitism.

Compatibility factors	**Produced by**	**Host**	**Reference**
HC toxin	Cochliobolus carbonum race 1	maize	(Walton and Panaccione 1993)
BZR toxin	Cochliobolus carbonum race 3	maize	(Xiao, Tsuge et al. 1992)
HV toxin	Cochliobolus victoriae	oats	(Wolpert and Macko 1989)
PC toxin	Periconia circinata	sorghum	(Macko, Stimmel et al. 1992)
Alternaria toxins	Alternaria alternata	various dicots	(Otani, Kohmoto et al. 1991)
spore germination fluid	Bipolaris oryzae	rice	(Xiao, Tsuda et al. 1990)
Extracellular protease	Pyrenopeziza brassicae	oilseed rape	(Ball, Ashby et al. 1991)
Extracellular protease	Pseudomonas lachrymans	cucumber	(Keen, Williams et al. 1967)
Oxalic acid	Sclerotinia sclerotiorum	Phaseolus vulgaris	(Godoy, Steadman et al. 1990)
FOL1 toxin	Fusarium oxysporum f. sp. lycopersici	tomato	(Sutherland and Pegg 1992)
Nod factors	Rhizobium spp.	legumes	(see text)
T-DNA	Agrobacterium tumefaciens	broad range	(Zambryski 1992)
30-kDa movement protein	tobacco mosaic virus	tobacco	(Jong and Ahlquist 1992)
Virulence factors	**Produced by**	**Host**	
HMT toxin	Cochliobolus heterostrophus race T	maize	(Braun, Siedow et al. 1990)
PTR toxin	Pyrenophora tritici-repentis	wheat	(Ballance, Lamari et al. 1989)
ET toxin	Exserohilum turcicum	maize	(Bashan and Levy 1992)
Pectic enzymes	Erwinia carotovora	potato	(Collmer and Keen 1986)
Pseudomonas toxins	Pseudomonas syringae	broad range	(Gross 1991)
Avirulence factors	**Produced by**	**Host**	
avrD product	Pseudomonas syringae pv. tomato	[soybean]	(Keen, Tamaki et al. 1990)
avrLr20 product	Puccinia recondita	wheat	(Jones and Deverall 1978)
avr9 product	Cladosporium fulvum	tomato	(Wit 1992)
Host factors	**Produced by**	**Parasite**	
Flavonoids	legumes	Rhizobium spp.	(see text)
Phenolics	broad range	Agrobacterium spp.	(Peters and Verma 1990)
Defense factors	**Produced by**	**Parasite**	
Cyclic hydroxamates	maize	Helminthosporium turcicum	(Couture, Routley et al. 1971)

compatibility factor from *Pseudomonas syringae* has been reported (Jakobek, Smith et al. 1993). The discovery of compatibility factors dates back to 1933 when Tanaka described AK-toxin, which permits *Alternaria kikuchiana* to infect Japanese pear (Scheffer 1989).

Virulence factors are only produced by pathogenic strains of parasites. Symbionts such as *Rhizobium* can produce virulence factors, in which case they are pathogenic (Giovanelli, Owens et al. 1973). Conversely, aggressive pathogens such as *Colletotrichum* can lose their pathogenicity without losing the ability to colonize their host, suggesting the loss of a virulence factor (Freeman and Rodriguez 1993). The observation that tabtoxin functions as a compatibility factor for *Pseudomonas syringae* BR2 on bean but as a virulence factor for *P. syringae* pv. *tabaci* 11528 on tobacco (Kinscherf, Coleman et al. 1991) underscores the importance of using functional roles to categorize determinants. It will be interesting to learn whether other virulence factors can also function as compatibility factors.

Avirulence factors have been characterized only recently. Production is governed by avirulence genes. Some factors, such as the *avr9* elicitor from *Cladosporium fulvum*, are encoded directly by avirulence genes (Wit 1992). Others, such as the *avrD* elicitor from *P. syringae* pv. *tomato*, are secondary products (Keen, Tamaki et al. 1990). These factors cause rapid host cell death and the induction of genes that produce defense factors. The relationships between host cell death, defense gene induction, and arrest of pathogen growth are unclear. Most elicitors lack specificity and do not function as avirulence factors.

Host factors have only been documented for infections that are caused by soil-borne bacteria. Flavonoids in the root exudates of legumes induce the *nod* genes of *Rhizobium* resulting in production of the Nod compatibility factors (Roche, Debelle et al. 1991; Spaink, Sheeley et al. 1991; Truchet, Roche et al. 1991; Brussel, Bakhuizen et al. 1992; Fisher and Long 1992). Similarly, phenolics secreted by host plants induce the *vir* genes of *Agrobacterium* resulting in the transfer of the T-DNA compatibility factor (Peters and Verma 1990). Given the importance of chemical signalling in the rhizosphere, it seems likely that host factors will be discovered that interact with many other classes of parasites. For example, the same isoflavones in soybean root exudates that induce the *nod* genes of *Bradyrhizobium japonicum* induce encystment and germination of *Phytophthora sojae* zoospores (Morris and Ward 1992).

Defense factors have received more study than factors in the other four categories but, ironically, they have the least direct experimental support. The strongest case in the literature for an inducible defense factor is based upon the disease of pea caused by *Nectria haematococca* mating population IV (VanEtten, Matthews et al. 1989). Genetic tests established that fungal virulence was quantitatively correllated with the production of pisatin demethylase, an enzyme that inactivates the pea-produced phytoalexin, pisatin. Pisatin inhibits fungal growth and is accumulated during infection. These observations strongly suggest that pisatin demethylase is a compatibility or virulence factor and, by inference, that pisatin is a defense factor. However, recent experiments reveal that disruption of the gene that encodes pisatin demethylase does not eliminate virulence (H. D. VanEtten, personal communication). This reopens the question: is pisatin a defense factor? A genetic study of phytoalexin production in *Arabidopsis* should eventually clarify matters but the results at this point are equivocal; one mutant that makes no camalexin retains full resistance but two others are only partially resistant (F. M. Ausubel, personal communication). Ectopic production of the grapevine phytoalexin, resveratrol, in tobacco diminished the incidence of disease (Hain, Reif et al. 1993). Overproduction of chitinase (typically, an inducible PR protein) enhanced resistance (Broglie, Chet et al. 1991). It is possible that none of these inducible factors are necessary for resistance but that each can be significant in some circumstances. We suspect that the restriction of lesion development in *compatible* interactions is mediated by inducible defense factors; perhaps this is their primary role.

Constitutive defense factors, while less studied, are more clearly established as significant determinants of resistance. The *Bx1* gene in maize is required for production of cyclic hydroxamates; non-producing mutants are more susceptible to *Helminthosporium turcicum* and *Ostrinia nubilalis* (Couture, Routley et al. 1971).

We encourage the reader to consider the myriad other factors that have been described to pass between host and parasite. Are there any that cannot be assigned to one of the five categories of determinants described here? Of particular importance are compatibility factors because these are the most fundamental determinants of whether or not parasitism can occur. Disrupting the action of compatibility factors should provide stable disease resistance insofar that resistance can be overcome only by the pathogen gaining a new function. There is no doubt that new functions (i.e., new compatibility factors) will evolve but this should take longer than it would to overcome resistance that is based upon avirulence factors. Pathogens can overcome this latter (gene-for-gene) resistance by simply ceasing to elaborate the avirulence factor. Greater efforts to identify compatibility factors are justified by the likelihood that their study will have the highest probability of producing durable disease resistance.

Literature references

Ball, A. M., A. M. Ashby, et al. (1991). "Evidence for the requirement of extracellular protease in the pathogenic interaction of Pyrenopeziza brassicae with oilseed rape." Physiological and Molecular Plant Pathology **38**: 147-161.

Ballance, G. M., L. Lamari, et al. (1989). "Purification and characterization of a host-selective necrosis toxin from Pyrenophora tritici-repentis." Physiological and Molecular Plant Pathology **35**(3): 203-214.

Bashan, B. and Y. Levy (1992). "Differential Response of Sweet Corn Cultivars to Phytotoxic Water-Soluble Compounds from Culture Filtrates of Exserohilum turcicum." Plant Disease **76**: 451-454.

Braun, C. J., J. N. Siedow, et al. (1990). "Fungal toxins bind to the URF13 protein in maize mitochondria and Escherichia coli." The Plant Cell **2**: 153-161.

Broglie, K., I. Chet, et al. (1991). "Transgenic Plants with Enhanced Resistance to the Fungal Pathogen Rhizoctonia solani." Science **254**: 1194-1197.

Brussel, A. A. N. v., R. Bakhuizen, et al. (1992). "Induction of Pre-Infection Thread Structures in the Leguminous Host Plant by Mitogenic Lipo-Oligosaccharides of Rhizobium." Science **257**: 70-72.

Collmer, A. and N. Keen (1986). The role of pectic enzymes in plant pathogenesis. Annual Review of Phytopathology. Palo Alto, Annual Reviews, Inc. 383-409.

Comstock, J. C. and R. P. Scheffer (1973). "Role of host-selective toxin in colonization of corn leaves by Helminthosporium carbonum." Phytopathology **63**: 24-29.

Couture, R. M., D. G. Routley, et al. (1971). "Role of cyclic hydroxamic acids in monogenic resistance of maize to Helminthosporium turcicum." Physiological Plant Pathology **1**: 515-521.

Fisher, R. F. and S. R. Long (1992). "Rhizobium-plant signal exchange." Nature **357**: 655-660.

Freeman, S. and R. J. Rodriguez (1993). "Genetic conversion of a fungal plant pathogen to a non-pathogenic, endophytic mutualist." Science **260**: 75-78.

Giovanelli, J., L. D. Owens, et al. (1973). "B-cystathionase. In vivo inactivation by rhizobitoxine and role of the enzyme in methionine biosynthesis in corn seedlings." Plant Physiology **51**: 492-503.

Godoy, G., J. R. Steadman, et al. (1990). "Use of mutants to demonstrate the role of oxalic acid in pathogenicity of Sclerotinia sclerotiorum on Paseolus vulgaris." Physiological and Molecular Plant Pathology **37**: 179-191.

Gross, D. C. (1991). Molecular and genetic analysis of toxin production by pathovars of Pseudomonas syringae. Annual Review of Phytopathology. Palo Alto, Annual Reviews, Inc. 247-278.

Hain, R., H.-J. Reif, et al. (1993). "Disease resistance results from foreign phytoalexin expression in a novel plant." Nature **361**: 153-156.

Jakobek, J. L., J. A. Smith, et al. (1993). "Suppression of bean defense responses by Pseudomonas syringae." The Plant Cell **5**: 57-63.

Johal, G. S. and S. P. Briggs (1992). "Reductase activity encoded by the HM1 disease resistance gene in maize." Science **258**: 985-987.

Jones, D. R. and B. J. Deverall (1978). "The use of leaf transplants to study the cause of hypersensitivity to leaf rust, Puccinia recondita, in wheat carrying the Lr20 gene." Physiological Plant Pathology **12**: 311-319.

Jong, W. D. and P. Ahlquist (1992). "A hybrid plant RNA virus made by transferring the noncapsid movement protein from a rod-shaped to an icosahedral virus is competent for systemic infection." Proc. Natl. Acad. Sci. USA **89**: 6808-6812.

Keen, N. T., S. Tamaki, et al. (1990). "Bacteria Expressing Avirulence Gene D Produce a Specific Elicitor of the Soybean Hypersensitive Reaction." Molecular Plant-Microbe Interactions **3**: 112-121.

Keen, N. T., P. H. Williams, et al. (1967). "Protease of Pseudomonas lachrymans in relation to cucumber angular leaf spot." Phytopathology **57**: 263-271.

Kinscherf, T. G., R. H. Coleman, et al. (1991). "Cloning and expression of the tabtoxin biosynthetic region from Pseudomonas syringae." Journal of Bacteriology **173**: 4124-4132.

Kuo, M., O. C. Yoder, et al. (1970). "Comparative specificity of the toxins of Helminthosporium carbonum and Helminthosporium victoriae." Phytopathology **60**: 365-368.

Macko, V., M. B. Stimmel, et al. (1992). "Structure of the host-specific toxins produced by the fungal pathogen Periconia circinata." Proc. Natl. Acad. Sci. USA **89**: 9574-9578.

Meeley, R. B., G. S. Johal, et al. (1992). "A biochemical phenotype for a disease resistance gene of maize." The Plant Cell **4**: 71-77.

Meeley, R. B. and J. D. Walton (1991). "Enzymatic detoxification of HC-toxin, the host-selective cyclic peptide from Cochliobolus carbonum." Plant Physiology **97**: 1080-1086.

Morris, P. F. and E. W. B. Ward (1992). "Chemoattraction of zoospores of the soybean pathogen, Phytophthora sojae, by isoflavones." Physiological and Molecular Plant Pathology **40**: 17-22.

Nelson, O. E. and A. J. Ullstrup (1964). "Resistance to leaf spot in maize." J Heredity **55**: 195-199.

Otani, H., K. Kohmoto, et al. (1991). Role of host-specifc toxins in the pathogenesis of Alternaria alternata. Molecular Strategies of Pathogens and Host Plants. New York, Springer-Verlag. 139-149.

Peters, N. K. and D. P. S. Verma (1990). "Phenolic compounds as regulators of gene expression in plant-microbe interactions." Molecular Plant-Microbe Interactions **3**: 4-8.

Roche, P., F. Debelle, et al. (1991). "Molecular Basis of Symbiotic Host Specificity in Rhizobium meliloti: nodH and nodPQ Genes Encode the Sulfation of Lipo-Oligosaccharide Signals." Cell **67**: 1131-1143.

Scheffer, R. P. (1989). "Ecological Consequences of Toxin Production by Cochliobolus and Related Fungi." NATO ASI Series **H27**: 285-300.

Scheffer, R. P. (1989). Host-specific toxins in phytopathology: origins and evolution of the concept. Host-specific toxins: recognition and specificity factors in plant disease. Tottori, Tottori University. 1-17.

Scheffer, R. P. and R. S. Livingston (1984). "Host selective toxins and their role in plant diseases." Science **223**: 17-21.

Scheffer, R. P., R. R. Nelson, et al. (1967). "Inheritance of toxin production and pathogenicity in Cochliobolus carbonum and Cochliobolus victoriae." Phytopathology **57**: 1288-1291.

Scheffer, R. P. and A. J. Ullstrup (1965). "A host-specific toxic metabolite from Helminthosporium carbonum." Phytopathology **55**: 1037-1038.

Spaink, H. P., D. M. Sheeley, et al. (1991). "A novel highly unsaturated fatty acid moiety of lipo-oligosaccharide signals determines host specificity of Rhizobium." Nature **354**: 125-130.

Sutherland, M. L. and G. F. Pegg (1992). "The basis of host recognition in Fusarium oxysporum f. sp. lycopersici." Physiological and Molecular Plant Pathology **40**: 423-436.

Truchet, G., P. Roche, et al. (1991). "Sulphated lipo-oligosaccharide signals of Rhizobium meliloti elicit root nodule organogenesis in alfalfa." Nature **351**: 670-673.

VanEtten, H. D., D. E. Matthews, et al. (1989). Phytoalexin detoxification: importance for pathogenicity and practical implications. Annual Review of Phytopathology. Palo Alto, Annual Reviews, Inc. 143-164.

Walton, J. D. and D. G. Panaccione (1993). "Host-Selective Toxins and Disease Specificity: Perspectives and Progress." Annual Review of Phytopathology **31**: in press.

Wit, P. J. G. M. d. (1992). Molecular characterization of gene-for-gene systems in plant-fungus interactions and the application of avirulence genes in control of plant pathogens. Annual Review of Phytopathology. Palo Alto, Annual Reviews, Inc. 391-418.

Wolf, S. J. and E. D. Earle (1991). "Effects of Helminthosporium carbonum race 1 toxin on host and non-host cereal protoplasts." Plant Science **70**: 127-137.

Wolpert, T. J. and V. Macko (1989). "Specific binding of victorin to a 100-kDa protein from oats." Proceedings of the National Academy of Sciences, U.S.A. **86**: 4092-4096.

Xiao, J., T. Tsuge, et al. (1992). "Further evaluation of the significance of BZR-toxin produced by Bipolaris zeicola race 3 in pathogenesis on rice and maize plants." Physiological and Molecular Plant Pathology **40**: 359-370.

Xiao, J. Z., M. Tsuda, et al. (1990). "Phytotoxins Produced by Germinating Spores of Bipolaris oryzae." Pytopathology **81**: 58-64.

Zambryski, P. C. (1992). Chronicles from the Agrobacterium-plant cell DNA transfer story. Annual Review of Plant Physiology and Plant Molecular Biology. Palo Alto, Annual Reviews, Inc. 465-490.

Oligogalacturonides Induce Proteinase Inhibitor Synthesis Via the Octadecanoid Pathway

Clarence A. Ryan, Barry McGurl, Gregory Pearce, Edward E. Farmer*, Javier Narvaez-Vasquez** and Martha Orozco-Cardenas**

Institute of Biological Chemistry
Washington State University
Pullman, WA 99164-6340
USA

INTRODUCTION

Proteinase inhibitor proteins are synthesized in leaves of several plant genera in response to localized and systemic signals that are released from sites of insect and pathogen attacks. Several chemical signals have been identified including oligouronides (14), chitosan fragments(15), an 18 amino acid polypeptide, called systemin (13), linolenic acid (4), and its derivative, jasmonic acid (3).

We have proposed a signal transduction pathway (4) (Figure 1) for the activation of proteinase inhibitor genes in which oligouronides and systemin are released by pathogen and insect attacks, respectively, and each interacts with a specific plasma membrane receptro, resultling in the release of linolenic acid. Evidence to date supports the pathway activated by wounding and systemin, but strong evidence for oligouronides activating the pathway has not been presented.

Several lines of evidence support a role for systemin in the pathway. Transgenic tomato plants expressing an antisense prosystemin gene, driven by the CaMV promoter, exhibit very weak systemic wound induction of Inhibitor I and II genes (10). When supplied with systemin, oligouronides, linolenic acid or jasmonic acid, the antisense plants synthesize proteinase inhibitors at the same level as wild type plants. Furthermore, several plant species have been shown to release of jasmonic acid in leaves in response to wounding (1,7). Promoter analyses of wound-inducible and jasmonate-inducible genes has revealed that the wound responsive elements and the jasmonate responsive elements are indistinguishable, lying within the same short nucleotide sequence (8,10).

The sites of inhibition of three potent inhibitors of the wound-inducible proteinase inhibitors have been determined using the various proposed intermediates of the pathway as inducers of synthesis of Inhibitors I and II in tomato plants. These

---

Present Addresses:
*Institut de Biologie et de Physiologie vegetales, Batiment de Biologie, CH-1015 Lausanne,Suisse.
**Unidad de Investigacion en Biotecnologia Agrijcola, Corporation Colombiana de Investigacion Agropecuaria, Apartado Aeareo 151123, El Dorado, Bogota, D. C. Colombia.

NATO ASI Series, Vol. H 81
Plant Molecular Biology
Edited by G. Coruzzi and P. Puigdomènech

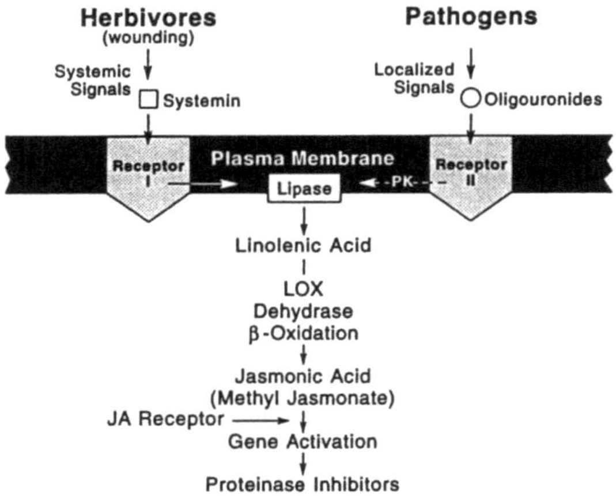

Figure I. Proposed model for the activation of proteinase inhibitor genes by systemin and oligouronides.

these inhibitors not only support the proposed signal transduction model, but provide evidence that oligouronides, like systemin, induce the expression of the two inhibitor genes via the octadecanoid pathway.The inhibitors are p-chloromercuribenzenesulfonic acid (PCMBS), diethyldithiocarbamic acid (DIECA) and acetylsalicylic acid (ASA). The inhibitor structures and their sites of inhibition in the pathway are shown in Figure 2.

INHIBITION OF THE PATHWAY BY PCMBS

PCMBS is a thiol reagent that is an inhibitor of active transport in plants (6). When placed on wounds on tomato leaf surfaces together with radioactively labeled systemin, PCMBS blocked transport of systemin into the nearby phloem and prevented transport out of the wounded leaf (11). Similarly, PCMBS supplied to young leaves severely inhibited the induction of proteinase inhibitors by wounding or systemin. However, PCMBS did not inhibit the induction of the proteinase inhibitors by oligouronides, linolenic acid or jasmonic acid (11). These results indicated that the chemical was blocking the signal transduction pathway between systemin production and linolenic acid release.

INHIBITION OF THE PATHWAY BY DIECA

DIECA, a potent reducing agent, strongly inhibits the induction of proteinase inhibitor synthesis in tomato leaves in response to wounding (5). When aqueous solutions of DIECA were supplied to excised, young tomato plants through their cut stems for 30 min and then wounded with a hemostat across the main vein of each leaf, DIECA inhibited the wound-induced accumulation of proteinase Inhibitor I in the leaves in a dose-dependent manner (5). Half maximal inhibition of induction by DIECA was at approximately 0.5 mM.

DIECA was tested to see if it would inhibit the induction of proteinase inhibitor synthesis by systemin, oligouronides, linolenic acid or jasmonic acid, four components of the proposed pathway (Figure 1). DIECA was supplied at 1.5 mM to excised young tomato plants through their cut petioles for 30 min, followed by solutions of each of the four inducing compounds, and the plants were incubated in light for 24 hr and assayed for Inhibitor I accumulation in the leaves. The inducing activity of Inhibitor I accumulation by systemin, oligouronides and linolenic were severely inhibited, but the inducing activity of jasmonic acid was unaffected. The extents of the inhibition of oligouronides and systemin by DIECA were comparable to the inhibition of wounding. This was consistent with the prediction from the model, and indicated that DIECA was exerting its inhibitory effects on the signalling pathway somewhere between linolenic acid and jasmonic acid (cf. Figure 2). DIECA does not inhibit soluble lipoxygenase activity present in the leaf cells nor does it inhibit commercial soybean lipoxygenase. It appears to be interacting specifically with 13S-hydroxylinolenic acid, the product of lipoxygenase action on linolenic acid (5).

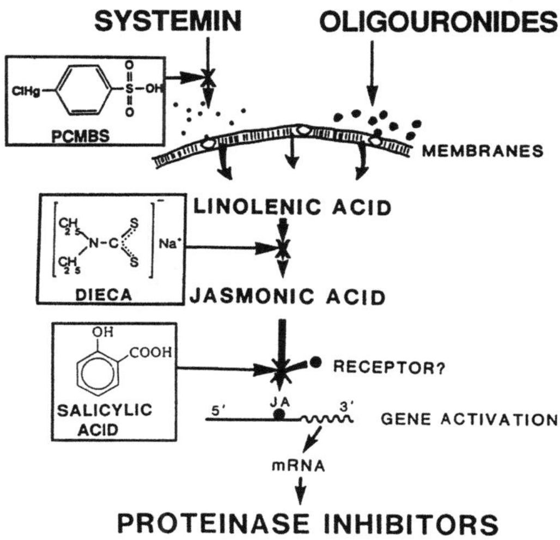

Figure 2. Sites of inhibition of PCMBS, DEICA and ASA

INHIBITION OF THE PATHWAY BY ASA

ASA, a primary signal for the systemic induction of PR proteins (16), was previously reported by Doherty and Bowles (2) to be an effective inhibitor of wound-induced and of oligouronide-induced proteinase inhibitor synthesis. The mode of inhibition was not known. We have now found that ASA inhibits the induction of proteinase inhibitors by systemin and by methyl jasmonate (12). Concentrations of I mM ASA essentially abolished the induction by methyl jasmonate. According to our proposed pathway (4), ASA would be inhibiting the pathway near the level of transcription (Figure 2). Since wounding and oligouronide induction of the proteinase inhibitors are already known to be inhibited by ASA (2), the cumulative data indicates that both oligouronides and systemin are signalling the activation of the proteinase inhibitor genes through the octadecanoid pathway.

The property of ASA as an inhibitor of the proteinase inhibitor induction pathway, while an inducer of the PR protein pathway, raises some interesting questions about the possible crosstalk between the two pathways. The proteinase inhibitors genes are induced very rapidly by wounding, whereas the PR proteins are induced much more slowly in response to pathogens. It is possible that the wounding by the insects provides an avenue of entry into the plant by pathogens, and during the period of pathogen incubation, while the levels of inhibitors throughout the plant are induced to protective levels, the increasing levels of SA induced by pathogens eventually shut down proteinase inhibitor synthesis and turn up PR protein synthesis to maximize protection against both insects and pathogens.

SUMMARY

The three inhibitors of proteinase inhibitor synthesis described here add to the evidence that both systemin and oligouronides utilize the octadecanoid pathway to activate the inhibitor genes. The sites of inhibition for each of the inhibitors has been localized to specific steps in the pathway by testing each with the various chemical components (inducers) of the proposed pathway, ie. systemin, oligouronides, linolenic acid, 13S-hydroxylinolenic acid, phytodienoic acid, and jasmonic acid. These inducers and the inhibitors provide useful biochemical markers to further study the biochemistry, molecular biology, and biochemical genetics of the pathway. Much research remains to be done, such as the biochemistry of prosystemin synthesis; the processing of prosystemin; the storage, release and transport of systemin; the nature of the receptor and the signal transduction pathway leading to intracellular linolenic acid production; and the fate of jasmonic acid and its role in gene transcription. It is anticipated that within the next few years that the components of the pathway and their regulation will be better understood and that the full extent of both polypeptide signalling and the octadecanoid pathway will become known.

ACKNOWLEDGEMENTS

This research was supported in part by Project #1791, College of Agriculture and Home Economics, and Grants DCB-9104542 and DCB-9117795 from the National Science Foundation.

LITERATURE CITED

1. Creelman, R. A., Tierney, M. L., and Mullet, J. E. 1992. Jasmonic Acid/Methyl Jasmonate Accumulate in Wounded Soybean Hypocotyls and Modulate Wound Gene Expression. Proc. Natl. Acad. Sci. USA. 89, 4938-4941.
2. Doherty, H. M., Selvendran, R. R., and Bowles, D. J. 1988. The Wound Response of Tomato Plants Can Be Inhibited by Aspirin and Related Hydroxybenzoic Acids. Physiol. and Mol. Plant Pathol. 33, 377-384.
3. Farmer, E. E. and Ryan, C. A. 1990. Interplant Communication: Airborne Methyl Jasmonate Induces Synthesis of Proteinase Inhibitors in Plant Leaves. Proc. Natl. Acad. Sci. USA. 87, 7713-7716.
4. Farmer, E. E. and Ryan, C. A. 1992. Octadecanoid jasmonate precursors activate the synthesis of wound-inducible proteinase inhibitors. The Plant Cell 4, 129-134.

5. Farmer, E. E., Walker-Simmons, M., and Ryan, C. A. Evidence for Hydroperoxidation in the Signalling Pathway for the Wound-Induction of Proteinase Inhibitors in Tomato Leaves. In Review.
6. Giaquinta, R. 1976. Evidence for Phloem Loading from the Apoplast. Chemical Modification of Membrane Sulfhydryl Groups. Plant Physiol. 57, 872-875.
7. Gundlach, H., Muller, J. J., Kutchan, T. M., and Zenk, M. H. 1992. Jasmonic Acid Is a Signal Transducer in Elicitor-Induced Plant Cell Cultures. Proc. Natl. Acad. Sci. USA. 89, 2389-2393.
8. Kim, S-R., Choi, J-L., Costa, M. A., and An, G. 1992. Identification of G-Box Sequence as an Essential Element for Methyl Jasmonate Response of Potato Proteinase Inhibitor II Promoter. Plant Physiol. 99, 627-631.
9. Mason, H. S., DeWald, D. B., and Mullet, J. E. 1993. Identification of a Methyl Jasmonate-Responsive Domain in the Soybean vspB Promoter. The Plant Cell. 5, 241-251.
10. McGurl, B., Pearce, G. Orozco-Cardenas, M., and Ryan, C. A. 1992. Structure, Expression and Antisense Inhibition of the Systemin Precursor Gene. Science 255, 1570-1573.
11. Narvaez-Vasquez, J., Orozco-Cardenas, M. L., and Ryan, C. A. Systemic Induction of Proteinase Inhibitor Synthesis Is Inhibited by PCMBS in Tomato Plants. In Review.
12. Narvaez-Vasquez, J. and Ryan, C. A. Acetylsalicylic acid Inhibits Synthesis of Proteinase Inhibitors in Tomato Leaves Induced by Systemin and Methyl Jasmonate. In Review.
13. Pearce, G., Strydom, D., Johnson S., and Ryan, C.A. 1991. A Polypeptide from Tomato Leaves Activates the Expression of Proteinase Inhibitor Genes. Science 253, 895-897.
14. Ryan, C.A. Oligosaccharide Signalling in Plants. 1987. Ann. Rev. Cell. Biol. 3, 295-317.
15. Walker-Simmons, M., Hadwiger, L., and Ryan, C.A. 1983. Chitosans and Pectic Polysaccharides Both Induce the Accumulation of the Antifungal Phytoalexin Pisatin in Pea Pods and Antinutrient Proteinase Inhibitors in Tomato Leaves. Biochem. Biophys. Res. Comm. 110, 194-199.
16. Yalpani, N., Lilverman, P., Wilson, M. , Kleier, D. A., and Raskin, I. 1991. Salicylic Acid Is a Systemic Signal and an Inducer of Pathogenisis-Related Proteins in Virus-Infected Tobacco. The Plant Cell 3, 809-818.

EXPRESSION OF GENES ENCODING THIONINS AND LIPID-TRANSFER PROTEINS. A COMBINATORIAL MODEL FOR THE RESPONSES OF DEFENSE GENES TO PATHOGENS.

Antonio Molina and Francisco García-Olmedo
Laboratorio de Bioquímica y Biología Molecular
E.T.S.Ingenieros Agrónomos. E-28040 Madrid, Spain

INTRODUCTION

Seeds normally accumulate high levels of proteins that are either toxic or inhibitory towards heterologous systems, including pathogens, pests and predators. This is the case of cereal kernels, where a substantial fraction of the non-storage proteins is represented by different families of these proteins (for reviews, see García-Olmedo et al.1989, 1992). The same types of proteins are often present in other tissues, either under normal development or under stress, including infection by pathogens. The thionins (THs) and the so-called non-specific lipid transfer proteins (LTPs) are two such families. Thionins were first reported in wheat endosperm (Balls et al., 1942) and their *in vitro* antipathogenic properties have been known for over 20 years (Fernadez de Caleya et al. 1972), while cereal LTPs were first reported in barley aleurone (Mundy and Rogers 1986) and we have recently shown that different members of this family can be isolated from other barley tissues and that they are potent growth inhibitors of bacterial and fungal plant pathogens (Molina and García-Olmedo, 1991; Molina et al.1993; Molina and García-Olmedo, submitted). In this report, we will update our knowledge of the molecular biology and inhibitory properties of barley THs and LTPs, and will describe the developmental and pathogen-induced expression of the different genes encoding these proteins. Also, a working model for the response of defense genes against infection by different pathogens will be proposed.

NATO ASI Series, Vol. H 81
Plant Molecular Biology
Edited by G. Coruzzi and P. Puigdomènech

STRUCTURAL TYPES AND THEIR GENETIC CONTROL

Thionins

Thionins are cysteine-rich polypeptides of about 5 kDa that have been found in a variety of tissues from different taxa (see García-Olmedo et al., 1989; 1992). Based on their similarity, known sequences can be grouped into at least 5 clusters or types (I-V), three of which (I,II,V) can be found in cereals such as wheat or barley.

The first type, which corresponds to the original purothionins from wheat endosperm (Balls et al. 1942), is 45 residues long, highly basic, with no acid residues, and has four disulfide bridges. There are one or two genes of type I per haploid genome in group-1 chromosomes of wheat and barley genomes. These genes encode precursor proteins in which the sequence of the mature protein is preceded by that of a signal peptide that is co-translationally processed and followed by that of a C-terminal, acidic peptide that undergoes a post-translational excision (Ponz et al. 1983; García-Olmedo et al. 1992).

Type-II thionins are less basic, with some acidic residues, and have an extra amino acid located in the central loop, as well as four disulfide bridges (Gausing 1987; Bohlmann and Apel 1987). Genes of this type, which seem to be present in 10-100 copies per haploid genome, have been located in chromosome 6H of barley and encode precursor proteins with the same characteristics as those of type I (Gausing 1987; Bohlmann and Apel 1987; Bohlmann et al.1988).

Type-V thionins are 37 residues long and have three disulfide bridges, two basic and two acidic residues, and a similar precursor structure as the previous types (Castagnaro et al. 1992). Genes of type V, one copy per haploid genome, have been located within a few kb of type-I genes in group-1 chromosomes in the wheat genomes (Castagnaro et al. 1992).

Lipid Transfer Proteins

Members of the LTP family have been characterized in barley

aleurone (Mundy and Rogers, 1986) and in green tissues, where we have recently reported a distinct subfamily of at least four members, which are structurally closer to LTPs isolated from other plant species than to the aleurone type (Molina et al.1993). Genes for these LTPs are located in chromosomes 3H (*Ltp2*, encoding protein Cw18; *Ltp4*, Cw21) and 7H (*Ltp3*, Cw20; Molina and García-Olmedo, submitted).

INHIBITORY PROPERTIES

Toxicity and Specificity

Both THs and LTPs are inhibitory and biocidal (at higher concentrations) towards bacterial and fungal pathogens. It has been shown that thionins are able to alter membrane permeability, indirectly inhibiting macromolecular biosynthesis (Carrasco et al. 1981), and to directly inhibit certain enzymes (Diaz et al. 1992 and unpublished), while the mechanism of action of LTPs remains unknown. As shown in Fig.1, different THs of types I and II show different activity towards a given pathogen, and the same is true for the barley LTP variants tested, but the relative ranking of the variants changes with the pathogen (i.e. *Clavibacter michiganensis* versus *Fusarium solani*).

The ability of THs to enhance resistance against certain pathogens *in vivo* has been recently reported (Carmona et al.1993).

Complementation and Synergism

Data in Fig.1 also show that while LTPs are about 20-fold more active than THs against the bacterial pathogen, they are somewhat less active than THs against the fungal one. Furthermore, we have found that the two types of proteins can act additively against the bacterium (not shown) and synergistically against the fungus (Fig.2). *Complementation*, as well as *additive* and *synergic* effects, might be relevant to explain interactions *in vivo*.

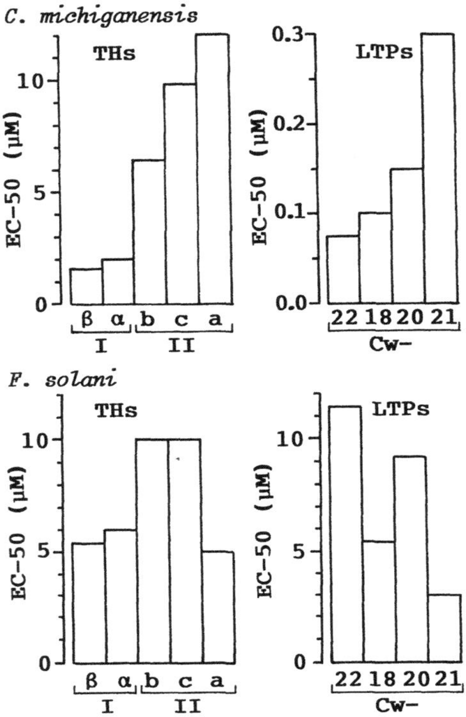

Fig.1. Growth inhibition of the bacterium *Clavibacter michiganensis* and the fungus *Fusarium solani* by TH and LTP variants from barley.

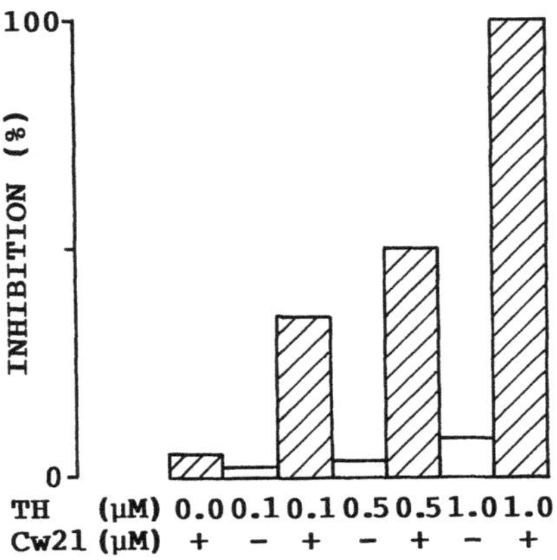

Fig 2. Synergism of THs (typeI) and LTPs (Cw21; 1µM) against *Fusarium solani*.

Resistance to Defense Proteins

Intraspecific variability of pathogen susceptibility to different defense proteins and the induction of resistance - including cross resistance - are being actively investigated by us, both to improve our understanding of compatible/incompatible interactions and because of the relevance of these properties when engineering plants for enhanced resistance. Examples of this intraspecific variability are shown in Fig.3.

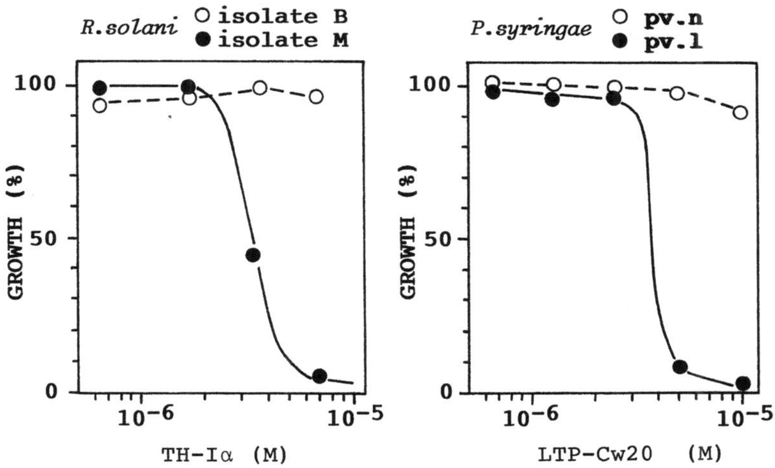

Fig.3. Differential *in vitro* resistance among pathogen strains.

EXPRESSION OF DEFENSE GENES

Normal Development

Expression patterns of three LTPs and of type-II THs are summarized in Table 1. All tissues tested, except endosperm, expressed the three LTPs investigated, and sequence-specific probes showed quantitative differences in the expression levels of the three genes. The probe used for the THs (type-II-specific) detected high levels of mRNA in etiolated leaves and lower levels in other tissues. Thionins of types I and V were restricted to endosperm (not shown). The LTPs described here are located in vascular tissues and in the outer cell layer of the exposed surfaces of the plant, and can be *washed* out with buffer.

Table 1. Expression patterns* of *Ltp* and *Th* genes in barley

TISSUE	(age)	GENE *Ltp2*	*Ltp3*	*Ltp4*	*Th*
Root	(young)	(+)	(+)	(+)	(+)
Leaf/dark	(young)	++	++	++	+++++
/light	(young)	++	++	+++	+
	(old)	+	+	(+)	(+)
Shoot apex		+	++	+	+++
Stem	(young)	+++	+++	+++++	+
	(old)	+	+	+	(+)
Spike/palea-lemma		(+)	++	+	(+)
/grain coats		(+)	+++	++	-
/rachis		(+)	++	+	(+)
/endosperm		-	-	-	-

* after overnight exposure; parenthesis indicate 5d exposure.

Responses to Abiotic Stimuli and to Pathogens

Young barley plants were exposed to different physical stresses and treated with different plant hormones and agrichemicals as summarized in Table 2. Steady-state mRNA levels of *Ltp* and *Th* genes were little affected by the physical treatments, except for moderate responses to salinity of two of the *Ltp* genes and the response to drought of *Th* genes. Abscisic acid increased the expression of both types of genes, and methyl jasmonate switched off *Ltp* genes, while markedly increasing the expression of *Th* genes. All other treatments had no detectable effects.

The current status of our analysis of the expression of these genes in response to pathogens is summarized in Table 3. Responses of the pathogenesis-related gene *Pr1* from barley (probe gift of W.Knogge) are also indicated.

Table 2. Responses (nX of basal levels) of *Ltp* and *Th* genes to abiotic stimuli (nd, not determined; -, no effect).

	GENES			
STIMULI	*Ltp2*	*Ltp3*	*Ltp4*	*Th*
Salinity	2X	-	2X	-
Cold	-	-	-	-
Drought	-	-	-	4X
Wounding	-	-	-	-
Methyl jasmonate	.05X	.06X	.12X	20X
Abscisic acid	5X	2X	3X	3X
Ethylene	-	-	-	-
Ethephon	-	-	-	nd
Salicylate	-	-	-	nd
Isonicotinic acid	-	-	-	nd

Table 3. Responses (nX of basal levels) of *Ltp* and *Th* genes to pathogens (-, no effect; C/I, compatible/incompatible).

		GENES				
PATHOGEN/CULTIVAR	C/I	*Ltp2*	*Ltp3*	*Ltp4*	*Th*	*Pr1*
*E.graminis**						
(Av6)/Pallas	I	3X	3X	9X	3X	6X
(vir6)/Pallas	C	3X	3X	8X	3X	6X
*R.secalis***						
/Atlas46	I	3X	-	10X	4X	10X
/Atlas	C	-	-	-	4X	-
/Turk	I	4X	-	16X	-	20X
/Hannchen	C	-	-	-	-	-
P.syringae						
(153)/Bomi	I	-	-	-	-	-
(DC3000)/Bomi	C	-	-	-	5X	4X

* filters from W.Knogge (Köln) and ** from L.Boyd (Norwich).

Different combinations of the genes investigated were affected in the different plant/pathogen interactions, and the extent and time-courses of the observed effects were also different. Thus, in the infection by *Erysiphe graminis*, both LTPs and THs were affected within the first few hours, before the progress of the compatible and the incompatible interactions can be distinguished microscopically from each other, according to Boyd (1993), whereas in the infection with *Rhyncosporium secalis*, LTPs were rapidly affected, while THs lagged behind by two days. Barley cv Turk, which failed to increase TH mRNA when challenged with *R.secalis,* was able to do so in response to some chemical treatments. In the case of the infection with *Pseudomonas syringae*, no hypersensitive reaction was detected in the incompatible interaction and none of the genes investigated was expressed above basal levels. This *basal resistance* was likely due to constitutive levels of defense proteins active against the pathogen.

A WORKING MODEL

The above observations suggest a number of features related to defense proteins that can be incorporated to current models of plant/pathogen interactions (Fig.4). These can be summarized as follows:

a) Different defense genes can be activated through totally or partially independent pathways and/or receptors, a combination of which are activated by a given pathogen in the same or in different sets of plant cells.

b) The elicited response might be *gratuitous* if the elicited proteins are not inhibitory (individually or in combination) against a particular pathogen, which is nevertheless *seen* by the plant.

c) A compatible interaction might result not only from failure of the plant to recognize a particular pathogenic strain, and to activate the appropriate defense gene(s), but also from the ability of the specific strain to resist inhibitor(s) to which other strains are susceptible.

d) Basal levels of defense proteins (for which additional roles in development are not excluded) are often above 1 x 10^{-5} moles/kg fresh weight, concentrations that should be sufficient to inhibit growth of many pathogens by a mechanism that does not imply a hypersensitive reaction.

e) The same defense proteins might be involved in all the different situations described above (b,c, and d).

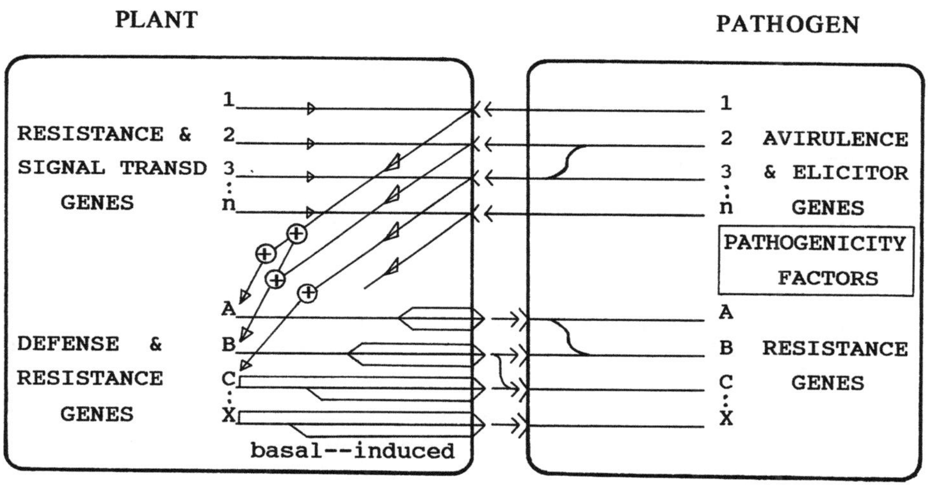

Fig.4. Schematic representation of activation of defense genes.

ACKNOWLEDGEMENTS. The collaboration of W.Knogge and coworkers (Kóln) and of L.Boyd and coworkers (Norwich), as well as financial support from the Comisión Interministerial de Ciencia y Tecnología (grant BIO90/0084) are gratefully acknowledged.

REFERENCES

Balls AK, Hale WS, Harris TH (1942) A crystalline protein obtained from a lipoprotein of wheat flour. Cereal Chem 19:279-288

Bohlmann H, Apel K (1987) Isolation and characterization of cDNAs coding for leaf-specific thionins closely related to the endosperm-specific hordothionin of barley (Hordeum vulgare L.). Mol Gen Genet 207:446-454

Bohlmann H, Clausen S, Behnke S, Giese H, Hiller C, Reiman-Philipp U, Schrader G, Barkholt V, Apel K (1988) Leaf-specific thionins of barley - a novel class of cell wall proteins toxic to plant-pathogenic fungi and possibly involved in the defense mechanism of plants. EMBO J 7:1559-1565

Boyd L (1993) Gene expression in barley in response to mildew infection. In "Engineering plants against pests and pathogens" Proc.Juan March Workshop, Madrid, Spain. p. 83.

Carmona MJ, Molina A, Fernandez JA, Lopez-Fando JJ, García-Olmedo F (1993) Expression of the α-thionin gene from barley in tobacco confers enhanced resistance to bacterial pathogens. Plant J 3:457-462

Carrasco L, Vazquez D, Hernandez-Lucas C, Carbonero P, García-Olmedo F (1981) Thionins: plant peptides that modify membrane permeability in cultured mammalian cells. Eur J Biochem 116:185-189

Castagnaro A, Maraña C, Carbonero P, García-Olmedo F (1992) Extreme divergence of a novel wheat thionin generated by a mutational burst specifically affecting the mature protein domain of the precursor. J Mol Biol 224:1003-1009

Diaz I, Carmona MJ, García-Olmedo F (1992) Effects of thionins on β-glucuronidase in vitro and in plant protoplasts. FEBS Lett 296:279-282

Fernadez de Caleya R, Gonzalez-Pascual B, García-Olmedo F, Carbonero P (1972) Susceptibility of phytopathogenic bacteria to wheat purothionins in vitro. Appl Microbiol 23:998-1000

García-Olmedo F, Rodriguez-Palenzuela P, Hernandez-Lucas C, Ponz F, Maraña C, Carmona MJ, Lopez-Fando JJ, Fernandez JA, Carbonero P (1989) The thionins: a protein family that includes purothionins, viscotoxins and crambins. Oxford Surveys Plant Mol Cell Biol 6:31-60

García-Olmedo F, Carmona MJ, Lopez-Fando JJ, Fernandez JA, Castagnaro A, Molina A, Hernandez-Lucas C, Carbonero P (1992) Characterization and analysis of thionin genes. In Genes involved in plant defense (Boller T, Meins F, eds) Plant Gene Research, pp 283-302. Springer-Verlag

Gausing K (1987) Thionin genes specifically expressed in barley leaves. Planta 171:241-246

Molina A, García-Olmedo (1991) Patent application P 9101258 (24.05.91) PCT/EP92/01130

Molina A, Segura A, García-Olmedo F (1993) Lipid transfer proteins (nsLTPs) from barley and maize leaves are potent inhibitors of bacterial and fungal plant pathogens. FEBS Lett 316:119-122

Mundy J, Rogers, JC (1986) Selective expression of a probable α-amylase/protease inhibitor in barley aleurone cells: comparison to the barley α-amylase/subtilisin. Planta 169:51-63

Ponz F, Paz-Ares J, Hernandez-Lucas C, Carbonero P, García-Olmedo F (1983) Synthesis and processing of thionin precursors in developing endosperm from barley (Hordeum vulgare L). EMBO J 2:1035-1040

Structure/Function Studies of Lea Proteins

Leon Dure III
Department of Biochemistry
The University of Georgia
Athens, GA 30602-7229
USA

Over 10 years ago several families of proteins that accumulate to high levels during the maturation phase of cotton embryogenesis were identified by 2D gel electrophoresis (Dure et al. 1981). These proteins could be induced to accumulate earlier in embryogenesis when immature embryos were dissected and incubated in dilute solutions of ABA (without ABA, the embryos germinate precociously). Subsequently, cDNA representatives of many of these protein families were obtained by subtractive hybridization and identified by hybrid arrested translation (Galau et al. 1986). To reflect the developmental period of high accumulation, these protein families were named Lea, mnemonic for late embryogenesis abundant; likewise *lea* genes and mRNAs. By 1988 a representative gene from each of six different *lea* families had been sequenced and their amino acid sequence deduced (Baker et al. 1988). In the last several years a large number of gene sequences have been reported that are homologous to the sequenced *lea* gene families. Many were uncovered in studies designed to find genes whose expression is up-regulated in response to water stress, salt stress and cold stress. In most all cases, the high expression of these genes was found to be up-regulated by exogenous ABA as well. Furthermore, the stress-induced expression of these genes was found initially in non-seed tissues.

These facts have led to the possibility that the expression of those *lea* genes that are also stress-induced in leaves, roots, etc., is regulated by the common aspect of seed maturation, water stress, salt stress and cold stress -- namely, the loss of intracellular water. The experimental findings also suggested that the expression of these genes is mediated by the plant growth regulator abscisic acid (ABA).

However it should not be concluded that all the genes up-regulated in expression in seed maturation, in environmental stress responses or by increases in intracellular

NATO ASI Series, Vol. H 81
Plant Molecular Biology
Edited by G. Coruzzi and P. Puigdomènech

ABA constitute a single battery of mutually inclusive genes. A more realistic view of the genes regulated by the inductive events given above is that there is a set of genes induced in common by all the inductive events but that this set is but a subset of the entire body genes up-regulated by a single inductive event. This notion is presented in Figure 1 below. Each ellipse of this rosette of genes represents genes up-regulated by the various events. The overlapping region of the ellipses represents those genes involved in the water loss response. The regions outside the region of commonality represent genes whose expression is peculiar to the individual stress, to ABA or due to seed developmental regulation.

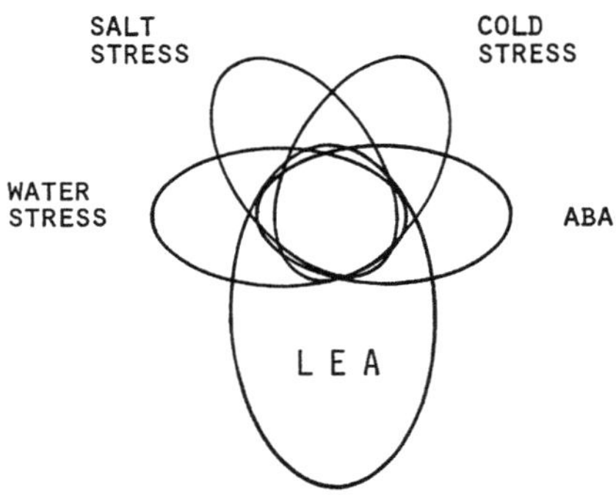

Figure 1. Diagram of overlapping sets of genes.

Clearly, there are ABA-inducible genes that are not part of the late-embryogenesis cohort of up-regulated genes, just as there should be cold-stress-induced genes that are not part of the water-stress response. The *kin* genes (Kurkela & Franck, 1990), which are not induced by other stresses nor expressed in seeds, are examples. However they appear to be ABA responsive. Although this diagram is not a quantitative representation of the number of genes involved, the *lea* cohort is shown larger than the others in order to emphasize that, by definition, this designation represents all genes that give rise to abundant proteins during the final stage of embryo development. Only a portion of these genes are likely to be stress-induced.

Table 1 gives an indication of the complexity of the response to water stress alone as reflected in the current literature. Genes for a number of enzymes have been found to be up-regulated in water-stressed tissues, many of which seem appropriate for osmotically equipping cells to resist water loss by increasing the intracellular pool of sugar alcohols and amino acid derivatives. The rationale for increases in other proteins with demonstrable or presumed functions is less clear and awaits a more detailed description of what cells do when their water content drops.

Gene Products Up-regulated by Water Stress

<u>Known or putative functions</u>

- enzymes for synthesis of sugar alcohols
- enzymes for synthesis of proline and glycine betaine
- thiol protease
- gly-rich cell water proteins
- integral membrane transport facilitator proteins
- anti-fungal proteins

<u>Unknown functions</u>

- Lea D-7 & homologs
- Lea D-11 & homologs (dehydrins and rab proteins)
- Lea D-19 & homologs (wheat Em)
- Lea D-113 & homologs

However, the majority of known genes induced to high expression by water stress give rise to proteins of unknown function. Furthermore, many of these genes increase in expression in response to salt and/or cold stress and are participants in the development program that brings about seed maturation. As mentioned before, it is conceivable that these genes are all up-regulated by the water loss aspect of all these stressed and developmental events. It should be recognized, however, that most of the data indicating the elevated levels of gene expression are based on northern blotting measurements of mRNA levels. As will be seen further along, the sequenced genes are, in most all cases, members of multigene families. Further, there is considerable size and sequence heterogeneity among family members of a given species. In fact in some cases, family members are linked only by the shared occurrence of a particular amino acid sequence motif. Northern blotting may suggest that the same gene is up-regulated by different stresses when, in actuality, different family members may be activated by each stress which are quite different overall in amino acid composition and sequence one from another. Whether these quite different proteins carry out the same cellular function is at present unknown. Establishing the function of the various Lea/stress induced proteins has been an on-going pursuit of this laboratory and is the theme of this presentation.

Predicting Protein Function by Determining Protein Structure

It may be possible in some cases to determine the function of certain proteins by manipulating their levels in cells and observing phenotypic characteristics. Gene number can be increased, gene expression negated by anti-sense DNA and the normal temporal/spatial aspects of gene expression altered by the substitution of natural promoters with promoters of different temporal/spatial specificity. However, in view of the large number of genes seemingly involved in an individual stress response or in a developmental program, and of the large number of members that constitute each gene family -- many of which share only regions of sequence homology -- this approach may generate phenotypes that are difficult to interpret. Furthermore, many phenotypic changes (e.g. increased desiccation tolerance) can only confirm the importance of a given gene in a stress response without defining on a molecular basis precisely what the protein product is doing in the stressed cell.

We have chosen to determine the 3D structure and the resulting biophysical/chemical properties of many of these proteins in hopes that this information will suggest functions that can be directly tested.

One initial step in this direction can be taken by examining the known sequences of a given protein family in order to establish regions of common chemical properties and regions of predicted common 2D structure. The predictive computer programs of Chou & Fasman (1978), Granier et al. (1978) and Argos (1989) have been used, and only those sequence domains that yield a common structural prediction from the three programs (helix, β sheet, β turn and random coil) are emphasized. The names of the several families of genes/proteins used throughout are those established for cotton (Galau et al. 1986).

Lea D-19 family of proteins

Twelve protein sequences from eight plant species have been determined for this group. It includes the long-known Em wheat protein. Although their sizes range from 83 to 153 amino acids, these proteins constitute the most homogeneous of the stress proteins of unknown function with regard to overall chemical properties. They are uniformly high in the charged amino acids glu, lys and arg and in gly. They are predicted to be over 50% random coil interspersed with a few random helical regions. There is a high degree of sequence homology among these proteins if chemically synonymous amino acid residues are considered identical.

The chief basis for the difference in size found in this group is the existence of a sequence that is repeated in some members, but not in others, as shown by dot matrix analysis of protein sequence (Dure 1993). The large amount of random coil predicted for the Em protein and its resultant high water-binding property led Lane and coworkers (McCubbin et al. 1985) to speculate that its function in wheat embryos is to resist total desiccation by tenaciously retaining water molecules as the seeds desiccate.

Lea D-113 family of proteins

Five representatives of this group of proteins representing four species have been sequences. Their sizes range from 88 to 173 residues. They are high in gly, thr and ala residues. The only sequence domain common to these proteins is found in the N-terminal 60-80 residues which are uniformly predicted to be a single long α helix. The remaining residues comprise entirely a putative random coil region high in gly and thr. The variation in length among these proteins reflects differences in the length of this random coil C-terminal tail. Curiously, within this tail no strict amino acid sequence homology is found. The occurrence of the abundant gly and thr residues follows no discernable pattern, suggesting that what is functionally important in this region in the lack of secondary structure itself.

A space-filling model of the cotton D-113, showing the N-terminal helix followed by the random coil tail is given in Figure 2. The gly and thr/ser residues are colored gold and yellow, respectively.

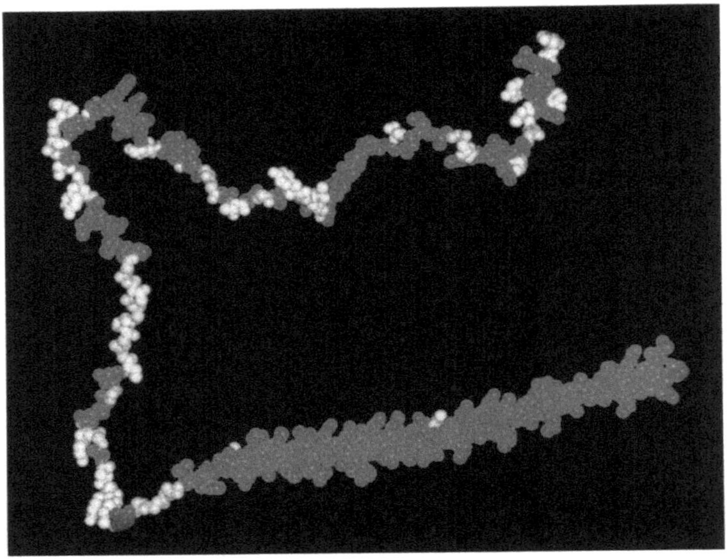

Figure 2. Predicted structure of cotton D-113 protein.

Lea D-11 family of proteins

This is the most diverse and curious group of the Lea/water-stress proteins found to date and also the most numerous. The sequence of over 30 representatives have been established, obtained from 13 plant species. Rice alone has furnished five different sequences.

Historically, sequence diversity among these proteins was not apparent. The first several sequences obtained suggested that the unifying sequence elements among these proteins were the presence of a stretch of 7-9 contiguous ser residues found roughly midway through the molecules, followed by a stretch of 14 residues that was repeated in most instances with high fidelity near the C-terminus of the molecules. These common sequence features were embedded in regions very high in gly and thr that lacked strict sequence homology, similar to the D-113 family. The gly/thr-rich regions are predicted to be principally random coils, whereas the repeating 14-mer is predicted to form about three turns of α helix bounded by β turns.

Subsequently, sequences were found that violated one or more of these unifying properties. A few sequences predicted proteins of over 500 residues. These invariably contained more than two copies of the 14-mer repeat. Other sequences lacked the poly-ser stretch. Most surprising of all, a few sequences revealed a reversal in the predicted dominant feature of secondary structure. In these instances the typical 20-30% gly content drops to 3-7% and the most abundant residues are glu and lys. Not surprisingly, these latter proteins are predicted to be mostly helical.

Figure 3 demonstrates how these sequence relationships can be visualized by dot matrix sequence analysis. In panel A the sequence of a barley protein of this group is compared with itself. In order to generate a dot, 4 out of a window of 6 amino acids must be gly, ser or thr. The high frequency of these residues is apparent from the large number of dots generated. The positions of the two 14-mer stretches (which are not high in gly, ser or thr) are conspicuous by the absence of dots. The position of the poly-ser stretch is also indicated on the matrix. Panel B shows a tomato sequence that contains three 14-mers elements compared with itself. In this case 4 out of 6 residues must be a charged residue (glu, asp, lys, arg) to generate a dot. Clearly the large number of dots indicate that this protein is an example of the subset of this group predicted to be largely helical in secondary structure. No dots appear in the poly-ser region. In panel C the barley and tomato proteins are compared with each other in the conventional manner. Amino acid homology is noted by dots when 5 residues out of 7 are identical. Only the poly-ser region and the 14-mer repeats (2 in the barley protein, 3 in the tomato protein) generate dots, indicating that only in these regions are these sequences related. Panel D shows that even among the proteins that are high gly/thr, strict sequence homology is often lacking when proteins from different genera are compared. Here the barley protein is compared with a protein from the resurrection plant *Craterostigma p.* in the conventional manner with 5 of 7 residue identity producing a dot. Other than the poly-ser tract and the two 14-mer elements present in each molecule, very little homology is noted elsewhere in the molecules.

From this it is apparent that the only common element found among this group of apparent homologs is the 14-mer unit. The "dehydrins" and "rob" proteins are members of this group. Table 2 summarizes these relationships.

In view of the diversity of properties displayed by these proteins, it is difficult to imagine that they perform a common function in water-stressed cells; a function that depends on the 14-mer stretch of residues -- unless that function be very non-specific. Although a few of the members of this group are predicted to be largely helical rather than informationless disordered structures, it is possible that a common function may

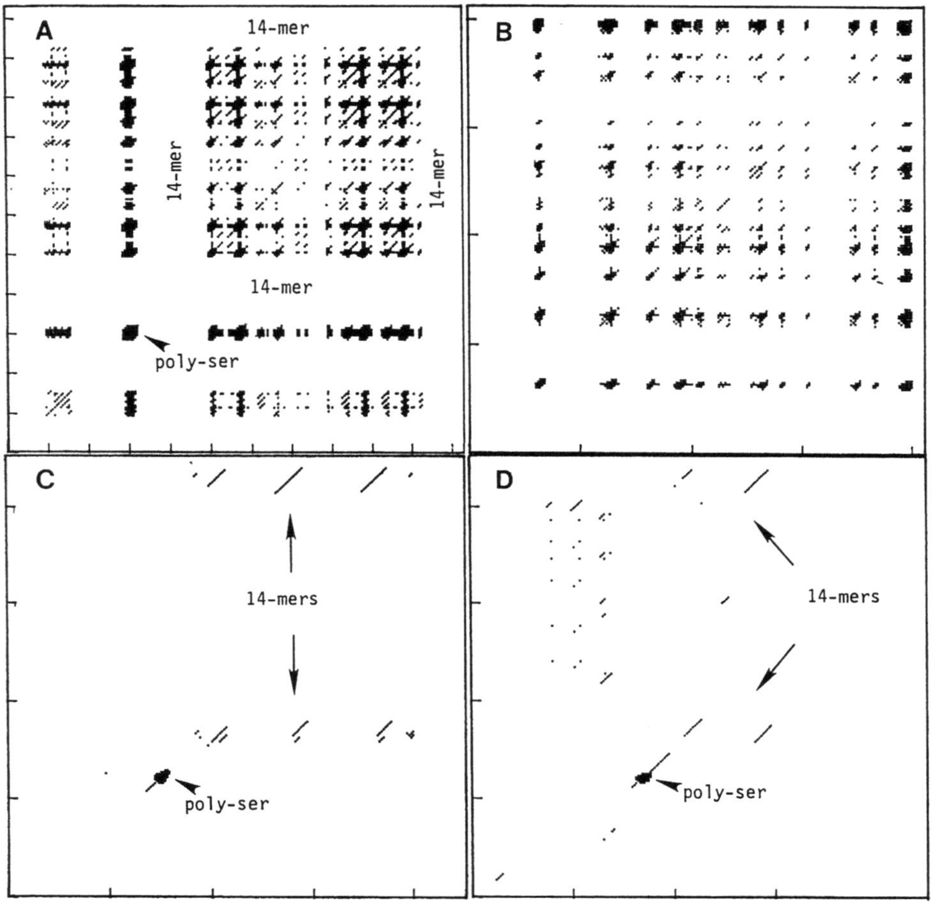
A
14-mer
14-mer
14-mer
14-mer
poly-ser
B
C
14-mers
poly-ser
D
14-mers
poly-ser

Figure 3

result from a lack of 3D structure by both types of proteins. Conversely, the occurrence of the 14-mer motifs in these diverse proteins may simply reflect a meaningless evolutionary curiosity.

Characteristics of D-11 Homologs

	Number
<u>Species providing sequences</u>	13
<u>Size range (amino acid residues)</u>	
140-200	20
201-300	8
300+	3
<u>poly ser domain</u>	
1	33
0	8
<u>14-mer domains</u>	
1	2
2	23
3	3
6	2
11	1
<u>gly/thr based random coil</u>	28
<u>glu/lys based α helix</u>	3

Lea D-7 family of proteins

This family has five members to date representing five plant species and range in size from 136 to 243 residues. These molecules share considerable sequence homology throughout, the wide range in size being totally attributable to the variable number of a contiguously tandem 11-mer repeat motif. The number of repeats range from 5 to 13, as seen in Table 3.

D-7 Family

	Amino Acids	# 11-mers	# hinges	% helix
Cotton	136	5 (55)	0	76
Carrot	163	6 (66)	1	69
Barley	213	9 (99)	2	69
Wheat	244	10 (110)	3	72
Brassica	243	13 (143)	1	65

The repeating unit models as an ideal amphiphilic α helix with apolar residues at positions 1,2, 5 and 9 forming a hydrophobic surface with a small right-handed twist. Negatively charged or amide residues at positions 3, 7 and 11 provide a very polar surface on the opposite side of the helix and lys residues at positions 6 and 8 are located at the interfaces of these surfaces. Position 10 contains a random assortment of residues, whereas position 4 is chiefly ala. Unsurprisingly, these repeating units are predicted to be helical. Since these units are mostly tandemly arranged, the length of these helices are large, involving many turns of the helix. Perhaps to prevent the formation of an overly long, rigid helix, the tandem repeat units are interrupted in all but the smallest protein (cotton) by one or more short stretches of residues whose sequence is unrelated to the repeating unit. These regions may be considered hinge regions separating the helical regions bècause they may allow the molecule to bend. There is unmistakable sequence homology between these family members in the N and C terminal regions bordering the repeating elements. A detailed analysis of the structure of the 11-mer repeating unit and its possible structural consequences has been published (Dure, 1993a).

Of all the Lea/water stress proteins of unknown function presented here, this family of proteins would seem the most likely to form and to function as an entity of prescribed 3D structure, probably based on the repeating units. Yet the function must allow for a wide range of molecular sizes because of the different numbers of 11-mer units involved. A likely higher-order structure would be a dimer formed by the interaction of two apolar surfaces of two helical monomeric proteins. Such a structure would be analogous to the coiled coil superhelix that forms from helices whose fundamental building blocks are the tandemly-repeating 7-mers of animal structural proteins. There is, however, a fundamental difference between the 7-mer-based coiled coil and an 11-mer-based coiled coil. The former is a left-handed superhelix, whereas the putative 11-mer-based structure would form a right-handed superhelix. No such structure has been found by protein chemists to date.

Many conceivable functions can be imagined for a superhelical protein with the properties of the D-7 family. Because of the number and position of the charged residues of the superhelical dimer, the possibility that the protein sequesters phosphate salts during desiccation, thus avoiding precipitation or crystallization of these salts, is an attractive idea. Experiments designed to test this possibility are being carried out.

As for the functions of the other Lea/water stress proteins that seem to be largely random coils, such functions would have to be very general and independent of size. Since random coil molecules carry little steric information, the role of these proteins would seem not to require any. A role as "reverse chaperones" preserving existing intracellular structures through random electrostatic binding to them, i.e. serving as analogs to hydration water, is one such possibility. Experiments designed to demonstrate such a role are also in progress.

References

Argos P (1989) Editbase, computer program. Purdue Research Foundation and USDA-ARS

Baker JC, Steele C, Dure L III (1988) Sequence and characterization of six Lea proteins and their genes from cotton. Plant Mol Biol 11:277-291

Chou PY, Fasman GD (1978) Prediction of the secondary structure of proteins from their amino acid sequence. Adv Enzymol 47:45-148

Dure L III, Greenway SC, Galau GA (1981) Developmental biochemistry of cottonseed embryogenesis and germination: XIV. Changing mRNA populations as shown by in vitro and in vivo protein synthesis. Biochemistry 20:4162-4168

Dure L III (1993a) Structural motifs in Lea proteins. In: Plant Responses to Cellular Dehydration During Environmental Stress. TJ Close, EA Bray (eds) (In press)

Dure L III (1993b) A repeating 11-mer amino acid motif and plant desiccation. The Plant Journal 3:363-369

Galau GA, Hughes DW, Dure L III (1986) Abscisic acid induction of cloned cotton late embryogenesis abundant (Lea) mRNAs. Plant Mol Biol 7:155-170

Granier J, Osguthorpe DJ, Robson B (1978) Analysis of the accuracy and implications of simple methods for predicting the secondary structure of globular proteins. J Mol Biol 120:97-120

Kurkala S, Franck M (1990) Cloning and characterization of a cold- and ABA-inducible *Arabidopsis* gene. Plant Mol Biol 15:137-144

McCubbin WD, Kay CM, Lane BG (1985) Hydrodynamic and optical properties of the wheat germ Em protein. Can J Biochem Cell Biol 63:803-811

GENE REGULATION BY ABSCISIC ACID AND WATER STRESS IN MAIZE

Miguel Angel Freire, M. Mar Alba, Anders B. Jensen, Mª Françoise Niogret, Adela Goday and Montserrat Pagès.
Departament de Genètica Molecular. Centre d'Investigació i Desenvolupament. C.S.I.C. Jorge Girona Salgado 18-26.
08034 Barcelona. Spain

INTRODUCTION

Abscisic acid (ABA) mediates a large number of physiological and developmental processes in plants such as embryo maturation, seed dormancy and the response of plants to osmotic stress (for a review see Zeevaart and Creelman 1988).

The response of plant cells to ABA includes modulation of specific mRNAs and proteins in different tissues (Skriver and Mundy 1990, Quatrano et al. 1992). Developmental studies in seeds and physiological studies on water-stress indicate that ABA controls quantitative and qualitative differences in the accumulation and repression of specific mRNAs and proteins. During embryogenesis the endogenous levels of ABA increase prior to desiccation of the embryo. The increase in endogenous ABA levels has been correlated with the developmental expression of ABA-responsive genes in mature embryos in different plant species. After germination ABA-responsive mRNAs rapidly decrease or totally disappear. In young embryos exogenously applied ABA induces the synthesis of ABA-responsive mRNAs but also represses the accumulation of specific messages.

When vegetative plant tissues are exposed to water stress, the level of endogenous ABA increases and dramatic changes in the concentration of mRNAs and proteins are also observed (Pla et al 1989). As desiccation progresses, the metabolic activity decreases and ABA-responsive mRNAs accumulate in water-stressed leaves (Gomez et al 1988, Mundy et al 1988). Some water stress-responsive mRNAs are identical to those induced during embryo development (Vilardell et al 1990). Dramatic changes in the

NATO ASI Series, Vol. H 81
Plant Molecular Biology
Edited by G. Coruzzi and P. Puigdomènech

concentration of mRNAs and proteins have also been described during rehydration of water-stressed tissues. Selective stabilization/ repression of the ABA-responsive mRNAs has been ascribed to both transcriptional (Marcotte et al 1988, Mundy et al 1990, Vilardell et al 1991) and post-transcriptional regulation (Quatrano et al 1991, Vilardell et al 1990).

TRANSCRIPTIONAL REGULATION

Although the molecular basis of ABA action on gene expression has been the subject of numerous studies, to date little progress has been made in the elucidation of the intermediates in the ABA signal transduction pathway. Many genes responsive to ABA (*rab*) are controlled at the transcriptional level. Functional dissection of *rab* promoters using transient expression assays in protoplasts has revealed a specific sequence element (ABRE) which appears to confer ABA responsiveness upon the reporter gene (Marcotte et al. 1989, Mundy et al. 1990, Pla et al. 1993). Moreover, leucine zipper-type proteins binding to the ABRE sequence have been identified (Guiltinan et al. 1990, Oeda et al. 1991). However, recent studies have shown that changes in ABA levels do not always accompany gene expression (Pla et al. 1989, Nordin et al. 1991, Espelund et al. 1992).

Our previous studies have shown that *rab* mRNAs are expressed during late embryo development of maize *viviparous* (*vp2* and *vp5*) mutants (Pla et al 1989; Pla et al 1991). These mutants are ABA-deficient and the *viviparous* embryos are devoid of desiccation. No induction of *rab* genes was obtained in vegetative tissues of *vp2* by water-stress treatment. These data seem to indicate that specific expression of *rab* genes in late embryogenesis does not depend on the presence of high levels of endogenous ABA in the tissues and seems to be relatively independent of embryo desiccation.

It has been reported that ABA-responsive gene promoters, the rice *rab* 16B and carrot DC8 promoter, are developmentally regulated during seed formation but are not active in vegetative organs of transgenic tobacco plants following water stress or ABA treatment (Yamaguchi-Shinozaki et al. 1990, Goupil et al. 1992). Our previous results showed that a *rab* 17/CAT transgene was developmentally regulated in transgenic tobacco plants, although the promoter activity was not studied in adult plants (Vilardell et al. 1991). The different ways of regulating *rab* genes in seeds or in plants may reflect either that heterologous promoters are deficient in specific *cis* elements required for ABA or osmotic stress regulation or they may be due to the absence of specific transcription factors in vegetative tissues of tobacco. These results strongly suggest that alternative pathways involving different molecular mechanisms are responsible for seed-specific expression and ABA-osmotic stress induction of *rab* genes. Indeed, preliminary evidence of tissue-specific factors involved in the regulation of *rab* genes has been recently obtained. Distinct regulatory protein factors with apparently similar affinity for the ABRE sequence have been detected in embryo and vegetative tissues of maize (Pla et al. 1993). Taken together, these data suggest that different levels of control are involved in the expression of *rab* genes.

OTHER LEVELS OF CONTROL

During ABA treatment, the relative importance of other levels of control such as post-transcriptional, post-translational and degradation rates affecting mRNA and protein amount is virtually unknown. Post-transcriptional and/or post-translational regulation of the ABA-responsive genes has been described.

```
ZeaMay:MA16       M-AAADVEYR-CFVGGL-AWATSNESLENAFASYGEILDSKVITDRETGRSRGFGFVTFSSENSMLDAIENMNGKELDGRNITVNQAQSR
SorBic:G-rich2    M-AAADVEYR-CFVGGL-AWATNNETLEQAFANFGQVIDSKVITDRETGRSRGFGFVTFSSEQSMLDAIENMNGKELDGRNITVNQAQSR
SorBic:G-rich1    ------------------------NSLHSAFSTYGEVLESKIILDRETQRSRGFGFVTFSSEEAMRSAIEGMNGKELDGANITVNEAQSR
SpiOle:28kD       ------EEAK-LFVGNL-PYDVDSEKLAGIFDAAGVVEIAEVIYNRETDRSRGFGFVTMSTVEEAEKAVELLNGYDMDGRQLTVNKAAPR
DauCar:G-rich     ------------------------------FGD---ITDSKIINDRETGRSRGFGFVTFKDEKSMRDAIEGMNGQELDGRNITVNEAQSR
NicPlu:30kD       ------PDLK-IFVGNL-LFSADSAALAELFERAGNVEMVEVIYDKLTGRSRGFGFVTMSSKEEVEAACQQFNGYELDGRALRVNSGPPP
NicPlu:31kD       ------EDLK-LFVGNL-PFSVDSAALAGLFERAGNVEIVEVIYDKLSGRSRGFGFVTMSTKEEVEAAEQQFNGYEIDGRAIRVNAGPAP
NicSyl:28kD       ------EDAK-LFVGNL-PYDIDSEGLAQLFQQAGVVEIAEVIYNRETDRSRGFGFVTMSTVEEADKAVELYSQYDLNGRLLTVNKAAPR
NicSyl:31kD       ------EDAK-LFVGNL-PYDVDSEGLARLFEQAGVVEIAEVIYNRDTDQSRGFGFVTMSTVEEAEKAVEMYNRYDVNGRLLTVNKAARR
NicSyl:33kD       ------EGGR-LYVGNL-PFSMTSSQLSEIFAEAGTVANVEIVYDRVTDRSRGFAFVTMGSVEEAKEAIRLFDGSQVGGRTVKVNFPEVP
DroMel:Hrb87F     NAGA---TVKKLFVGGLR-DDHDEECLREYFKDFGQIVSVNIVSDKDTGKKRGFAFIEFDDYDPVDKIILQKTHS-IKNKTLDVKKAIAK
XenLae:hnRNPA1a   GAHL---TVKKIFVGGIK-EDTEEDHLREYFEQYGKIEVIEIMTDRGSGKKRGFAFVTFEDHDSVDKIVIQKYHT-VNNHNSQVRKALSK
XenLae:hnRNPA1b   GAHL---TVKKIFVGGIK-EDTEEDHLREYFEQYGKIEVIEIMTDRGSGKKRGFAFVTFEDHDSVDKIVIQKYHT-VNNHNSQVRKALSK
XenLae:Nucleol.   ENKKERDS-RTLFVKNI-PYSTTVEELQEIFEN---AKDIRIPTGKD-GSNKGIAYVEFSNEDEANKALEEKQGAEIEGRSIFVDFTGEK
GalGal:Nucleol.   ENKKERDA-RTLFVKNL-PYRVTEDEMKNVFEN---ALEVRLVLNKE-GSSKGMAYIEFKTEAEAEKALEEKQGTEVDGRAMVIDYTGEK
MesAur:Nucleol.   --KKVRAA-RTLLAKNL-SFNITEDELKEVFED---ALEIRLV-SQD-GKSKGIAYIEFKSEADAEKNLEEKQGAEIDGRSVSLYYTGEK
MusMus:CArG-BFA   --------VKKIFVGGLNPEA-TEEKIREYFGQFGEIEAIELPIDPKLNKRRGFVFITFKEEDPVKKVLEKKFHT-VSGSKCEIKVAQPK
HomSap:hnRNPA-B   --------VKKIFVGGLNPESPTEEKIREYFGEFGEIEAIELPMDPKLNKRRGFVFITFKEEEPVKKVLEKKFHT-VSGSKCEIKVAQPK
HomSap:U1-70kD    DPNAQGDAFKTLFVARV-NYDTTESKLRREFEVYGPIKRIHMVYSKRSGKPRGYAFIEYEHERDMHSAYKHADGKKIDGRRVLVDVERGR

Consensus                    LFVGNL          L   F  FG I                RGFGFV T          A          I G     I
```

Figure 1. The amino acid RNA binding domains of RNP-CS type proteins.

A detailed study of ABA regulation of the Em gene in wheat embryos and in vegetative tissues has shown that ABA may confer selective stability/ translatability on Em mRNA (Quatrano et al.1991). Studies of the regulation of the *rab* 17 gene in embryos and callus cells has shown that post-translational mechanisms regulate the accumulation of the protein (Vilardell et al. 1990). Comparable levels of mRNA are induced specifically in these two tissues by ABA. Highly phosphorylated *rab*17 protein is accumulated in embryos (Goday et al 1988). However, *rab* 17 protein is poorly synthesized in callus cells. All these data indicate that different levels of control regulate mRNA stability and protein accumulation during ABA response.

THE RNA-BINDING 16 kDa GLYCINE-RICH PROTEIN OF MAIZE

We have previously described a 16 kDa glycine-rich protein (MA16) in maize, which is involved in a variety of ABA and stress responses in the plant (Gomez et al 1988).

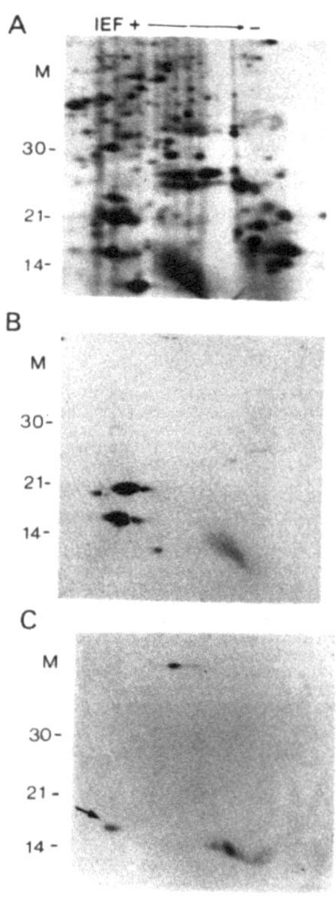

Figure 2. Two-dimensional electrophoretic resolution of translation products from : A) poly A^+ RNA of mature embryos.
B) Hybrid-selected pMAH9 complementary mRNA sequences. C) Full-length cDNA transcript labelled with an arrow.

The sequence of MA16 predicts that it is an RNA-binding protein, since it contains a ribonucleoprotein consensus sequence-type RNA-binding domain (CS-RBD) (Mortenson and Dreyfuss 1988). Figure 1 shows the alignment of MA16 with other plant proteins. Each of the proteins contains two perfect RNA recognition motifs.

The cDNA clone (pMAH9) was obtained by differential screening of a maize dry embryo cDNA library, using cDNA synthesized from an immature embryo poly A^+ RNA with or without ABA treatment. MA16 mRNA was shown to accumulate in maize embryos, roots and leaves after ABA treatment (Gomez et al 1988, Lobreaux et al 1992). MA16 mRNA was also induced in maize leaves upon wounding (Gomez et al 1988), and by mercuric chloride (Didierjean et al 1992), and iron (Lobreaux et al 1992) treatment.

IDENTIFICATION OF MA16

The identity of the protein encoded by pMAH9 cDNA was established by a hybridization-selection/translation assay. Mature embryo mRNAs were hybridized to pMAH9, translated **in vitro**, and the protein products were resolved by two-dimensional gel electrophoresis (figure 2). Unambiguous identification of the protein encoded by pMAH9 was hindered, however, by the presence of other translation products in the two-dimensional gels. Recently, the sequence of a maize glycine-rich protein showing about 75% homology with MA16 has been described (Didierjean et al 1992). The protein CHEM2 has a predicted molecular mass of 15.5 kDa and an isoelectric point of 7.46. Therefore, it is a strong candidate for the cross-hybridization observed in the hybrid/ selection experiment. To distinguish between these polypeptides, **in vitro** transcription/ translation of the full length cDNA was performed. A single protein of 16 kDa and pI 6 was produced which closely matched the predicted molecular mass (15,427 Da) and pI(5,7) of the protein. Further confirmation of the identity of this translation product was obtained by immunoprecipitation with an antiserum (α-19) directed against a 19 amino-acid synthetic peptide, (Gly)3-Tyr-(Gly)2-(Arg)2-(Gly)5-Tyr-(Gly)4, derived from the predicted protein sequence (Ludevid et al 1992). Interestingly , this antibody does not recognize

MA16 by immunoprecipitation of *in vivo* labelled protein extracts. The α-19 antiserum was directed against a synthetic peptide from the glycine-rich domain of the protein. Several RNA-binding proteins contain a glycine-rich domain. It has been proposed that the glycine-rich domain is involved in protein-protein interactions and/or in modulating the interaction of the RNA-binding domain with the polynucleotides (Dreyfuss et al 1988). The lack of accessibility of the glycine-rich- epitope of MA16 to the α-19 antibody may indicate that MA16 specifically interacts with RNAs in the cell.

As a first step to elucidate the function of MA16 within the cell, we produced an antibody that reacted specifically with the MA16 protein synthesized *in vivo*. A β-galactosidase- MA-16 fusion protein was expressed in E. coli, purified and injected into rabbits, yielding a polyclonal antiserum α-MA16. By means of immunoprecipitation of *in vivo* labelled protein extracts and western blot analysis of proteins from mature embryos, α-MA16 specifically recognized MA16 as shown in figure 3.

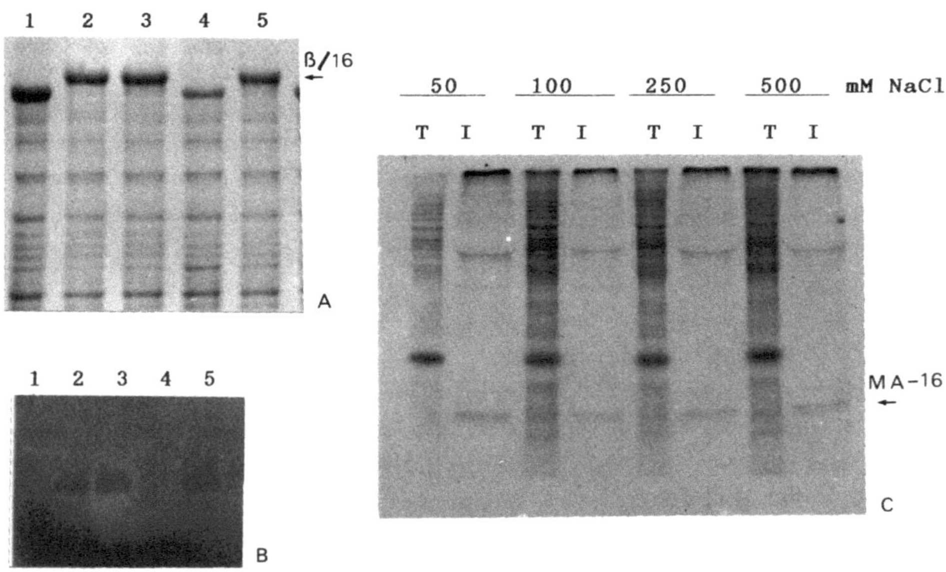

Figure 3. Bacterially expressed β-galactosidase/ MA16 (A), and immunodetection with α-19 (B). Immunoprecipitation of *in vivo* labelled proteins from maize embryos with α-MA16 (C).

HYPOTHETICAL FUNCTION OF MA16

The predicted RNA-binding property of the MA16 protein was assessed by using ribohomopolymer-binding assays (Ludevid et al 1992). The observation that MA16 contains CS-RBD and binds to poly(G) and poly(U) at high salt concentrations demonstrates that, by analogy with other known CS-RBD RNA-binding proteins, for example PABP and several hnRNP proteins (Dreyfuss et al 1988b), this protein is a "bona fide" RNA-binding protein. RNA-binding proteins play important roles in many aspects of RNA metabolism. As a first step in gaining insight into the function of MA16, we analyzed the sequence similarities between MA16 and RNA-binding proteins reported in the literature.

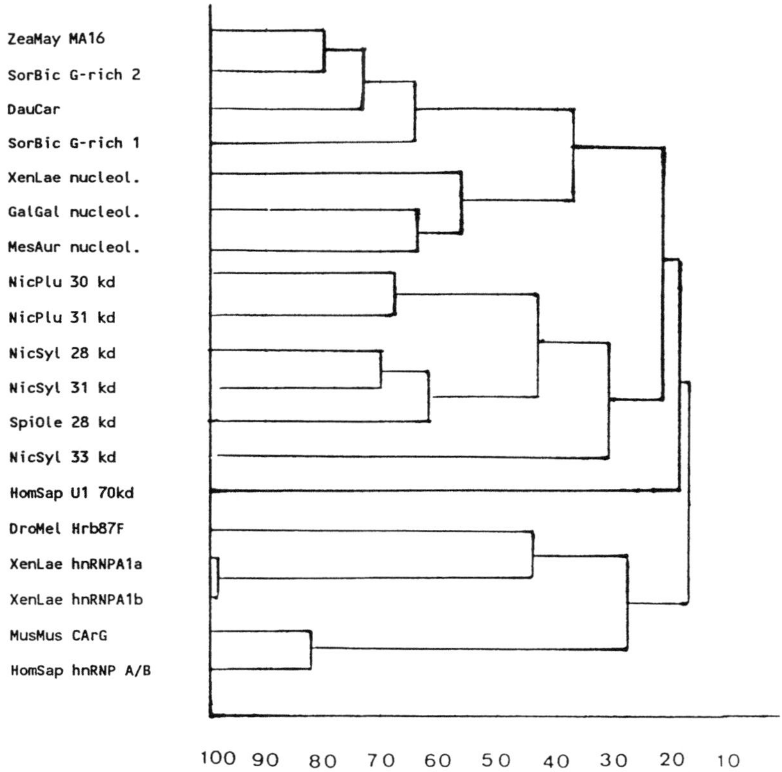

Figure 4. Dendogram showing the relationship between different proteins. Distance along the matrix is proportional to the differences between sequences.

The data shown in the dendogram (figure 4) indicate that the sorghum (Cretin et al 1991) and carrot (Sturm 1992) proteins may be the homologues of the maize protein. Significant homologies were found to tobacco (Mieszczak et al 1992) and spinach RNA-binding proteins (Schuster et al 1991) and to mammalian proteins (Kenan et al. 1991) such as nucleolin and the A1 protein of the heterogeneous nuclear ribonucleoprotein complex. Whether these similarities indicate a functional relationship beyond that due to the CS-RBDs themselves remains to be ascertained.

Current evidence indicates that RNA molecules undergo diverse metabolic processes as a result of developmental or environmental cues, although the molecular mechanisms are largely unknown. Considering the similarity of MA16 to other known RNA-binding proteins and that MA16 is involved in general stress responses, the maize protein may play a role in the maturation of specific RNAs in stress response. Further elucidation of the function of MA16 as well as identification of the MA16 associated RNAs will contribute to the understanding of the molecular mechanisms of ABA action in post-transcriptional regulation of gene expression.

ACKNOWLEDGEMENTS

This work was supported by grant BIO91-0546 from CICYT.

REFERENCES

Cretin, C. Puigdomenech, P (1990) Glycine-rich RNA-binding proteins from *Sorghum vulgare*. Plant Mol. Biol. 15, 783.

Didierjean, L. Frendo, P. Burkard, G. (1992) Stress responses in maize: Sequence analysis of cDNAs encoding glycine-rich proteins. Plant Mol. Biol. 18:847-849.

Dreyfuss, G. Swanson M.S. Piñol-Roma S (1988a). Heterogeneous nuclear ribonucleoprotein particles and the pathway of mRNA formation. Trends Biochem. Sci. 13, 86-91.

Dreyfuss, G.Philipson L Mattaj I.W. (1988b) Ribonucleoprotein particles in cellular processes. The J. of Cell Biol. 106, 1419-1425.

Espelund, M. Saeboe-Larssed, S. Hughes, D.W. Galau, G.A. Larsen, F.Jakobsen, K.S. (1992) Late embryogenesis-abundant genes encoding proteins with different numbers of hydrophilic repeats are regulated differentially by abscisic acid and osmotic stress. The Plant Journal 2 241-252.

Goday, A. Sanchez-Martinez D. Gomez, J. Puigdomenech, P Pages M (1988). Gene expression in developing Zea mays embryos: Regulation by abscisic acid of a highly phosphorylated 23

to 25 kD group of proteins. Plant Physiol. 88, 564-569.
Gomez, J. Sanchez-Martinez, D. Steifel, V. Rigau, J. Puigdomenech, P. Pages, M. (1988) A gene induced by the plant hormone abscisic acid in response to water stress encodes a glycine-rich protein. Nature 334, 262-264.
Goupil, P. Hatzopoulos, P. Franz, G Hempel, F.D. You, R. Sung, Z.R. (1992). Transcriptional regulation of a seed-specific carrot gene DC8. Plant Mol. Biol. 18, 1049-1063.
Guiltinan, M.J. Marcotte, W.R. and Quatrano, R.S. (1990). A plant leucine zipper protein recognizes an abscisic acid response element. Science. 250, 267-270.
Kenan, D.J. Query, Ch.C. Keene, J.D.(1991). RNA recognition: towards identifying determinants of specificity. TIBS 16 214-220.
Lobreaux, S. Hardy, T. Briat, J.F.(1993). Abscisic acid is involved in the iron-induced synthesis of maize ferritin. EMBO J 12: 651-657.
Marcotte, W.R. Bayley, Ch.C. and Quatrano, R.S. (1988). Regulation of a wheat promoter by abscisic acid in rice protoplasts. Nature 335, 454-457.
Mieszczak, M. Klahre, U. Levy, J.H. Goodall, G.J. Fillipowicz, W. (1992) Multiple plant RNA binding proteins identified by PCR. Mol. Gen. Genet. 234, 390-400.
Mortenson, E. Dreyfuss, G. (1989). RNP in maize protein. Nature 337, 312.
Mundy, J. Chua, N.H. (1988) Abscisic acid and water stress induce the expression of a novel rice gene. EMBO J 7:2279-2286.
Mundy, J. Yamaguchi-Shinozaki, K. Chua, N.H. (1990). Nuclear proteins bind conserved elements in the abscisic acid-responsive promoter of a rice RAB gene. Proc. Natl. Acad. Sci. USA 87, 1406-1410.
Nordin, K. Heinno, P. Palva E.T. (1991). Separate signal pathways regulate the expression of a low-temperature-induced gene in Arabidopsis thaliana (L.) Heynh. Plant Mol. Biol. 16, 1061-1071.
Oeda, K. Salinas, J. Chua, N.H. (1991). A tobacco bZip transcription activator (TAF-1) binds to a G-box-like motif conserved in plant genes. EMBO J. 10, 1793-1802.
Pla, M. Goday, A. Vilardell J. Gomez J. and Pages, M.(1989). Differential regulation of ABA-induced 23-25 kDa proteins in embryo and vegetative tissues of the viviparous mutants of maize. Plant Mol. Biol.13, 385-394.
Pla, M. Gomez, J. Goday, A. Pagès, M. (1991). Regulation of the abscisic acid-responsive gene *rab* 28 in maize *viviparous* mutants. Mol. Gen. Genet. 230, 394-400
Pla, M. Vilardell, J. Guiltinan, M.J. Marcotte, W.R. Niogret, M.F. Quatrano, R.S. Pagès, M. (1993) The cis-regulatory element CCACGTGG is involved in ABA and water-stress responses of the maize gene *rab* 28. Plant Mol. Biol. 21 259-266.
Quatrano, R.S. Guiltinan, M.J. Marcotte, W.R. (1992). Regulation of gene expression by abscisic acid. In Control of Plant Gene Expression (Ed. D-PS Verma) The Telford Press, Caldwell N.J. (in press).
Schuster, G. Gruissem, W. (1991) Chloroplast mRNA 3'end processing requires a nuclear-encoded RNA-binding protein. EMBO J 8, 1493-1502.
Skriver, K. and Mundy, J. (1990). Gene expression in response to

abscisic acid and osmotic stress. Plant Cell 2, 503-512.
Sturm, A. (1992). A wound-inducible glycine-rich protein from *Daucus carota* with homology to single-stranded nucleic acid-binding proteins. Plant Physiol. 99, 1689-1692.
Vilardell, J. Goday, A. Freire, M.A. Torrent, M. Martinez, M.C. Torne, J.M. and Pages, M. (1990). Gene sequence, developmental expression, and protein phosphorylation of RAB-17 in maize. Plant Mol. Biol. 14, 423-432.
Vilardell, J. Mundy. J. Stilling, B. Leroux, B. Pla, M. Freyssinet, G. Pages, M. (1991). Regulation of the maize *rab* 17 gene promoter in transgenic heterologous systems. Plant Mol. Biol.17 985-993.
Yamaguchi-Shinozaki, K. Mino, M. Mundy, J. and Chua, N.H. (1990). Analysis of an ABA-responsive rice gene promoter in transgenic tobacco. Plant Mol. Biol. 15, 905-912.
Zeevaart, J.A.D. Creelman, R.A. (1988). Metabolism and physiology of abscisic acid. Ann. Rev. Plant Physiol. Plant Mol. Biol. 39, 439-473.

Analysis and regulation of gene expression in the resurrection plant *Craterostigma plantagineum*

Dorothea Bartels, Pekka Heino, Don Nelson, Detlef Michel, Antonella Furini, Giovanni Bernacchia, Riccardo Velasco, Renza Roncarati, Ralph Elster, Gerhard Schwall, Josefa Alamillo, Francesco Salamini
Max-Planck-Institut für Züchtungsforschung
Carl-von-Linné-Weg 10
D-50829 KÖLN: Germany

Introduction

Plant responses to water loss are very complex and depend on the severity of the water loss as well as on the developmental stage and the physiological condition of the affected plants. To study responses to protoplastic dehydration at the molecular level we initiated research with the objective to identify gene products which may contribute to water stress tolerance. The unique ability of resurrection plants to withstand severe water loss greater than 90% make them a suitable system to study water stress tolerance: upon rewatering these plants recover quickly from the stress (Gaff 1971, Bartels et al. 1990). We have chosen *Craterostigma plantagineum* as a representative of the resurrection plants for our studies. An attractive feature of this resurrection plant is that we can analyse desiccation tolerance in undifferentiated callus tissue, too. Callus withstands rapid dehydration if it is treated with the plant hormone abscisic acid (ABA) prior to the drying treatment (Bartels et al. 1990). This makes two experimental systems available from the same plant for the isolation of molecular components relevant to desiccation. Initially we isolated a large number of cDNA clones which are induced upon dehydration and/or ABA treatment (Bartels et al. 1990, Piatkowski et al. 1990, Bartels et al. 1992). The purpose of this chapter is to summarize some characteristic structural and regulatory features of the expressed genes.

NATO ASI Series, Vol. H 81
Plant Molecular Biology
Edited by G. Coruzzi and P. Puigdomènech

Results and Discussion

Analysis of cDNA clones: Initially we isolated a large number of cDNA clones which were induced during dehydration using a simple differential screening approach. A characteristic feature of these clones is their rapid induction during the first hours of water loss. The majority of the abundant cDNA clones could be classified into eleven sequence-related families based on cross hybridization studies. DNA sequence information is available on all the abundant cDNA clones. As a result of the sequence analysis the majority of the transcripts are related to Lea (= late embryogenesis abundant) type genes (Galau et al. 1986). Lea genes are characterized by the following features: abundantly expressed in mature embryos, transcripts can be prematurely induced by ABA in embryos or they may be responsive to osmotic stress imposed on vegetative tissues (Skriver and Mundy 1990, Bray 1991). The deduced proteins are usually hydrophilic and several contain regions which most likely exist as α-helices. According to their sequence motifs a number of different groups of Lea genes have been defined (Baker et al. 1988, Dure et al. 1989, Dure 1993). Desiccation induced transcripts from *Craterostigma* can be assigned to at least seven different types of Lea genes. An overview is given in Table 1. The degree of homology varies: conservation may be restricted to particular sequence motifs and sequence structures (e.g. pcC11-45, Bartels et al., unpublished) or genes can be highly identical like the recently discovered homology between pcC27-45 (Piatkowski et al. 1990) and the cotton Lea 14 gene (Galau et al. 1993); a comparison of the deduced amino acid sequence is given in Fig. 1. The high expression of these different Lea type genes in dehydrated vegetative tissues of *Craterostigma* suggests that similar metabolic processes are occurring in the maturation phase of seeds to which desiccation tolerance is restricted during the life cycle of higher plants.

```
                 .         .         .         .         .
  1 MAQLMNKAKNFVAEKVANVFKPKASVEDVDLKDVGRHGITYLTRICVENP
    |.||::|||:||.:||||:.||.|||.|||||.|:|..:.| .:::|.||
  1 MSQLLEKAKDFVVDKVANIKKPEASVSDVDLKHVSRECVEYGAKVSVSNP

                 .         .         .         .         .
 51 YSASIPVGEIKYTLKSAGRVIVSGNIPDPGSLKGNDKTMLEPAIKVPHSA
    || |||:.||.|.::||||.|.||.||||||||:.|.|||:.::|||..
 51 YSHSIPICEISYNFRSAGRGIASGTIPDPGSLKASDTTMLDVPVKVPYNI

                 .         .         .         .         .
101 LVSLIKDIGADMDIDYVLELGLVVDLPVIGNFTIPLSHKGEMKLPGLSDI
    ||||:|||||| |||| |||||.:|||::||||||||:|||:|||.||||
101 LVSLVKDIGADWDIDYELELGLTIDLPIVGNFTIPLSQKGEIKLPTLSDI

151 F
    |
151 F
```

Fig. 1: Alignment of the amino acids deduced from pcC27-45 (*Craterostigma*) and the Lea 14 expressed in cotton embryos; the sequences are 65.5% identical and 81% similar

<u>Functional aspects</u>: The function of these Lea gene products is up to now unknown, but the tools to investigate functional aspects are available. Hypotheses for the role of Lea proteins during dehydration have been proposed by Dure et al. 1989 and Dure 1993. It is remarkable that the Lea type genes are not only conserved among higher plants but immunologically related proteins have also been observed in lower eucaryotes like e.g. in cyanobacteria (Close and Lammers 1993). Overexpression of three different *Craterostigma* proteins in tobacco did not lead to altered phenotypes of the transformed tobacco plants (Iturriaga et al. 1992). This suggests that these proteins do not interfere with the normal metabolism and makes it unlikely that they encode proteins with enzymatic properties. Characteristic for the D11 group is that nearly all genes assigned to this group contain a stretch of serines. The serine residues in the D11-maize proteins expressed in embryos are phosphorylated (Goday et al. 1988). By *in vivo* labelling and immunoprecipitation it was shown that also the homologous proteins in *Craterostigma* leaves are phosphorylated (Fig. 2). Whether the phosphorylation is essential for the function of

the proteins is unclear; it is also not known whether the Lea D11 proteins are always phosphorylated. The study of mutated proteins will answer these questions. All the *Craterostigma* proteins encoded by the cDNAs listed in Table 1 were analysed by *in vivo* labelling whether they may occur in phosphorylated forms. With the exception of pcC6-19 and the related pcC27-04 none of the other proteins was found to be phosphorylated.

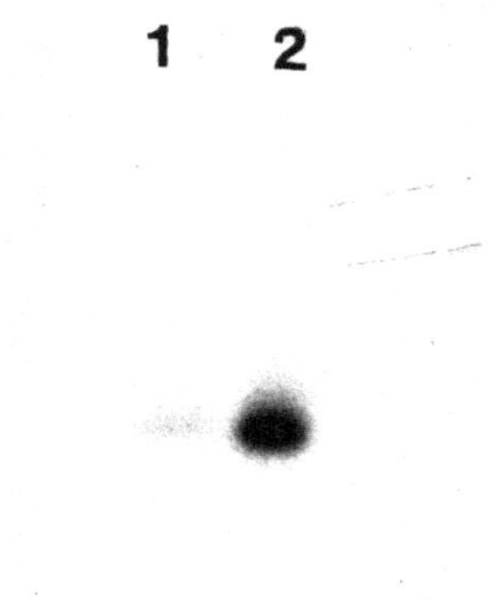

Fig. 2: Phosphorylation of pcC6-19 induced by ABA in *Craterostigma* leaves. Leaf samples were incubated in 500 µl 10 mM pipes buffer pH 6.8 containing 10^{-4} M ABA for 2.5 h before 800 µCi anorganic ^{32}P-labelled phosphate was added. The leaves were labelled for 4 hours, then the soluble proteins were extracted and immunoprecipitated with antiserum directed against pcC6-19 encoded proteins. The precipitated proteins were separated on a SDS-PAGE gel and visualized by autoradiography. (Lane 1 represents samples not treated with ABA and lane 2 contains the ABA-treated protein extracts.)

Regulation of gene expression: All the desiccation induced genes from *Craterostigma* can also be induced by ABA. In the callus the ABA treatment is connected with the tolerant phenotype. An important aspect of this research is to understand the regulatory cascade starting with the dehydration stress and resulting in gene expression most likely involving ABA in the signalling chain. One approach to

Table 1: **Lea type genes inducible in Craterostigma leaves by drying or ABA**

Craterostigma cDNA clone	homologous Lea type gene	characteristic features	References[1]
pcC6-19	cotton Lea D11	genes in many plants identified, contain stretch of serines and lysine rich motif	Bray 1992 Baker et al. 1988 Skriver & Mundy 1990
pcC34-62	cotton Lea D7 brassica pLea 76		Baker et al. 1988 Harada et al. 1989
pcC163	cotton Lea D29		Baker et al. 1988
pcC26	cotton Lea D113		Baker et al. 1988
pcC27-45	cotton Lea 14		Galau et al. 1993
pcC11-24	*Arabidopsis* LTI 65 " " *rd29*	also cold inducible only partial homology	Nordin et al. 1993 Yamaguchi-Shinozaki & Shinozaki 1993
pcC3-06	DC8 carrot wheat group 3 Lea barley ABA inducible gene		Franz et al. 1989 Curry & Walker-Simmons 1993 Hong et al. 1988

[1]The references for the pcC *Craterostigma* cDNA clones are given in Bartels et al. 1990, Piatkowski et al. 1990 and Bartels et al. 1993, in press.

tackle these questions is the isolation of promoters in order to identify important regulatory elements. Promoters from three different genes (CDeT27-45, CDeT6-19, cDeT11-24) were selected for this research. A functional analysis of the three promoters should reveal common regulatory principles. The experimental approach for this analysis is the construction of various promoter GUS fusions which are tested in a homologous transient expression assay (transfected callus protoplasts) for ABA responsiveness or which are transferred into other plants and analysed for their abilities to drive GUS expression.

For the CDeT27-45 promoter a 5' upstream region was identified which is important in conferring ABA inducibility to the promoter in transient assays (Michel et al. 1993). The importance of this fragment could be further substantiated by DNA protein binding studies (Nelson and Bartels, submitted).

The activity of this promoter is being analysed in tobacco, *Arabidopsis* and *Craterostigma* plants. A remarkable observation was that in tobacco this promoter is highly active only in embryo tissues of maturing tobacco seeds (Michel et al. 1993). The promoter could not be activated in tobacco leaves with ABA treatments. This was surprising and suggests that some signalling intermediates are not active in the tobacco plant. The situation appears to be different in *Arabidopsis*: like in tobacco the highest activities are measured in seeds, but in young seedlings of transformed plants the promoter can be induced by ABA. This ABA responsiveness, however, is lost during development and in mature leaves the promoter cannot be triggered with ABA (Furini and Bartels, unpublished). To assure that the promoter contains all the elements necessary to respond to dehydration or ABA the analysis of the transgenic *Craterostigma* plants is in progress. The promoter of the CDeT6-19 genes shows some pronounced differences, compared with CDeT27-45: 1. in transgenic tobacco, *Arabidopsis* and *Craterostigma* plants constitutive expression was measured (Michel, Furini, Bartels unpublished), 2. in tobacco the promoter could be activated by ABA or drying (Michel and Bartels unpublished). The ABA-mediated expression of pcC6-19 requires no *de novo* protein synthesis in contrast to pcC27-45 (Bartels et al. 1992, Bartels unpublished), this supports the notion of a partially different activation pathway.

Conclusion

Analysis of gene expression and the promoter analysis available up to now suggests that pathway leading to desiccation tolerance in the resurrection plant *Craterostigma plantagineum* is very similar to the ones found for seeds in higher plants. This suggests that the genetic information necessary for desiccation tolerance in vegetative tissues is possibly available in many higher plants. Desiccation tolerance is a multifactorial phenomenon and it might as one part require the expression of an array of different Lea genes; however, it is expected that also genes involved in the synthesis of osmolytes and membrane channel proteins are essential.

References

Baker JC, Steele C, Dure III L. (1988) Sequence and characterization of 6 Lea proteins and their genes from cotton. Plant Mol Biol 11:277-291.

Bartels D, Hanke C, Schneider K, Michel D, Salamini F (1992) A desiccation-related Elip like gene from the resurrection plant *Craterostigma plantagineum* is regulated by light and ABA. EMBO J 11:2771-2778.

Bartels D, Schneider K, Terstappen G, Piatkowski D, Salamini F (1990) Molecular cloning of ABA-modulated genes from the resurrection plant *Craterostigma plantagineum* which are induced during desiccation. Planta 181:27-34.

Bartels D, Velasco R, Schneider K, forlani F, furini A, Salamini F (1993) Resurrection plants as model systems to study desiccation tolerance in higher plants. In: Mabry T, Nguyen H, Dixon R eds, Applications and prospects of biotechnology for arid and semi-arid lands. In press.

Bray E (1991) Regulation of gene expression by endogenous ABA during drought stress. In: Abscisic acid physiology and biochemistry (Davies WJ, Jones HG, eds), Bios Scientific Publishers, Oxford, UK.

Close TJ, Lammers PJ (1993) An osmotic stress protein of cyanobacteria is immunologically to plant dehydrins. Plant Physiol 101:773-779.

Curry J, Walker-Simmons MK (1993) Unusual sequence of group 3 LEA (II) mRNA inducible by dehydration stress in wheat. Plant Mol Biol 21:907-912.

Dure III L (1993) A repeating 11-mer amino acid motif and plant desiccation. The Plant Journal 3:363-369.

Dure III L, Crouch M, Harada J, Ho T-H, Mundy J, Quatrano R, Thomas T, Sung ZR (1989) Common amino acid sequence domains among the Lea proteins of higher plants. Plant Mol Biol 12:475-486.

Franz G, Hatzopoulos P, Jones TJ, Krauss M, Sung ZR (1989) Molecular and genetic analysis of an embryonic gene, DC 8 from *Caucus carota* L. Mol Gen Genet 218:143-151.

Gaff DF (1971) Desiccation tolerant flowering plants in Southern Africa. Science 174:1033-1034.

Galau GA, Hughes DW, Dure III L (1986) Abscisic acid induction of cloned cotton late embryogenesis abundant (Lea) mRNAs. Plant Mol Biol 7:155-170.

Galau GA, Wang H, Hughes DW (1993) Cotton Lea5 and Lea14 encode atypical late embryogenesis-abundant proteins. Plant Physiol 101:695-696.

Goday A, Sanchez-Martinez D, Gómez H, Puigdomènech P, Pagès M (1988) Gene expression in developing *Zea mays* embryos: Regulation by abscisic acid of a highly phosphorylated 23- to 25-kD group of proteins. Plant Physiol 88:564-569.

Harada JJ, DeLisle AJ, Baden CS, Crouch ML (1989) Unusual sequence of an abscisic acid-inducible mRNA which accumulates late in *Brassica napus* seed development. Plant Mol Biol 12:395-401.

Hong B, Uknes SJ, Ho T-hD (1988) Cloning and characterization of a cDNA encoding a mRNA rapidly induced by ABA in barley aleurone layers. Plant Mol Biol 99:495-506.

Iturriaga G, Schneider K, Salamini F, Bartels D (1992) Expression of desiccation-related proteins from the resurrection plant *Craterostigma plantagineum* in transgenic tobacco. Plant Mol Biol 20 20:555-558.

Michel D, Salamini F, Bartels D, Dale P, Baga M, Szalay A (1993) Analysis of a desiccation and ABA-responsive promoter isolated from the resurrection plant *Craterostigma plantagineum*. The Plant Journal 4 (in press).

Nordin K, Vahala T, Palva ET (1993) Differential expression of two related, low-temperature-induced genes in *Arabidopsis thaliana* (L.) Heynh. Plant Mol Biol 21:641-653.

Piatkowski D, Schneider K, Salamini F, Bartels D (1990) Characterization of five abscisic acid-responsive cDNA clones isolated from the desiccation-tolerant plant

Craterostigma plantagineum and their relationship to other water-stress genes. Plant Physiol 94:1682-1688.
Skriver K, Mundy J (1990) Gene expression in response to abscisic acid and osmotic stress. The Plant Cell 2:503-512.
Yamaguchi-Shinozaki K, Shinozaki K (1993) Characterization of the expression of a desiccation-responsive *rd29* gene of *arabidopsis thaliana* and analysis of its promoter in transgenic plants. Mol Gen Genet 236:331-340.

GENE EXPRESSION AND PLANT RESPONSES TO WATER DEFICIT

John E. Mullet
Department of Biochemistry and Biophysics
Texas A&M University
College Station, Texas 77843
United States

Introduction

A large percentage of the world's crops are exposed to chronic or sporadic periods of drought (Boyer, 1982). In addition, even in the absence of drought, the timing and amount of rainfall are primary determinants of crop selection and yield (Ludlow and Muchow, 1990; Boyer, 1982). Boyer (1982) estimated that 44.7% of U.S. soils are subject to low water availability and drought accounted for the highest proportion of insurance indemnities for crop losses in the U.S. from 1939 to 1978. Improved drought tolerance in crop species would result in improved yield stability.

One approach to improve crop performance in water limited environments is to select for genotypes which have improved yield in this environment. This approach has proved partially successful but difficult due to the variability of rainfall and the polygenic nature of drought tolerance. An additional approach to improve drought tolerance involves identification and selection of traits which contribute to drought avoidance, drought tolerance, or water use efficiency. The list of potentially important traits is extensive because water deficit can negatively affect nearly all stages of the plant's life cycle. In addition, responses to drought vary depending on plant history, duration and severity of water deficit and compounding biotic or abiotic stress. A partial list of potentially important traits includes; matching plant phenology to available water supply, seedling dessication tolerance, water extraction efficiency (i.e., root/shoot ratios, root system), water use efficiency (i.e., conductance, stomatal control, cuticle characteristics), osmotic adjustement, and modulation of partitioning to maintain reproductive structures during late season drought. Most of these traits are complex and our understanding or their control limited.

NATO ASI Series, Vol. H 81
Plant Molecular Biology
Edited by G. Coruzzi and P. Puigdomènech

The list of traits contributing to drought tolerance suggests two different strategies to identify genes important for drought tolerance. First, genotypes, which in field conditions exhibit drought tolerance, can be characterized in order to understand the physiological and biochemical basis of tolerance. Crosses between succeptible and tolerant lines can be constructed to determine the genetic complexity of tolerance. Furthermore, application of RFLP or related genetic marker approaches will allow the genes controlling traits contributing to drought tolerance to be mapped. For example, Rosenow (Rosenow, 1987) found that sorghum lines with the "stay green" phenotype (i.e., B35) exhibit improved post-flowering drought tolerance compared to the normal senescent lines (i.e., Tx7000). Physiological studies indicated that the basis of this response may be continued photosynthesis in the "stay green" lines which provides carbon to developing reproductive structures. Sorghum was found to have the unusual characteristic of maintaining open stomata once flowering occurs even in water limiting conditions. Other studies indicated that cytokinin levels remain elevated in "stay green" lines and that root growth continued for longer in these genotypes compared to senescent lines. Recently, Dr. Rosenow, Dr. Nguyen and Dr. Kumaravadivel have established 160 F6 lines from a cross of B35 and Tx7000. The F6 lines have been characterized for segregation of "stay green" and drought tolerance. We are presently applying RFLP and RAPD markers to map the DNA loci controlling this trait. Eventually, Dr. Rosenow will use this information to help build new genotypes which have improved drought tolerance.

In the example described above, it would be desirable to clone the genes which control "stay green". This would allow more precise gene tracking when constructing genotypes, allow greater genotypic variability to be generated throught transgenic plants and provide a better understanding of the biochemical basis of the response. At present, isolation of genes such as the ones controlling "stay green" from sorghum is difficult. A more general limitation of this approach is that variation selected in the field (or that generated by mutation) most often identifies genes wich regulate several downstream processes because these genes cause the largest changes in phenotype. For example, mutations which affect plant water relations or seed dessication tolerance most often alter ABA responsive genes which, if modified, would have a smaller impact on plant's phenotype.

The study of genes controlling known biochemical pathways and the isolation of genes which are induced in response to water deficit provides a complementary approach to the one described above. Once isolated, biochemical and genetic (over expression, antisense) approaches can be applied to determine the role a specific gene plays in plant responses to water deficit. In addition, the relationship between inducible genes and genes controlling traits conferring drought tolerance in field conditions can be established. As a first step towards this latter objective, we are mapping dehydration inducible genes on the sorghum genetic map (RFLP). This data will be related to information on DNA loci which contribute to sorghum drought tolerance under field conditions.

Isolation of genes responsive to water deficit

We (Guerrero et al.,1990; Creelman and Mullet, 1991) and others (i.e., Yamaguchii-Shinozaki et al.,1992; Mundy and Chua,1998; Cohen and Bray,1990; Bartels et al.,1990; Claes et al.,1990) have used differential cDNA screening to identify genes which show altered expression in plants exposed to water deficit. Choice of starting material and water deficit condition determines the types of genes obtained from this approach. We have isolated cDNAs corresponding to genes with water deficit modulated expression from pea leaves (Guerrero et al.,1990) and the soybean hypocotyl zone of cell elongation (Creelman and Mullet, 1991). In addition, we have examined global changes in gene expression in roots (Creelman et al., 1990). In both cases (pea leaves and soybean stems) we subjected the plants to defined water deficits which were mild. Pea leaves were dehydrated until they lost 10% of their fresh weight. Leaves were then incubated at 100% relative humidity to prevent further water loss and changes in gene expression were monitored for up to 4 hrs. This treatment reduced cell turgor to near zero but did not subject cells to great loss of cell volume or severe negative osmotic potentials. Soybean seedlings were grown in darkness in 100% relative humidity in well watered conditions and then transferred to vermiculite having a water potential of -0,3 MPa (still at 100% relative humidity). This relatively mild water deficit causes stem growth to cease for up to 20hrs followed by gradual recovery. Root growth after a very short period of inhibition, continues at control rates (Creelman et al., 1990). Changes in water status, growth, and cell extensibility in these soybean seedlings have been extensively studied by Boyer and colleagues (i.e., Steudle and Boyer, 1985; Meyer and Boyer, 1981; Bozarth et al., 1987).

In both peas and soybeans, we first established that mRNA populations were altered in specific ways by the water deficit treatment and then isolated cDNAs corresponding to the modulated mRNA (Guerrero and Mullet, 1988; Mason et al.,1988). The results of these studies have been related to changes in abscisic acid which accumulates in dehydrated tissue (Creelman et al., 1990; Mason et al.,1988) and to changes which may reflect altered metabolic status or osmotic adjustments involved in cell adaptation to water limiting conditions.

Water deficit, abscisic acid and lipoxygenase

Water deficit is known to alter hormone levels (rev., Davies et al., 1986). In particular, ABA levels increase within 30 min following reduction of cell turgor (Guerrero and Mullet, 1988). The accumulation of ABA in water limited plants has numerous functions including inhibition of stomatal opening (although probably not initial closure) (i.e., Cummins et al., 1971), differential inhibition of shoot growth (Creelman et al., 1990), maintenance of root growth (Saab et al., 1990), modulation of ion transport (Biddington and Bearman, 1982), water permeability (Markhart, 1984) and induction of ABA responsive genes (rev. Dure et al., 1989). ABA responsive genes include the dehydrins which encode soluble, boiling soluble proteins which may play a role in protecting plant tissues from dehydration induced damage (Close et al. 1989; Dure et al.,1989). The dehydrins (lea's) accumulate during seed maturation and in vegetative tissue of the resurrection plant which tolerates severe dehydration (Bartels et al.,1990). This suggests that some ABA inducible genes help protect of plants from damage during relatively severe water loss. In our experiments using mild water deficits, relatively few of the changes in gene expression were caused by ABA. However, ABA accumulates in the tissues we subjected to mild water deficit to a level sufficient to contribute to differential inhibition of shoot vs root growth (Creelman et al., 1990).

Induction of ABA biosynthesis in dehydrated tissues requires protein synthesis and can be inhibited by pretreatment with inhibitors of transcription (Guerrero and Mullet, 1986) or translation (Guerrero and Mullet, 1986). The induction of ABA biosynthetic capacity in response to dehydration is rapid ocurring within 30 mins post-loss of cell turgor (i.e., Guerrero and Mullet, 1988). ABA which accumulates in water deficient plants is synthesized *de novo* and is derived from an oxygenated carotenoid such as violaxanthin (Robichaud et al., 1980). Evidence has accumulated that a non-heme dioxygenase is involved in ABA biosynthesis probably mediating the conversion of oxygenated carotenoids to xanthoxin which is processed further to ABA (Creelman et al., 1992). The dioxygenase may be the rate limiting step in ABA biosynthesis and subject to regulation in water deficient plants. This idea is based on the large pool of oxygenated carotenoids present in plants (Gamble and Mullet, 1986; Rock and Zeevaart,1991) and the lack of change in the rate of conversion of xanthoxin to ABA in extracts from control and water stressed tissues (Sindu and Walton, 1987). Recent studies have shown that induction of ABA biosynthesis in response to tissue dehydration can be blocked by inhibitors of lipoxygenase (Creelman et al.,1992) and that *lox* mRNA levels rapidly increase when pea leaves are dehydrated to zero turgor (Bell and Mullet, 1991).

Changes in gene expression related to altered growth

Growth is extremely sensitive to water deficit. This is not surprising because water deficit can reduce cell turgor which is required for growth. In addition, water movement into growing zones can be disrupted by changes in water status which reduce water potential gradients from the xylem to growing zones. Gene expression is growing zones changes in response to water deficit. In soybean, the polysome content of the stem growing zone is reduced in response to water deficit. RNA populations also change in response to water deficit (Mason et al., 1988). Analysis of cDNAs revealed that genes encoding tubulin and actin were down regulated in stem growing zones in response to water deficit (Creelman and Mullet, 1991). This makes sense because cell growth in stems stops in water limited plants reducing the requirement for tubulin and actin. Expression of cell wall proteins is also altered in the stem zone of cell elongation (Creelman and Mullet, 1991). This may relate to changes in cell wall extensibility which occur under these conditions. Interestingly, HRGPs which may make the cell wall non-extensible and genes encoding glycine rich proteins, are down regulated in the stem growing zone whereas PRP-1 is up regulated. Perhaps these changes make the wall less extensible but not irreversibly so.

Altered transport, metabolism, osmotic adjustment and gene expression

Hanson and Hitz (1982) reviewed metabolic limitations which occur as a consequence of water deficit and benefits derived from adjustments allowing continued growth and development in water limited environments. I think it is useful to consider changes in gene expression induced by water deficit within this same context. For example, water deficit leads to stomatal closure and reduction of photosynthesis. In addition, reduced transpiration results in a decrease of nitrate arrival to the leaf which eventually limits amino acid biosynthesis. Reduction of cell turgor is also known to change transport through plasmodesmata as well as other membrane channels (rev., Oparka, 1990). These observations suggest that water deficit could elicit changes in gene expression which help cells to adjust to these altered metabolic conditions.

Alterations in transport, cell metabolism and the need for osmotic adjustment in response to water deficit may help explain increased expression of three genes we isolated from pea leaves. One of these genes encodes a thiol protease which has high homology to a vacuolar thiol protease from soybean including a signal sequence and putative vacuolar targeting domain (Kalinski et al., 1992). One rationale for induction of a vacuolar thiol protease in water limited plants is to release amino acids from vegetative storage proteins located in the vacuole. The amino acids could subsequently be utilized in protein biosynthesis, or accumulation of glycine betaine or proline. A second water deficit inducible gene we have characterized encodes a protein with homology to aldehyde dehydrogenases (Guerrero et al., 1990). While the substrate for this putative enzyme is unknown, the protein shows homology to betaine aldehyde dehydrogenase (Weretilnyk and Hanson, 1990). A third water deficit inducible gene (*trg*31 [turgor responsive gene], formerly named cDNA 7a) encodes a membrane protein containing 6 membrane spanning domains (Guerrero et al.,1990). When this gene was first sequenced in 1989, it showed homolgy to *nod*26, a peribacteroid membrane protein (Miao et al.,1992), *glp*F, the glycerol facilitator protein from E.*coli* (Sweet et al., 1990) and MIP26, a membrane intrinsic protein of the bovine lens (Paul and Goodenough, 1983). In recent years several additional homologous proteins/genes have been isolated including; TIP (tonoplast intrinsic protein), *bib* (found in drosophila), *tob*RB7 (root specific gene in tobacco) and *fps*1 (complements *fdp*1 in yeast) (Paul and Goodenough, 1983).

The cDNA corresponding to *trg*31 was isolated by differentially screening a cDNA library from pea leaf mRNA for genes induced in response to reduced cell turgor (Guerrero et al.,1990). Subsequently it was shown that mRNA hybridizing to *trg*31 accumulated rapidly when excised stems or leaves were dehydrated or when hydroponically grown seedlings were exposed to 0.6 M mannitol (Guerrero et al.,1990). Run-on nuclear transcription assays showed that the increase in *tgr*31 mRNA was paralleled by increased transcription at least in the initial phase of accumulation. In non-stressed 10 day old seedlings, *trg*31 mRNA was low in leaves and stems but relatively high in roots (Guerrero et al.,1990). In addition, some induction could be obtained when seedlings were treated with ABA or after heat shock. Interestingly, a similar gene was isolated by Yamaguchi-Shinozaki et al. in 1992 by differentially screening an Arabidopsis cDNA library for dehydration inducible genes (Yamaguchi-Shinozaki et al.;1992). In addition, we have recently isolated a cDNA from sorghum which encodes a protein with 84% sequence similarity to pea *trg*31. Expression of the sorghum gene is also induced when seedlings are subjected to water deficit. These results indicate that *trg*31 may play a general role in plant responses to water deficit.

I. REFERENCES

Bartels, D., Schneider, K., Terstappen, G., Piatkowski, D., Salamini, F. (1990) Molecular cloning of abscisic acid-modulated genes which are induced during dessication of the resurrection plant *Craterostigma plantagineum*. Planta 181:27-34.

Bell, E., Mullet, J.E. (1991) Lipoxygenase gene expression is modulated in plants by water deficit, wounding, and methyl jasmonate. Mol. Gen. Genet. 230:456-462.

Biddington, N.L., Dearman, A.S., (1982) The effect of abscisic acid on root and shoot growth of cauliflower plants. In: Plant Growth Regulation, Vol. 1:15-24, The Hague, Dr. W.Junk Publishers.

Boyer, J.S.(1982) Plant productivity and environment. Science 218:444-448.

Bozarth, C.S., Mullet, J.E., Boyer, J.S. (1987) Cell wall proteins at low water potentials. Plant Physiol. 85:261-267.

Claes, B., Dekeyser, R., Villarroel, R., Van den Bulcke, M., Bauw, G., Van Montagu, M., Caplan, A. (1990) Characterization of a rice gene showing organ-specific expression in response to salt stress and drought. The Plant Cell 2:19-27.

Close, T.J., Kortt, A.A., Chandler,P.M. (1989) A cDNA-based comparison of dehydration-induced proteins (dehydrins) in barley and corn. Plant Mol.Biol. 13.95-108.

Cohen, A., Bray, E.A. (1990) Characterization of three mRNAs that accumulate in wilted tomato leaves in response to elevated levels of endogenous abscisic acid. Planta 182:27-33.

Creelman, R.A., Bell, E., Mullet, J.E. (1992) Involvement of a lipoxygenase-like enzyme in abscisic acid biosynthesis. Plant Physiol. 99:1258-1260.

Creelman, R.A., Mason, H.S., Bensen, R.J., Boyer, J.S., Mullet, J.E. (1990) Water deficit and abscisic acid cause differential inhibition of shoots *versus* root growth in soybean seedlings. Plant Physiol. 92:205-214.

Creelman, R.A., Mullet, J.E. (1991) Water deficit modulates gene expression in growing zones of soybean seedlings. Analysis of differentially expressed cDNAs, a new b-tubulin gene, and expression of genes encoding cell wall proteins. Plant Mol. Biol. 17:591-608.

Cummins, W.R., Kende, H., Raschke, K. (1971) Specifity and reversibility of the rapid stomatal response to abscisic acid. Planta 99:347-351.

Davies, W.J., Metcalfe, J., Lodge, T.A., da Costa, A.R. (1986) Plant growth substances and the regulation of growth under drought. Aust. J. Plant Physiol. 13:105-125.

Dure, L., Crouch, M., Harada, J., Ho, T.H.D., Mundy, J., Quatrano, R., Thomas, T., Sung, Z.R. (1989) Common amino acid sequence domains among the LEA proteins of higher plants. Plant Mol. Biol. 12:475-486.

Gamble, P.E., Mullet, J.R. (1986) Inhibition of carotenoid accumulation and abscisic acid biosynthesis in fluridone-treated dark-grown barley. Eur. J. Biochem. 160:117-121.

Guerrero, F., Mullet, J.E. (1986) Increased abscisic acid biosyntesis during plant dehydration requires transcription. Plant Physiol. 80:588-591.

Guerrero, F.D., Jones, J.T., Mullet, J.E. (1990) Turgor-responsive gene transcription and RNA levels increase rapidly when pea shoots are wilted. Sequence and expression of three inducible genes. Plant Mol. Biol. 15:11-26.

Guerrero, F.D., Mullet, J.E. (1988) Reduction of turgor induces rapid changes in leaf translatable RNA. Plant Physiol. 88:401-408.

Hanson, A.D., Hitz., W.D. (1982) Metabolic responses of mesophytes to plant water deficits. 33:163-203.

Hong, B., Barg, R., Ho, T.H.D. (1992) Developmental and organ-specific expression of an ABA-and stress-induced protein in barley. Plant Mol. Biol. 18:663-674.

Kalinsky, A., Melroy, D.L., Dwivedi, R.S., Herman, E.M. (1992) A soybean vacuolar protein (P34) related to thiol proteases is synthesized as a glycoprotein precursor during seed maturation. J. Biol. Chem. 267:12068-12076.

Ludlow, M.M., Muchow, R.C. (1990) A critical evaluation of traits for improving crop yields in water-limited environments. Advances in Agronomy 43:107-153.

Markhart, A.H.III (1984) Plant Physiol. 74:81-83.

Mason, H.S., Mullet, J.E., Boyer, J.S. (1988) Polysomes, messenger RNA, and growth in soybean stems during development and water deficit. Plant Physiol. 86:725-733.

Meyer, R.F., Boyer, J.S. (1981) Osmoregulation, solute distribution, and growth in soybean seedlings having low water potentials. Planta 151:482-489.

Miao, G.-H., Hong, Z., Verma, D.P.S. (1992) Topology and phosphorylation of soybean nodulin-26, an intrinsic protein of the peribacteriod membrane. J. Cell Biol. 118: 481-490.

Miller, K.J., Kennedy, E.P., Reinhold, V.N. (1986) Osmotic adaptation by gramnegative bacteria: possible role for periplasmic oligosaccharides. Science 231:48-51.

Mullet, J.E. (1990) Reversible inhibition of hypocotyl growth in soybean seedings exposed to water deficit. Plant Gene Transfer 249-256.

Mundy, J., Chua, N.-H. (1988) Abscisic acid and water- stress induce the expression of a novel rice gene. EMBO J. 7:2279-2286.

Oparka, K.J. (1990) What is phloem unloading? Plant Physiol. 94:393-396.

Paul, D.L., Goodenough, D.A. (1983) In vitro synthesis and membrane insertion of bovine MP26, an integral protein from lens fiber plasma membrane. J. Cell Biol. 96:636-638.

Robichaud, C.S., Wong, J., Sussex, I.M. (1980) Developmental Genetics 1:325-330.

Rock, C.D., Zeevart, J.A.D. (1991) The *aba* mutant of *Arabidopsis thaliana* is impaired in epoxy- carotenoid biosynthesis. Proc. Natl. Acad. Sci. USA. 88: 7496-7499.

Rosenow, D.T. (1987) in Food and Grain Production in Semi-Arid Africa

Saab, I.N., Sharp, R.E., Pritchard, J., Voetberg, G.S. (1990) Increases endogenous abscisic acid maintains primary root growth and inhibits shoot growth of maize seedlings at low water potentials. Plant Physiol. 93:1329-1336.

Sindu, R.K., Walton, D.C. (1987) The conversion of xanthoxin to abscisic acid by cell-free preparations from bean leaves. Plant Physiol. 85:916-921.

Steudle, E., Boyer, J.S. (1985) Hydraulic resistance to radical water flow in growing hypocotyl of soybean mesured by a new pressure-perfusion technique. Planta 164:189-200.

Sweet, G., Gandon, C., Voegele, R., Wittekind, N., Beuerl, J., Truniger, V., Lin, E.C.C., Boss, W. (1990) Glycerol facilitator of *Escherichia coli:* Cloning of *glpF* and identification of the *glpF* product. J. Bacteriol. 172:424-430.

Weretilnyk, E.A., Hanson, A.D. (1990) Molecular cloning of a plant betaine-aldehyde dehydrogenase, an enzime implicated in adaptation to salinity and drought. Proc. Natl. Acad. Sci. USA 87:2745-2749.

Yamaguchi-Shinozaki, K., Koizumi, M., Urao, S., Shinozaki, K. (1992) Molecular cloning and characterization of 9 cDNAs for genes that are responsive to dessication in *Arabidopsis thaliana:* Sequence analysis of one cDNA clone that encodes a putative transmembrane channel protein. Plant Cell Physiol. 33:217-224.

The Molecular Biology of Fruit Ripening

Steve Picton, Julie. E. Gray and Don Grierson
University of Nottingham, AFRC Research Group in Plant Gene Regulation
Department of Physiology and Environmental Science
Sutton Bonington Campus, Loughborough
Leics LE12 5RD
UK

Introduction

The past 15 years has witnessed an explosion in the number of research papers utilising techniques of molecular biology to answer questions about the developmental regulation of gene expression in higher plants. This paper will draw on research in the authors' laboratory utilizing the tomato (*Lycopersicon esculentum*) as a model plant and the process of fruit ripening as a clearly defined developmental system which has been used to examine the regulation and biochemical function of specific changes in gene expression [for review see 9]. We shall provide an overview of the development of techniques for isolation, identification and subsequent manipulation of expression of individual genes implicated in the process of fruit ripening and show how manipulation of individual genes is of use both in fundamental and applied research on food crops.

Ripening of fleshy fruits is characterised by changes that lead to an increase in the appeal and attractiveness of the fruit to potential consumers, thus ensuring subsequent seed dispersal. In tomato the most obvious of these are the changes in colour, flavour and aroma that attract the consumer and textural changes that make the fruit more edible (**Figure 1**). These changes are caused by alterations in the biochemical processes occurring in the fruit cells; they appear to affect all cellular compartments and are regulated in a highly specific manner. A number of enzymes change in activity or are synthesized *de novo* during ripening and the whole process requires changes in the expression of specific genes. Some "house-keeping" mRNAs remain at a constant level throughout the process, whereas other mRNAs, required during early fruit development, show a decline and new mRNAs are transcribed from genes specifically required for the fruit ripening process (**Figure 1**),[35, 22, 2, 21]. We have concentrated on genes that encode mRNAs that accumulate in a ripening-specific manner and illustrate how the isolation and *in planta* manipulation of such genes has allowed us to define the biochemical role of their encoded products and has led to the production of transgenic fruit with improved commercial characteristics [9].

NATO ASI Series, Vol. H 81
Plant Molecular Biology
Edited by G. Coruzzi and P. Puigdomènech

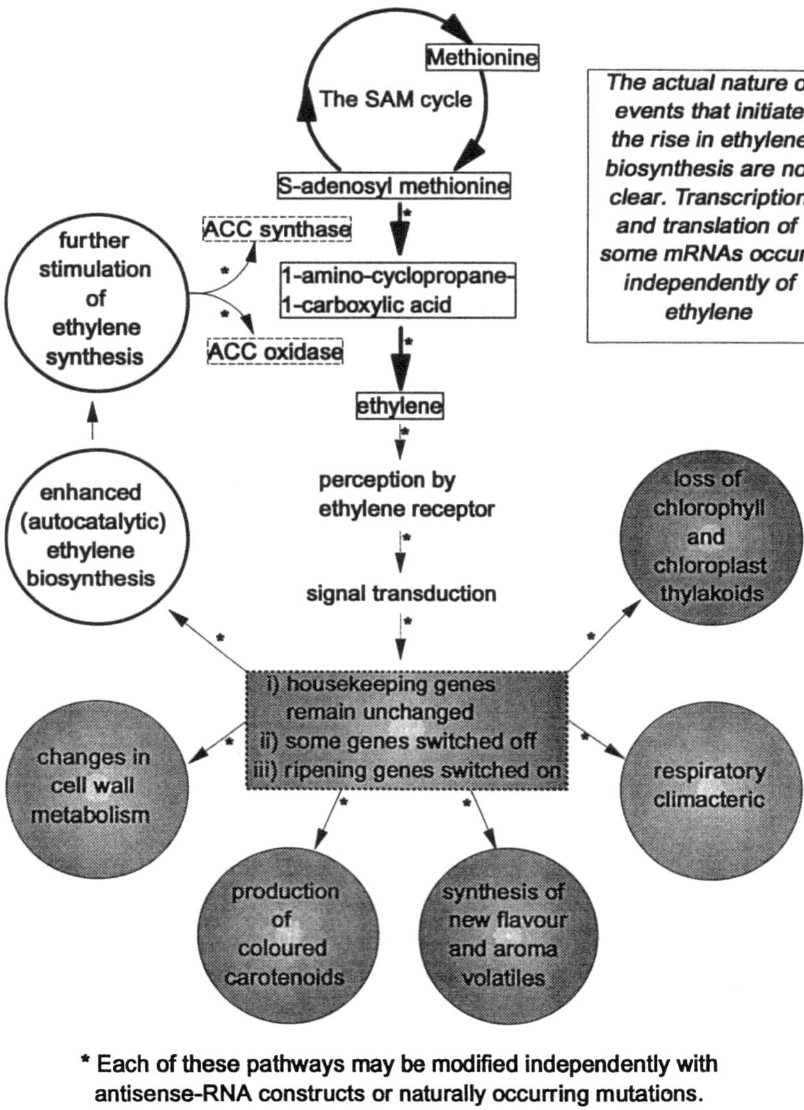

Figure 1. **Schematic representation of the interrelationships between ethylene biosynthesis, alterations in gene expression and the ripening of tomato fruit.**

The Tomato Plant and Fruit Ripening

The tomato plant provides the molecular biologist with several inherent advantages. The plant is easily propagated all year round and is a self-pollinating diploid with a relatively small genome and an extensive chromosome map. Many ripening mutants have been mapped, introduced into near isogenic lines and are available from germplasm resource centres. Tomato explants can be modified genetically utilising *Agrobacterium* mediated transformation [17, 24, 37, 34, 3, 4, 8, 12, 13, 25, 20, 7], thus allowing the stable introduction of novel genes or the down regulation of endogenous ripening genes.

Tomato fruit belong to the class of fruit that display a climacteric pattern of ripening [1], that is, the onset of the ripening process is accompanied by a dramatic rise in respiration rate and an increase in ethylene evolution. The onset of the respiratory rise can be stimulated by the plant growth regulator ethylene, which is thus implicated in the regulation of ripening of climacteric fruits [45]. Addition of ethylene to pre-climacteric fruit stimulates the ripening process whilst its removal or interference with its perception or action can delay ripening. Application of ethylene to mature, but pre-climacteric, tomato fruit leads to the appearance of a number of new mRNAs thought to be required for the ripening events [23, 21]. The pattern of accumulation of individual mRNAs has been widely studied and supports the hypothesis that ripening requires the co-ordinated expression of a number of new genes and that ethylene plays a role in this process by regulation of the expression of specific ripening genes [9].

Construction and screening of fruit-derived cDNA libraries has led to the isolation of ripening clones [9] and confirmed that ripening is determined by developmentally regulated changes in gene expression. These ripening clones provided the starting material for the manipulation of single genes in transgenic plants in order to determine the biochemical and physiological role of their products in the ripening process.

Plant Gene Manipulation using Antisense RNA

It has been demonstrated that it is possible to alter gene expression *in planta* by insertion of a recombinant gene in such an orientation that it produces an "antisense" RNA transcript (**Figure 2**) [37, 43]. The stable integration and expression of such a construct causes a targeted down regulation of the endogenous homologous plant gene [37, 43, and reviews 9 and 19]. The exact mechanism that leads to a dramatic reduction in the level of the mRNA from the endogenous gene is, at present, unknown but due to the target specificity of the antisense RNA it is thought that base pairing and creation of rapidly degraded RNA:RNA duplexes may function in the observed action of the antisense RNA. The antisense approach allows the modification of expression of individual ripening genes utilizing cDNAs cloned from the tissues in which the endogenous gene is normally expressed [37, 38, 39, 3, 34, 4, 13, 25, 29, 12, 7]. This creates a targeted mutation, facilitating the identification of the biochemical role of the gene product. Furthermore, the transgenic fruit in which a single gene is modified in this way, potentially have altered ripening characteristics [9, 32, 29, 25, 26].

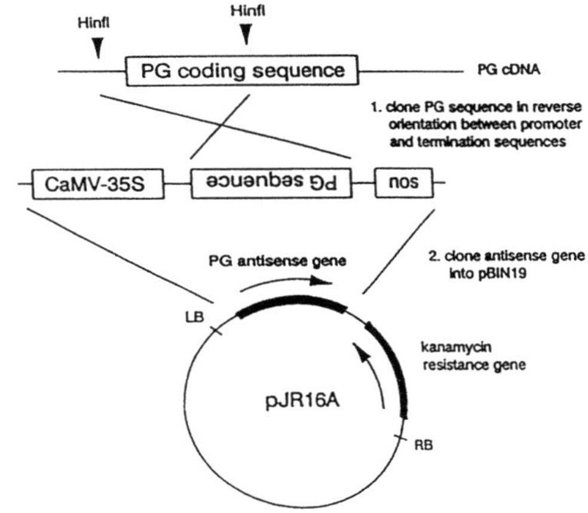

Figure 2.Construction of the polygalacturonase (PG) antisense gene. A *Hin*F1 fragment from the 5'-end of the PG cDNA was inserted in the reverse orientation between the promoter and terminator sequences of a plant transformation vector. Following *Agrobacterium* mediated transformation, the DNA between the left border (LB) and right border (RB) becomes integrated into the plant genome. The expression of the kanamycin resistance gene allows selection of transformed explants. When the integrated PG sequence is transcribed an "antisense" PG RNA is produced .

The Tomato Polygalacturonase Gene

Polygalacturonase (PG), a cell wall solubilizing enzyme which acts on the pectin fraction of the cell wall [41, 6], had been previously thought to be a major determinant in the softening of tomato fruits during ripening [15, 16] and this stimulated research on cloning the gene. The PG protein was purified from tomato fruit and shown, *in vitro*, to cause dissolution of the middle lamella of unripe pericarp tissue [41,6]. A partial amino acid sequence of the purified PG polypeptide was determined to aid in subsequent identification of a PG cDNA.

A ripe tomato fruit cDNA library was constructed and screened against probes prepared from unripe fruit in order to isolate 19 groups of "ripening-related" cDNAs [35]. DNA sequencing of some of these clones and comparison of the derived amino acid sequence with that obtained from the PG protein led to the identification of a cDNA, TOM 6, as encoding the polygalacturonase gene [10, 11]. Clones encoding tomato PG were also identified by other groups and the actual PG gene itself was also isolated and sequenced [33, 29 and review 9].

A PG antisense RNA construct was prepared (**Figure 2**) and stably integrated into transgenic Ailsa Craig tomato plants in order to examine the role of the enzyme in fruit ripening. The transgenic plants showed a reduction in level of both PG mRNA transcripts and the PG protein [37]. Consistent with the gene-specific mechanism of action of antisense RNA, other ripening parameters examined such as ethylene evolution and pigment accumulation were unaffected in the transformed fruit (**Figure 3**). The primary transgenic plants [37] were self-pollinated and the antisense gene was shown to be inherited in a normal Mendelian fashion and gave rise to a population of plants bearing zero, one or two antisense genes [39]. Plants with no inheritance of the antisense gene showed normal PG mRNA and protein levels whilst plants inheriting two copies of the transgene showed a reduction of PG activity to less than 1% of that observed in wild-type controls [39]. Despite the dramatic reduction of PG activity in the initial experiments, the transgenic fruit showed no significant change in compressibility, often used as an index of softening. Examination of the level of pectin degradation in the transgenic fruit however showed a reduction in the level of pectin depolymerization in a transgene-dependant manner [39]. Expression of a PG antisense gene in other tomato cultivars has been reported to increase fruit firmness [34, 29] and can sometimes be detected in Ailsa Craig fruit (GE Hobson, personal communication). Taken together, these results clearly demonstrated that the previously held view that PG was the major determinant of fruit softness was not correct and indicated that other enzymes also play an important role. Despite this, the transgenic fruit with PG antisense genes were shown to have altered rheological properties [32] and are being exploited commercially for improving the processing qualities of fruit extracts. These transgenic fruit also show an increased resistance to mechanical damage and problems associated with the splitting of fruit and subsequent post-harvest secondary fungal infections (See Figure 2, reference 9).

In parallel studies, the derived amino acid sequence of a previously identified pectinesterase (PE) enzyme, also implicated in cell wall metabolism, was used to screen a tomato fruit cDNA library and isolate a tomato fruit PE cDNA. This cDNA has also been utilised to produce transgenic antisense plants in which the PE gene is specifically down regulated. Analysis of these plants has shown that the fruit have greatly reduced PE activity, however the fruit continue to ripen normally [12]. Similar PE antisense fruit have also been examined elsewhere [42].

The results of both PG [37, 38, 39, 34, 29] and PE [12, 42] antisense experiments have demonstrated that the softening of tomatoes most likely results from an array of changes in cell wall metabolism, brought about by a range of different enzymes rather than being the result of a single key determinant. Further experiments are underway, utilising specific natural ripening mutations, to clone other enzymes that may be involved in the complex metabolism of cell walls that contributes to the softening of the fruit during the ripening process.

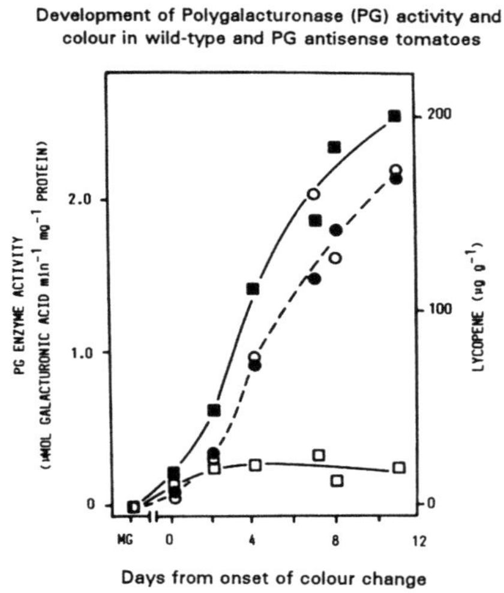

Figure 3. Accumulation of the polygalacturonase enzyme and the carotenoid pigment lycopene in control and PG antisense fruit. PG enzyme activity (wild-type, ■ ; PG antisense, □) and lycopene accumulation (wild-type, ●; PG antisense ○) were measured in fruit prior to the onset of ripening (the mature green stage, MG) and at intervals after the onset of colour change.

Fruit Pigmentation and Manipulation of Colour Change with Antisense RNA

The ripening of fruits is often associated with changes in pigmentation, the tomato being an obvious example, with the fruit progressing from dark green to a red colour during the course of normal ripening. These changes are associated with the breakdown and loss of chlorophylls and the increased accumulation of coloured carotenoid compounds, particularly lycopene and β-carotene. Following DNA sequencing of other ripening-related cDNAs and comparison of the sequences with those in nucleic acid databases, it was shown that the fruit clone, TOM 5 [35, 28], had significant homology with the *crtB* genes of *Erwinia herbicolor* and *Rhodobacter capsulatus*. These genes are believed to code for enzymes responsible for the formation of the carotenoid phytoene and it was postulated that the homologous tomato gene may be involved in the production of carotenoid pigments during fruit ripening.

In order to define a biochemical role for the cloned gene in the ripening fruit, an antisense TOM 5 construct was transformed into tomato plants. These TOM 5 antisense plants produced normal fruit, however when fully ripe the fruit remained pale yellow in colour and flowers borne on such plants were observed to have reduced yellow pigmentation [4, 7]. Further examination of the transgenic fruit showed that they were greatly impaired in their ability to synthesize phytoene.

Lycopene and β-carotene, responsible for the deep red colour of ripe fruit, were almost undetectable, whilst other parameters such as PG accumulation were unaffected. This experiment clearly demonstrated that TOM 5 encoded the enzyme phytoene synthase [4]. This example again shows that a combined biochemical and molecular approach can lead to the identification of specific genes involved in the ripening process and illustrates the power of the antisense approach for identifying the biochemical role of a gene product.

The isolation of genes involved in carotenoid biosynthesis also has commercial applications. As well as the down-regulation of endogenous genes by antisense RNA, it has now proved possible to insert a gene in a "sense" orientation in order to complement a natural mutation, shown by the elegant experiments using TOM 5 to restore lycopene synthesis in the *yellow flesh* tomato mutation (**Figure 4**) [7], or to increase the accumulation of lycopene by over-expression of the TOM 5 sense construct in the natural wild-type background (Fray, unpublished). Such over production yields fruit with enhanced colouration and may be of great commercial value to producers of processed tomato products such as pastes, sauces and soups.

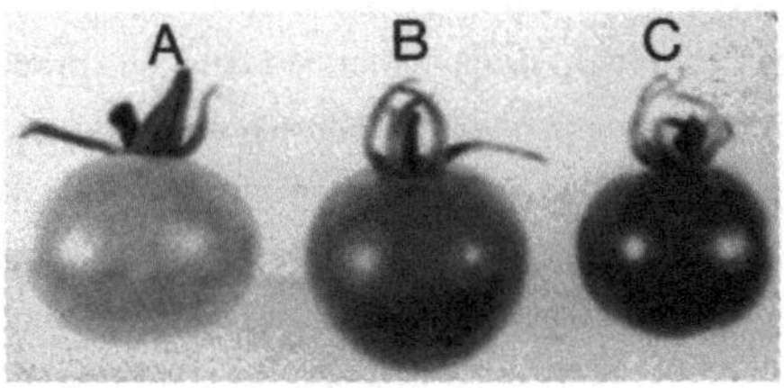

Figure 4. **Complementation of a defective phytoene synthase gene in the *yellow flesh* mutant.** The *yellow flesh* mutant fails to accumulate the carotenoid pigment, lycopene, due to a defective phytoene synthase gene. The fruit thus turn yellow when fully ripe rather than red.Three lines of the *yellow flesh* mutant transformed with a construct encoding a sense phytoene synthase RNA show A, (yellow fruit) no complementation; (B) partial or (C) full complementation (red fruit) of the mutant phenotype.

Manipulation of ethylene biosynthesis in climacteric tomato fruit utilizing antisense RNA techniques

As discussed in the introductory section of this chapter, the plant growth regulator, ethylene, has a major role in both the onset and continued progression of ripening in climacteric fruits such as tomato (**Figure 1**). The ability to manipulate genes involved in the biosynthesis of ethylene in fruit may help define the way in which ethylene regulates ripening. Furthermore, plants in which ethylene biosynthesis has been modified could be expected to have an altered ripening phenotype that may prove of commercial interest.

The pathway of ethylene biosynthesis has been elucidated, proceeding from the S-adenosylmethionine (SAM) cycle via the intermediate 1-aminocyclopropane-1-carboxylic acid (ACC) to ethylene [45]. The two key enzymes in the pathway are ACC synthase, catalysing the conversion of SAM to ACC and ACC oxidase (formerly the ethylene-forming enzyme), that catalyses the conversion of ACC to ethylene (**Figure 1**).

The approach taken in our laboratory to identify cDNAs with possible involvement in ethylene biosynthesis was to screen a fruit cDNA library to identify members whose mRNA showed increased accumulation both during the ripening of fruit **and** following the wounding of fruit and leaf tissue [35, 36], all situations involving a dramatic increase in the evolution of ethylene. This approach identified the clone, TOM 13, which encoded an mRNA with accumulation characteristics consistent with involvement in ethylene production [36]. To test the role of the TOM 13 mRNA, an antisense RNA construct based on the TOM 13 cDNA was transformed into tomato plants and ethylene production assayed in both transgenic fruit and wounded leaf tissue [13]. Endogenous TOM 13 mRNA accumulation was severely inhibited in ripening transgenic fruit and ethylene production was substantially reduced in both the ripening fruit (**Figure** 5) and wounded leaves [13]. As with the PG antisense plants, primary transformants were self-pollinated to give rise to a segregating population of plants containing zero, one or two copies of the antisense gene. It was demonstrated that both TOM 13 mRNA accumulation and ethylene production were inhibited in a transgene-dosage dependant manner and it was therefore proposed that the TOM 13 gene product may encode one of the two key enzymes in higher plant ethylene biosynthesis, ACC synthase or ACC oxidase [13]. Further analysis of the TOM 13 antisense plants showed much reduced ACC oxidase activity suggesting that the TOM 13 product encoded at least part of the elusive ACC oxidase system. Final proof of the biochemical identity of the TOM 13 encoded product came following the expression of a full length reconstructed TOM 13 cDNA in *S. cerevisiae* and the production of ethylene from its immediate precursor, ACC, in transformed yeast cultures [14]. In another experiment, an elicitor induced ACC oxidase, with a high level of homology to the TOM 13 cDNA was cloned from tomato cell cultures and expressed in *Xenopus* oocytes, again conferring the ability on the cell line to convert ACC to ethylene [40]. Once the nature of the ACC oxidase was predicted from the deduced amino acid sequence of TOM 13, it became a relatively simple matter to solubilise and characterise the enzyme from tomato fruit [18, 44].

This was the first ever use of antisense techniques to assign a biochemical role to an unidentified gene and the same approach was used subsequently to identify the carotenoid pigment gene, phytoene synthase, described in the previous section .

Examination of homozygous ACC oxidase plants showed that the fruit developed normally and began to ripen at a similar chronological stage to wild-type controls [13]. Transgenic fruit left attached to the vine showed accumulation of red pigmentation over several weeks but measurement of pigment levels revealed reduced lycopene accumulation [5, 26]. The transgenic fruit also demonstrated an increased resistance to over-ripening and shrivelling following extensive storage either on or off the plant [26]. Interestingly, when picked and allowed to ripen off the vine, the low-ethylene fruit never ripened fully, but several aspects of ripening could be at least partially restored by supplying them with ethylene [26]. Plants with these genetic modifications are now being examined in field scale trials in order to assess if such a quality is commercially exploitable. An additional phenotypic characteristic of the ACC oxidase antisense plants was a delay in the onset of foliar senescence of approximately two weeks [26]. In separate experiments, transgenic plants were produced in which ethylene synthesis was greatly inhibited with an ACC synthase (**See Figure 1**) antisense gene [30, 31, 25]. These tomatoes failed to ripen either on or off the vine but ripening was largely restored by picking the fruit and treating with ethylene [25].

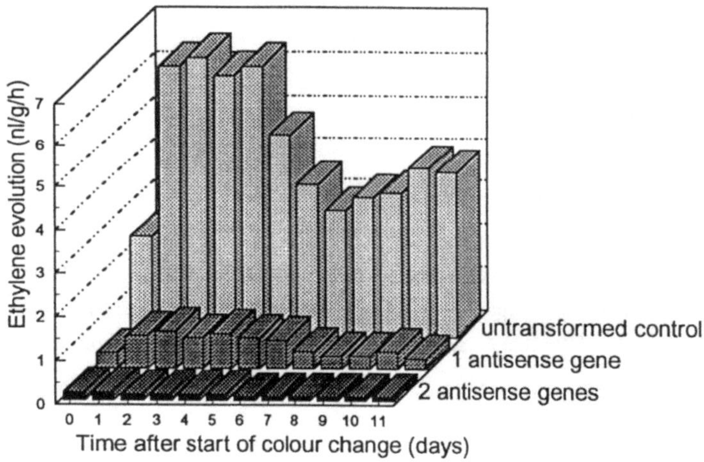

Figure 5. **Ethylene evolution from ACC oxidase antisense fruit.** Evolution of ethylene from wild-type and ACC oxidase antisense fruit containing one or two copies of the antisense gene was measured after the onset of colour change.

Conclusions and Future Development

The experiments outlined here, spanning some ten years of research, demonstrate that the goal of cloning, identifying and manipulating specific genes in order to alter plant developmental pathways is possible. Some of the experiments have allowed us to redefine potential roles of previously isolated enzymes, as is the case with PG. The results of the PG and PE antisense experiments now point to the softening process of tomato fruit as being a complex process involving more than one enzyme; nevertheless, PG antisense fruit have improved processing qualities and such engineered fresh or processed fruit products may become available to the consumer in the USA and Europe later this decade.

The use of antisense RNA as a diagnostic tool to define specific gene function has proved a great success, with genes encoding the ethylene biosynthetic enzyme, ACC oxidase [13, 14], and the carotenoid biosynthesis enzyme, phytoene synthase [4, 7] being identified. Further, the production of plants with low levels of ethylene production [25, 13, 20, 26], has allowed a more detailed examination of the role of this plant growth regulator in the control of fruit ripening and other developmental processes such as foliar senescence [26] and abscission [20]. Such genetic modification may also prove of great commercial value.

Recently, the use of the extreme tomato ripening mutant, *ripening inhibitor (rin)* has led to isolation of further cDNAs encoding unknown biochemical functions associated with ripening [27] and these will be examined using a similar antisense RNA approach to that outlined above. It now becomes clear that a more detailed understanding of specific biochemical pathways that lead to the ripening of fruit will be required in order to facilitate the targeting, isolation and manipulation of specific genes involved in the process.

Acknowledgements:

Work in the authors laboratory was supported by a grant from the Agricultural and Food Science Research Council to Don. Grierson. All work was carried out under Ministry of Agriculture, Food and Fisheries licences. We thank Lee Whotton for assistance in the preparation of Figure 1.

References:

1. Biale JB, Young RE: Respiration and ripening in fruits-retrospect and prospect. In: Friend J,Rhodes MJC (eds) Recent Advances in the Biochemistry of Fruit and Vegetables. pp 1-39 Academic Press, London (1981).
2. Biggs MS, Harriman RW, Handa AK: Changes in gene expression tomato fruit ripening. Plant Physiol.81: 395-403 (1986).
3. Bird CR, Smith CJS, Moreau P, Bevan MW, Bird AS, Hughes S, Morris PC, Grierson D, Schuch W: The tomato polygalacturonase gene and ripening specific expression in transgenic plants. Plant Mol.Biol. 11: 651-662 (1988).
4. Bird CR, Ray JA, Fletcher JD, Boniwell JM, Bird AS, Teulieres C, Blain I, Bramley PM, Schuch W: Using antisense RNA to study gene function: Inhibition of carotenoid biosynthesis in transgenic tomatoes. Bio-technology 9:635-639 (1991).
5. Bouzayen M, Hamilton AJ, Picton S, Barton S, Grierson D: Identification of genes for the ethylene-forming enzyme and inhibition of ethylene synthesis in transgenic plants using antisense genes. Biochem.Soc.Trans. 20:76-79 (1992).
6. Crookes PR, Grierson D: Ultrastructure of tomato fruit ripering and the role of polygalacturonase isoenzymes in cell wall degradation. Plant Physiol. 72:1088-1093 (1983).
7. Fray R, Grierson D: Identification and genetic analysis of normal and mutant phytoene synthase genes of tomato by sequencing, complementation and co-supression. Plant Mol. Biol. In Press (1993).
8. Giovanoni JJ, DellaPenna D, Bennet AB, Fischer RL: Expression of a chimeric polygalacturonase gene in transgenic *rin (ripening inhibitor)* tomato fruit results in polyuronide degradation but not fruit softening. Plant Cell. 1:53-63 (1989).
9. Gray J, Picton S, Shabeer J, Grierson D: Molecular biology of fruit ripening and its manipulation with antisense genes.Plant Mol.Biol. 19:69-87 (1992).
10. Grierson D, Maunders MJ, Slater A, Ray J, Bird CR, Schuch W, Holdsworth MJ, Tucker GA, Knapp JE: Gene Expression during tomato ripening. Phil.Trans.R.Soc.Lond. B 314: 399-410 (1986).
11. Grierson D, Tucker GA, Keen J, Ray J, Bird CR, Schuch W: Sequencing and identification of a cDNA clone for tomato polygalacturonase. Nucl. Acids. Res. 14:8595-8603 (1986).
12. Hall LN, Tucker GA, Smith CJS, Watson CF, Seymour GB, Bundick Y, Boniwell JM, Fletcher JD, Ray JA, Schuch W, Bird CR, Grierson D. Antisense inhibition of pectin esterase gene expression in transgenic tomatoes. The Plant Journal. 3:121-129 (1993).
13. Hamilton AJ, Lycett GW, Grierson D: Antisense gene that inhibits synthesis of the hormone ethylene in transgenic plants. Nature. 346: 284-287 (1990).
14. Hamilton AJ, Bouzayen M, Grierson D: Identification of a tomato gene for the ethylene forming enzyme by expression in yeast. Proc.Natl.Acad. Sci. USA. 88: 7434-7437 (1991).
15. Hobson GE: Polygalacturonase in normal and abnormal tomato fruit. Biochem.J. 92: 324-332 (1964).
16. Hobson GE: The firmmess of tomato fruit in relation to polygalacturonase activity. J. Hort.Sci. 40:66-72 (1965).
17. Horsch RB, Fry JE, Hoffman NL, Eiochholtz D, Rogers SG, Frayley RT: A simple and general method for transferring genes into plants. Science. 227:1229-1231 (1985).
18. John P: How plant molecular biologists revealed a surprising relationships between two enzymes, which took an enzyme out of a membrane where it was not located, and put it into the soluble phase where it could be studied. Plant Mol. Biol.Rep. 9:192-192 (1991).
19. Jorgensen R: Altered gene expression in plants due to *trans* interactions between homologous genes. TIBTECH. 8:340-344 (1990).
20. Klee HJ, Haiford MB, Kretzner KA, Barry GF, Kishmore GM: Control of Ethylene synthesis by expression of a bacterial enzyme in transgenic tomato plants. The Plant Cell. 3:1187-1193 (1991).
21. Lincoln JE, Cordes S, Read E, Fischer RL: Regulation of gene expression by ethylene during *Lycopersicon* esculentum (Tomato) fruit development. Proc.Natl.Acad.Sci. USA 84:2793-2797 (1987).

22. Mansson PE, Hsu D, Stalker D: Characterisation of fruit specific cDNAs from tomato. Mol. Gen. Genet. 200:356-361 (1985).
23. Maunders MJ, Holdsworth MJ, Slater A, Knapp JE, Bird CR, Schuch W, Grierson D : Ethylene stimulates the accumulation of ripening-related mRNAs in tomatoes. Plant Cell Envir.10:177-184 (1987).
24. McCormick S, Neidermeyer J, Fry J, Barnasson A, Horsch A, Frayley R: Leat disc transformation of cultivated tomato (*L.esculentum*) using *Agrobacterium tumefaciens*. Plant Cell Rep. 5:81 (1986).
25. Oeller PW, Wong LM, Taylor LP, Pike DA, Theologis A: Reversible inhibition of tomato fruit senescence by antisense 1-aminocyclopropane-1-carboxylate synthase. Science. 254: 437-439 (1991).
26. Picton S, Barton SL, Bouzayen M, Hamilton AJ, Grierson D: Altered fruit ripening and leaf senescence in tomatoes expressing an antisense ethylene-forming enzyme transgene. The Plant Journal. 3:469-481 (1993).
27. Picton S, Gray J, AbuBaker U, Lowe A, Grierson D: cDNA cloning and characterisation of novel ripening-related mRNAs with altered patterns of accumulation in the *ripening inhibitor (rin)* tomato ripening mutant. Plant Mol.Biol. Submitted for publication (1993).
28. Ray J, Bird C, Maunders M, Grierson D, Schuch W: Sequence of pTOM 5, a ripening related cDNA from tomato. Nucl. Acids Res. 15:10587 (1987).
29. Redenbaugh K, Hiatt W, Martineau B, Kramer M, Sheehy R, Sanders R, Houck C, Emlay D: Safety Assessment of Genetically Engineered fruits and Vegetables. A Case study of the Flavr Savr™ Tomato. CRC Press, Boca Ratón, USA. (1992).
30. Sato T, Theologis A: Cloning the mRNA encoding 1-amino-cyclopropane-1-carboxylate synthase, the key enzyme for ethylene biosynthesis in plants. Proc.Natl.Acad. Sci.USA. 86:6621-6625 (1989).
31. Sato T, Oeller PW, Theologis A: The 1-aminocyclopropane-1-carboxylate synthase of *Curcubita*. J. Biol.Chem. 266:3752-3759 (1991).
32. Schuch W, Hobson G, Kanczler J, Tucker G, Robertson D, Bright S, Bird C: Improvement of tomato fruit quality through genetic engineering. HortScience. 26:1517-1520 (1991).
33. Sheehy RE, Pearson J, Brady CJ, Hiatt WR: Molecular characterization of tomato fruit polygalacturonase. Mol. Gen.Genet. 208:30-36 (1987).
34. Sheehy RE, Kramer M, Hiatt WR: Reduction of polygalacturonase activity in tomato fruit by antisense RNA. Proc.Natl. Acad. Sci. USA. 85:8805-8809 (1988).
35. Slater A, Maunders MJ, Edwards K, Schuch W, Grierson D: Isolation and characterization of cDNA clones for tomato polygalacturonase and other ripening related proteins. Plant Mol.Biol. 5:137-147 (1985).
36. Smith CJS, Slater A, Grierson D: Rapid appearance of mRNA correlated with ethylene synthesis encoding a protein of MW 35,000. Planta. 168:94-100 (1986).
37. Smith CJS, Watson CF, Ray J, Bird CR, Morris PC, Schuch W and Grierson D:Antisense RNA inhibition of polygalacturonase gene expression in transgenic tomatoes. Nature. 334:724-726 (1988).
38. Smith CJS, Watson CF, Bird CR, Ray J, Schuch W, Grierson D: Expression of a truncated tomato polygalacturonase gene inhibits expression of the endogenous gene in transgenic plants. Mol.Gen.Genet. 224:477-481 (1990).
39. Smith CJS, Watson CF, Morris PC, Bird CR, Seymour GB, Gray JE, Arnold C, Tucker GA, Schuch W, Harding S, Grierson D: Inheritance and effect on ripening of antisense polygalacturonase genes in transgenic tomatoes. Plant Mol.Biol. 14:369-379 (1990).
40. Spanu P, Reinhart D, Boller T: Analysis and cloning of the ethylene-forming enzyme from tomato by functional expression of its mRNA in *Xenopus laevis* oocytes. EMBO.J. 10:2007-2013 (1991).
41. Themmen APN, Tucker GA, Grierson D: Degradation of isolated tomato cell walls by purified polygalacturonase *in vitro*. Plant Physiol. 69:122-124 (1982).
42. Tieman DM, Harriman RW, Ramamohan G, Handa AK: An antisense pectinmethylesterase gene alters pectin chemistry and soluble solids in tomato fruit. The Plant Cell. 4:681-687 (1992).
43. van der Krol AR, Lenting PE, Veenstra J, van der Meer IM, Koes RE, Gerats AGM, Mol JNM and Stuitje AR: An anti-sense chalcone synthase gene in transgenic plants inhibits flower pigmentation. Nature. 333:866-869 (1988).

44. Ververidis P, John P: Complete recovery *in vitro* of ethylene forming enzyme activity. Phytochemistry. 30:725-727 (1991).
45. Yang SF, Hoffmann NE: Ethylene biosynthesis and its regulation in higher plants. Ann. Rev. Plant Physiol. 35:155-189 (1984).

ALTERED PATTERNS OF PROTEIN SYNTHESIS IN *opaque-2* MAIZE ENDOSPERM

Jeffrey E. Habben[a], Gloverson L. Moro[a], Mauricio A. Lopes[a], Etti Or[a], Bruce Hamaker[b] and Brian A. Larkins[a]

[a]Department of Plant Sciences, University of Arizona, Tucson, AZ 85721 USA
[b]Department of Food Science, Purdue University, West Lafayette, IN 47907 USA

INTRODUCTION

The search for maize genotypes with an increased level of lysine, an essential amino acid, led to the discovery that certain mutations, such as *opaque-2* and *floury-2*, result in seeds with enhanced nutritional value for monogastric animals[1,2]. These mutations cause a reduction in synthesis of the prolamin proteins, or zeins, and an increase in other endosperm proteins, the non-zeins. Typically, zeins account for around 60% of the endosperm protein. Because zeins contain virtually no lysine or tryptophan, their abundance is an important factor contributing to the inferior nutritional quality of maize seed protein. Although *opaque-2* and *floury-2* mutants were found to have significantly improved nutritional value, their soft endosperms, which cause kernel chipping and breakage, insect susceptibility, and inferior food processing, ultimately prevented these mutants from becoming commercially useful[3].

Early studies with *opaque-2* revealed that in certain genetic backgrounds regions of the endosperm acquired a vitreous phenotype, similar to that of normal seeds. The genes that condition this response were called *opaque-2* "modifiers". Although the modifier genes behave as genetically complex traits, plant breeders at the International Maize and Wheat Improvement Center (CIMMYT) recognized their potential for converting the soft endosperm of *opaque-2* mutants to seeds with a vitreous phenotype and a high lysine content. Through a complex back-crossing and recurrent selection procedure, they developed a large number of high lysine maize varieties known as Quality Protein Maize (QPM). While QPM has not yet reached commercial production, its improved protein and seed quality are becoming well documented[4,5], and there is increasing commercial interest in utilizing this germplasm.

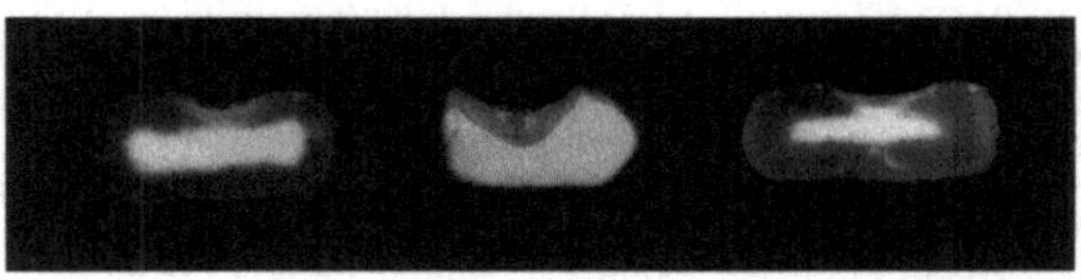

Fig. 1. Cross-section of Tuxpeno normal (A), Tuxpeno *opaque-2* (B), and Tuxpeno modified-*opaque-2* (C) maize kernels.

NATO ASI Series, Vol. H 81
Plant Molecular Biology
Edited by G. Coruzzi and P. Puigdomènech

BIOCHEMICAL EFFECTS OF *opaque-2* MODIFIERS

The primary phenotypic change conditioned by *opaque-2* modifiers is the conversion of starchy endosperm to a hard, vitreous texture (Fig. 1). This alteration in phenotype is associated with the enhanced synthesis of one of the prolamin storage proteins, the gamma-zein. Typically, QPM genotypes contain between 3- and 4-times more gamma-zein than unmodified *opaque-2* mutants[6]. Their enhanced content of gamma-zein is illustrated in Figure 2, which shows a comparison of the zein and non-zein proteins in Tuxpeno normal, *opaque-2*, and modified-*opaque-2* (QPM). For this comparison, we extracted protein from equal amounts of endosperm flour and separated it by SDS-PAGE. Compared to the normal genotype, there is a reduction in 22-kD and 19-kD alpha-zein proteins in both *opaque-2* mutants. However, only in the modified *opaque-2* mutant is there a major increase in the 27-kD gamma-zein. This analysis also illustrates that per unit of flour there are more non-zein proteins in the *opaque-2* mutants than in the normal genotype. Based on one dimensional SDS-PAGE, the differences in non-zein proteins between normal and *opaque-2* genotypes are quantitative rather than qualitative and correspond to a group of polypeptides between 40-kD and 75-kD.

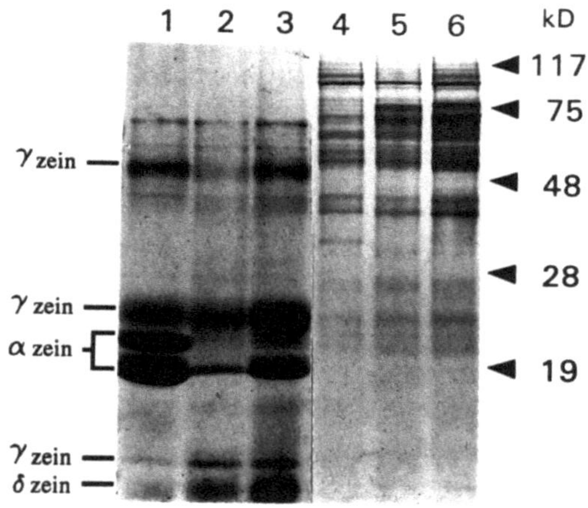

Fig. 2. SDS-PAGE separation of zein and non-zein proteins from Tuxpeno normal, Tuxpeno *opaque-2*, and Tuxpeno modified-*opaque-2* maize endosperm. Proteins were extracted with a high pH, borate buffer containing SDS and 2-mercaptoethanol. Non-zeins were separated from zeins by incubating the protein extract in a final concentration of 70% ethanol. Each lane contained protein from 1.5 mg of flour. Lane 1, Tuxpeno, zeins; Lane 2, Tuxpeno *opaque-2*, zeins; Lane 3, Tuxpeno modified-*opaque-2*, zeins; Lane 4, Tuxpeno, non-zeins; Lane 5, Tuxpeno *opaque-2*, non-zeins; Lane 6, Tuxpeno modified-*opaque-2*, non-zeins. The position and molecular mass of protein markers are shown on the right.

The reduction in alpha-zeins, which contain no lysine, and the increase of non-zeins, which contain on the average 4% to 5% lysine[7], contribute to an increased percentage of lysine in *opaque-2* and QPM grain (Table I). A comparison of the protein in 100 mg of endosperm flour from Tuxpeno normal, *opaque-2* and modified *opaque-2* reveals that the reduction in alpha-zein content caused by the *opaque-2* mutation is partially offset in QPM by the activity of the modifier genes, which cause an increase in gamma-zein synthesis. The absolute content of lysine in the *opaque-2* and modified *opaque-2* mutants is nearly the same, and both contain more than double the lysine of the normal genotype. However, the increased content of gamma-zein in the modified mutant results in a somewhat lower percentage of lysine in the QPM when expressed on a protein basis. Thus, the QPM phenotype is a compromise between normal and *opaque-2*.

Table I. Amount of Protein and Lysine from Endosperm Flour of Tuxpeno normal, Tuxpeno *opaque-2* , and Tuxpeno modified-*opaque-2*. Protein levels were determined by Kjeldahl analysis and lysine amounts by amino acid analysis.

Genotype	mg protein/100 mg flour	mg lys/100 mg flour	mg lys/100 mg protein
Tuxpeno	12.1	0.11	0.90
Tuxpeno *opaque-2*	7.2	0.26	3.56
Tuxpeno modified-*opaque-2*	10.2	0.25	2.49

The enhanced synthesis of the 27-kD gamma-zein protein in modified *opaque-2* mutants is accompanied by a nearly equal increase in the synthesis of gamma-zein mRNA[8]. This suggests that the products of the modifier genes impact the expression of the genes encoding gamma-zein. Although we do not know the mechanism by which this occurs, our recent genetic evidence indicates there are at least two modifier genes that operate through a *trans*-acting mechanism[9]. Experiments are in progress to more specifically define the molecular mechanism(s) through which the modifiers alter the expression of the gamma-zein genes.

CHARACTERIZATION OF THE LYSINE-CONTAINING ENDOSPERM PROTEINS

The gene encoding the Opaque-2 protein has been isolated and shown to be a transcriptional regulatory protein that is responsible for the expression of some alpha-zein genes[10], as well as a gene encoding a ribosome inactivating protein[11]. While this activity accounts for the reduced alpha-zein content in *opaque-2* mutants, it does not explain the increased synthesis of so many non-zein proteins. It has been suggested that the increased amount of non-zeins is a compensatory response to the reduction in zein synthesis[12]. However, if this was the case one would predict that the increased synthesis of gamma-zein in QPM would be offset by a reduction in non-zein proteins, and this does not happen (Fig. 2). Therefore, it appears that the enhanced synthesis of non-zein proteins is a direct consequence of the *opaque-2* mutation and not simply a nitrogen sink effect.

The non-zein proteins are the principal source of lysine in the kernel[7]. Knowledge of their origin and nature would provide a basis for studying the inheritance of these nutritionally important proteins and might provide an explanation for their coordinately increased synthesis in *opaque-2* mutants. To investigate the origin of these proteins, we homogenized developing endosperm tissue in a low ionic-strength isotonic buffer and separated the cellular components by sucrose gradient centrifugation. This technique allow us to distinguish four fractions: soluble proteins that floated on top of the gradient, a layer of endomembranes that banded between a density of 1.06 and 1.13 g/mL, a layer of mitochondria that banded between a density of 1.13 and 1.19 g/mL, and a layer of protein bodies that banded between a density of 1.19 and 1.26 g/mL. These four fractions accounted for 76%, 2%, 4%, and 18% of the non-zein protein and 82%, 7%, 3%, and 7%, respectively, of the protein-bound lysine. Thus, the majority of the lysine-containing proteins were recovered as soluble proteins by this technique.

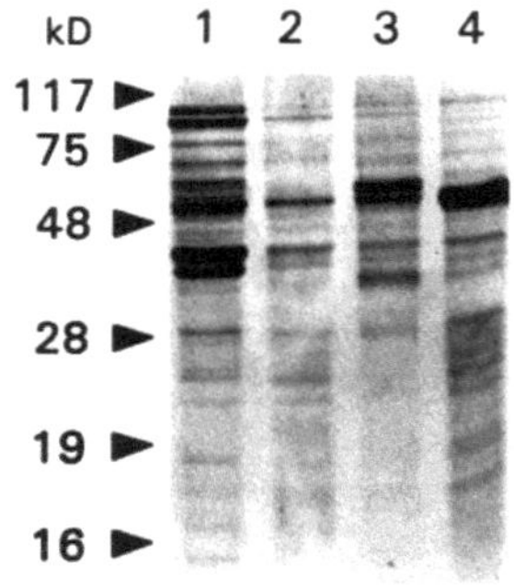

Fig. 3. SDS-PAGE separation of non-zein proteins from sucrose gradient fractions. Initially, 28-DAP W64A *opaque-2* endosperm tissue was disrupted in a low-ionic strength buffer, the extract strained through Miracloth and the starch grains allowed to sediment. The tissue and starch grains were back-extracted with fresh buffer and the supernatants pooled and separated by centrifugation in discontinuous sucrose gradients. Samples from four regions were analyzed for organelle separation by determining the distribution of the marker enzymes Cyt *c* reductase and Cyt *c* oxidase. Each lane contained 80 μg of non-zein protein from the soluble fraction (Lane 1), the endomembranes (Lane 2), the mitochondria (Lane 3), and the protein bodies (Lane 4). The position and molecular mass of protein markers are shown on the left. (From Habben et al. [7])

SDS-PAGE of the non-zein proteins from these four fractions revealed a number of qualitative differences, reflecting differences in organellar composition (Fig. 3). The complexity of proteins floating on the gradient was similar to that of the total non-zein fraction from mature kernels (cf. Fig. 2, lanes 4,5 and 6 and Fig. 3, lane 1). Thus, it appears that the majority of the abundant non-zeins are soluble proteins in the developing kernel. There were many similar proteins between the endomembrane fraction and the protein body layer, which is consistent with their common origin[13]. Each of these fractions contained a prominent polypeptide of 59-kD, which was very abundant in the protein body fraction though less so in the endomembranes. The profile of proteins from the mitochondrial fraction was distinct, indicating that these organelles were relatively well separated from other cellular components by gradient centrifugation. This conclusion was also supported by assays for the marker enzymes cytochrome C oxidase and cytochrome C reductase, which showed essentially little enzyme activity associated with proteins floating on the gradient[7]. Thus, while there may be a small amount of cross-contamination of organelles between the four gradient fractions, they appear to be adequately separated.

In order to further characterize the major non-zeins and measure their content, we immunized rabbits with proteins from the soluble fraction and obtained a polyclonal antiserum. When the antibodies were tested by immunoblotting with total non-zein proteins from mature seed, all of the major polypeptides were found to react. Furthermore, the intensity of the reaction reflected the abundance of the proteins. When we tested proteins from equal amounts of flour from Tuxpeno normal, *opaque-2,* or modified *opaque-2*, the antibody reaction was proportional to the amount of non-zein protein in each genotype (Fig. 4). Therefore, it appears that this antiserum may be useful to estimate the amount of non-zein proteins among different genotypes.

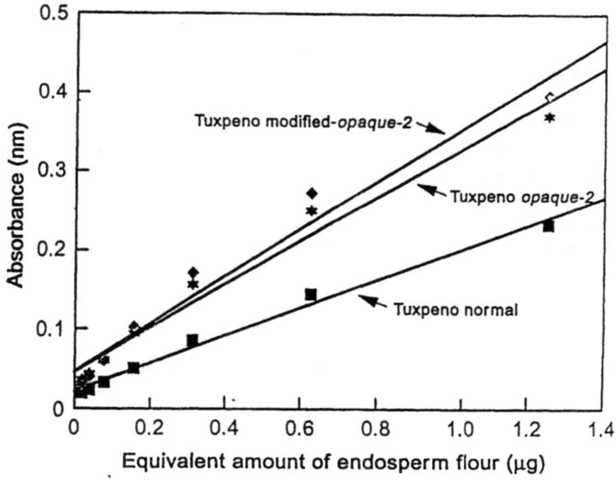

Fig. 4. ELISA analysis of non-zein proteins from endosperm flour of Tuxpeno normal, Tuxpeno *opaque-2*, and Tuxpeno modified-*opaque-2*. The data for all three slopes had a R^2 greater than 0.96.

As one approach to identify the major non-zein proteins, we used the complex antiserum to screen an endosperm cDNA expression library. From the nucleotide sequence of the clones obtained, we deduced the amino acid sequence of the protein and compared it to those in the protein data base. In our initial screen, approximately 60% of the cDNA clones we isolated shared sequences with known genes. Approximately 30% were zein sequences, while the remainder corresponded to a variety of proteins, many of which were enzymes such as sucrose synthase, nucleoside diphosphate kinase, and trypsin and chymotrypsin inhibitors (Table II).

Table II. Isolated Maize cDNA Clones with Sequence Identity to Known Genes. The clones were isolated from an endosperm-specific cDNA library by immunoscreening with complex antibodies raised against the soluble protein fraction. (From Habben et al.[7])

Clone	Description[a]	No. of clones[b]	Amino Acid Identity[c]	Organism Match
RSP033	α-glucan phosphorylase	1	86 (73)	*Ipomoea batatas*
RSP145	Aspartate aminotransferase	2	95 (70)	*Panicum miliaceum*
RSP111	Chymotrypsin inhibitor-2	3	69 (42)	*Hordeum vulgare*
RSP037	Elongation factor	2	75 (54)	*Arabidopsis thaliana*
RSP027	Maize catalase	1	55 (51)	*Zea mays*
SPF004	Nucleoside diphosphate kinase	7	78 (51)	*Spinacia oleracea*
RSP030	Sucrose binding protein	1	100 (16)	*Glycine max*
RSP118	Sucrose synthase	5	100 (64)	*Zea mays*
RSP021	Trypsin inhibitor	9	94 (76)	*Zea mays*
RSP081	60S ribosomal protein	1	44 (107)	*Homo sapiens*

[a] Identification is based on the best sequence identity to known genes.
[b] Number of isolates from 79 randomly sequenced clones.
[c] Amino acid identities are the percentage of match and the length of match (in parentheses).

Most of these proteins are involved in the major biosynthetic pathways of the endosperm, such as carbohydrate synthesis (alpha-glucan phosphorylase), amino acid biosynthesis (aspartate aminotransferase), and protein synthesis (60S ribosomal subunit protein).

We have not yet matched these clones to specific polypeptides in the SDS gels, although this work is in progress. We believe that by characterizing this complex mixture of proteins through this approach we will be able to identify most of the major non-zein proteins and account for the major sources of lysine in the kernel. These clones will also provide tools to investigate the expression of the corresponding genes and characterize the mechanism(s) responsible for the coordinate enhanced expression of certain genes in developing *opaque-2* endosperm.

ACKNOWLEDGMENTS

This research was supported by grants from the National Institutes of Health (GM36970)(BAL) and Pioneer Hi-Bred International (BAL and BH), and graduate fellowships from CNPq/RHAE, Brazil (GM and MAL).

REFERENCES

1. E.T. Mertz, L.S. Bates and O.E. Nelson, Mutant gene that changes protein composition and increases lysine content of maize endosperm, *Science* 145: 279 (1964).

2. O.E. Nelson, E.T. Mertz and L.S. Bates, Second mutant gene affecting amino acid pattern of maize endosperm proteins, *Science* 150: 1469 (1965).

3. E. Villegas, S.K. Vasal and M. Bjarnason, Quality Protein Maize-what is it and how it was developed, *In* "Quality Protein Maize," E.T. Mertz, ed. American Association of Cereal Chemists, St. Paul, MN (1992) p. 27.

4. R. Bressani, V. Benavides, E. Acevedo and M.A. Ortiz, Changes in selected nutrient contents and protein quality of common and Quality Protein Maize during rural tortilla preparation. *Cereal Chem.* 67: 515 (1990).

5. R. Magnavaca, Quality Protein Maize development in Brazil, *In* "Quality Protein Maize, " E.T. Mertz, ed., American Association of Cereal Chemists, St. Paul, MN (1992) p. 98.

6. J.C. Wallace, M.A. Lopes, E. Paiva and B.A. Larkins, New methods for extraction and quantification of zeins reveals a high content of gamma-zein in modified *opaque*-2 maize, *Plant Physiol.* 92: 191 (1990).

7. J.E. Habben, A.W. Kirleis and B.A. Larkins, The *opaque-2* mutation causes a pleiotropic increase of lysine-containing proteins in maize endosperm, *Plant Physiol.* 101: (In press) (1993).

8. K.B. Geetha, C.R. Lending, M.A. Lopes, J.C. Wallace and B.A. Larkins, *opaque*-2 modifiers increase gamma-zein synthesis and alter its distribution in maize endosperm, *Plant Cell* 3: 1207 (1991).

9. M.A. Lopes, T. Helentjaris and B.A. Larkins, Genetic mapping of *opaque-2* modifier genes in Quality Protein Maize (In preparation)(1993).

10. R.J. Schmidt, F.A. Burr, M.J. Aukerman and B. Burr, Maize regulatory gene *opaque*-2 encodes a protein with a "leucine zipper" motif that binds to zein DNA, *Proc. Natl. Acad. Sci. USA* 87: 46 (1990).

11. S. Lohmer, M. Maddaloni, M. Motto, N. DiFonzo, H. Hartings, F. Salamini and R.D. Thompson, The maize regulatory locus *Opaque*-2 encodes a DNA-binding protein which activates the transcription of the *b-32* gene, *EMBO J* 10: 617 (1991).

12. E.T. Mertz, Case histories of existing models, *In* "Genetic Improvement of Seed Proteins, " National Academy of Sciences, Washington, D.C. (1976) p. 57.

13. B.A. Larkins and W.J. Hurkman, Synthesis and deposition of zein in protein bodies of maize endosperm, *Plant Physiol.* 62: 256 (1978).

bZIP Proteins Regulating Gene Expression in Maize Endosperm

Robert J. Schmidt, Leonard D. Pysh,
Mariena Ketudat, Ronald L. Parsons,
and Gisela Hoschek
Department of Biology, 0116
Center for Molecular Genetics
University of California, San Diego
La Jolla, California 92093-0116
U.S.A.

Introduction

A major focus of the research in my laboratory has been directed towards understanding the regulation of seed storage protein gene expression in maize. More specifically, we have concentrated our efforts on elucidating the mechanism by which one particular regulatory protein, Opaque-2 (O2), affects expression of specific members of the zein family of seed storage protein genes. What has made this system of gene regulation approachable is the many years of genetic and biochemical studies that preceded the molecular analysis (reviewed in Motto et al. 1989, and Schmidt 1992). This earlier work had already suggested that O2 was a transcriptional activator and had identified likely target genes under O2 control. Due to this pioneering research, rapid progress towards understanding the basic interactions between O2 and its target genes was possible once the *O2* gene was cloned.

The *O2* gene and cDNA have been cloned by our laboratory (Schmidt et al. 1987, Schmidt et. al. 1990) and independently by another laboratory (Motto et al. 1988, Hartings et al. 1989). The deduced amino acid sequence contains a conserved basic leucine zipper DNA-binding motif. The presence of this motif suggested that O2 was capable of binding DNA sequences in the predicted target genes. This was consistent with the predictions of the genetic analyses which suggested that O2 was a transcriptional activator. Although a number of plant bZIP transcription factors have been identified in recent years (Katagiri and Chua, 1992), O2 is distinguished by the fact that its

NATO ASI Series, Vol. H 81
Plant Molecular Biology
Edited by G. Coruzzi and P. Puigdomènech

function has been dissected by genetic means. Analysis of a number of o2 mutants has helped elucidate O2 function and identify likely target genes.

O2 regulates the expression of certain members of the zein seed storage protein gene family: namely, those belonging to the 22-kD zein class (see Motto et al. 1989, Schmidt et al. 1992). In addition, it regulates the expression of a gene, *b-32*, that encodes an abundant protein that appears to play a role in plant defense (Lohmer et al. 1991, Bass et al. 1992). There is also evidence to suggest that O2 may regulate an enzyme that functions in endosperm amino acid biosynthesis (Brochetto-Braga et al. 1992).

I. Cloning and Characterization of the O2 Gene

O2 is the only plant bZIP gene isolated by genetic means. Transposon tagging led to the cloning of *O2* (Schmidt et al. 1987, Motto et al. 1988). A number of *Spm*-tagged alleles were obtained by Schmidt, Burr and Burr (1987) from a single experiment involving about a half million progeny. At least one of these alleles contained the autonomous 8.3 kb element, while other mutants were due to the insertion of deletion derivatives of *Spm* (Schmidt et al. 1987, Aukerman and Schmidt 1993). In another transposon tagging of *O2*, Motto et al. (1988) obtained an *Ac* insertion from a population of about 760,000 progeny. Subsequent isolation and sequencing of these tagged *O2*

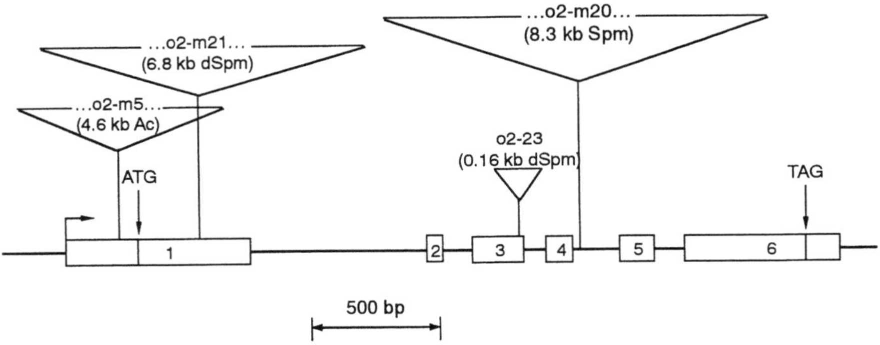

Figure 1. Diagram of the O2 gene showing intron, exons (numbered boxes) and insertion sites of transposon-tagged alleles obtained during transposon mutagenesis of the O2 gene. The initiating (ATG) and stop (TAG) codons are indicated along with the transcription start (bent arrow).

alleles showed the *O2* gene to be composed of 6 exons spanning about 4.5 kb (Figure 1). Analyses of RNA accumulation showed a tight correlation between O2 expression and zein gene expression (Schmidt et al. 1988). O2 expression appears restricted to developing endosperm tissue.

II. Functional Domains of O2

A number of important functional domains have been identified in O2. The portions of the O2 protein that are needed for DNA binding, dimerization, nuclear targeting and transcriptional activation have been identified, and their locations within a linear representation of the O2 protein are shown in Figure 2.

DNA Binding

The deduced amino acid sequence indicates the presence of a basic leucine zipper (bZIP) DNA-binding motif. The bZIP motif is a distinctive structural domain consisting of two subdomains: a sequence of basic amino acids, which contact the DNA, and a heptameric repeat of leucines or similarly hydrophobic amino acids, which form a dimerization interface (reviewed in Lamb and McKnight 1991). The bZIP motif in O2 is capable of interacting

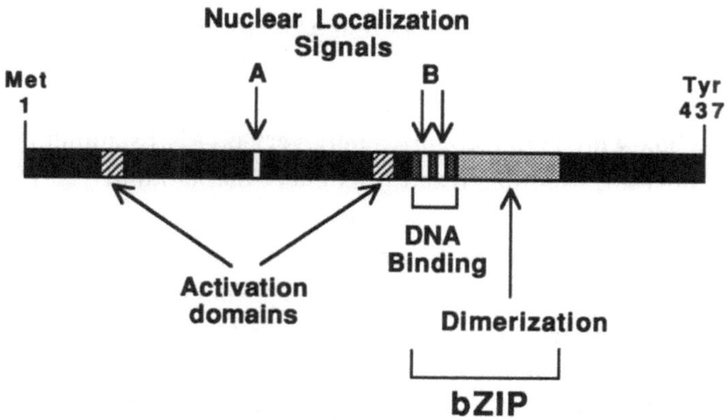

Figure 2. A diagram of functional domains within the O2 protein. This cartoon portrays a linear O2 protein with relative positions of various domains indicated. The first (Met) and last (Tyr) amino acids of the predicted amino acid sequence are shown.

with a specific sequence in the promoter of 22-kD zein genes. This sequence is located about 300 base pairs upstream of the initiating codon and just 3' of a highly conserved sequence element (prolamin box) found in all zein genes (Figure 3A). The 22-kD zein genes encode one of the more abundant subclasses of the zein storage protein family (reviewed in Motto et al. 1989, Schmidt 1992).

DNase I footprinting and *in vitro* DNA-binding analyses showed that O2 binds an imperfect palindrome in the 22-kD zein gene promoter (Schmidt et al. 1990, 1992). This sequence, 5'-TCCACGTAGA-3' (O2 binding site or O2 box), is similar to the sequence recognized by a variety of other plant bZIP proteins (Katagiri and Chua 1992, Schmidt 1992). The bZIP domain alone is sufficient for recognition of this sequence. As few as 96 amino acid residues spanning the basic domain and first four repeats of the leucine zipper are sufficient for recognition of the O2 box in 22-kD zein gene promoters (Schmidt et al. 1990). Likewise, a single point mutation in an absolutely conserved residue of the basic domain results in a loss of DNA binding by the mutant protein and a concomitant decrease in zein gene expression (Aukerman et al. 1991). Taken together, these data indicate that O2 regulates zein gene expression through a direct interaction with the target genes and not indirectly through the activation of another regulatory protein.

The O2 binding site has been found only in 22-kD zein gene promoters and not in the promoters of other members of the zein gene family. This observation is consistent with the selective decrease in accumulation of 22-kD zein transcripts and, ultimately, 22-kD zein proteins observed in *o2* mutants. An inspection of the sequences in a number of 22-kD zein clones, however, shows that several have a mutated O2 box (Figure 3B). One of these, pML1, has a mutated box which O2 recognizes only very poorly (Schmidt et al. 1992). Extrapolating from this result, it is likely that the other 22-kD zein clones containing altered O2 box sequences will also fail to bind the O2 protein. As discussed elsewhere (Schmidt 1992), these genes are probably inactive "pseudogenes," a feature characteristic of this gene family.

Another gene under the control of *O2*, called *b-32*, contains binding sites for the O2 protein that are different from those found in the 22-kD zein gene promoter. Lohmer et al. (1991) identified five sites (Table 1) by DNase I footprinting that were protected by large amounts of a bacterially expressed O2 protein. These are likely to be relatively low affinity sites as oligonucleotides based on these individual sites fail to produce shifted complexes when incubated with O2 in electrophoretic mobility shift assays (R. Thompson, personal communication). Perhaps the juxtaposition of five relatively weak sites is required for binding of O2 to the *b-32* promoter.

A

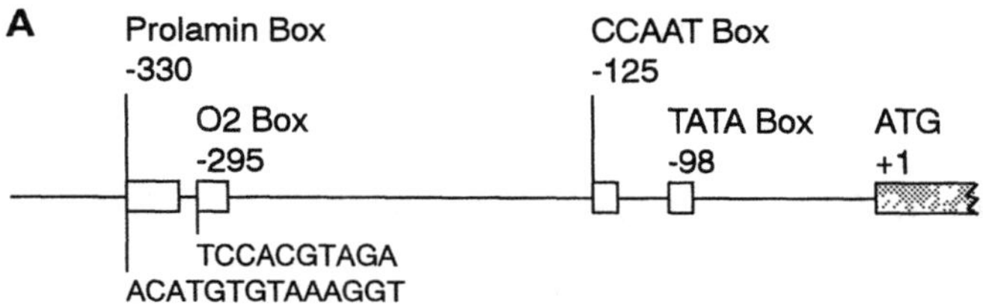

B

```
                         O2 Box
 -305                                      -274
5'-TGCATGTCATTCCACGTAGAT-AAAAGAATTC-3'  22Z-1
5'-TGCATGTCATTCCACGTAGATGAAAAGAATTC-3'  22Z-2
5'-TGCATTTCATTCCACGTAGAT-AAAAGAGTTC-3'  22Z-3
5'-TGCATGTCATTCCACGTAGATGAAAAGAATTC-3'  22Z-4
5'-TGCATGTCATTCCACATAGATTAAAAAAAATC-3'  pML1
5'-TCGATTTCATTCCACGTAGATAAAAAGAGTTC-3'  pML2
5'-TGCATGTCA-TCCAAGTATATGAAAATAATTC-3'  Z7
5'-TGCATCTCATTCCACGTAGATAAAAAGAGTTC-3'  zA1
5'-TGCATGTCATTCCTCATGTATGGAAAGAATTC-3'  gZ22.8
5'-TGCATCTCA-TCCAAGTATATCAAAAGAATTC-3'  ψgZ22.8
5'-TGCATGTCATTCCACGTAGATGAAAAAA-TTC-3'  ZSF4C1
5'-TGCATGTCATTCCACGTAGATGAAAAGAATTC-3'  ZSF4C2-C5
```

Figure 3. A) Conserved sequence elements within the promoter of 22-kD zein genes. **B)** Comparison of the O2 box and neighboring sequences in different 22-kD zein genes. The ACGT core of the imperfect palindromic binding site (O2 box) is missing in several 22-kD zein genomic clones. Sequences are from Schmidt et al. 1992 (22Z-1 to 4), Wandelt and Feix 1989 (pML1 and 2), Kridl et al. 1982 (Z7), Spena et al. 1982 (zA1), Thompson et al. 1992 (gZ22.8 and ψgZ22.8), and Liu and Rubenstein 1992 (ZSF4C1 to C5).

The presence of apparently different O2 binding sites in the promoters of 22-kD zeins and *b-32* is surprising. Although O2 can bind these sites *in vitro*, possibly one or both of these promoters are bound *in vivo* by heterodimeric complexes between O2 and another bZIP protein. Such heterodimers may bind strongly to sites that O2 homodimers bind only weakly. Alternatively, O2 may be able to recognize a wide spectrum of sites. A recent study (Izawa et al. 1993) indicates that O2 does have a greater tolerance than a number of other plant bZIP proteins for variations surrounding an ACGT core binding site. The results from this study predict that the 5'-TCCACGTAGA-3' site in 22-kD zein genes should be a high affinity site whereas the 5'-GTTGACGTGA-3' site in *b-32* is predicted to be a low affinity site.

Table 1. Sequences Recognized by the O2 Protein

O2 Binding Site	Gene	Position*	Designation
5'-TCCACGTAGA-3'	*22Z-4*	-197	O2 Box
5'-GTTGACGTGA-3'	*b-32*	-210	B5
5'-GATGACATGA-3'	"	-167	B4
5'-GATGATGTGG-3'	"	-143	B3
5'-GATGATATGG-3'	"	-117	B2
5'-GATGACATGG-3'	"	-68	B1

* indicates position relative to the TATA box of the respective gene

Finally, O2 will activate expression of the distantly related 22-kD-like coixin genes from *Coix lacryma-jobi* (Job's tears) (P. Arruda, unpublished). Coix, maize, tripsacum and sorghum belong to the same grass tribe, *Andropogoneae*. Recently, Ottoboni et al. (1993) demonstrated by sequence comparison that the 22-kD zeins and 25-kD coixin genes share high sequence identity throughout the promoter and coding regions. DNA-binding studies using ^{32}P-labelled coixin promoter fragments incubated with a β-gal/O2 fusion protein indicate that O2 recognizes a sequence within the first 250 bp 5' of the initiation codon (P. Arruda and R. Schmidt, unpublished). Although the promoters from these genes are very similar to 22-kD zein genes, they do not contain the O2 box found in *22Z-1* through *22Z-4* of Schmidt et al. (1992). It will be interesting to know what sequence in these promoters is being recognized by O2. An O2 homologue has been identified in *Coix* (L. Ottoboni, A. Leite and P. Arruda, unpublished) and its bZIP domain is nearly identical to the bZIP domain of O2.

Transcriptional Activation

DNA-binding by O2 *in vitro* shows that the protein can recognize the target gene promoter, but it does not indicate whether O2 activates transcription. Using yeast as a heterologous system we have demonstrated that O2, in the absence of other maize proteins, can activate expression of a *lacZ* reporter gene containing a multimer of the O2 binding site (Schmidt et al. 1992). Likewise, in transient assays performed on maize endosperm suspension cells, O2 activates a *GUS* reporter gene containing the 22-kD zein gene promoter. In this same system it will not, however, activate transcription from the promoter of the distantly related 27-kD zein gene, unless the cryptic O2 box in the 27-kD promoter (5'-TTTACGTAGA-3') is converted to the O2 box found in 22-kD zein gene promoters (Ueda et al. 1992). Curiously, although O2 activates gene expression from a promoter containing a single O2 box in maize, in yeast only a multimer of the O2 box is sufficient for activation by O2. In yeast, neither the *22Z-4* promoter driving *lacZ* nor a monomer or dimer of the O2 box could provide activation by O2 (M. Ketudat and R. Schmidt, unpublished). This suggests that, although O2 can activate expression in a heterologous eukaryote, the mechanism by which it interacts with the O2 binding site may differ from the one being utilized in maize endosperm cells.

Although the individual O2 binding sites in the *b-32* promoter appear to bind O2 only weakly, the intact promoter responds to O2 expressed from the *CaMV* 35S promoter in transient assays performed in tobacco protoplasts (Lohmer et al. 1991). As with our analysis in yeast cells, where O2 is expressed from the constitutive phosphoglycerate kinase promoter, the transient assays in tobacco cells rely on high levels of O2 expression to obtain activation. Because of the high levels of expression, it is difficult to know how physiologically relevant these studies are to events occurring in endosperm cells where levels of O2 protein may be significantly lower and interactions between O2 and other maize proteins may play a significant role in the activation of these two different genes. These studies do demonstrate that O2 has the capacity to activate transcription upon binding to these recognition sequences, and both systems are being used to define domains in O2 important to activation. From such studies it appears that O2 contains two acidic activation domains, one near the amino terminus and another near the basic domain (Figure 2).

Nuclear Localization

The sequences in O2 that are important for nuclear targeting have been recently determined. By immunolocalization Varagona et al. (1991)

demonstrated that O2 is localized to the nucleus in maize endosperm cells as well as in transgenic tobacco. In order to determine the domains of O2 that direct this protein to the nucleus, an alternative monocot system was developed. Using onion epidermal cells as a convenient monocot for transient assays of nuclear targeting, Varagona et al. (1992) used O2::GUS fusions to localize the targeting signals to two short sequences within the O2 protein (Fig. 2). The most efficient of the two is contained within the basic domain and consists of a bipartite signal. Thus, the basic domain is bifunctional, being important to DNA binding and nuclear localization. These two functions are separable, as a DNA-binding deficient mutant of O2 (see below) is still localized to the nucleus (R. Varagona and N. Raikhel, unpublished). The second nuclear localization signal is located between residues 101 and 135 and has the structure of a simian virus 40 nuclear targeting signal. A bipartite domain appears to be the more common nuclear targeting signal among plant proteins (Raikhel 1992).

III. Does O2 Binding/Activation Depend on Other Regulatory Proteins?

Our analysis of an EMS-induced mutant of *O2*, *o2-676* (Aukerman et al. 1991), suggested that in maize endosperm cells, O2 can interact with another protein to bind zein gene promoters. The mutant protein o2-676 will not bind the target site in 22-kD zein gene promoters under any conditions tested. Likewise, when o2-676 is expressed in yeast under conditions where the normal O2 activates expression, the mutant protein produces background levels of activity (M. Ketudat and R. Schmidt, unpublished). We conclude that this mutant form of O2 is incapable of binding to the target site sequence on its own. However, in endosperm, this mutant is capable of activating zein gene expression at about 50% of wild-type levels. This contrasts markedly with the levels of expression (about 10% of wild-type) seen in endosperm from an *o2* null allele.

How is it that the o2-676 protein can partially activate zein gene expression in maize endosperm cells, yet the protein is incapable of recognizing the target site sequence in *in vitro* binding assays and in experiments performed *in vivo* using yeast cells? DNA binding by o2-676 in maize endosperm cells may be dependent on an association with another (or other) maize proteins. This may involve a protein that forms a heterodimer with o2-676 to recognize the O2 box, or it may involve protein-protein contacts between o2-676 and another protein that binds to an adjacent site.

In support of the former model, the fact that the O2 box is not a perfect palindrome suggests that a heterodimeric complex could better contact this site than a homodimeric complex. Also, it is clear that other endosperm proteins besides O2 can bind the O2 box in 22-kD zein genes. We have shown that at least two shifted complexes can be detected in mobility shift assays using the labelled O2 box incubated with proteins from a nuclear-enriched fraction derived from o2 null or o2-676 mutant endosperms (Schmidt et al. 1992, and Figure 4). These proteins bind specifically, as the shifted complexes do not occur when a mutant site is used. It is possible that one or more of the proteins in these complexes forms heterodimeric complexes with O2 that bind the O2 box. An efficient *in vivo* labelling technique for endosperm proteins coupled with antibody precipitation using an anti-O2 antibody would help resolve this question. Likewise, screening an endosperm cDNA expression library with a labelled O2 box should identify recombinant clones expressing proteins that can bind this site. Recombinant O2 clones could be avoided by making such a library from an o2 null line.

In support of the latter model, there is a highly conserved *cis*-element that lies just 20 nucleotides upstream of the O2 box. This *cis*-element is often referred to as the prolamin box due to its presence in all prolamin genes from maize as well as in the prolamins of other species (see Thompson and Larkins, 1989). A nuclear factor does bind to this site (Maier et al. 1987, Quayle and Feix 1992). Considering the close proximity of the O2 box and the prolamin box, protein-protein interactions may occur that mutually stabilize the association of these proteins on the promoter. This stabilization may be sufficient to allow the mutant o2-676 protein to interact (albeit only weakly) with the O2 box. The isolation of the prolamin-box binding factor is requisite to testing this model. Identifying recombinant lgt11 clones expressing the prolamin box binding protein should also be possible by library screening, provided this binding factor does not represent an obligate heterodimeric complex.

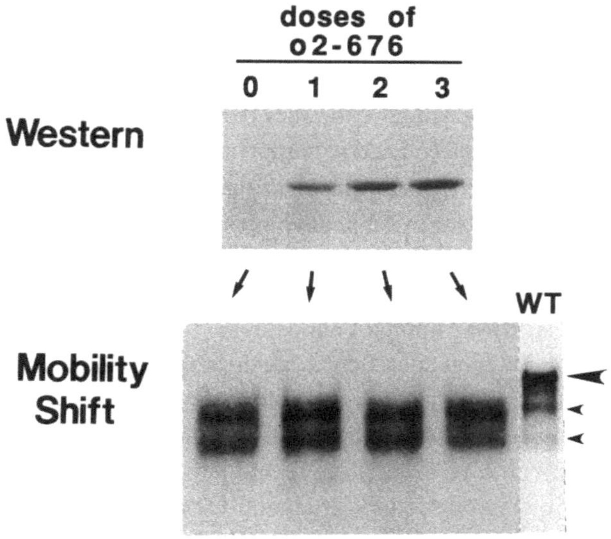

Figure 4. Analysis of O2 box binding proteins in endosperm nuclear extracts from an *o2-676* by *o2* null allelic series. Top panel is a protein blot of the nuclear extracts reacted with O2 antiserum. Numbers along the top of the figure indicate the genome equivalents of o2-676 in the triploid endosperm. In the lane with zero genomic equivalents of o2-676 (i.e. a homozygous o2 null mutant) there is no O2 protein to react with the antibodies. The lower panel is a mobility shift assay using a labelled O2 box oligonucleotide incubated with the four different nuclear extracts analyzed in the upper panel. The unbound probe is not included in the figure and only the shifted complexes are shown. The lane labelled "WT" shows the wild-type mobility shift pattern from a separate experiment. The large arrowhead indicates the uppermost complex which is present only in wild-type *O2* extracts (see Schmidt et al. 1992), whereas the two smaller arrowheads indicate the O2 independent complexes seen in extracts from *o2* null and the DNA binding mutant *o2-676*.

IV. An Endosperm bZIP Protein that Interacts with O2

We have recently identified a bZIP protein from maize endosperm that interacts with O2 (Pysh et al. 1993). This protein, OHP1, shares significant homology with O2 through the bZIP domain (76% identity) but differs significantly from O2 both 3' and 5' of this DNA binding motif. OHP1 has a predicted molecular weight of 42,300, which is slightly smaller than the O2 protein (MW 47,100). As determined by mobility shift assays using *in vitro* translated O2 and OHP1, OHP1 forms heterodimers with O2 that bind the O2 box in 22-kD zein genes. The affinity of the heterodimer for the O2 site appears equal to the affinity of either the O2 homodimer or the OHP homodimer (Pysh et al. 1993). The significance of these observations to O2 function and zein gene expression remains to be elucidated. Recent experiments indicate that OHP1 protein is considerably less abundant than is O2, at least during the mid-stage (20 days after pollination) of endosperm development. A comparison of protein levels at other time points in endosperm development has not yet been made. From studies of RNA expression during endosperm development it appears that OHP1 message initiates earlier and peaks ahead of the O2 transcript (L. Pysh and R. Schmidt, unpublished). If RNA levels accurately indicate protein levels, then OHP1 protein appears earlier than O2 and decreases during the peak of O2 production.

How might we fit these observations into a possible role of OHP1 in zein gene expression? One piece of information that is critical to forming models of OHP1 function is to know whether DNA binding by OHP1 can result in transcriptional activation. We have obtained inconclusive results with regards to addressing this question. Our preliminary results suggest that OHP1 can behave as an activator, although we suspect that this occurs only when very high levels of OHP1 are being expressed. Relative to O2, OHP1 appears to be a weak activator. Considering the apparent low levels of OHP1 relative to O2 in endosperm, transcriptional activation of zein genes is likely to be more influenced by O2 than OHP1.

So what role could OHP1 be playing? Could OHP1 be the bZIP protein that was previously speculated to interact with the o2-676 mutant protein, thus generating partial activation of zein genes? This seems unlikely since attempts to demonstrate DNA binding dimers of o2-676 and OHP1 have failed (data not shown).

One possibility we are considering is that OHP1 is a repressor of zein gene expression. The role of OHP1 may be to keep zein genes quiescent early in endosperm development by binding to the O2 box and negatively regulating zein transcription. The role of O2 would be two-fold: first to titrate away the OHP1 by forming dimers that displace and sequester OHP1, and, second, to bind the O2 box to provide high levels of zein gene expression. If this scenario were true, one would expect OHP1 to be present early in endosperm development (which it is) and to be expressed at much lower levels than O2 (which it appears to be) to allow a quick titration of OHP1 by O2. This model also provides an explanation for the observed o2-676 phenotype. The model must account for the fact that, *in vivo*, o2-676 partially activates zein gene expression, yet the mutant protein appears incapable of binding to the O2 box based on *in vitro* binding assays and *in vivo* studies in yeast cells. These observations can be explained by invoking the model of OHP1 function proposed above. The o2-676 protein maintains a wild-type dimerization motif which would allow it to titrate the OHP1 repressor from the zein promoter. This may cause partial activation of zein gene expression by simply removing the repressor from the promoter. However, since the defective basic domain of o2-676 prevents its binding to the O2 box, this mutant can never induce normal levels of zein gene expression. This model is testable and further experiments are in progress.

Another possibility is that O2 and OHP1 act cooperatively to regulate the expression of target genes other than zeins. If this occurs, it most likely happens early in endosperm development when apparent levels of OHP1 and O2 are most similar. Determining the role of OHP1 in maize development will be complicated by the fact that the locus encoding OHP1 is duplicated or triplicated in some maize inbreds, while existing as a single locus in others (L. Pysh and R. Schmidt, unpublished).

V. Summary

The analyses to date have shown that the product of the O2 gene is an endosperm specific DNA binding protein that plays an important regulatory role in seed storage protein gene expression and in the expression of a gene, *b-32,* that likely plays a role in defense against pathogens. The regulation of these different genes by O2 is dependent on the recognition of diverse sequence elements in the promoters of 22-kD zein genes and *b-32*. Not surprisingly, evidence from molecular and genetic analyses of O2 and various mutants (discussed above) implicate other endosperm proteins in this process. Understanding the precise mechanism by which O2 exerts it effect on these genes will depend on the isolation and characterization of these other factors.

The existence of endosperm mutants besides *o2* that affect seed storage protein expression implicate several other loci in this process (see Motto et al. 1989). Molecular approaches to gene isolation as well as traditional transposon tagging should facilitate the goal of understanding how these regulatory products fit into the overall scheme of maize seed development.

Acknowledgments

This work was supported by grants to R.J.S. from the National Institutes of Health (GM41286) and the McKnight Foundation and to L.D.P. by the Howard Hughes Medical Institute Predoctoral Program. We thank K. Archer for assistance with figures.

References

Aukerman MJ, Schmidt RJ, Burr B, Burr FA (1991) An arginine to lysine substitution in the bZIP domain of an *opaque-2* mutant in maize abolishes specific DNA binding. Genes Dev 5:310-320

Aukerman, MJ, Schmidt RJ (1993) A 168 bp derivative of *Suppressor-mutator/Enhancer* is responsible for the maize *o2-23* mutation. Plant Molec Biol 21:355-362

Bass HW, Webster C, O'Brian GR, Roberts JKM, Boston RS (1992) A maize ribosome-inactivating protein is controlled by the transcriptional activator *Opaque-2*. Plant Cell 4:225-234

Brochetto-Braga, MR, Leite A, Arruda P (1992) Partial purfication and characterization of lysine-ketoglutarate reductase in normal and opaque-2 maize endosperms. Plant Physiol 98:1139-1147

Hartings H, Maddaloni M, Lazzaroni N, Di Fonzo N, Motto M, Salamini F, Thompson R (1989) The *O2* gene which regulates zein deposition in maize endosperm encodes a protein with structural homologies to transcriptional activators. EMBO J 8:2795-2801

Izawa T, Foster R, Chua N-H (1993) Plant bZIP protein DNA binding specificity. J Mol Biol 230:(in press)

Katagiri F, Chua N-H (1992) Plant transcription factors: present knowledge and future challenges. Trends Genet 8:22-27

Kridl JC, Vieira J, Rubenstein I, Messing J (1982) Nucleotide sequence analysis of a zein genomic clone with a short open reading frame. Gene 28:113-118

Lamb P, McKnight SL (1991) Diversity and specificity in transcriptional regulation: The benefits of heterotypic dimerization. Trends Biochem Sci 16:417-422

Liu C-N, Rubenstein I (1992) Molecular characterization of two types of 22 kilodalton α-zein genes in a gene cluster in maize. Mol Gen Genet 234:244-253

Lohmer S, Maddaloni M, Motto M, Di Fonzo N, Hartings H, Salamini F, Thompson RD (1991) The maize regulatory locus *Opaque-2* encodes a DNA-binding protein which activates the transcription of the *b-32* gene. EMBO J 10:617-624

Maier U, Brown JWS, Toloczyki C, Feix G (1987) Binding of a nuclear factor to a consensus sequence in the 5' flanking region of zein genes from maize. EMBO J 6:17-22
Motto M, Di Fonzo N, Hartings H, Maddaloni M, Salamini F, Soave C, Thompson RD (1989) Regulatory genes affecting maize storage protein synthesis. Oxford Surveys of Plant Molecular and Cell Biology 6:87-114
Motto M, Maddaloni M, Ponziani G, Brembilla M, Marotta R, Di Fonzo N, Soave C, Thompson R, Salamini F (1988) Molecular cloning of the *o2-m5* allele of *Zea mays* using transposon marking. Mol Gen Genet 212:488-494
Ottoboni LMM, Leite A, Yunes JA, Targon ML, de Souza Filho GA, Arruda P (1993) Sequence analysis of 22 kDa-like α-coixin genes and their comparison with homologous zein and kafirin genes reveals highly conserved protein structure and regulatory elements. Plant Mol Biol 21:765-778.
Pysh L, Aukerman MJ, Schmidt RJ (1993) OHP1: A Maize bZIP Protein that Interacts with Opaque-2. Plant Cell 5:227-236
Quayle TJA, Hetz W, Feix G (1991) Characterization of a maize endosperm culture expressing zein genes and its use in transient transformation assays. Plant Cell Reports 9:544-548
Raikhel N (1992) Nuclear targeting in plants. Plant Physiol 100:1627-1632
Schmidt RJ (1992) Opaque-2 and zein gene expression. In: Verma DP (ed) Control of plant gene expression, CRC Press, London, pp. 337-355
Schmidt RJ, Burr FA, Aukerman MJ, Burr B (1990) Maize regulatory gene *opaque-2* encodes a protein with a "leucine-zipper" motif that binds to zein DNA. Proc Natl Acad Sci USA 87:46-50
Schmidt RJ, Burr FA, Burr B (1987) Transposon tagging and molecular analysis of the maize regulatory locus *opaque-2*. Science 238:960-963
Schmidt RJ, Ketudat M, Aukerman MJ, Hoschek G (1992) Opaque-2 is a transcriptional activator that recognizes a specific target site in 22-kD zein genes. Plant Cell 4:689-700
Spena A, Viotti A, Pirrotta V (1982) A homologous repetitive block structure underlies the heterogeneity of heavy and light chain zein genes. EMBO J1:1589-1594
Thompson GA, Siemieniak DR, Sieu LC, Slightom, JL, Larkins, BA (1992) Sequence analysis of linked maize 22 kDa α-zein genes. Plant Mol Biol 18:827-833
Ueda T, Waverczak W, Ward K, Sher N, Ketudat M, Schmidt RJ, Messing J (1992) Mutations of the 22- and 27-kD zein promoters affect transactivation by the Opaque-2 protein. Plant Cell 4:701-709
Varagona MJ, Schmidt RJ, Raikhel NV (1991) Monocot regulatory protein Opaque-2 is localized in the nucleus of maize endosperm and transformed tobacco plants. Plant Cell 3:105-113
Varagona, M.J., R.J. Schmidt and N.I. Raikhel. (1992). Nuclear localization signal(s) required for nuclear targeting of the maize regulatory protein, Opaque-2. Plant Cell 4:1213-1227
Wandelt C, Feix G (1989) Sequence of a 21 kd zein gene from maize containing an in-frame stop codon. Nucl Acids Res 17:2354

Regulation of the Expression of Genes Coding for Proline-Rich Proteins in Maize

José A. Martínez-Izquierdo, Marcelo Menossi, Virginia Stiefel, Nora García, and Pere Puigdomènech.
Departamento de Genética Molecular, CID-CSIC. Jordi Girona, 18, 08034 Barcelona, Spain.

Introduction

Plant proline-rich proteins can be classified into several families depending on its function or structure (José and Puigdomènech, 1993). One of these families are the hydroxyproline-rich glycoproteins (HRGPs) where a large proportion of prolines are hydroxylated post-translationally and contain a variable proportion of glycosylated residues. HRGPs are one of the main extractable components of cell walls from both dicotyledonous (where they are often called extensins) and monocotyledonous species. Another group of genes coding for proline-rich proteins (PRPs) have been better characterized in soybean (Hong *et al.*, 1987). Immunological methods have shown that maize HRGPs (Ludevid *et al.*, 1990) and soybean PRPs (Ye *et al.*, 1991) are located in the cell wall. In general, all these proteins are supposed to have a structural role in the wall (Cassab and Varner, 1988). Other proteins containing domains rich in proline, such as maize HyPRP, have also been described (José and Puigdomènech, 1992) but nothing is yet known about their location in the cell.

Extensins are developmentally regulated in a tissue-specific manner (Memelink *et al.*, 1987; Ye *et al.*, 1991) and they can be induced by fungal infection and wounding (Corbin *et al.*, 1987). In some cases, genes coding for HRGPs show very specific patterns of expression as it is the case for a gene mainly expressed in lateral root initiation in tobacco (Keller and Lamb, 1989). PRPs are encoded by a number of genes whose mRNAs are accumulated preferentially in distinct plant tissues, for instance, they are expressed in defined stages of hypocotyl development (Hong *et al.*, 1989).

NATO ASI Series, Vol. H 81
Plant Molecular Biology
Edited by G. Coruzzi and P. Puigdomènech

HRGP genes have a relatively simple genomic structure in the cereal species studied so far. They are probably represented by one gene copy per haploid genome in maize (Stiefel *et al.*, 1990), sorghum (Raz *et al.*, 1991) and rice (Caelles *et al.*, 1992). However, in dicotyledonous species HRGPs are encoded by multigenic families, and each particular gene has a defined pattern of expression. All the cereal HRGP genes analyzed until now are also simple in terms of gene structure with no introns interrupting their coding regions, although in all cases they have an intron (Chen and Varner, 1985; Raz *et al.*, 1992) in the 3'-nontranslated region. Presumably a number of essential regulatory boxes of the promoter of the maize HRGP gene are contained in the first 600 bp upstream from the initiation translation codon as it has been shown in transient expression experiments (Tagu *et al.*, 1992).

Accumulation of HRGP mRNA in meristematic tissues of maize

The expression of the *HRGP* gene parallels histone mRNA accumulation in tissues rich in dividing cells, such as coleoptile nodes, plumule, root apex and calli (Stiefel *et al.*, 1988, 1990; Ludevid *et al.*, 1990), being very low in adult tissues, in particular in elongating and differentiated tissues of root, leaf and stem. The mRNA coding for the maize HRGP is transiently accumulated in provascular cells in germinating embryos, young leaves and roots. This property can be better observed by using *in situ* hybridization (Stiefel *et al.*, 1990; Ruiz-Avila *et al.*, 1991). For instance in young roots the accumulation of HRGP mRNA in the provascular cells can be observed in Figure 1. In flower meristems the pattern of expression of the maize *HRGP* gene also coincides with the regions having a high cell division activity (Figure 2).
The analysis of the expression patterns of genes in certain cell types of the immature embryo may be interesting to understand the cell types present at different stages of the morphogenetic stages of plant embryogenesis and as a way to identify genes that regulate these processes. The expression of *HRGP* is only detectable at the protein and mRNA levels in the embryo axis, being its mRNA barely

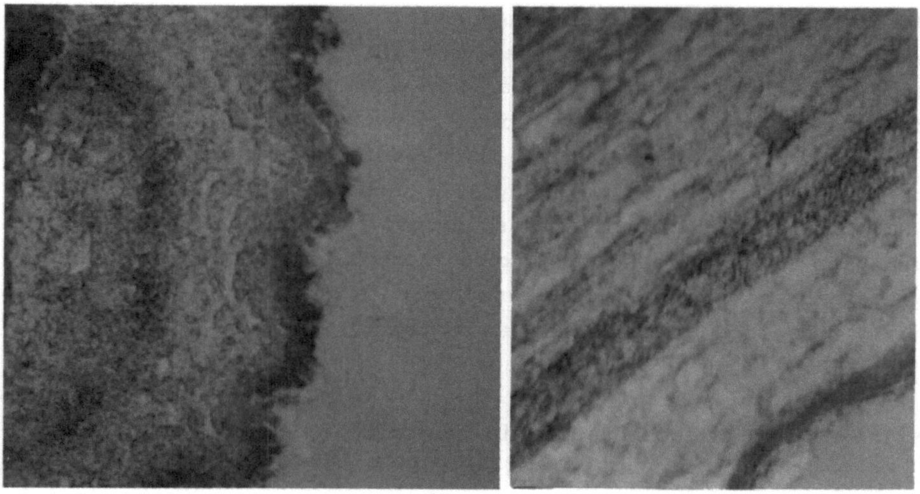

Figure 1. HRGP mRNA accumulation in the provascular cilinder of maize roots. The *in situ* hybridization was carried out on fixed paraffin sections with a ^{35}S-labelled probe.

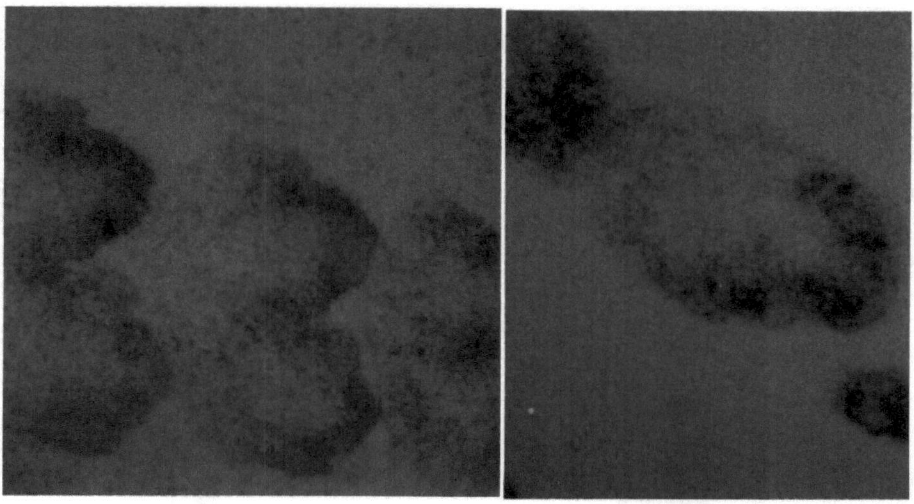

Figure 2. HRGP mRNA accumulation in maize floral meristems observed by *in situ* hybridization. The floral meristem sections were hybridized with an HRGP cDNA probe labelled with ^{35}S nucleotides.

detectable in the scutellum (Ruiz-Avila *et al.*, 1991, 1992). In the embryo axis, the *HRGP* mRNA preferentially accumulates in dividing and vascular cells in a similar way to histone H4 mRNA, although this one is restricted to defined groups of cells. This fact has been interpreted as indicating that both mRNA have a different rate of turnover during the cell cycle (Ruiz-Avila *et al.*, 1992). The *HRGP* transient expression associated to vascular systems in vegetative tissues and its expression in provascular systems in young embryos gives to *HRGP* gene the role of useful marker for the early formation of the vascular systems.

Maize HRGP mRNA accumulation in response to fungal elicitors

It has been reported that some *HRGP* genes from dicots (mainly carrot, bean and tomato) are activated in response to mechanical wounding, ethylene treatment and fungal infection of some plant tissues (Cooper, 1988). Not only the infection of the plant by fungi but also the treatment of plant cells with elicitors derived from fungal cell walls is able to activate *HRGP* genes (Corbin *et al.*; 1987, Lamb *et al.*, 1989). Although, maize *HRGP* gene is activated by wounding and ethylene (Ludevid *et al.*, 1990; Tagu *et al.*, 1992), activation by fungi or their elicitors has not been so far proved.

Fungal elicitor treatment of suspension-cultured cells of the maize line Black Mexican Sweet (BMS) caused an increase in *HRGP* mRNA in an elicitor dose dependent manner (Figure 3).

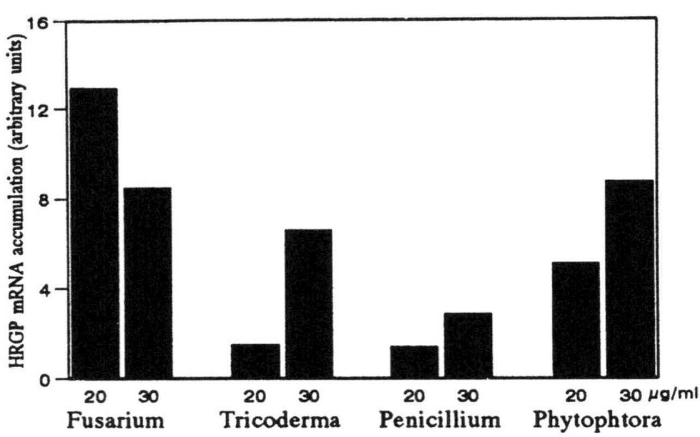

FIGURE 3: *HRGP* mRNA accumulation in response to fungal elicitors. BMS maize cells growing exponentially were challenged with fungal elicitors during 8 h and collected. The elicitors were obtained from fungal cell walls by conventional methods, except for *P. megasperma* where a semipurified fraction was used (Ayers *et al.*, 1976). As controls BMS cells treated with elicitors 0 h or non-treated cells 8 h were used. *HRGP* mRNA bands observed in the different treatments in the RNA blot autoradiograms were quantitized by densitometric scanning.

The increment in mRNA accumulation seems to depend on the fungus species, being the highest increment when BMS cells were challenged with 20 μg/ml of elicitor from *Fusarium monilliforme*, a natural pathogen of maize (Kuchareck *et al.*, 1966). Not even a more purified fraction of elicitors from *Phytophtora megasperma* (Ayers *et al.*, 1976) at 30 μg/ml was able to produce a similar increment (30% less) than *F. moniliforme* elicitors (Figure 3). The activation of the *HRGP* gene by the elicitors of the non-pathogenic fungi of maize probed, namely, *Tricoderma*, *Penicillium* and *Phytophtora* genus is not an unexpected result, because it has been shown that the response of plant to elicitors is often non-specific and in some cases dose dependent. Interestingly enough, the higher increment in the accumulation of *HRGP* mRNA produced by elicitors of a natural pathogen of maize (*F. monilliforme*) suggests some degree of specific response of the plant besides the non-specific one observed with the other fungi. To prove this hypothesis, elicitors of other fungi, and in particular other natural pathogens of maize, should be investigated.

Transient expression analysis of HRGP promoter in immature embryos by using particle bombardment.

In order to identify *cis*-acting regulatory elements responsible for the complex pattern of HRGP expression present in the *HRGP* promoter, several chimeric genes were constructed by fusing 5' promoter deletions with a constant 3' end in the nucleotide +16 (in the coding region) with the *Gus* reporter gene, in phase. These translational fusions may in general be preferred to transcriptional ones, because the former ones probably better reflect the natural situation of the gene, taking into account that transcriptional regulatory sequences around the translational codon that have been recently described.

The plasmid with 699 bp promoter was able to direct the expression of the *Gus* gene (Figure 3) to comparable levels, although lower than the two strong constitutive promoters, those corresponding to the CaMV 35S and rice actin gene (results not shown). Removal of 194 bp in p505 construction only reduced the GUS activity by 14%, however further deletion of 208 bp in construction p297 resulted in a drastic reduction of 39%, suggesting that those 208 bp contain elements important for quantitative enhancement of transcription. In the last construction tested, where another 142 bp were deleted, the level of GUS activity driven by the 159 bp promoter, the activity although 20% lower than p297, was still significatively higher that the basal level, suggesting that this minimal promoter has still relatively strong quantitative *cis*-acting elements. The relatively high levels of expression that the latter two constructions still show, contrast with the very low levels, slightly higher than basal ones, shown by equivalent transcriptional constructions (Tagu *et al.*, 1992), pointing out to the presence of quantitative *cis*-acting elements around the ATG start codon. The difference among the chimeric genes in terms of known putative regulatory boxes, is the absence in p159 of a putative ethylene box, that differs in one nucleotide from the box implicated in ethylene regulation found in several promoters of bean chitinase genes (Broglie *et al.*, 1989). This fact makes p159 and the other constructions interesting to study *HRGP* regulation by ethylene.

The approach used here, with only one reporter gene to assess the transient expression of *HRGP* promoter by bombardment, with the appropriate statistical analysis, gives reproducible results with SEM ranging from 2.5 to 11.3% of the corresponding averages (Figure 4). This could be an alternative strategy to the use of a second plasmid, with another reporter gene under the control of a standard promoter as an internal control to evaluate transfection efficiency. The use of two promoters with two reporter genes results, in some instances, in experimental artifacts (Farr and Roman, 1993).

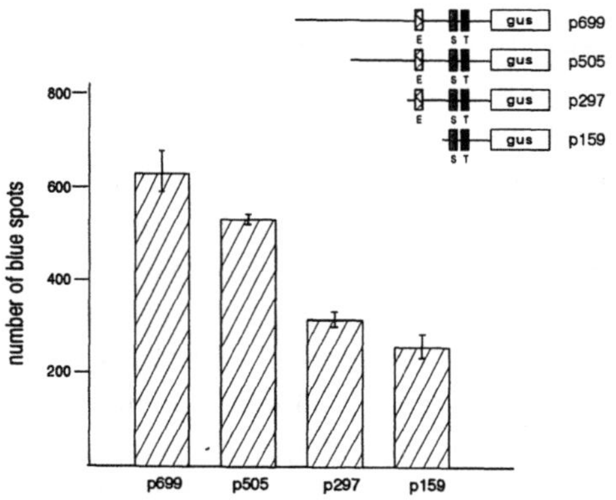

Figure 4: Transient expression analysis of *HRGP* promoter in immature embryos by using particle bombardment. The chimeric constructions depicted in the top-right part of the figure, are translational *HRGP* promoter fusions at nucleotide +16 (A of the ATG start codon is +1) with the *Gus* reporter gene and NOS 3' downstream regulatory signals. Numbers represent the length of the promoter up to the ATG start codon. E, S and T boxes stand for putative ethylene (Broglie *et al.*, 1989), *Shrunken* (Maas *et al.*, 1990) and TATA boxes. The different DNA constructions have been introduced into 14 DAP immature embryos of W64A maize line, by bombarding with 1 μm diameter gold particles coated with DNA. Embryos were excised and placed on MS0 medium-containing dishes 24 h before bombarding. 24 h after bombarding embryos were transferred into dishes with a X-gluc staining solution, incubated for 18 h and the developed blue spots were counted under a magnifying glass. Graphed values represent the average of the number of blue spots from at least 3 different bombarded plates (9 embryos per plate) ± SEM. The observed differences among *HRGP* constructions are statistically significative, as determined by the Student t method.

Discussion

Plant HRGP genes have a well-defined pattern of expression as observed by both RNA blot analysis and *in situ* hybridization. Their mRNA has been shown to be accumulated in regions of the plant rich in dividing cells and particularly in provascular cells (Ye et al.,

1991). They have also been shown to be induced by mechanical stress and pathogen infection (Corbin et al., 1987). In dicotyledonous species this effect appears to be separated in specific genes having defined functions either in development or in defense reactions. In the species of cereals studied so far the proteins equivalent to dicot HRGP both in their extraction properties, cell wall localization and sequence appear to be a more simple system. This is the case at least for maize, sorghum and rice. In these species the gene coding for a highly repetitive proline-rich protein that is modified post-translationally in part of its proline residues (Kieliszewski et al., 1990) has been cloned and at least in maize the same gene appears to code for a mRNA that is developmentally regulated and induced both by wounding and by ethylene (Ludevid et al., 1990; Tagu et al., 1992).

In the present report we show that the *HRGP* gene can be induced in maize cell suspensions by elicitors prepared from different fungal species. It is interesting to note that there is no specific correlation between specificity and the induction of the gene indicating that HRGP accumulation is part of a general reaction of the plant against mechanical stress or external attack. The mRNA corresponding to HRGP is accumulated at specific cell layers both in young roots and in floral meristems, confirming that HRGP is a useful probe to detect regions active in cell division. Finally it appears that the gene elements directing the expression of the maize HRGP gene are concentrated in the 500 bp proximal to the gene as observed by microparticle bombardment. The analysis of this region may yield elements important for the different types of expression described.

ACKNOWLEDGMENTS

The authors acknowledge the continuous support of Plan Nacional de Investigación Científica y Técnica (grant BIO-91/0648) and Dirección General de Investigación Científica (grant PB90-0129). The technical assistance of Dr. Pilar Fontanet is greatly acknowledged. M.M. is recipient of a fellowship from CNPq and N.G. from ICI.

REFERENCES

Ayers, A.R., Ebel, J., Valent, B., Albersheim, P.: Host-pathogen interactions. X, Fractionation and biological activity of an elicitor isolated from the mycelial walls of *Phytophtora megasperma* var. *sojae*. Plant Physiol 57: 760-765, 1976.

Broglie, K.E., Biddle, P., Cressman, R., Broglie, R., 1989: Functional analysis of DNA sequences responsible for ethylene regulation of a bean chitinase gene in transgenic tobacco. Plant Cell 1: 599-607.

Caelles, C., Delseny, M., Puigdomènech, P., 1992 : The hydroxyproline -rich glycoprotein gene from *Oryza sativa*. Plant Mol. Biol. **18**: 617-619.

Cassab, G.I., Varner, J.E., 1988: Cell wall proteins. Annu. Rev. Plant Physiol. Plant Mol. Biol. **39**: 321-353.

Chen, J.A, Varner, J.E., 1985: An extracellular matrix protein in plants: characterization of a genomic clone for carrot extensin. EMBO J. **4**: 2145-2151.

Collinge, D.B., Slusarenko, A..J., 1987: Plant gene expression in response to pathogens. Plant Mol. Biol. **9**:389-410.

Cooper, J.B.: Cell wall extensin genes. In: Temporal and spatial regulation of plant genes (Plant Gene Research). Verma, D.P.S. & Goldberg, R.B. (eds.). Springer-Verlag, Vienna, 1988, pp. 235-251.

Corbin, D.R., Sauer, N., Lamb, C.J., 1987: Differential regulation of a hydroxyproline-rich glycoprotein gene family in wounded and infected plants. Mol. Cell. Biol. 7: 4337-4344.

Farr, A., Roman, A.: A pitfall of using a second plasmid to determine transfection efficiency. Nucleic Acid Res. **20**: 920, 1993.

Hong, J.C., Nagao, R.T., Key, J.L., 1987: Characterization and sequence analysis of a developmentally regulated putative cell wall protein gene isolated from soybean. J. Biol. Chem. **262**: 8367-8376.

Hong, J.C., Nagao, R.T., Key, J.L., 1989: Developmentally regulated expression of soybean proline-rich cell wall protein genes. Plant Cell **1**: 937-943.

José, M., Puigdomènech P., 1993: Structure and expression of genes coding for structural proteins of the plant cell wall. New Phytol. **124**, In press.

José, M., Ruiz-Avila, L., Puigdomènech, P., 1992: A maize embryo-specific gene encodes a proline-rich hydrophobic protein. Plant Cell **4**: 413-423.

Keller, B., Lamb, C.J., 1989: Specific expression of a novel cell wall hydroxyproline-rich glycoprotein gene in lateral root initiation. Genes Dev. **3**: 1639-1646.

Kieliszewski, M.J., Leykam, J.F., Lamport, D.T.A., 1990: Structure of the threonine-rich extensin from *Zea mays*. Plant Physiol. **92**: 316-326.

Kuchareck, T.A., Kommedahl, T.: Kernel infection and corn stalk and rot caused by *F. moniliforme*. Phytopathology **56**: 983-984, 1966.

Lamb, C.J., Lawton, M.A., Dron, M., Dixon, R.A., 1989: Signals and transduction mechanisms for activation of plant defenses against microbial attack. Cell **56**: 215-224.
Ludevid, M.D., Ruiz-Avila, L., Vallés, M.P., Stiefel, V., Torrent, M., Torné, J.M., Puigdomènech, P., 1990: Expression of genes for cell-wall proteins in dividing and wounded tissues of *Zea mays* L. Planta **180**: 524-529.
Maas, C., Schaal, S., Werr, W.: A feedback control element near the transcription start site of the maize *Shrunken* gene determines promoter activity. EMBO J. **9**: 3447-3452, 1990.
Memelink, J., Hoge, J.H.C., Schilperoort, R.A., 1987: Cytokinin stress changes the developmental regulation of several defence-related genes in tobacco. EMBO J. **6**: 3579-3583.
Raz, R., Crétin, C., Puigdomènech, P., Martínez-Izquierdo, J.A., 1991: The sequence of hydroxiproline-rich glycoprotein gene for *Sorghum vulgare*. Plant Mol. Biol. **16**: 365-367.
Raz, R., Josè, M., Moya, A., Martínez-Izquierdo, J.A., Puigdomènech, P., 1992: Different mechanisms generating sequence variablility are revealed in distinct regions of the hidroxyproline-rich glycoprotein gene from maize and related species. Mol. Gen. Genet. **233**: 252-259.
Ruiz-Avila, L., Ludevid, M.D., Puigdomenech, P., 1991: Differential expression of a hydroxyproline-rich cell wall protein gene in embryonic tissues of Zea mays L. Planta **184**: 130-136.
Ruiz-Avila, L., Burgess, R., Stiefel, V., Ludevid, M.D., Puigdomènech, P., 1992: Accumulation of cell wall hydroxyproline-rich glycoprotein mRNA is an early event in maize embryo cell differentiation. Proc. Natl. Acad. Sci. USA **89**: 2414-2418.
Stiefel, V., Pérez-Grau, L., Albericio, F., Giralt, E., Ruiz-Avila, L., Ludevid, M.D., Puigdoménech, P., 1988: Molecular cloning of cDNAs encoding a putative cell wall protein from *Zea mays* and immunological identification of related polypeptides. Plant Mol. Biol. **11**, 483-493.
Stiefel, V., Ruíz-Avila, L., Raz, R., Vallés, M.P., Gómez, J., Pagés, M., Martínez-Izquierdo, J.A., Ludevid, M.D., Langdale, J.A., Nelson, T., Puigdomènech, P., 1990: Expression of a maize cell wall hydroxyproline-rich glycoprotein gene in early leaf and root vascular differentiation. Plant Cell **2**: 785-793.
Tagu, D., Walker, N., Ruiz-Avila, L., Burgess, S., Martínez-Izquierdo, J.A., Leguay, J.-J., Netter, P., Puigdomènech, P., 1992: Expression of a maize HRGP gene by wounding and ethylene. Functional analysis of the promoter by transient expression using microprojectile bombardment. Plant Mol. Biol. **20**: 529-538.
Ye, Z.-H., Song, Y.-R., Marcus, A., Varner, J.E., 1991: Comparative localization of three classes of cell wall proteins. Plant J. **1**: 175-183.

THE PLANT MITOCHONDRIAL TRANSFER RNAs HAVE THREE DIFFERENT GENETIC ORIGINS

André Dietrich, Laurence Maréchal-Drouard and Jacques-Henry Weil
Institut de Biologie Moléculaire des Plantes du CNRS, Université Louis Pasteur,
12 rue du Général Zimmer, 67084 Strasbourg, France.

INTRODUCTION

In mammals and fungi, the mitochondrial (mt) genome contains 22-24 transfer RNA (tRNA) genes and the corresponding tRNA population is sufficient to allow the incorporation of all 20 amino acids into proteins, thanks to a number of unusual codon-anticodon interactions which enable this apparently incomplete tRNA set to read all 61 sense codons (Maréchal-Drouard et al 1992).

In plant mitochondria, the situation is much more complex. A first set of tRNA genes located in the mt genome, are probably inherited from the endosymbiotic ancester of the mitochondrion, and are therefore called "native" or "genuine". A second set of tRNA genes, also found in the mt genome, have 95-100% similarity to their chloroplast counterparts, and probably result from chloroplast DNA insertions into the mt genome during evolution ("promiscuous DNA"); some of these "chloroplast-like" tRNA genes are expressed in plant mitochondria to yield functional tRNA species. But together the "native" and the "chloroplast-like" tRNAs are not sufficient to allow mt protein synthesis, so that plant mitochondria need a third set of tRNAs, which are encoded in the nuclear genome and must be imported from the cytosol into the mitochondria. The plant mitochondrial tRNA population is therefore a mixture of species which have three different genetic origins.

1) mt tRNAs encoded by the plant mt genome

In wheat and maize, where the most systematic studies were performed, only about 16 different tRNA genes seem to be present in the mt genome. In wheat (Joyce et al 1988a) only 16 *Hind*III restriction fragments carrying putative tRNA genes were characterized, and the almost perfect stoechiometry observed in the hybridization signals' intensities argued against the presence of more than one tRNA gene per restriction fragment. In fact, most of the wheat mt tRNA genes studied show no close physical linkage with each other. In maize (Sangaré et al 1990b) 16 different RNA genes were identified (table I) and these genes appear to be scattered over the mt genome. Out of these 16 tRNA genes, 6 turned out to be present in chloroplast DNA insertions and were shown to be expressed in maize mitochondria. Some of the wheat or maize mt tRNA genes are duplicated (Joyce and Gray 1988, Joyce et al 1988b), but many of them are single copy (Joyce and Gray 1989a, Sangaré et al 1990b).

NATO ASI Series, Vol. H 81
Plant Molecular Biology
Edited by G. Coruzzi and P. Puigdomènech

On the mt genome of a dicot, *Petunia hybrida*, 17 different tRNA genes have been identified (table I) and shown to be scattered. These genes include 5 chloroplast-like genes (Weber-Lotfi et al 1992).

Table I : "Native" and "chloroplast-like" tRNAs or tRNA genes in maize, wheat, potato and petunia mitochondria. (a) Number of tRNA genes in the mitochondrial genome; (b) tRNAs identified in mitochondria; (c) tRNA encoded with a methionine anticodon (CAU) and becoming an isoleucine-specific species after post-transcriptional modification of the C (Weber et al 1990); nd=not determined. Data are taken from Sangaré et al (1990b), Joyce and Gray (1989b), Maréchal-Drouard et al (1990a) and Weber-Lotfi et al (1992).

Amino acid	Anticodon	Genetic Origin	Maize(a)	Wheat(b)	Potato(b)	Petunia(a)
Asp	GUC		1	-	+	-
Cys	GCA		-	-	+	1
Gln	UUG		1	+	+	1
Glu	UUC		2	+	+	1
Gly	GCC		-	-	+	1
Ile	CAU(c)		1	+	+	1
Lys	UUU	"Native"	1	+	+	1
*f*Met	CAU	Mitochondrion-	1	+	+	2
Phe	GAA	encoded	-	-	+	1
Pro	UGG		2	+	+	1
Ser	GCU		1	+	+	1
Ser	UGA		1	+	+	2
Tyr	GUA		1	+	+	1
Val	nd		-	-	+	-
Val	nd		-	-	+	-
Asn	GUU		2	+	+	1
Cys	GCA		1	+	-	-
His	GUG	"Chloroplast-like"	1	-	+	1
*m*Met	CAU	Mitochondrion-	1	+	+	3
Phe	GAA	encoded	1	+	-	-
Ser	GGA		-	+	+	1
Trp	CCA		1	+	+	1

It should be pointed out however that the tRNA genes present in the mt genome of wheat, maize or petunia (including the "chloroplast-like" genes) only correspond to 12-14 amino acids, if one considers their anticodons and assumes that in plant mitochondria the "universal" genetic code is used (there is so far no evidence for any deviation), while a minimal number of 23 tRNAs with different anticodons would be necessary to read the 61 sense codons, taking into account the wobble (Crick 1966) and "two out of three" (Lagerkvist 1978, Samuelson et al 1983) hypotheses.

Table II : "Native" and "chloroplast-like" mitochondrial tRNA genes sequenced in higher plants. (a) The tRNA encoded by this maize mitochondrial gene could in fact become an isoleucine tRNA after post-transcriptional modification of the C in the first position of the anticodon (see Weber et al 1990).

Aminoacid	Anticodon	Genetic origin	Plant species	References
Asp	GUC		Maize	Parks et al 1985
			Wheat	Joyce et al 1988a
Cys	GCA		Tomato	Izuchi & Sugita 1989
Gln	UUG		Wheat	Joyce & Gray 1988
			Wheat	Joyce & Gray 1989a
Glu	UUC		*Arabidopsis*	Chen et al 1989
			Maize	Sangaré et al 1989b
			Soybean	Wintz et al 1987
			Wheat	Gualberto et al 1989a
Gly	GCC		Lupine	Bartnik & Borsuk 1986
			Oenothera	Binder et al 1990
Ile	CAU		Potato	Weber et al 1990
Lys	UUU		Maize	Sangaré et al 1989a
			Oenothera	Binder et al 1990
*f*Met	CAU	"Native"	Lupine	Borsuk et al 1986
		Mitochondrion-	Maize	Parks et al 1984
		encoded	*Oenothera*	Gottschalk & Brennicke 1985
			Soybean	Grabau 1987
			Wheat	Gray & Spencer 1983
*m*Met (a)	CAU		Maize	Parks et al 1984
Phe	GAA		*Oenothera*	Binder et al 1990
Pro	UGG		Maize	Sangaré et al 1990a
			Wheat	Runeberg-Roos et al 1987
			Wheat	Joyce et al 1988a and b
Ser1	GCU		*Arabidopsis*	Chen et al 1989
			Maize	Wintz et al 1988b
			Oenothera	Binder et al 1990
			Wheat	Joyce et al 1988a
Ser2	UGA		Maize	Sangaré et al 1989c
			Wheat	Joyce et al 1988a
Tyr	GUA		*Arabidopsis*	Chen et al 1989
			Bean	Bird et al 1989
			Maize	Sangaré et al 1989a
			Oenothera	Binder et al 1990
			Tomato	Izuchi et al 1990
			Wheat	Joyce et al 1988a
Asn	GUU		Bean	Bird et al 1989
			Lupine	Karpinska & Augustyniak 1988
			Tomato	Izuchi et al 1990
			Wheat	Joyce & Gray 1989c
Cys	GCA		Maize	Wintz et al 1988b
			Wheat	Joyce & Gray 1989b
His	GUG		Maize	Iams et al 1985
			Oenothera	Binder et al 1990
			Sunflower	Ceci et al 1989
*m*Met	CAU	"Chloroplast-like"	*Arabidopsis*	Wintz et al 1988a
		Mitochondrion-	Maize	Sangaré et al 1989a
		encoded	Soybean	Wintz et al 1988a
Phe	GAA		Wheat	Joyce & Gray 1989b
Ser3	GGA		*Oenothera*	Schuster & Brennicke 1987
			Wheat	Joyce & Gray 1989b
Trp	CCA		Maize	Leon et al 1989
			Oenothera	Schuster et al 1988
			Wheat	Maréchal et al 1987

Table III : "Native", "chloroplast-like" and "cytosol-like" mitochondrial tRNAs identified in higher plants. (a) This maize mitochondrial tRNA could in fact become an isoleucine tRNA after post-transcriptional modification of the C in the first position of the anticodon (see Weber et al 1990); nd = not determined.

Aminoacid	Anticodon	Genetic origin	Plant species	References
Asp	GUC		Potato	Maréchal-Drouard et al 1990a
Cys	GCA		Potato	Maréchal-Drouard et al 1990a
Gln	UUG		Potato	Maréchal-Drouard et al 1990a
			Wheat	Joyce & Gray 1989b
Glu	UUC		Potato	Maréchal-Drouard et al 1990a
			Wheat	Joyce & Gray 1989b
Gly	GCC		Potato	Maréchal-Drouard et al 1990a
Ile1	L*AU		Potato	Weber et al 1990
Lys	UUU		Potato	Maréchal-Drouard et al 1990a
			Wheat	Joyce & Gray 1989b
*f*Met	CAU		Bean	Maréchal et al 1986
		"Native"	Potato	Maréchal-Drouard et al 1990a
		Mitochondrion-	Wheat	Joyce & Gray 1989b
*m*Met (a)	CAU	encoded	Wheat	Joyce & Gray 1989b
Phe	GAA		Bean	Maréchal et al 1985b
			Potato	Maréchal-Drouard et al 1990a
Pro	UGG		Bean	Runeberg-Roos et al 1987
			Potato	Maréchal-Drouard et al 1990a
			Wheat	Joyce & Gray 1989b
Ser1,2	GCU, UGA		Potato	Maréchal-Drouard et al 1990a
			Wheat	Joyce & Gray 1989b
Tyr	GUA		Bean	Maréchal et al 1985c
			Potato	Maréchal-Drouard et al 1990a
			Wheat	Joyce & Gray 1989b
Val1,2	nd		Potato	Maréchal-Drouard et al 1990a
Asn	GUU		Potato	Maréchal-Drouard et al 1990a
			Wheat	Joyce & Gray 1989b
Cys	GCA		Wheat	Joyce & Gray 1989b
His	GUG		Potato	Maréchal-Drouard et al 1990a
*m*Met	CAU		Bean	Maréchal et al 1986
		"Chloroplast-like"	Potato	Maréchal-Drouard et al 1990a
		Mitochondrion-	Wheat	Joyce & Gray 1989b
Phe	GAA	encoded	Wheat	Joyce & Gray 1989b
Ser3	GGA		Potato	Maréchal-Drouard et al 1990a
			Wheat	Joyce & Gray 1989b
Trp	CCA		Bean	Maréchal et al 1985a
			Potato	Maréchal-Drouard et al 1990a
			Wheat	Joyce & Gray 1989b
Ala	IGC		Potato	Maréchal-Drouard et al 1990a
Arg1,2	ICG, NCU		Potato	Maréchal-Drouard et al 1990a
Gly	GCC		Wheat	Joyce & Gray 1989b
Ile2	nd		Potato	Maréchal-Drouard et al 1990a
Leu1	C*AA		Bean	Green et al 1987
		"Cytosol-like"	Potato	Maréchal-Drouard et al 1990b
		Nucleus-encoded	Wheat	Joyce & Gray 1989b
Leu2	NAG		Bean	Maréchal-Drouard et al 1988
Leu3	nd		Bean	Maréchal-Drouard et al 1988
Leu4	NAA		Bean	Maréchal-Drouard & Guillemaut 1988
Leu2,3,4,5	nd		Potato	Maréchal-Drouard et al 1990a
Thr1,2	nd		Potato	Maréchal-Drouard et al 1990a
Val	GAC		Wheat	Joyce & Gray 1989b

A number of tRNA genes are clearly missing from all higher plant mt genomes studied so far (tables I and II), for instance those for alanine, arginine, leucine and threonine tRNAs. But the number and the identity of the missing mt tRNA genes seem to vary between monocots and dicots. For example no $tRNA^{Gly}$ gene has been found in the wheat or maize mt genome (Joyce et al 1988a, Sangaré et al 1990b), whereas a $tRNA^{Gly}$ gene has been identified in lupine (Bartnick and Borsuk 1986), *Oenothera* (Binder et al 1990) and petunia (Weber-Lotfi 1992) mitochondria.

The mt genome of *Chlamydomonas reinhardtii* represents an extreme case, as only 3 tRNA genes, corresponding respectively to codons CAA/G(Gln), AUG(Met) and UGG(Trp) have been found in the 15.8 kbp mt DNA of this unicellular green alga (Boer and Gray 1988, Michaelis et al 1990). The liverwort (*Marchantia polymorpha*) mt DNA represents the opposite extreme case, as 29 tRNA genes, encoding 27 different tRNA species, have been located after sequencing of the complete (184 kbp long) mt DNA. Two of these 29 genes (coding for $tRNA^{Met}_{f}$ and $tRNA^{Tyr}$) are duplicated, and none of them is "chloroplast-like" (Oda et al 1992). One is tempted to conclude that this mt DNA codes for a complete set of tRNAs, however 2 tRNA genes have not been found, namely a gene coding for a AUU/C-decoding $tRNA^{Ile}$ and a gene coding for a ACA/G-decoding $tRNA^{Thr}$.

The occurrence of "chloroplast-like" tRNA genes in the plant mt genome results from the presence of promiscuous chloroplast DNA which has been inserted into the plant mt DNA in the course of evolution (Stern and Lonsdale 1982). The isolation from bean mitochondria of a mature "chloroplast-like" $tRNA^{Trp}$ clearly distinguishable from its chloroplast counterpart (Maréchal et al 1985a) was the first evidence for the expression of a "chloroplast-like" tRNA gene in plant mitochondria. Sequencing of the two bean $tRNAs^{Trp}$ also showed that a "chloroplast-like" mt tRNA can differ from its chloroplast counterpart in the modified nucleotide content, suggesting differences in the post-transcriptional modification systems between mitochondria and chloroplasts (Maréchal et al 1985a).

As shown in tables III and IV and fig. 1, 6 "chloroplast-like" tRNAs have been identified in wheat mitochondria, and 5 in potato mitochondria. While 4 tRNAs are common to the two plants, namely $tRNA^{Met}_{m}$, $tRNA^{Asn}$, $tRNA^{Trp}$ and $tRNA^{Ser}$(GGA), 2 tRNAs are only present in wheat mitochondria ($tRNA^{Cys}$ and $tRNA^{Phe}$) and 1 is specific to potato mitochondria ($tRNA^{His}$).

Differences in the identity of "chloroplast-like" tRNAs exist not only between wheat and potato, but also between the two monocots, wheat and maize: As shown in tables I and II, a "chloroplast-like" $tRNA^{Ser}$(GGA) gene is present in the wheat but not in the maize mt genome, whereas a $tRNA^{His}$ gene of the chloroplast type is present in the maize mt genome, but has not been found in the wheat mt genome. More studies would be needed to determine when, during evolution, these transfers of genetic information from the chloroplasts to the mitochondria have occurred.

Table IV : Identification and genetic origin of 31 potato mitochondrial tRNAs. Correspondence with the known wheat mitochondrial tRNAs, as taken from Joyce and Gray (1989b). The genetic origin of the identified species has been determined by hybridization to nuclear or mitochondrial DNA; nd = not determined.

Aminoacid	Anticodon	Genetic origin	Counterpart in wheat
Asp	GUC		-
Cys	GCA		-
Gln	UUG		+
Glu	UUC		+
Gly	GCC		-
Ile1	L*AU		-
Lys	UUU	"Native"	+
*f*Met	CAU	Mitochondrion-	+
Phe	GAA	encoded	-
Pro	UGG		+
Ser1	GCU		+
Ser2	UGA		+
Tyr	GUA		+
Val1	nd		-
Val2	nd		-
Asn	GUU		+
His	GUG	"Chloroplast-like"	-
*m*Met	CAU	Mitochondrion-	+
Ser3	GGA	encoded	+
Trp	CCA		+
Ala	IGC		-
Arg1	ICG		-
Arg2	NCU		-
Ile2	nd		-
Leu1	C*AA	"Cytosol-like"	+
Leu2	nd	Nucleus-encoded	-
Leu3	nd		-
Leu4	nd		-
Leu5	nd		-
Thr1	nd		-
Thr2	nd		-

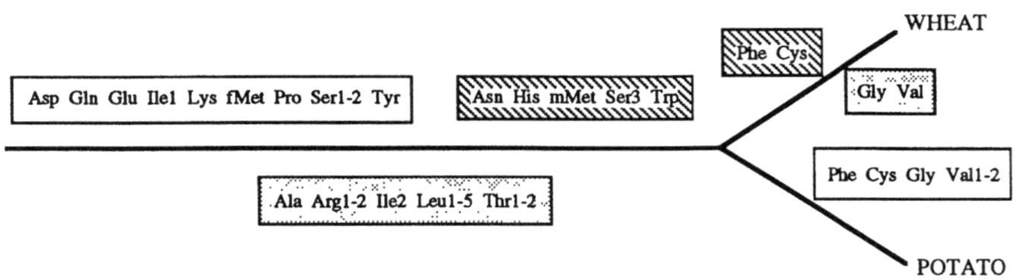

Figure 1 : Differences in the genetic origin of mt tRNAs between potato and wheat. tRNA isoacceptors are numbered according to table IV. "Native" mitochondrion-encoded tRNAs, "Chloroplast-like" mitochondrion-encoded tRNAs and "cytosol-like" nucleus-encoded tRNAs are presented in open, hatched and dotted boxes, respectively.

2) mt tRNAs encoded by the plant nuclear genome

As discussed above, a number of tRNA genes are missing from the plant mt genomes studied so far. That the corresponding tRNAs are coded for by the nuclear genome was first suggested by the presence of "cytosol-like" tRNAs in plant mitochondria, and then demonstrated by experiments showing that these mt tRNAs hybridize to nuclear DNA and not to mt DNA (Maréchal-Drouard et al 1988). Among these tRNAs, in the case of bean, there are 4 $tRNAs^{Leu}$, 3 of which have been completely sequenced and shown to differ from their cytosolic counterparts only by the presence of a Gm at position 18 instead of a G (Green et al 1987, Maréchal-Drouard and Guillemaut 1988, Maréchal-Drouard et al 1988). In potato also, the mt $tRNA^{Leu}$(C*AA) only differs from its cytosolic counterpart by the presence of a Gm at position 18 (Maréchal-Drouard et al 1990b).

Out of the 31 mt tRNAs identified in potato, 11 hybridize only to nuclear DNA (Maréchal-Drouard et al 1990a), namely the 5 $tRNAs^{Leu}$, the 2 $tRNAs^{Thr}$, the 2 $tRNAs^{Arg}$, the $tRNA^{Ala}$ and a $tRNA^{Ile}$ (table IV). But the identity of the nucleus-encoded tRNAs can differ from species to species, especially between monocots and dicots, as shown for instance by a comparison between wheat and potato (table IV and fig. 1): While mt $tRNA^{Gly}$ and $tRNA^{Val}$ are "native" tRNAs coded for by mt DNA in potato (Maréchal-Drouard et al 1990a), these two tRNAs are "cytosol-like" and are presumably encoded by nuclear DNA in wheat (Joyce and Gray 1989b). The differences in the nucleus-encoded tRNAs and also in the "chloroplast-like" tRNAs observed among plants (table III), especially between monocots and dicots (fig. 1), suggest that genetic information transfer and rearrangement has continued after segregation of species and that the evolutionary flux is still in progress.

Direct proof for the import of a nucleus-encoded tRNA was obtained by *in vivo* experiments (Small et al 1992): After introduction of the bean nuclear $tRNA^{Leu}$ (C*AA) gene (Green et al 1987) into potato protoplasts by electroporation, the presence of this foreign bean tRNA could be detected, using a specific oligo-nucleotide probe, in the cytosol and inside the mitochondria of regenerated transgenic potato plants.

But one does not know the mechanisms which allow the selective import of some of the nucleus-encoded tRNAs into the mitochondria (fig. 2). Are these tRNAs imported as precursors (as in the case of the proteins imported into mitochondria) or as mature molecules ? Is the tRNA naked during its import, or is it associated with a protein, such as the corresponding mt aminoacyl-tRNA synthetase (which is also encoded in the nucleus and imported into the mitochondria) or a tRNA methylase (for instance the enzyme which specifically methylates bean $tRNAs^{Leu}$ at position 18) ? The *in vivo* import of heterologous tRNAs into the mitochondria of transgenic plants (Small et al 1992) may help identifying the structural features of the tRNA which are essential for the import process (as constructs using altered tRNA genes can be used), but an *in vitro* sytem would certainly be extremely useful to study the mechanisms of tRNA import.

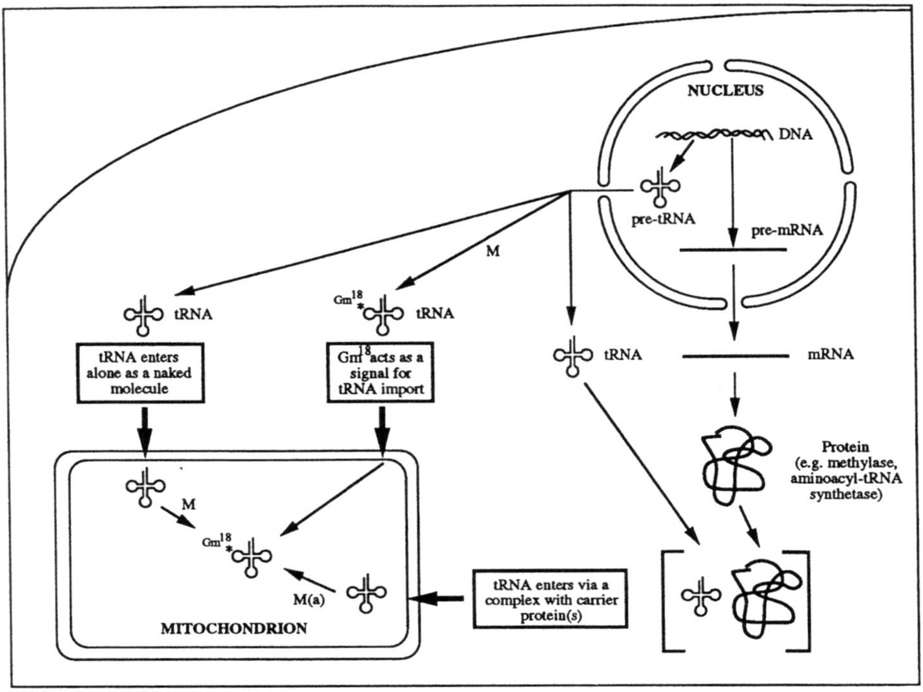

Figure 2 : Hypothetical mechanisms which might allow the selective import of some of the nucleus-encoded tRNAs into plant mitochondria. M = methylation; M(a) = methylation could occur inside or outside the mitochondria.

3) Plant mitochondrial protein synthesis involves tRNAs of different genetic origins

The plant mt tRNA population appears to consist of molecules of nuclear, plastid and genuine ("native") mitochondrial origins. That these tRNA species all participate in protein synthesis in the mitochondria is supported by several arguments: i) they are processed to their mature size and contain modified nucleotides, ii) they are good substrates for the corresponding aminoacyl-tRNA synthetases and for the tRNA nucleotidyl transferase, iii) no nucleus-encoded or "chloroplast-like" tRNA appears to have a genuine ("native") counterpart (table III), in other words there is apparently no redundancy, which suggests that all these tRNA species, whatever their genetic origin, are necessary for plant mt protein synthesis. This may place some constraints on the evolution of the sequences of mitochondria-encoded tRNAs and may account, at least in part, for the fact that plant mt tRNAs have canonical primary and secondary structures, as opposed to metazoan mt tRNAs (e.g. Okimoto and Wostenholme 1991).

There is only one amino acid, namely isoleucine, for which two potato mt isoacceptor tRNAs have been characterized, one encoded by the mt genome, the other by the nuclear genome (see table IV). These two tRNAs^{Ile} decode different synonymous codons. Interestingly the "native" (mitochondrial DNA-encoded) one has a L*AU anticodon (L* is a derivative of lysidine, which is a modified cytidine) and is encoded by a mt gene possessing a methionine-specific (CAU) anticodon (Weber et al 1990). This is reminiscent of what happens in *E. coli*, where a tRNA(CAU) accepting methionine and recognizing the AUG methionine codon is converted (upon post-transcriptional modification of C into L) into a tRNA(LAU) accepting isoleucine and recognizing the AUA isoleucine codon (Muramatsu et al 1988).

CONCLUSION

While in mammals all mt tRNAs are encoded in the mt genome, in yeast one nucleus-encoded tRNA^{Lys} is imported into mitochondria, but this tRNA does not seem to participate in mt protein synthesis (Martin et al 1979, Tarassov and Entelis 1992). In plants, one of the most intriguing feature of the mt tRNA population is the fact that these tRNAs have three genetic origins, which underlines the complexity of the genetic information transfers between the various plant cell compartments and of the mitochondrial gene divergence during evolution. The fact that higher plant mitochondria contain more tRNAs than mitochondria of other organisms, together with the occurrence of messenger RNA editing (Covello and Gray 1989, Gualberto et al 1989b, Hiesel et al 1989) allows plant mitochondria to use the universal genetic code. It should be pointed out that the import into mitochondria of nucleus-encoded tRNAs occurs not only in plants, but also in protozoa (e.g. Suyama 1986, Hancock and Hajduk 1990).

The elucidation of the mechanisms involved in tRNA import is of great importance not only for the understanding of this process, but also because it might provide ways of targeting other nucleic acids into mitochondria, circumventing the currently unsolved problem of plant mitochondria transformation.

Acknowledgments : The authors wish to thank Prof. P. Guillemaut (IUT Louis Pasteur, Strasbourg, France) and Dr. I. Small (INRA Center, Versailles, France) for their contribution to this work.

REFERENCES

Bartnik E, and Borsuk P (1986) A glycine tRNA gene from lupine mitochondria. Nucleic Acids Res. **14:** 2407

Binder S, Schuster W, Grienenberger JM, Weil JH and Brennicke A (1990) Genes for tRNA^{Gly}, tRNA^{His}, tRNA^{Lys}, tRNA^{Phe}, tRNA^{Ser} and tRNA^{Tyr} in *Oenothera* mitochondrial DNA. Curr. Genet . **17**:353-358

Bird S, Duncker B, Garber P and Bonen L (1989) Nucleotide sequence of the bean mitochondrial DNA region containing the $tRNA^{Asn}$ and $tRNA^{Tyr}$ genes. Nucleic Acids Res. **17**:4379

Boer PH and Gray MW (1988) Transfer RNA genes and the genetic code in *Chlamydomonas reinhardtii* mitochondria. Curr. Genet. **14**:583-590

Borsuk P, Sirko A and Bartnik E (1986) The methionine tRNA gene from lupine mitochondria. Nucleic Acids Res. **14**:7508

Ceci LR, Ambrosini M, Siculella L and Gallerani R (1989) Location of a single $tRNA^{His}$ gene on the master chromosome of sunflower mitochondrial DNA. Plant Science **61**:219-225

Chen HC, Wintz H, Weil JH and Pillay DTN (1989) Three mitochondrial tRNA genes from *Arabidopsis thaliana*: evidence for the conversion of a $tRNA^{Phe}$ gene into a $tRNA^{Tyr}$ gene. Nucleic Acids Res. **17**:2613-2621

Covello PS and Gray MW (1989) RNA editing in plant mitochondria. Nature **341**: 662-666

Crick FHC (1966) Codon-anticodon pairing : the wobble hypothesis. J. Mol. Biol. **19**:548-555

Gottschalk M and Brennicke A (1985) Initiator methionine tRNA in *Oenothera* mitochondria. Curr. Genet. **9**:165-168

Grabau EA (1987) Cytochrome Oxydase subunit II gene is adjacent to an initiator methionine tRNA gene in soybean mitochondrial DNA. Curr. Genet. **11**:287-293

Gray MW and Spencer DF (1983) Wheat mitochondrial DNA encodes a eubacterial-like initiator methionine transfer RNA. FEBS Lett. **161**:323-327

Green AG, Maréchal L, Weil JH and Guillemaut P (1987) A *Phaseolus vulgaris* mitochondrial $tRNA^{Leu}$ is identical to its cytoplasmic counterpart : sequencing and *in vivo* transcription of the gene corresponding to the cytoplasmic $tRNA^{Leu}$. Plant Mol. Biol. **10**:13-19

Gualberto JM, Domon C, Weil JH and Grienenberger JM (1989a) Nucleotide sequence of the wheat mitochondrial $tRNA^{Glu}$(UUC) gene. Nucleic Acids Res. **17**:3586

Gualberto JM, Lamattina L, Bonnard G, Weil JH and Grienenberger JM (1989b) RNA editing in wheat mitochondria results in the conservation of protein sequences. Nature **341**:660-662

Hancock K and Hajduk SL (1990) The mitochondrial tRNAs of *Trypanosoma brucei* are nuclear-encoded. J. Biol. Chem. **265**:19208-19215

Hiesel R, Wissinger B, Schuster W and Brennicke A (1989) RNA editing in plant mitochondria. Science **246**:1632-1634

Iams KP, Heckman JE and Sinclair JH (1985) Sequence of histidyl tRNA present as a chloroplast insert in mtDNA of *Zea mays*. Plant Mol. Biol. **4**:225-232

Izuchi S and Sugita M (1989) Nucleotide sequence of a tomato mitochondrial $tRNA^{Cys}$(GCA) gene. Nucleic Acids Res. **17**:1248

Izuchi S, Terachi T, Sakamoto M, Mikami T and Sugita M (1990) Structure and expression of tomato mitochondrial genes coding for $tRNA^{Cys}$(GCA), $tRNA^{Asn}$(GUU) and $tRNA^{Tyr}$(GUA) : a native $tRNA^{Cys}$ gene is present in dicot plants but absent in monocot plants. Curr. Genet. **18**:239-243

Joyce PBM and Gray MW (1988) Nucleotide sequence of a wheat mitochondrial glutamine tRNA gene. Nucleic Acids Res. **16**:1210

Joyce PBM, Spencer DF, Bonen L and Gray MW (1988a) Genes for $tRNA^{Asp}$, $tRNA^{Pro}$, $tRNA^{Tyr}$ and two $tRNAs^{Ser}$ in wheat mitochondrial DNA. Plant Mol. Biol. **10**:251-262

Joyce PBM, Spencer DF and Gray MW (1988b) Multiple sequence rearrangements accompanying the duplication of a $tRNA^{Pro}$ gene in wheat mitochondrial DNA. Plant Mol. Biol. **11**:833-843

Joyce PBM and Gray MW (1989a) Nucleotide sequence of a second glutamine tRNA gene in wheat mitochondrial DNA. Nucleic Acids Res. **17**:4885

Joyce PBM and Gray MW (1989b) Chloroplast-like transfer RNA genes expressed in wheat mitochondria. Nucleic Acids Res. **17**:5461-5476

Joyce PBM and Gray MW (1989c) Aspartate and asparagine tRNA genes in wheat mitochondrial DNA: a cautionary note on the isolation of tRNA genes from plants. Nucleic Acids Res. **17**:7865-7878

Karpinska B and Augustyniak H (1988) An asparagine tRNA gene from lupine mitochondria. Nucleic Acids Res. **16**:6239

Lagerkvist U (1978) "Two out of three": an alternative method for codon reading. Proc. Natl. Acad. Sci. USA **75**:1759-1762

Leon P, Walbot V and Bedinger P (1989) Molecular analysis of the linear 2.3 kb plasmid of maize mitochondria: apparent capture of tRNA genes. Nucleic Acids Res. **17**:4089-4099

Maréchal L, Guillemaut P, Grienenberger JM, Jeannin G and Weil JH (1985a) Sequence and codon recognition of bean mitochondria and chloroplast $tRNAs^{Trp}$: evidence for a high degree of homology. Nucleic Acids Res. **13**:4411-4416

Maréchal L, Guillemaut P, Grienenberger JM, Jeannin G and Weil JH (1985b) Structure of bean mitochondrial $tRNA^{Phe}$ and localization of the $tRNA^{Phe}$ gene on the mitochondrial genome of maize and wheat. FEBS Lett. **184**:289-293

Maréchal L, Guillemaut P and Weil JH (1985c) Sequences of two bean mitochondrial $tRNAs^{Tyr}$ which differ in the level of post-transcriptional modification and have a prokaryotic-like large extra-loop. Plant Mol. Biol. **5**:347-351

Maréchal L, Guillemaut P, Grienenberger JM, Jeannin G and Weil JH (1986) Sequences of initiator and elongator methionine tRNAs in bean mitochondria. Plant Mol. Biol. **7**:245-253

Maréchal L, Runeberg-Roos P, Grienenberger JM, Colin, J, Weil JH, Lejeune B., Quetier F and Lonsdale DM (1987) Homology in the region containing a $tRNA^{Trp}$ gene and a (complete or partial) $tRNA^{Pro}$ gene in wheat mitochondrial and chloroplast genomes. Curr. Genet. **12**:91-98

Maréchal-Drouard L and Guillemaut P (1988) Nucleotide sequence of a bean mitochondrial leucine tRNA and of its cytoplasmic counterpart. Re-examination of bean mitochondrial and cytoplasmic $tRNA^{Leu}1$ sequences. Nucleic Acids Res. **16**:11812

Maréchal-Drouard L, Weil JH and Guillemaut P (1988) Import of several tRNAs from the cytoplasm into the mitochondria in bean *Phaseolus vulgaris*. Nucleic Acids Res. **16**:4777-4788

Maréchal-Drouard L, Guillemaut P, Cosset A, Arbogast M, Weber F, Weil JH and Dietrich A (1990a) Transfer RNAs of potato (*Solanum tuberosum*) mitochondria have different genetic origins. Nucleic Acids Res. **18**:3689-3696

Maréchal-Drouard L, Neuberger M, Guillemaut P, Douce R, Weil JH and Dietrich A (1990b) A nuclear-encoded potato (*Solanum tuberosum*) mitochondrial $tRNA^{Leu}$ and its cytosolic counterpart have identical nucleotide sequences. FEBS Lett. **262**: 170-172

Maréchal-Drouard L, Dietrich A and Weil JH (1992) Adaptation of transfer RNA population to codon usage in cellular organelles. In: Hatfield DL, Lee BJ and Pirtle RM (eds). Transfer RNA in Protein Synthesis. 436 pp., Boca Raton: CRC Press: 125-140

Martin RP, Schneller JM, Stahl AJC and Dirheimer G (1979) Import of nuclear deoxyribonucleic acid-coded lysine-accepting transfer ribonucleic acid (anticodon CUU) into yeast mitochondria. Biochemistry **18**:4600-4604

Michaelis G, Vahrenholz C and Pratje E (1990) Mitochondrial DNA of *Chlamydomonas reinhardtii:* The gene for apocytochrome b and the complete function and map of the 15.8 kb DNA. Mol. Gen. Genet. 223:211-216

Muramatsu T, Nishihawa K, Nemoto F, Kuchino Y, Nishinura S, Miyazawa T and Yokoyama S (1988) Codon and amino acid specificities of a transfer RNA are both converted by a single post-transcriptional modification. Nature **336**:179-181

Oda K, Yamato K, Ohta E, Nakamura Y, Takemura M, Nozato N, Akashi K, Kanegae T, Ogura Y, Kohchi T and Ohyama K (1992) Gene organization deduced from the complete sequence of liverwort *Marchantia Polymorpha* mitochondrial DNA: a primitive form for plant mitochondrial genome. J. Mol. Biol. **223**:1-7

Okimoto R and Wostenholme DR (1991) A set of tRNAs that lack either the TYC arm or the dihydrouridine arm : towards a minimal tRNA adaptor. EMBO J. **9**:3405-3411

Parks TD, Dougherty WG, Levings CS III and Timothy DH (1984) Identification of two methionine transfer RNA genes in the maize mitochondrial genome. Plant Physiol. **76**:1079-1082

Parks TD, Dougherty WG, Levings CS III and Timothy, DH (1985) Identification of an aspartate transfer RNA gene in maize mitochondrial DNA. Curr. Genet. **9**:517-519

Runeberg-Roos P, Grienenberger JM, Guillemaut P. Maréchal L, Gruber V, Weil JH (1987) Localization, sequence and expression of the gene coding for $tRNA^{Pro}$(UGG) in plant mitochondria. Plant Mol. Biol. **9**:237-246

Samuelson T, Elias P, Lustig F, Axberg T, Fölsch G, Akesson B and Lagerkvist U (1983) Aberrations of the classic codon reading scheme during protein synthesis *in vitro*. J. Biol. Chem. **255**:4583-4588

Sangaré A, Lonsdale DM, Weil JH and Grienenberger JM (1989a) Sequence analysis of the $tRNA^{Tyr}$ and $tRNA^{Lys}$ genes and evidence for the transcription of a chloroplast-like tRNAMet in maize mitochondria. Curr. Genet. **16**:195-201

Sangaré A, Weil JH and Grienenberger JM (1989b) Nucleotide sequence of a maize mitochondrial $tRNA^{Glu}$(UUC) gene. Nucleic Acids Res. **17**:5837

Sangaré A, Weil JH and Grienenberger JM (1989c) Nucleotide sequence of a maize mitochondrial $tRNA^{Ser}$(UGA) gene. Nucleic Acids Res. **17**:7979

Sangaré A, Weil JH and Grienenberger JM (1990a) Maize and wheat mitochondrial $tRNA^{Pro}$ coding regions have similar sequences but different organizations. Biochim. Biophys. Acta **1049**:96-98

Sangaré A, Weil JH, Grienenberger JM, Fauron C and Lonsdale D (1990b) Localization and organization of tRNA genes on the mitochondrial genomes of fertile and male sterile lines of maize. Mol. Gen. Genet. **223**:224-232

Schuster W and Brennicke A (1987) Plastid, nuclear and reverse transcriptase sequences in the mitochondrial genome of *Oenothera* : is genetic information transferred between organelles via RNA ? EMBO J. **6**:2857-2863

Schuster W, Grienenberger JM, Weil JH and Brennicke A (1988) Comparison of $tRNA^{Trp}$ and $tRNA^{Pro}$ gene sequences in the chloroplast and mitochondrial genomes of *Oenothera*. Nucleic Acids Res. **16**:7737

Small I, Maréchal-Drouard L, Masson J, Pelletier G, Cosset A, Weil JH and Dietrich A (1992) *In vivo* import of a normal or mutagenized heterologous transfer RNA into the mitochondria of transgenic plants : towards novel ways of influencing mitochondrial gene expression ? EMBO J. **11**:1291-1296

Stern DB and Lonsdale DM (1982) Mitochondrial and chloroplast genomes of maize have a 12-kilobase DNA sequence in common. Nature **299**:698-702

Suyama Y (1986) Two dimensional polyacrylamide gel electrophoresis analysis of *Tetrahymena* mitochondrial tRNA. Curr. Genet. **10**:411-420

Tarassov IA and Entelis NS (1992) Mitochondrially-imported cytoplasmic $tRNA^{Lys}$(CUU) of *Saccharomyces cerevisiae*: in vivo and in vitro targetting systems. Nucleic Acids Res. **20**:1277-1281

Weber F, Dietrich A, Weil JH and Maréchal-Drouard L (1990) A potato mitochondrial isoleucine tRNA is coded for by a mitochondrial gene possessing a methionine anticodon. Nucleic Acids Res. **18**:5027-5030

Weber-Lotfi F, Maréchal-Drouard L, Folkerts O, Hanson M and Grienenberger JM (1992) Localization of tRNA genes on the *Petunia hybrida* 3704 mitochondrial genome. Plant Mol. Biol., in press.

Wintz H, Chen HC and Pillay DTN (1987) Nucleotide sequence of a soybean mitochondria tRNAGlu(TTC) gene. Nucleic Acids Res. **15**:10588

Wintz H, Chen HC and Pillay DTN (1988a) Presence of a chloroplast-like elongator tRNA^{Met} gene in the mitochondrial genomes of soybean and *Arabidopsis thaliana*. Curr. Genet. **13**:255-260

Wintz H, Grienenberger JM, Weil J H and Lonsdale DM (1988b) Localization and nucleotide sequence of two tRNA genes and a tRNA pseudogene in the maize mitochondrial genome: evidence for the transcription of a chloroplast gene in mitochondria. Curr. Genet. **13**:247-254

Expression of the plant mitochondrial genome

Stefan Binder, Axel Brennicke, Lutz Grohmann, Rudolf Hiesel, Volker Knoop, Wolfgang Schuster
Institut für Genbiologische Forschung Berlin GmbH, Ihnestrasse 63, 14195 Berlin, FRG

1. Introduction

The highly variable structure of the plant mitochondrial genome imposes several conditions upon expression and correct assembly of reading frames (Gray et al. 1992; Bonen et al. 1991). Plant mitochondrial genomes contain between 10 and 100 times the coding capacity of animal mitochondrial DNA (Ward et al. 1981), but the extra amount of DNA is only in part attributable to genuine extra genes (Lonsdale 1989). Most of the genome expansions are due to non-coding sequences, such as introns, transcribed and non-transcribed spacers between reading frames and sequence duplications (Gray 1992; Wissinger et al. 1991). These duplications occur without apparent attention to coding information, and recombination of gene fragments in numerous instances results in different pseudo genes (Wissinger et al. 1991). Another factor of sequence expansion is provided by inserted chloroplast and nuclear sequences, which in most instances are non-functional (Schuster and Brennicke 1988). Some tRNA genes of plastid origin, however, appear to be transcribed in the mitochondrial compartment (Joyce and Gray 1989).

Apparently spontaneous or by many factors inducible recombinations have been found in many higher plant species to readily restructure the entire mitochondrial genome. For example, passage through tissue culture can lead to individual regenerants with largely altered genome structures. From these recombined genomes presumably only functional arrangements will be maintained and will survive, suggesting that the observed viable structures represent just the tip of the iceberg of *in vivo* recombination events (Lonsdale 1989).

NATO ASI Series, Vol. H 81
Plant Molecular Biology
Edited by G. Coruzzi and P. Puigdomènech

The large number of different "surviving" genome arrangements suggests a certain flexibility in the regulatory signals and their arrangements in transcription units. In the following we will discuss specific features of gene expression in plant mitochondria with respect to transcription initiation, RNA editing, splicing and the resulting implications for the transfer process of mitochondrial genetic information to the nuclear genome during the course of the endosymbiotic evolution.

2. Transcription of the mitochondrial genome

There must be numerous transcription initiation sites present on the genome, from which transcripts are initiated to ensure correct expression in spite of the many recombination events. Several of these intraorganellar sequence recombinations have been observed to occur very close to essential reading frames, separating potential promoter regions from the genes to be expressed. These observations imply that transcription initiation sites are of a general nature in plant mitochondrial genomes, i.e. they are interchangeable between different genes. Recombination events have for example duplicated a potential promoter region in the mitochondrial genome of *Oenothera* resulting in identical transcription initiation sites for two genes, each encoding a different subunit of complex IV of the respiratory chain (Hiesel et al. 1987).

Consensus sequences surrounding transcription initiation sites have been identified in the monocots wheat and maize (Gray et al. 1992; Lonsdale 1989) and have been confirmed in these two plants by faithful transcription initiation *in vitro* by mitochondrial lysates (Hanic-Joyce and Gray 1991; Rapp and Stern 1992). Analysis of transcription initiation sites in the dicot plants *Oenothera* and soybean (Brown et al. 1991) further substantiated the flexible structure of these loose consensus sequences that most likely act as transcriptional promoters (Binder and Brennicke 1993). In dicot plants a promoter consensus sequence was derived that at first sight differs from the promoter consensus conserved in monocots. Only a core of four nucleotides appears to be conserved between the two angiosperm lineages, suggesting that the promoter regions recognized by the mitochondrial RNA polymerase and/or its specific co-factors have evolved slightly differently in both lineages of higher plants. In *Oenothera* potential promoter sequences have been found conserved between protein coding genes and a ribosomal RNA coding gene further supporting the interchangeability of transcription initiation sites also between different classes of genes (Binder and Brennicke 1993). Transcription initiation in plant mitochondria shows similarity to animal and fungal mitochondrial - and interestingly also to the T7 and T3 bacteriophage-transcription systems with respect to the relative positions of the conserved sequence and the first transcribed nucleotide (Gray et al. 1992; Hanic-Joyce and Gray 1991; Rapp and Stern 1992). Unlike prokaryotic and eukaryotic transcription systems, where the conserved consensus sequences are located several nucleotides upstream of the first transcribed nucleotide, the mitochondrial conserved sequences include the first transcribed nucleotide and even extend several nucleotides into the transcribed sequence. This positioning results in very similar 5'-termini of the different mitochondrial transcripts, which in the dicot plant *Oenothera* usually start with the dinucleotide 5'-GA-3' (Binder and Brennicke 1993). Further *in vitro* analyses are required to determine in detail the connection between promoter activities and nucleotide identities in the promoter sequences.

3. RNA editing

The currently most enigmatic posttranscriptional alteration of the transcribed mRNA molecules in plant mitochondria is observed as C to U and U to C nucleotide transitions by RNA editing (Bonen 1991; Covello and Gray 1989; Gray et al. 1992; Gualberto et al. 1989; Hiesel et al. 1989). These nucleotide identity switches in most instances result in distinct alterations of the information content of the mRNA molecules. The protein specified by the edited mRNA differs in many instances considerably from the protein predicted from the genomic information content in the mitochondrial DNA. RNA editing in plant mitochondria affects almost all of the so far investigated protein coding transcripts (Wissinger et al. 1992). The predicted protein sequences are altered up to 15 % by RNA editing, suggesting that the properties of the edited mRNA specified protein products differ considerably from those predicted by the genomic sequence. In some instances functionally essential transitions of otherwise highly conserved amino acid positions have been described for RNA editing events in plant mitochondria. These observations suggest that without RNA editing no functional protein can be synthesized from the genomic sequence (Covello and Gray 1990; Hernould et al. 1993). The C to U (and U to C) alterations in plant mitochondria have been found to introduce translational start (Chapdelaine and Bonen 1992), and stop codons, but also the elimination of stop codons such as UAA to CAA alterations has been observed (Gray et al. 1992; Hiesel et al. 1993; Wissinger et al. 1992).

Current research is focussed on the identification of the biochemical processes involved in this C to U and U to C RNA editing process and on the elucidation of the underlying specificity determinants that decide which nucleotide is to be altered by RNA editing. The biochemical process could be a straightforward deamination or amination of the C or U nucleotide in the mRNA (Gray et al. 1992; Wissinger et al. 1992). Editing could alternatively invoke a transglycosylation event without interrupting the phosphate backbone of the nucleic acid, whereas the also not excluded complete nucleotide exchange would interrupt the nucleic acid molecule during the editing process. This last process is rather unlikely to occur in plant mitochondria, since intermediates should have been detected by the various analyses of primer extension and cDNA cloning. However, no preference for such transcript termini at RNA editing sites has ever been observed. Specificity determinants for RNA editing sites could reside in trans-acting factors, such as guide RNA molecules analogous to those known from trypanosome mitochondria or alternatively could be encoded in cis in the mRNA molecule itself by elements in the secondary or tertiary structure. Such intramolecular determinants are, however, difficult to imagine in transcripts, where 50 or more RNA editing sites are concentrated in reading frames of 400 nucleotides, which does not leave much variability in the localization of the specificity determinants in the mRNA molecule itself. The development of in vitro editing systems

(Araya et al. 1992) is expected to further our understanding of the RNA editing process in plant mitochondria.

The origin of RNA editing in the evolutionary history of plants is as yet unclear. A survey of all major groups of land plants has shown that RNA editing is present in mitochondria of all land plants, including ferns and the *Lycopodiatae* (Hiesel et al. 1993). In bryophyte mitochondria, however, so far no evidence for RNA editing has been reported. It is currently still open, whether RNA editing evolved after the separation of the bryophytes from the other land plants or whether RNA editing is a more ancient trait that has been lost only in bryophytes, but has been functionally maintained in the other groups of land plants. In the extant green algae discussed as related to the ancestors of land plants, the *Charophyceae*, no editing has been observed, but cannot be generally excluded as yet due to the small number of species and genes sampled to date.

4. Cis- and trans-splicing in plant mitochondria

To date about 20 introns have been identified in higher plant mitochondrial genes, all of them belonging to the organellar group II family (Gray et al. 1992). The presence of some of these introns has been reported to be species-specific, as exemplified by the differences in the structure of the coxII gene, which contains either none, one or two introns (Lippok et al. 1992), or the nad4 gene, which is interrupted by either 2, 3 or 4 introns (Gass et al. 1992). The structures of other genes with up to 4 introns, however, are conserved during the evolution of higher plant.

Three genes coding for subunits one, two and five of the NADH-dehydrogenase of the respiratory chain are each encoded by 5 exons, which are not located in cis in a linear arrangement, but are distributed around the genome in distant positions (Bonen 1991; Binder et al. 1992; Chapdelaine and Bonen 1991; Knoop et al. 1991; Wissinger et al. 1991). Presumably by several of the recombination events very common in plant mitochondrial genomes have the exons of the nad1, nad2 and nad5 genes been scattered to three different locations. The 5 exons of the nad2 gene, for example, have been disrupted once between exons B and C and are now located in two different regions of the genome (Binder et al. 1992). The connection between the different RNA molecules that have to be linked at their exon borders to form one continuous open reading frame is presumably mediated by the group II intron parts flanking the trans-located exon borders. These partial introns can be folded into the complete secondary structure models conserved for group II introns with the interruption located in domain IV. Similar trans-splicing group II introns have also been indentified in chloroplasts of higher plants for the rps12 and in *Chlamydomonas* for the psaA genes (see Bonen 1991 for references).

In plant mitochondria these trans-splicing genes have presumably been created from once cis-arranged genes that were interrupted in the group II intron structure at sites that have no deleterious effect on splicing efficiency. The probabilities of two molecule reactions required for such trans-splicing events to occur *in vivo* in comparison to the efficiencies of intramolecular cis-splicing reactions are as yet unclear. The intermolecular trans-splicing, however, seems to be very efficient with respect to intron specificity, since so far no mis-spliced cDNAs have been reported. At least five of the six trans-splicing events identified todate appear to be well conserved during plant evolution (Nugent and Palmer 1993), suggesting that these interruptions were once-upon-a-time events that occurred early in higher plant evolution. These disrupted gene structures may have been retained due to their advantage over a long inflexible linear cis-structure in the fluid surrounding provided by the highly variable mitochondrial genome.

Fig. 1: Features of gene expression in plant mitochondria

A) Some of the sequence regions repeated in the mitochondrial genomes of higher plants are active in homologous recombination (arrowheads) resulting in the formation of several smaller sub-genomic molecules when present as direct repeats. Some of the genes in plant mitochondria have been split by recombination events and portions of these genes are now located in distant regions of the genome (brackets).

B) A continous reading frame of exons A, B, C, D has to be assembled from mRNA molecules independently transcribed from the plant mitochondrial promoter indicated at the beginning of the arrows for the two schematically shown genomic regions. Mature mRNAs are assembled from the different transcripts by trans-splicing of interrupted group II introns (represented by hatched regions). Regions indicated by the dotted circles are high-lighted in part C to show specific sequence alterations introduced by post-transcriptional modification.

C) RNA editing in plant mitochondria alters the information content of the mRNA post-transcriptionally. The primary RNA transcript synthesized from the genomic sequence is altered by numerous C to U conversions. Within open reading frames the encoded amino acid sequence is altered whenever non silent codon positions are changed. This examplary region was assembled to show events ocurring in the first and second codon positions as single or double editing events leading to alteration of the encoded amino acid identity and to show a C to U editing event introducing a translational stop codon in the open reading frame. This hypothetical sequence example shows how serine codons can by single or double editing events be altered to code for three different amino acids other than serine after RNA editing. The left hand part presents several observed RNA editing events in different group II intron sequences of plant mitochondria. Some of these events, as for example elimination of the AC mismatches in domains 1 and 6 of plant mitochondrial introns, are postulated to be essential for correct folding and therefore also for assembly of a splicing-competent intron structure. In such instances RNA editing may be essential for splicing of this intron and may therefore be involved in regulating translation in plant mitochondria.

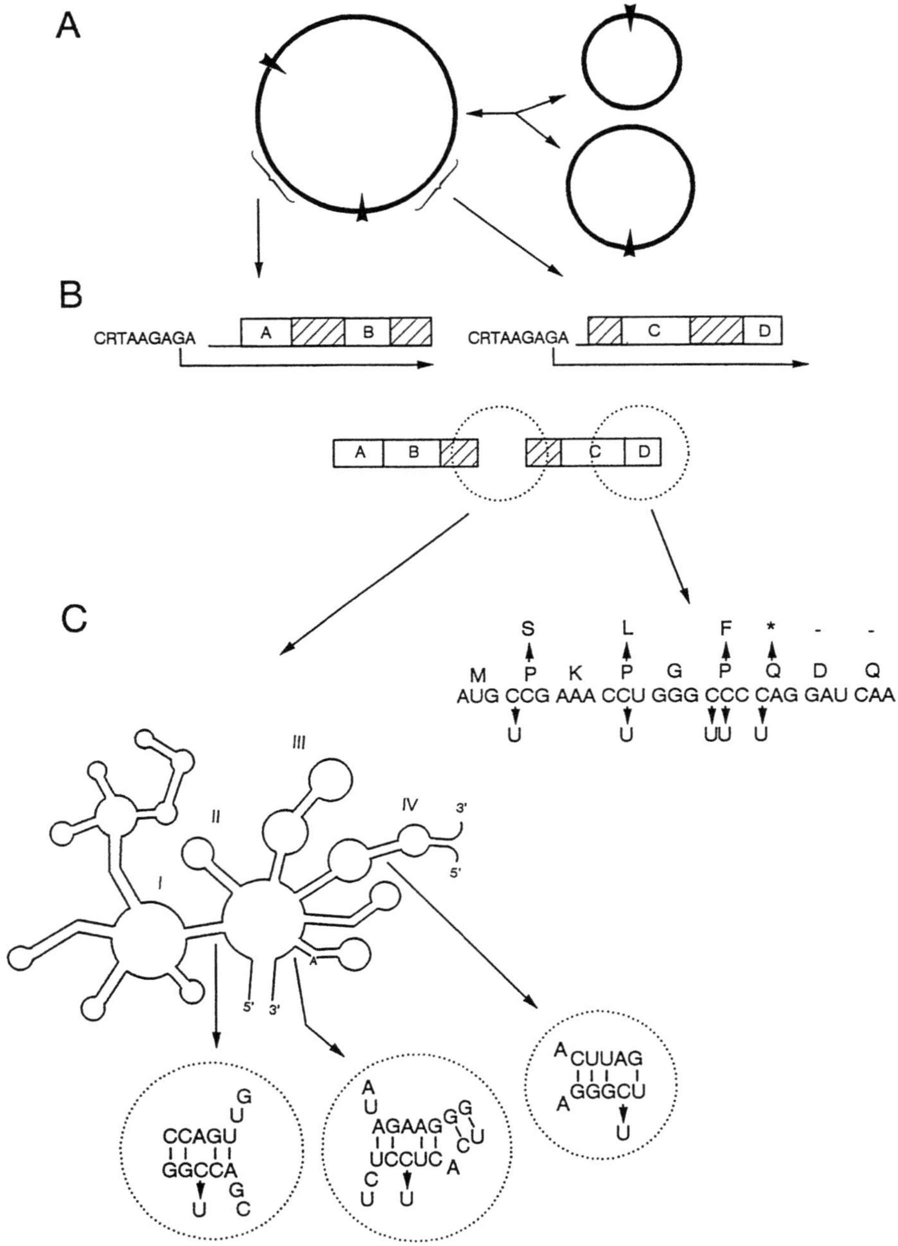
A
B
CRTAAGAGA
A
B
CRTAAGAGA
C
D
A
B
C
D
C
S L F * - -
M P K P G P Q D Q
AUG CCG AAA CCU GGG CCC CAG GAU CAA
U U UU U
I
II
III
IV
3'
5'
5'
3'
A
CCAGU
GGCCA
G
U
G
C
U
A
U
AGAAG
UCCUC
G
G
U
C
A
C
U
U
A
CUUAG
GGGCU
A
U

5. Editing and splicing

The comparison of the exon RNA editing status of spliced versus unspliced transcripts has provided the to date best evidence for the post-transcriptional nature of the RNA editing process (Sutton et al. 1991; Yang et al. 1991). The observation that spliced exons are usually edited at more potential editing positions than unspliced precursors suggests that editing and splicing are independent processes of transcript maturation.

An interdependence between editing and splicing is, however, suggested by the observation of several editing sites in intron sequences. Here some editing events are required to form AU pairs from AC mismatches in base pairing stems in order to allow energetically favoured helix pairings (Binder et al. 1992; Wissinger et al. 1991). Particularly editing events in short stems which have been shown in analyses in yeast introns to be essential for *in vivo* and *in vitro* splicing provide circumstancial evidence for a requirement of editing before splicing competent introns are generated. RNA editing may thus be involved in controlling the excision of introns in some instances.

Such regulatory functions of RNA editing in providing translatable messages including the more obvious requirement for editing in open reading frames may provide a rationale for the existence of RNA editing in plant mitochondria which may thus add an additional level of regulation for gene expression.

6. Translocated mitochondrial genes are expressed in the nucleus

In higher plant mitochondria not only the genome structures vary, but several examples of divergent gene contents between different plant species have also been reported (Brennicke et al. 1993). While in the legume family the coxII gene has in some species been transferred to the nuclear genome (Nugent and Palmer 1991; Covello and Gray 1992) in *Oenothera* the rps12 gene coding for ribosomal protein S12 has been transferred from the mitochondrial genome to the nuclear compartment (Grohmann et al. 1992). In both these instances the functional mitochondrial gene has been partially or completely lost and the protein product is now synthesized from the activated nuclear gene. Several transition stages of this transfer are still manifest in the legume family, where for example in soybean an apparently intact yet not transcribed *coxII* gene has been identified in the mitochondrial genome besides the active nuclear copy (Covello and Gray 1992). In cowpea, on the other hand, the mitochondrial copy has been lost completely (Nugent and Palmer 1991). In *Oenothera* the mitochondrial genome has retained a partial copy of the rps12 gene, although the relevant coding function has completely been transferred to the nuclear genome (Grohmann et al. 1992).

Such functional gene transfers from the mitochondrial compartment to the nuclear genome have to overcome several problems. Presumably any mitochondrial group II intron has to be removed during the transfer to the nuclear genome. Additionally the need for RNA editing in the genomic sequence integrated in the nucleus has to be eliminated, since at present no evidence for RNA editing in nuclear encoded transcripts has been observed in higher plants. Analysis of the nuclear genomic copies indeed shows that the nucleotide sequences conform in all instances to the edited sequences of mitochondrial mRNAs (Nugent and Palmer 1991; Covello and Gray 1992; Grohmann et al. 1992). This observation suggests that a fully edited and spliced, mature mitochondrial mRNA may have been used as the starting molecule from which transfer of the nucleic acid information to the nuclear genome has been initiated via a reverse transcription step (Schuster and Brennicke 1987). The nuclear copies of the transferred mitochondrial genes must have gained information for expression such as promoters and translational initiation signals as well as presequences necessary to target the protein product back to the mitochondrial compartment (Brennicke et al. 1993).

7. Conclusion

The expression of plant mitochondrial genes shows several unique features not observed in other organelles. Frequent C to U and occasional U to C RNA editing occurs in exon as well as intron sequences and may be used for regulating gene expression on the RNA level in higher plant mitochondria. RNA editing in introns may be required for correct assembly of cis- as well as trans-splicing introns. RNA editing appears to be an old process that most likely has been established during the evolution of land plants. The frequent recombinations in plant mitochondria require flexible regulatory signal arrangements, such as general promoter structures, for example, that are interchangeable between different genes. Comparisons of the gene content in plant mitochondria have revealed differences in the intracellular locations of several genes. The observed different stages of the endosymbiontic sequence transfer allow more definite conclusions about the individual requirements and probabilities ruling these gene transfer processes.

References

Araya A, Domec C, Begu D and Litvak S (1992) An in vitro system for the editing of ATP synthase subunit 9 mRNA using wheat mitochondrial extracts. Proc. Natl. Acad. Sci. USA 89:1040-1044

Binder S, Marchfelder A, Brennicke A and Wissinger B (1992) RNA editing in trans-splicing intron sequences of nad2 mRNA in *Oenothera* mitochondria. J. Biol. Chem. 267:7615-7623

Binder S and Brennicke A (1993) Transcription initiation sites in mitochondria of *Oenothera berteriana.* J. Biol. Chem. 268:7849-7855

Bonen L (1991) The mitochondrial genome: so simple yet so complex. Curr. Opinion Genet. Dev. 1:515-522

Brennicke A, Grohmann L, Hiesel R, Knoop V and Schuster W (1993) The mitochondrial genome on its way to the nucleus: different stages of gene transfer in higher plants. FEBS Lett. 325:140-145

Brown GG, Auchincloss AH, Covello PS, Gray MW, Menassa R and Singh M (1991) Characterization of transcription initiation sites on the soybean mitochondrial genome allows identification of a transcription-associated sequence motif. Mol. Gen. Genet. 228:345-355

Chapdelaine Y and Bonen L (1991) The wheat mitochondrial gene for subunit 1 of the NADH dehdrogenase complex: a trans-splicing model for this gene-in-pieces. Cell 65:465-472

Covello PS and Gray MW (1989) RNA editing in plant mitochondria. Nature 341:662-666

Covello PS and Gray MW (1992) Silent mitochondrial and active nuclear genes for subunit 2 of cytochrome c oxidase (cox2) in soybean: evidence for RNA-mediated gene transfer. EMBO J. 11:3815-3820

Covello PS and Gray MW (1990) RNA sequence and the nature of the Cu_A-binding site in cytochrome c oxidase. FEBS Lett. 2685-7

Gass DA, Makaroff CA and Palmer JD (1992) Variable intron content of the NADH dehdrogenase subunit 4 gene of plant mitochondria. Curr Genet 21: 423-430

Gray MW (1992) The endosymbiont hypothesis revisited. Int. Rev. Cytol. 141: 233-357

Gray MW, Hanic-Joyce PJ and Covello PS (1992) Transcription, processing and editing in plant mitochondria. Annu. Rev. Plant Physiol. Plant Mol. Biol. 43:145-175

Grohmann L, Brennicke A and Schuster W (1992) The mitochondrial gene encoding ribosomal protein S12 has been translocated to the nuclear genome in *Oenothera.* Nucleic Acids Res. 20:5641-5646

Gualberto JM, Lamattina L, Bonnard G, Weil JH and Grienenberger J-M (1989) RNA editing in wheat mitochondria results in the conservation of protein sequences. Nature 341:660-62

Hanic-Joyce PJ and Gray MW (1991) Accurate transcription of a plant mitochondrial gene in vitro. Mol. Cell. Biol. 11: 2035-2039

Hernould M, Suharsono S, Litvak S, Araya A and Mouras A (1993) Male-sterility induction in transgenic tobacco plants with an unedited atp9 mitochondrial gene from wheat. Proc. Natl. Acad. Sci. USA 90:2370-2374

Hiesel R, Combettes B and Brennicke A (1993) Evidence for RNA editing in mitochondria of all major groups of land plants except the *Bryophyta.* Proc. Natl. Acad. Sci. USA in press

Hiesel R, Schobel W, Schuster W and Brennicke A (1987) The cytochrome oxidase subunit I and subunit III genes in *Oenothera* mitochondria are transcribed from identical promoter sequences. EMBO J. 6:29-34

Hiesel R, Wissinger B, Schuster W and Brennicke A (1989) RNA editing in plant mitochondria. Science 246:1632-1634

Joyce PBM and Gray MW (1989) Chloroplast-like transfer RNA genes expressed in wheat mitochondria. Nucleic Acids Res. 17:5461-76

Knoop V, Schuster W, Wissinger B and Brennicke A (1991) Trans splicing integrates an exon of 22 nucleotides into the nad5 mRNA in higher plant mitochondria. EMBO J. 10:3483-3493

Lippok B, Brennicke A and Wissinger B (1992) The coxII gene in carrot mitochondria contains two introns. Mol. Gen. Genet. 232:322-327

Lonsdale DM (1989) The plant mitochondrial genome. In: Stump PK and Conn EE (eds) The Biochemistry of Plants: A Comprehensive Treatise. Academic Press, New York 15: 229-295

Nugent JM and Palmer JD (1991) RNA-mediated transfer of the gene coxII from the mitochondrion to the nucleus during flowering plant evolution. Cell 66:473-481

Nugent JM and Palmer JD (1993) Evolution of gene content and gene organization in flowering plant mitochondrial DNA: a general survey and further studies on coxII gene transfer to the nucleus. In: Brennicke A and Kück U (eds) Plant mitochondria. VCH Weinheim: 163-170

Rapp WD and Stern DB (1992) A conserved 11 nucleotide sequence contains an essential promoter element of the maize mitochondrial atp 1 gene. EMBO J. 11:1065-1073

Schuster W and Brennicke A (1988) Interorganellar sequence transfer: plant mitochondrial DNA is nuclear, is plastid, is mitochondrial. Plant Science 54:1-10

Schuster W and Brennicke A (1987) Plastid, nuclear and reverse transcriptase sequences in the mitochondrial genome of *Oenothera:* Is genetic information transferred between organells via RNA? EMBO J. 6:2857-2863

Sutton CA, Conklin PL, Pruitt KD and Hanson MR (1991) Editing of pre-mRNAs can occur before cis- and trans-splicing in *Petunia* mitochondria. Mol. Cell. Biol. 11:4274-77

Ward BL, Anderson RS and Bendich AJ (1981) The mitochondrial genome is large and variable in a family of plants (*Cucurbitaceae*). Cell 25:793-803

Wissinger B, Brennicke A and Schuster W (1992) Regenerating good sense: RNA editing and trans splicing in plant mitochondria. Trends Genet. 8:297-328

Wissinger B, Hiesel R, Schobel W, Unseld M, Brennicke A and Schuster W (1991) Duplicated sequence elements and their function in plant mitochondria. Z.Naturforsch. 46c:709-716

Wissinger B, Schuster W and Brennicke A (1991) Trans splicing in *Oenothera* mitochondria: nad1 mRNAs are edited in exon and trans splicing group II intron sequences. Cell 65:473-482

Yang AJ and Mulligan RM (1991) RNA editing intermediates of cox2 transcripts in maize mitochondria. Mol. Cell. Biol. 11:4278-4281

REGULATION OF PLASTID GENE EXPRESSION

WILHELM GRUISSEM
Department of Plant Biology, University of California
Berkeley, California 94720, USA

INTRODUCTION

Complete and partial sequencing of plastid DNAs from several plants has revealed the existence of 120 to 135 genes encoded in the plastid genome (1, 2, 3, 4, 5). The genes encode proteins involved in photosynthesis, proteins required for transcription and translation, potential proteins with homology to the mitochondrial NADH oxidoreductase, and a number of open reading frames (ORFs) for proteins of unknown functions which share homology with proteins in *E.coli*. Most of the plastid genes are organized into polycistronic transcription units in higher plants, which reflects the prokaryotic origin of chloroplasts. Some of the transcription units encode only proteins for photosynthetic electron transport complexes, or genes involved in transcription and translation. The organization of genes for proteins with related functions into transcription units may allow these groups of genes to be co-regulated and could facilitate the coordinated production of subunits for a single protein complex. Other transcription units contain genes for photosynthetic proteins together with genes for tRNAs or ribosomal proteins.

The temporal, spatial and environmental control of plastid functions therefore requires a detailed understanding of the mechanisms by which the expression of the mono- and poly-cistronic genes is regulated. Insights into the molecular control of plastid genes will also provide a better understanding of the role of nuclear gene products in the regulation of plastid gene expression. This review provides a brief outline of the mechanisms by which plastid gene expression can be regulated at transcriptional, post-transcriptional, translational and post-translational levels.

TRANSCRIPTIONAL REGULATION OF PLASTID GENES

The molecular structure, subunit composition and regulation of plastid RNA polymerase is currently not well understood. There is still considerable controversy as to the origin and number of distinct or related RNA polymerase activities that contribute to the transcription of plastid genes. The biochemical characterization of the plastid RNA polymerase (RNAP) and related factors has been carried out in a number of vascular plants and unicellular algae, including maize, spinach, pea, mustard, *Chlamydomonas* and *Euglena*. A consensus emerges from these studies of an enzyme composed of multiple subunits with different molecular masses (Figure 1). It is not known at present if all proteins detected in the highly purified enzyme preparations are part of the multisubunit enzyme complex or co-purify with the RNAP activites until their precise functions have been assigned. Also, none of the highly purified soluble RNA polymerase activities have been demonstrated convincingly to retain specificity for initiation at selected plastid promoter sites. In contrast, less purified enzyme preparations can correctly initiate transcription at chloroplast promoters with good efficiency (6, 7). These partially fractionated chloroplast RNA polymerase activities have been utilized to define the DNA sequences required for transcription initiation for rRNA, mRNA and tRNA genes (8, 9, 10).

NATO ASI Series, Vol. H 81
Plant Molecular Biology
Edited by G. Coruzzi and P. Puigdomènech

Heterologous DNA probes from *E. coli* genes for RNAP subunits and computer-aided searches of plastid DNA sequences have been applied successfully to identify plastid genes that share homology with the DNA sequences coding for the a, b, and b′ subunits of *E. coli* RNAP. A relationship between subunits of the purified or partially fractionated plastid RNAP activites and the *rpo* genes has been established by immunological cross-reaction and direct N-terminal amino acid sequencing of specific proteins (11, 12, 13, 14). Together, these results can be regarded as strong evidence that the plastid *rpo* genes contribute to a functional RNA polymerase in plastids.

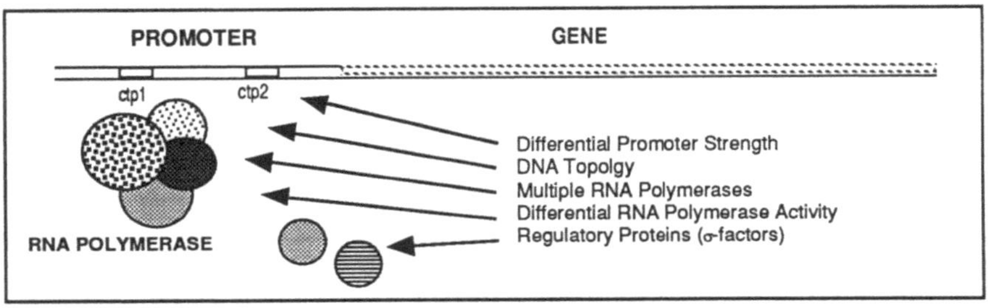

Figure 1: Transcription regulatory mechanisms

Recent analysis of the plastid genome in the parasitic plant *Epifagus virginiana* has shown, however, that all of the genes for photosynthetic proteins and the putative NADH oxidoreductase have been deleted (15). Of the genes for transcriptional and translational functions all of the *rps* and *rpl* genes as well as most *trn* genes are retained, but the *rpoB/C1/C2* transcription unit has been deleted from the plastid genome (16). Transcription of the remaining plastid genes, however, can still be detected. This result is consistent with a nuclear-encoded plastid RNAP activity. Previously published results of chloroplast RNA accumulation in mutant or heat-induced ribosome-deficient plastids (17, 18) could also be considered as further evidence for a nuclear-encoded plastid RNA polymerase. RNAP activity generally increases early during leaf development and light-induced chloroplast development and decreases later in development (19, 20). Because the increase in light-induced RNAP activity parallels or precedes the accumulation of 16S rRNA and mRNAs from *rpoB* and different ribosomal protein genes early in chloroplast development, it is possible that a nuclear-encoded RNAP is initally required for the synthesis of the plastid-encoded RNAP subunits to build up high levels of RNAP activity.

Comparative analysis of DNA sequences (21) located upstream of the 5' ends from primary RNA transcripts shows considerable similarity to the *E.coli* promoter consensus DNA sequence (22). Examination of wild type and mutant plastid promoter DNA sequences in vitro using a partially fractionated chloroplast RNAP activity (6, 7) has confirmed their function to direct transcription initiation of tRNA and protein genes (23, 24, 25, 26, 27, 28, 29). The promoter region of most plastid transcription units consists of two DNA sequence blocks (ctp1 = "-35" and ctp2 = "-10") in which certain bases are invariant, and when deleted or mutated alter transcription initiation efficiency (23, 24, 25, 26, 27, 28). However, not all plastid genes that are efficiently transcribed in vitro have a conserved *E.coli*-like ctp1-ctp2 promoter organization. Some tRNA genes such *trnR1* and *trnS1* lack the typical ctp1-ctp2 promoter organization, but can be efficiently transcribed by a soluble RNAP activity with less than 10 bp upstream of the tRNA coding region (30). Two other plastid promoter regions have recently been identified in the maize *petE* (31) and barley as well as tobacco *psbD-psbC* (32, 33, 34) transcription units that appear to be activated by white or blue light. DNA sequences upstream of the blue light-inducible barley mRNA do not contain the ctp1-ctp2 promoter organization typical for other plastid genes, but are highly conserved among several monocot and dicot plants (35).

The abundance of plastid RNAs can vary substantially for different genes and as a function of developmental processes and envrionmental signals (8, 36, 37, 38). Differences in RNA accumulation are controlled in part at the transcriptional level, but the mechanisms that govern the differential transcription of plastid genes are currently not well understood. At present, the data strongly suggest that promoter strength is a principal regulatory factor that determines the differences in transcription rates for a large set of transcription units in the plastid genome. However, control of transcription rates strictly based on promoter strength does not generally apply to all plastid transcription units, because modulation of transcription rates have also been reported for certain genes during chloroplast development in spinach (e.g., *trnK, rpl2*; (19, 39)), barley (*psbA*; (40); *psbC-psbD* (34)), and maize (*petE* (31)), and as a result of different light qualities (e.g., *psaA*; (41)) or carotenoid-deficiencies (42). These changes in promoter activity could be controlled by regulatory DNA-binding proteins or different s-factors which bind to the catalytically active core of RNA polymerase to determine the specificity of transcription initiation. Proteins that may belong to a group of chloroplast s-factors have recently been isolated from mustard chloroplasts. These proteins confer DNA-binding activity on *E.coli* core RNAP in a mobility shift assay (43, 44). Their function for the chloroplast RNAP has not been established, but these proteins could be regulatory factors that alter the specificity of RNA polymerase for different promoter regions. Regulatory RNAP subunits in combination with specific DNA binding proteins and the RNAP core (or a different RNAP enzyme) may also form alternate transcription initiation complexes at light-activated promoter regions (31, 32, 33) or promoter regions that differ from the typical ctp1-ctp2 organization (30, 45, 46).

POST-TRANSCRIPTIONAL CONTROL OF PLASTID GENE EXPRESSION

It has now been demonstrated that processing and stability of mRNAs in plastids and mitochondria are also important determinants that affect the expression of their genes (47). For example, the conversion of proplastids or etioplasts into photosynthetically active chloroplasts is generally accompanied by transitory or stable increases in pools of mRNAs for protein subunits of the thylakoid membrane complexes (8, 36, 37, 47, 48, 49). Also, environmental factors such as light quality or high light in combination with carotenoid-deficient plastids affect the pools of specific plastid RNAs. For example, photosystem I-sensitizing red light selectively decreases the level of the *psaA* mRNA, which encodes the photosystem I chlorophyll a-apoprotein (41, 50). This reduced *psaA* mRNA level in red light is established through a combination of a lower *psaA* transcription rate and increased turnover of the mRNA (41).

Following transcription, most polycistronic transcripts in chloroplast undergo extensive RNA processing to yield tRNAs, ribosomal RNAs and monocistronic mRNAs (reviewed in (8)). Thus, any of the specific processing events could have regulatory implications for the function of the RNAs. Processing of polycistronic transcripts occurs normally in all plastid types but the rate of certain processing events may be decreased in non-leaf plastids (51). Fully processed mono- or dicistronic RNAs generally represent the predominant pool of chloroplast RNAs or polysomal mRNAs, and it is likely that processing of polycistronic transcripts increases the translational competence of the resulting monocistronic mRNAs. Several mRNAs for photosynthetic proteins have multiple 5' ends which, at least for some mRNAs, result from specific cleavage of the primary transcript in chloroplasts (31, 34, 52, 53, 54) and in non-photosynthetic plastids (55) that can also be demonstrated in vitro (29, 56, 57). Although the significance of 5' end processing is unknown, it may provide a mechanism for translational control or differential stabilization of the mRNA.

Most chloroplast mRNAs contain a uniqe inverted repeat (IR) sequence in their 3' untranslated region that can fold into a stable stem/loop structure (58). These sequences are highly conserved for the same gene between evolutionary divergent plants, but differ between genes in the chloroplast genome (59). *In vitro* studies with partially fractionated chloroplast extracts have shown that chloroplast 3' IR sequences do not generally function as transcription terminators (57), although chloroplast RNA polymerase can efficiently terminate in vitro at certain tRNA genes that lack 3' IR sequences (57, 60), or at known bacterial terminator regions (61). By impeding 3' exonucleases, these chloroplast 3' stem/loop structures could act as stabilizing elements that increase the half-life of the mature mRNA. When 3' stem/loop structures are removed from their location in the 3' untranslated region of mRNAs either by deletion mutagenesis (62) or by endonucleolytic cleavage (63), the resulting RNAs are rapidly degraded *in vitro*. The most direct evidence for a stabilizing function of the 3' IR sequence was obtained from experiments in transformed *Chlamydomonas* chloroplasts, in which partial or complete deletion of the *atpB* 3' IR results in a 60% to 80% loss of the mRNA (64). It is therefore resonable to conclude that 3' IR sequences act as cis-acting elements critical for the stability of mRNAs in chloroplasts from higher plants and algae.

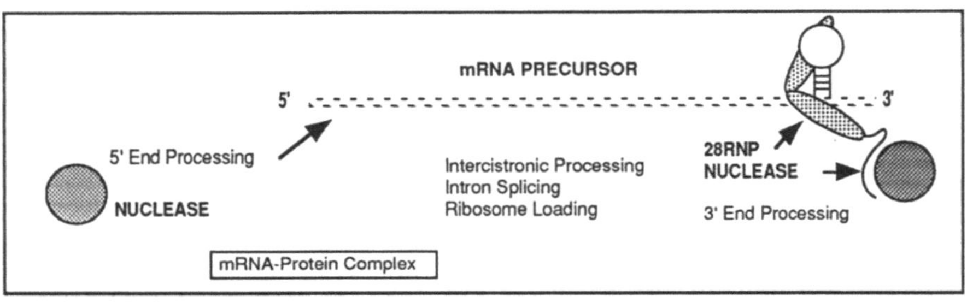

Figure 2: Regulatory mechanisms involved in plastid RNA processing

Studies with translation inhibitor-treated leaves and *Chlamydomonas* mutants have now provided infomation which shows that there is also an effect of translation on chloroplast mRNA turnover. The block of translation in leaf chloroplasts with lincomycin reduces the decay of *psbA* and *rbcL* mRNAs, and chloramphenicol accelerates the decay of the two mRNAs but with different kinetics (65). These results suggest that translation of *psbA* and *rbcL* mRNAs may be critical for initiating/facilitating their differential turnover. Consistent with this view are results from root amyloplasts and the *Chlamydomonas* chloroplast ribosome-deficient mutant *cr-1*, in which mRNAs for photosynthetic proteins are stable and accumulate, although they are specifically excluded from the polysomal RNA fraction (55, 66). Also,*Chlamydomonas* chloroplast mutants in which nonsense or frameshift mutations close to the 5' end of the coding region cause premature translation termination of *rbcL* mRNA (67), or different nuclear mutants that affect translation of *atpA* mRNA (68), *psbC* mRNA (69) and *psbD* mRNA (70), show either normal or increased levels of the transcripts. It is likely, therefore, that the turnover rate of chloroplast mRNAs is determined by a combination of factors, such as inherent stability of the RNA molecule, RNA-protein interactions, ribosome association, rate of translation elongation, and nucleolytic attack, which together influence the half-life of indvidual mRNas.

The biochemical approach designed to identify factors which function in the processing and stabilization of plastid mRNA and tRNA 3' ends has resulted in the identification (58, 71), purification and cloning of chloroplast-localized nuclear-encoded proteins which are critical for this process in vitro (72). Experiments using chloroplast processing extracts in combination with RNA mobility shift and UV crosslinking assays have established that specific RNA-protein complexes form with the different 3' ends of *petB, rbcL, psbA* (58) and *trnK* (73). FPLC and RNA affinity chromatography of spinach chloroplast proteins in combination with mRNA 3' end processing assays has resulted in the purification and cloning of two RNA-binding proteins of 28 kDa (28RNP; (72)) and 24 kDa (24RNP; S.Abrahamson and W.Gruissem, unpublished results) with conserved RNA-binding domains (47). The nuclear-encoded proteins are required for correct plastid mRNA 3' end processing *in vitro*, but the purified proteins do not have nuclease activity (72). These RNA-binding proteins may interact with nuclease(s) to direct the specific mRNA 3' end processing. In the absence of the 28RNP or 24RNP, the mRNA 3' end precursors are not processed and are degraded (72). Thus, precise processing of plastid mRNA 3' end precursors in which the nuclear-encoded RNA-binding proteins participate may be necessary to establish stable mRNA 3' ends. Similar RNA-binding proteins with molecular masses range from 28 kDa to 33 kDa and which are similar to the 28RNP and 24RNP from spinach have been isolated from tobacco (74, 75). All of the nuclear-encoded plastid proteins have two conserved RNA-binding domains which are found in RNA-binding proteins such as the U1A snRNP or *Drosphila sex-lethal* protein (76, 77, 78) The chloroplast RNA-binding proteins differ from the animal proteins in their N-terminal domain, however, which is very acidic and may function in protein-protein interaction.

TRANSLATIONAL CONTROL OF PROTEIN ACCUMULATION

Studies with different plants have shown that plastid mRNAs for photosynthetic proteins can accumulate to high levels in plants kept in darkness (8, 79), but the corresponding proteins are either not detectable, or accumulate to low levels. In many of these studies, however, it has been difficult to distinguish between arrest of translational elongation or rapid turnover of the protein because of the extended time of labeling assays in vitro or in isolated plastids.

Binding of co-factors is often required for the protein to accumulate. The importance of co-factors for the translation or stability of their cognate proteins was recently described for chlorophyll apoproteins in wild-type and mutant barley (80, 81, 82) and the *Chlamydomonas* y-1 mutant which fails to synthesize chlorophyll in the dark (83). The mRNA levels for the chlorophyll apoproteins are already high in dark-grown barley seedlings (81) and the *y-1* mutant (83). They are associated with thylakoid-bound polysomes in barley (84), but the proteins are not detectable either because their translation is arrested or they are rapidly degraded (81). Chloroplast-encoded chlorophyll apoprotein synthesis is rapidly induced in plastids from illuminated barley plants without changes in their mRNA levels or significant redistribution of the polysomes and is concomitant with the synthesis of chlorophyll (85). Accumulation of the photosystem II D1 and D2 chlorophyll apoproteins after illumination is delayed in *y-1* compared to *Chlamydomonas* wild type cells, which is consistent with their translation and/or stability being coupled to chlorophyll synthesis (83). The dependence of chlorophyll apoprotein accumulation on chlorophyll synthesis has

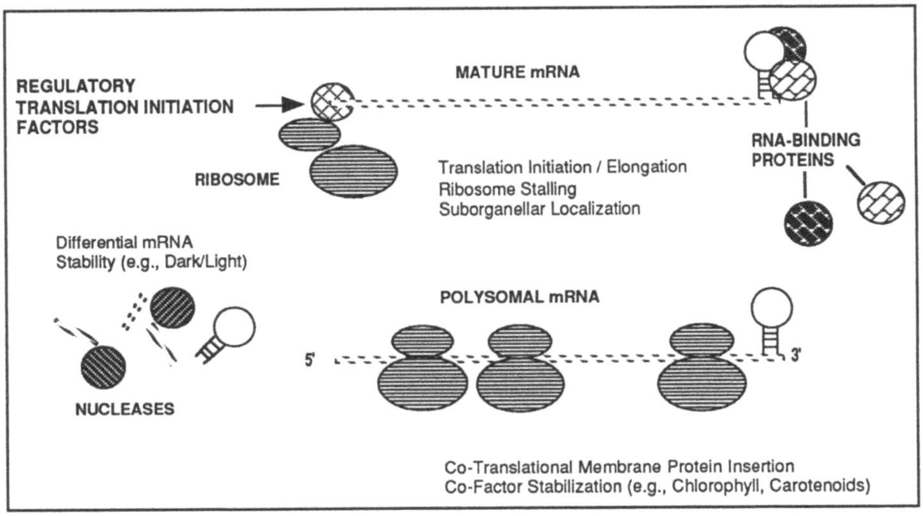

Figure 3: Translational control of mRNA stability

been demonstrated directly using lysed barley etioplasts supplemented with chlorophyll precursors (86), which has confirmed that the two events appear to be temporally linked. Insertion of chlorophyll into arrested of nascent apoproteins may be facilitated by ribosome pausing and stabilization of the translation intermediates, as was suggested for the synthesis of the D1 protein in barley (87, 88).

POST-TRANSLATIONAL CONTROL OF PLASTID GENE EXPRESSION

Many proteins of multisubunit complexes in chloroplasts are rapidly degraded when unable to assemble. This was first demonstrated for ribulose-1,5-bisphosphate carboxylase in *Chlamydomonas* (89), and subsequently extended in *Chlamydomonas* and other plants to proteins for photosystem I, photosystem II, the cytochrome b_6/f complex, and the ATP synthethase complex. Rapid pulse-labeling of proteins in *Chlamydomonas* has shown that the synthesis of certain subunits may be coordinated (90, 91), but the mechanisms that regulate the coordinated synthesis of two or more subunits are presently not know.

Several studies such as those described above have established the presence of proteolytic activites in chloroplasts (92). A general ATP-dependent protease that appears to be associated with thylakoid membranes and that degrades thylakoid proteins as well as stromal proteins has been identified and partially purified (93, 94), although it is not known if this activity consists of one non-specific protease or multiple specific proteases. A protease that appears to be specific for the turnover of the protein subunits from the photosystem II water-oxidation complex has been identified (95), suggesting that multiple, specific proteolytic enzymes may be associated with the thylakoid membrane system. Recently, genes for proteins have been identified in plants with sequences similar to Clp proteins (ClpA, ClpB, ClpP), which are part of ATP-dependent serine protease complexes that catalyze initial rate-limiting steps in proteolysis and which are present in all prokaryotic and eukaryotic cells examined (96). ClpA is an ATPase which activates an ATP-dependent ClpP endoprotease. ClpB has been identified as a heat-shock protein in *E.coli*, yeast and mammalian cells that may regulate proteolysis

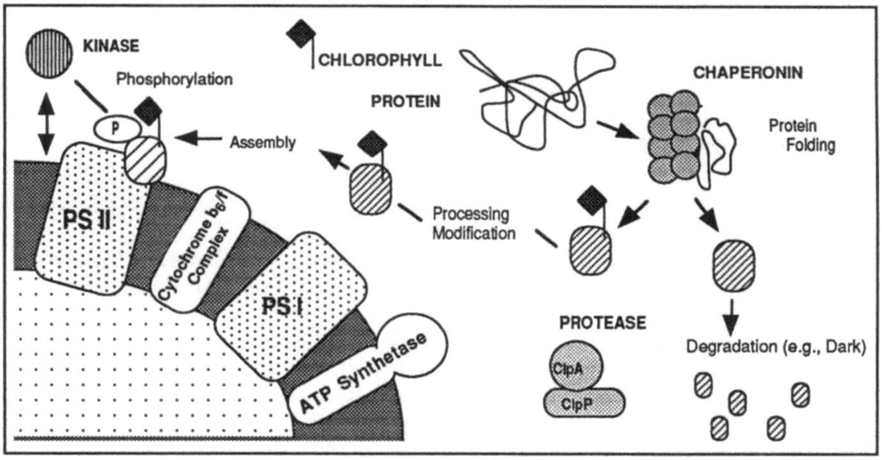

Figure 4: Control of protein assembly and degradation

but could also also protect proteins from heat-induced denaturation. The gene for the clpP endoprotease is encoded by the chloroplast genome (97, 98) and is retained in the truncated plastid genome of the parasitic plant *Epifagus* (16). Homologs of ClpA have also been identified in plants (99), but the protein appears to be encoded in the nuclear genome and transported into the chloroplast (92). The nuclear-encoded ClpA and chloroplast-encoded ClpP could be part of the ATP-dependend serine protease complex that is associated with the thylakoid membrane and degrades soluble and thylakoid membrane proteins (94). The precise role of the proteolytic chloroplast enzymes and the control of their activities in the degradation of unassembled subunits, or turnover of subunits in membrane and soluble protein complexes, has not been established. The nuclear location of the gene for the chloroplast ClpA homolog could point to a potentially interesting regulation of proteolysis in which the nucleus has an important control function. Moreover, the activity of ClpA/ClpP-type protease complexes could be rapidly adjusted by modification of the ClpA regulatory subunit (e.g., phosphorylation, ATP levels) in response to environmental (e.g., light) or developmental signals which, at least in part, could explain the rapid stabilization of photosynthetic proteins after illumination of dark-grown seedlings.

CONCLUSIONS

Together, this brief review illustrates that the temporal and spatial regulation of plastid gene expression is complex and involves several steps at which control can be exerted. The extent to which plastid gene expression is controlled at the transcriptional, post-transcriptional, translational, or post-translational level depends on several developmental and environmental factors. Considering the complexity of regulatory events in plastids, it is reasonable to speculate that most, if not all, of the potentially regulatory processes are controlled by genes in the nuclear genome. The combination of genetic, molecular, and biochemical approaches will help to identify many of the regulatory proteins involved in plastid gene expression over the next few years.

REFERENCES

1. K. Ohyama, *et al.*, *Nature* **322**, 572-574 (1986).
2. K. Shinozaki, *et al.*, *EMBO J.* **5**, 2043-2049 (1986).
3. M. Sugiura, *Bot. Mag. Tokyo* **100**, 407-436 (1987).
4. K. Umesono, H. Ozeki, *Trends Genet.* **3**, 281-287 (1987).
5. J. Hiratsuka, *et al.*, *Mol. Gen. Genet.* **217**, 185-194 (1989).
6. W. Gruissem, B. M. Greenberg, G. Zurawski, R. B. Hallick, in *Methods in Enzymology. Plant Molecular Biology* A. Weissbach, H. Weissbach, Eds. (Academic Press, Orlando, 1986), pp. 253-270.
7. E. M. Orozco, J. E. Mullet, L. Hanley-Bowdoin, N.-H. Chua, in *Methods in Enzymology* A. Weissbach, H. Weissbach, Eds. (Academic Press, Inc., Orlando, 1986), pp. 232-253.
8. W. Gruissem, in *The Biochemistry of Plants* A. Marcus, Eds. (Academic Press, San Diego, 1989), pp. 151-191.
9. L. Bogorad, I. K. Vasil, in *Cell Culture and Somatic Cell Genetics of Plants.* (Academic Press, Inc., San Diego, 1991),
10. G. L. Igloi, H. Kossel, *Crit. Rev. Plant Sci.* **10**, 525-558 (1992).
11. M. C. Little, R. B. Hallick, *J. Biol. Chem.* **263**, 14303-14307 (1988).
12. S. Purton, J. C. Gray, *Mol. Gen. Genet.* **217**, 77-84 (1989).
13. J. Hu, L. Bogorad, *Proc. Natl. Acad. Sci. USA* **87**, 1531-1535 (1990).
14. J. Hu, R. F. Troxler, L. Bogorad, *Nuc. Acids Res.* **19**, 3431-3434 (1991).
15. C. W. dePamphilis, J. D. Palmer, *Nature* **348**, 337-339 (1990).
16. C. W. Morden, K. H. Wolfe, C. W. dePamphilis, J. D. Palmer, *EMBO J.* **10**, 3281-3288 (1991).
17. W. Bünger, J. Feierabend, *Planta* **149**, 163-169 (1980).
18. A. Siemenroth, *et al.*, *Planta* **153**, 547-555 (1981).
19. X. W. Deng, W. Gruissem, *Cell* **49**, 379-387 (1987).
20. B. J. Baumgartner, J. C. Rapp, J. E. Mullet, *Plant Physiol.* **89**, 1011-1018 (1989).
21. L. Hanley-Bowdoin, N.-H. Chua, *Trends Biochem. Sci.* **12**, 67-70 (1987).
22. D. K. Hawley, W. R. McClure, *Nuc. Acids Res.* **11**, 2237-2255 (1983).
23. W. Gruissem, *et al.*, *Cell* **35**, 815-828 (1983).
24. W. Gruissem, G. Zurawski, *EMBO J.* **4**, 3375-3383 (1985).
25. W. Gruissem, G. Zurawski, *EMBO J.* **4**, 1637-1644 (1985).

26. W. Gruissem, G. Zurawski, in *Molecular Form and Function of the Plant Genome* L. van Vloten-Doting, G. S. P. Groot, T. C. Hall, Eds. (Plenum Publishing, New York, 1985), pp. 199-210.
27. G. Link, *EMBO J.* **3**, 1697-1704 (1984).
28. D. Bradley, A. A. Gatenby, *EMBO J.* **4**, 3641-3648 (1985).
29. L. Hanley-Bowdoin, E. M. Orozco, N.-H. Chua, *Mol. Cell. Biol.* **5**, 2733-2745 (1985).
30. W. Gruissem, *et al.*, *Nuc. Acids Res.* **14**, 7541-7556 (1986).
31. J. Haley, L. Bogorad, *Plant Cell* **2**, 323-333 (1990).
32. P. E. Gamble, J. E. Mullet, *EMBO J.* **8**, 2785-2794 (1989).
33. W. B. Yao, *et al.*, *Nuc. Acids Res.* **17**, 9583-9591 (1989).
34. T. B. Sexton, D. A. Christopher, J. E. Mullet, *EMBO J.* **9**, 4485-4494 (1990).
35. D. A. Christopher, M. Kim, J. E. Mullet, *Plant Cell* **4**, 785-798 (1992).
36. J. E. Mullet, *Annu. Rev. Plant Physiol. Plant Mol. Biol.* **39**, 475-502 (1988).
37. W. Gruissem, *Cell* **56**, 161-170 (1989).
38. W. Gruissem, in *Physiology, Biochemistry, and Genetics of Nongreen Plastids* C. D. Boyer, J. C. Shannon, R. C. Hardison, Eds. (American Society of Plant Physiologists, Rockville MD, 1989), pp. 227-240.
39. X.-W. Deng, *et al.*, *J. Biol. Chem.* **262**, 9641-9648 (1987).
40. R. R. Klein, J. E. Mullet, *J. Biol. Chem.* **265**, 1895-1902 (1990).
41. X.-W. Deng, *et al.*, *Plant Cell* **1**, 645-654 (1989).
42. J. C. Tonkyn, X.-W. Deng, W. Gruissem, *Plant Physiol.* **99**, 1406-1415 (1992).
43. S. Bülow, G. Link, *Plant Mol. Biol.* **10**, 349-357 (1988).
44. K. Tiller, A. Eisermann, G. Link, *Eur. J. Biochem.* **198**, 93-99 (1991).
45. G. S. Hudson, *et al.*, *J. Mol. Biol.* **200**, 639-654 (1988).
46. H. Neuhaus, A. Scholz, G. Link, *Curr. Genet.* **15**, 63-70 (1989).
47. W. Gruissem, G. Schuster, *in Control of Messenger RNA Stability* G. Brawerman and J. Belasco, Eds. (Academic Press, Inc., San Diego,1993), pp. 329-365.
48. W. Gruissem, *et al.*, *Trends Genet.* **4**, 258-263 (1988).
49. G. Link, in *The Photosynthetic Apparatus: Molecular Biology and Operation. Cell Culture and Somatic Cell Genetics of Plants.* L. Bogorad, I. K. Vasil, Eds. (Academic Press, Inc., San Diego, 1991), pp. 365-394.
50. R. E. Glick, *et al.*, *Proc. Natl. Acad. Sci.USA* **83**, 4287-4291 (1986).
51. A. Barkan, *Plant Cell* **1**, 437-445 (1989).
52. L. D. Crossland, S. R. Rodermel, L. Bogorad, *Proc. Natl. Acad. Sci. USA* **81**, 4060-4064 (1984).
53. J. E. Mullet, E. M. Orozco, N.-H. Chua, *Plant Mol. Biol.* **4**, 39-54 (1985).
54. P. E. Gamble, T. B. Sexton, J. E. Mullet, *EMBO J.* **7**, 1289-1297 (1988).
55. X.-W. Deng, W. Gruissem, *EMBO J.* **7**, 3301-3308 (1988).
56. P. Westhoff, *et al.*, *Mol. Gen. Genet.* **199**, 290-299 (1985).
57. D. B. Stern, W. Gruissem, *Cell* **51**, 1145-1157 (1987).
58. D. B. Stern, H. Jones, W. Gruissem, *J. Biol. Chem.* **264**, 18742-18750 (1989).
59. G. Zurawski, M. T. Clegg, *Annu. Rev. Plant Physiol.* **38**, 391-418 (1987).
60. J. C. Tonkyn, W. Gruissem, *Mol. Gen. Genet. In press* (1993).
61. L.-J. Chen, E. M. J. Orozco, *Nuc. Acids Res.* **16**, 8411-8431 (1988).
62. D. B. Stern, W. Gruissem, *Plant Mol. Biol.* **13**, 615-625 (1989).
63. C. Hsu-Ching, D. B. Stern, *J. Biol. Chem.* **266**, 24205-24211 (1991).
64. D. B. Stern, E. R. Radwanski, K. L. Kindle, *Plant Cell* **3**, 285-297 (1991).
65. P. Klaff, W. Gruissem, *Plant Cell* **3**, 517-529 (1991).

66. X.-Q. Liu, *et al., Plant Mol. Biol.* **12**, 385-394 (1989).
67. R. J. Spreitzer, *et al., Proc. Natl. Acad. Sci. USA* **82**, 5460-5464 (1985).
68. D. Drapier, J. Girard-Bascou, F.-A. Wollman, *Plant Cell* **4**, 283-295 (1992).
69. J.-D. Rochaix, *et al., EMBO J.* **8**, 1013-1021 (1989).
70. M. R. Kuchka, S. P. Mayfield, J.-D. Rochaix, *EMBO J.* **7**, 319-324 (1988).
71. J. Nickelsen, G. Link, *Mol. Gen. Genet.* **228**, 89-96 (1991).
72. G. Schuster, W. Gruissem, *EMBO J.* **10**, 1493-1502 (1991).
73. J. Nickelsen, G. Link, *Nuc. Acids Res.* **17**, 9637-9648 (1989).
74. Y. Li, M. Sugiura, *EMBO J.* **9**, 3059-3066 (1990).
75. L. Ye, *et al., Nuc. Acids Res.* **19**, 6485-6490 (1991).
76. R. J. Bandziulis, M. S. Swanson, G. Dreyfuss, *Genes Dev.* **3**, 431-437 (1989).
77. I. W. Mattaj, *Cell* **57**, 1-3 (1989).
78. D. J. Kenan, C. C. Query, J. D. Keene, *Trends. Genet.* **16**, 214-220 (1991).
79. R. G. Herrmann, P. Westhoff, J. Alt, J. Tittgen, N. Nelson, in *Molecular Form and Function of the Plant Genome* L. van Vloten-Doting, G. S. P. Groot, T. C. Hall, Eds. (Plenum Press, New York, 1985), pp. 233-256.
80. K. Kreuz, K. Dehesh, K. Apel, *Eur. J. Biochem.* **159**, 459-467 (1986).
81. R. R. Klein, J. E. Mullet, *J. Biol. Chem.* **261**, 11138-11145 (1986).
82. R. R. Klein, P. E. Gamble, J. E. Mullet, *Plant Physiol.* **88**, 1246-1256 (1988).
83. P. Malnoe, S. P. Mayfield, J.-D. Rochaix, *J. Cell Biol.* **106**, 609-616 (1988).
84. R. R. Klein, H. S. Mason, J. E. Mullet, *J. Cell Biol.* **106**, 289-301 (1988).
85. J. E. Mullet, P. G. Klein, R. R. Klein, *Proc. Natl. Acad. Sci. USA* **87**, 4038-4042 (1990).
86. L. A. Eichaker, *et al., J. Biol. Chem.* **265**, 13566-13571 (1990).
87. P. E. Gamble, R. R. Klein, J. E. Mullet, in *Photosynthesis* W. R. Briggs, Eds. (Alan R. Liss, Inc., New York, 1989), pp. 285-298.
88. J. Kim, P. G. Klein, J. E. Mullet, *J. Biol. Chem.* **266**, 14931-14938 (1991).
89. G. W. Schmidt, M. L. Mishkind, *Proc. Natl. Acad. Sci. USA* **80**, 2632-2636 (1983).
90. K. H. Jensen *et al., J. Cell Biol.* **103**, 1315-1325 (1986).
91. C. de Vitry, *et al., . Cell Biol.* **109**, 991-1006 (1989).
92. K. Keegstra, L. J. Olsen, S. M. Theg, *Annu. Rev. Plant Physiol. Plant Mol. Biol.* **40**, 471-501 (1989).
93. M. L. Mishkind, *et al., Curr. Topics Plant Biochem. Physiol.* **4**, 34-50 (1985).
94. X.-Q. Liu, A. T. Jagendorf, *Plant Physiol.* **81**, 603-608 (1986).
95. T. Kuwabara, K. Shinohara, *Curr. Res. Photosynth.* **III**, 743-746 (1990).
96. C. Squires, C. L. Squires, *J. Bacteriol.* **174**, 1081-1085 (1992).
97. J. C. Gray, S. M. Hird, T. A. Dyer, *Plant Mol. Biol.* **15**, 947-950 (1990).
98. M. R. Maurizi, *et al., J. Biol. Chem.* **265**, 12546-12552 (1990).
99. S. Gottesman, *et al., Proc. Natl. Acad. Sci. USA* **87**, 3513-3517 (1990).

POST-TRANSCRIPTIONAL REGULATION OF CHLOROPLAST GENE EXPRESSION IN CHLAMYDOMONAS

Jean-David Rochaix, Michel Goldschmidt-Clermont,
Caroline Monod and William Zerges

Departments of Molecular Biology and Plant Biology
University of Geneva, Geneva, Switzerland

INTRODUCTION

The biosynthesis of the photosynthetic apparatus in higher plants and eukaryotic algae occurs through the concerted action of two genetic systems located in the nucleus and chloroplast respectively. While some subunits of the photosynthetic complexes are encoded by the chloroplast genome and translated on 70S ribosomes, other subunits are encoded by nuclear genes, translated on cytosolic 80S ribosomes and imported into the chloroplast where they form functional complexes.

An attractive feature of the photosynthetic apparatus is that its functional state in vivo can be easily monitored through simple fluorescence measurements. Mutants affected in the activity of photosystem I or photosystem II display altered fluorescence properties and can therefore be recognized easily. Indeed, numerous photosynthetic mutants have been isolated from higher plants and eukaryotic algae using fluorescence screening (Bennoun and Delepelaire, 1982; Bishop, 1982; Miles, 1982).

The green unicellular alga Chlamydomonas reinhardtii has been used intensively for this type of analysis for several reasons. First, its photosynthetic function is dispensable when the cells are provided with a reduced carbon source such as acetate. Second, this alga has a well defined sexual cycle, and it can grow vegetatively either in a haploid or diploid state. Because it can be manipulated like a microorganism, extensive genetic analysis is feasible. Third, efficient nuclear and chloroplast transformation procedures are available (for review cf. Harris, 1989).

The analysis of several nuclear photosynthetic mutants of C. reinhardtii has revealed the existence of a surprisingly large number of nuclear encoded factors that appear to be required for the expression of defined chloroplast genes. These factors appear to act at several levels: RNA processing, RNA

NATO ASI Series, Vol. H 81
Plant Molecular Biology
Edited by G. Coruzzi and P. Puigdomènech

stability, translation and most likely assembly of photosynthetic complexes (Rochaix, 1992). Here we review recent results obtained with some of these mutants.

RNA PROCESSING: TRANS-SPLICING OF THE psaA RNA

The psaA gene encodes one of the reaction center polypeptides of photosystem I. In C. reinhardtii this gene has an unusual structure: it consists of three exons that are widely scattered on the chloroplast genome (Kück et al., 1987, fig. 1). These exons are flanked by consensus sequences of group II introns and they are transcribed independently so that the mature psaA RNA is assembled through two splicing reactions in trans (Kück et al., 1987; Choquet et al., 1988). A fourth chloroplast locus, tscA, that maps at a different position from either of the three exons encodes a small RNA that is required for the first trans-splicing reaction between exons 1 and 2 (Goldschmidt-Clermont et al., 1991).

A large number of mutants affected in a psaA mRNA maturation have been isolated. These mutants fall into 3 classes based on the psaA precursor RNAs they accumulate: mutants unable to splice exons 1 and 2 and exons 2 and 3 (class B); mutants unable to splice exons 2 and 3, but able to splice exons 1 and 2 (class A) and mutants unable to splice exons 1 and 2, but able to splice exons 2 and 3 (class C). These mutations have been grouped into 14 nuclear complementation groups: 5 for class A, 2 for class B and 7 for class C (Goldschmidt-Clermont et al., 1990). Because several of these complementation groups contain a single allele it is probable that many other nuclear genes are required for the maturation of psaA RNA. Whether class B mutations specifically affect psaA splicing is not known since no other group II introns have yet been identified in the chloroplast genome of C. reinhardtii. It is clear however that class A and C mutations act specifically since they affect only one of the two psaA trans-splicing reactions.

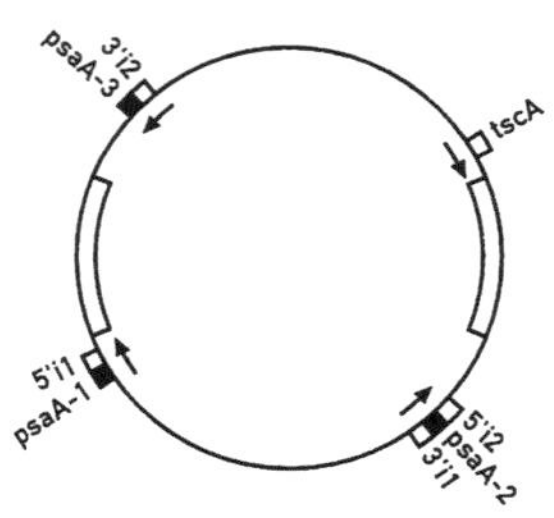

Figure 1. Split structure of psaA of C. reinhardtii. The three exons psaA-1, -2, -3 and their flanking intron sequences (5'i1, 3'i1, 5'i2, 3'i2) and tscA are indicated on the circular chloroplast genome (not drawn to scale). Arrows indicate the direction of transcription. The two large boxes represent the two segments of the chloroplast inverted repeat.

We have proposed a model in which the tscA RNA provides the missing link between the 5' and 3' parts of the first psaA intron (fig. 2, Goldschmidt-Clermont et al., 1991). The tscA RNA completes the group II catalytic core of this intron by base pairing with the 5' part of intron 1 thereby forming presumably domain I and also with the 3'part of intron 1 to form the stem of domain IV. The first psaA intron therefore consists of three discrete parts. This model agrees with several observations. Transformants containing large insertions within the peripheral loops of domains II and III are still capable of producing mature psaA RNA although the maturation of the psaA mRNA is slower. It is likely that these peripheral insertions do not markedly affect the catalytic core of the intron. The only insertion which interferes with splicing maps within the stem of domain IV where the tscA RNA pairs with the 3' part of intron 1 (Goldschmidt-Clermont et al., 1991). Recently the tscA locus has been identified in two other Chlamydomonas species, C. gelatinosa and C. zebra. (M. Turmel and C. Lemieux, unpublished results). In these two species the psaA intron can also be folded in the same way as in C. reinhardtii with a tripartite structure.

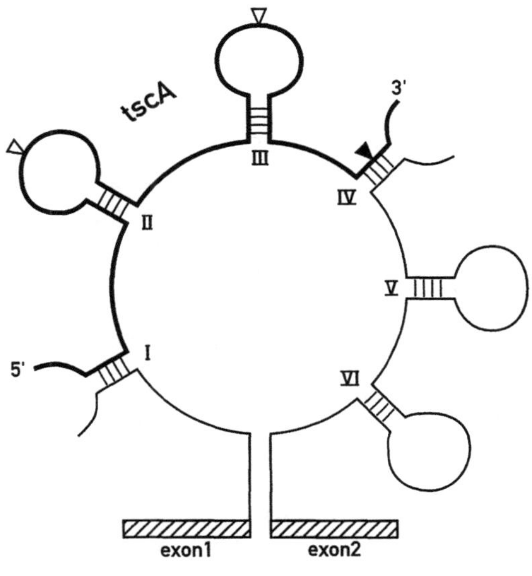

Fig. 2 Discontinuous structure of intron 1 of psaA. The tscA RNA is represented by a thick line. Its 5' end pairs with the 5' part of intron 1 (which follows exon 1) to form domain 1. The tscA RNA completes the catalytic core by forming domains II and III and it pairs with the 3' part of intron 1 (followed by intron 2) to form domain IV. The insertion within tscA that blocks splicing is represented by a closed triangle; two other insertions which do not inhibit splicing are indicated by open triangles.

Considering the large number of nuclear psaA trans-splicing mutants isolated, it is possible that some of the mutations act in an indirect way. This has indeed been shown for one of the class C nuclear mutants that lacks tscA RNA and that accumulates instead a larger RNA precursor that includes tscA. This mutant appears therefore to be affected in the processing of the tscA precursor RNA.

Nuclear transformation in C. reinhardtii can be used as a method for tagging genes. This is because in most cases the transforming DNA appears to integrate randomly within the nuclear genome (Debuchy et al., 1989; Kindle, 1990). Accordingly, transformation of the arg7cw15 strain with the pARG7.8 plasmid containing the gene of argininosuccinate lyase was used to generate $4x10^3$ ARG^+ transformants. From 14 high fluorescence photosynthetic mutants identified two were affected in trans-splicing. Genetic crosses showed co-segregation of the pARG7.8 plasmid marker with the mutation in one of the transformants. The tagged locus was cloned from a λ library of this mutant. A single copy probe from the locus was then used to isolate a wild-type allele clone from a cosmid library. Subcloning of the cosmid revealed that a 8 kb genomic DNA fragment can rescue the original mutation by transformation confirming that the corresponding gene has been isolated.

CHLOROPLAST RNA STABILITY

The levels of many plastid RNAs vary considerably during chloroplast development and plastid differentiation. These changes are caused mostly by the variable stability of individual plastid RNA (Deng and Gruissem, 1987; Deng et al., 1987; Mullet and Klein, 1987). A detailed understanding of plastid RNA stability is therefore important for understanding chloroplast biogenesis. We have chosen a genetic approach to this problem and isolated and characterized two nuclear mutants affected in psbD and psbB RNA accumulation.

The nuclear mutant nac2-26 is unable to accumulate detectable amounts of psbD RNA although in vivo pulse labeling of RNA shows that psbD is transcribed at similar levels as in wild-type (Kuchka et al. 1989). In this mutant psaA message accumulates as in wild-type although the second exon of psaA is located close to psbD and cotranscribed with it. While the psaA exon 2 precursor accumulates stably in psaA splicing mutants, it is rapidly degraded in double mutants for nac2-26 and trans-splicing indicating that the nuclear encoded factor defined by the nac2-26 mutation is required not only for stabilizing psbD mRNA, but also when it is part of the psaA exon 2 precursor (Kuchka et al. 1989). To identify the target site for the trans-acting factor involved in RNA stability on the RNA, a chimeric gene was constructed in which the psbD promoter and 5'psbD untranslated region were fused to the reporter gene aadA. When expressed in the chloroplast the aadA product aminoglycoside adenyl transferase (AAD) confers spectinomycin resistance (Goldschmidt-Clermont, 1991). The chimeric gene was inserted into the chloroplast genome of the photosynthetic mutant Fud50 using the atp-int vector (Goldschmidt-Clermont et al., 1991) and selecting for phototrophic growth. Transformants of mating type$^+$ were then crossed to the nac2-26 mutant and tetrads were scored for photosynthetic function and spectinomycin resistance. Because of the uniparental inheritance of the chloroplast genome from the mating-type + parent, all members of the tetrad will inherit the chimeric gene whereas the nuclear mutation will segregate 2:2. Analysis of numerous tetrads revealed in all cases cosegregation of the photosynthetic deficiency and spectinomycin sensitivity, thus indicating that the 5'untranslated region of psbD which consists of only 79 nucleotides is sufficient to induce degradation of the chimeric RNA in a nac 2-26 nuclear background. These results were confirmed by Northern analysis of the RNAs extracted from the members of 2 tetrads: loss of psbD RNA cosegregated with photosynthetic deficiency and spectinomycin sensitivity. Similar experiments using chimeric genes with the 3'untranslated region of psbD revealed that in this case the chimeric RNA is not degraded in the nuclear mutant background. These results show that the 5'untranslated region of psbD plays an important role in psbD message stability presumably by interacting directly or indirectly with nuclear-encoded factors.

Another nuclear mutant 222E from C. reinhardtii which is defective in photosystem II activity fails to accumulate specifically psbB transcripts and displays an altered transcription pattern downstream of psbB (Monod et al., 1992). A similar mutant has been characterized by Sieburth et al. (1991). The 5' and 3' termini of the psbB transcript have been mapped to 35 bases upstream of the initiation codon and approximately 600 bases downstream of the stop codon. To identify the target site for the trans-acting factor involved in psbB RNA stability that is presumably affected in the 222E mutant, a chimeric gene with the psbB 3'untranslated region fused to atpB was constructed and the same strategy as outlined above was followed. In this case the chimeric RNA was not degraded in the nuclear background of 222E, indicating that the 3'untranslated region is not sufficient to induce RNA degradation. Possibly the 5'untranslated region may be involved.

TRANSLATION: TWO NUCLEAR ENCODED FUNCTIONS ARE REQUIRED FOR TRANSLATION OF THE CHLOROPLAST psbC GENE

We have shown previously that the synthesis of the photosystem II core polypeptide P6, also called CP43, is affected in two nuclear mutants F34 and F64 belonging to distinct complementation groups (Rochaix et al., 1989). To test whether these mutations act at the level of translation or whether they promote rapid degradation of the P6 polypeptide we have constructed chimeric genes in which the psbC promoter and 5'untranslated region were fused to the aadA reporter gene. These genes were introduced into the chloroplast genome by particle gun mediated transformation and the transformants were analyzed using the same strategy as outlined above for the mutants affected in RNA stability. The chloroplast transformants produced AAD as determined by both spectinomycin resistance and the detection of the AAD polypeptide by Western blot analysis. AAD expression was not detected, however in sibling strains (also transformed with the chimeric gene) carrying either the F34 or the F64 nuclear mutations. Therefore, since the wild-type products of the genes defined by the F34 and F64 mutations are required for the ability of the psbC 5'untranslated region to direct translation of AAD, these nuclear encoded functions appear also to be required for the translation of the psbC message rather than for stability of P6.

In order to characterize the trans-acting factors mediating the F34 and F64 requirements for psbC translation, in vitro RNA protein binding experiments were performed. They revealed that chloroplast extracts from wild-type cells contain two protein species of 95 and 31 kDa which bind to an inverted repeat within the psbC 5'untranslated leader sequence as determined by UV crosslinking experiments. Chloroplast protein extracts from the F64 nuclear mutant contain, in addition to the 95 and 31 kDa species, novel binding activities suggesting that F64 activates repressors of psbC translation.

CONCLUSIONS

The interactions between the nucleus and the chloroplast appear to be rather complex. In all cases examined several nuclear encoded functions are required in post-transcriptional steps for expressing a given chloroplast gene, at least in C. reinhardtii. Since the chloroplast genome contains over hundred individual genes, the expression of this genome may require the participation of a very large number of nuclear genes. A list of nuclear mutations affecting chloroplast gene expression is given in Table 1. This list is likely to grow considerably during the next years.

A striking feature is that in most cases these mutations affect the expression of specific genes. Does this imply the existence of specific trans-acting factors ? This may not necessarily be the case as specific effects may result from the relatively unspecific binding of factors to unique cis elements that are located in strategically important positions within the 5'untranslated leader regions. It is well documented that the relative positions of short stem loop structures within the 5'untranslated leader region of the yeast gene GCN4 play a crucial role in the control of its expression (Abastado et al., 1991). Our preliminary results reveal a lack of specificity for several RNA binding activities that are altered in some of the mutants.

An understanding of the molecular basis of these chloroplast posttranscriptional processes will require the isolation of some of these nuclear encoded factors that are affected in the mutants. In this respect gene tagging appears to be a promising approach. However the highly efficient nuclear transformation method opens the possibility of cloning genes by complementation with wild-type genomic libraries. Isolation and characterization of the factors altered in these mutants is therefore likely to occur soon.

Table I. Nuclear mutation affecting the expression of chloroplast genes in C. reinhardtii

chloroplast gene	RNA stability	RNA processing	Translation protein turnover	Reference
psbA			1	Girard-Bascou et al. 1992
psbB			1-2	Jensen et al. 1986
				Sieburth et al. 1991
				Monod et al. 1992
psbC	1			Sieburth et al. 1991
psbC			2	Rochaix et al. 1989
psbD	1			Kuchka et al. 1989
psbD			2	Kuchka et al. 1988
psaA		14		Goldschmidt-Clermont et al. 1990
psaB			2	Girard-Bascou et al. 1987
atpA	1			Drapier et al. 1992
atpA			1	Drapier et al. 1992
atpB	1			Drapier et al. 1992
rbcL	1			H. Beasley, L. Mets, unpublished results

ACKNOWLEDGEMENTS

We thank N. Roggli for drawings. This work was supported by grant No 31-34014.92 from the Swiss National Fund.

REFERENCES

Abastado, J.P. Miller, P.F., Jackson, B.M. and Hinnebusch, A.G. (1991) Mol. Cell. Biol. 11, 486-496.

Bennoun, P. and Delepelaire, P. (1982) In Methods in chloroplast molecular biology. Edelman, M. et al. eds. pp. 25-38. Elsevier Biomedical Press.

Bishop, N.I. (1982) In Methods in chloroplast molecular biology. Edelman, M. et al. eds pp. 51-63, Elsevier Biomedical Press.

Choquet, Y., Goldschmidt-Clermont, M., Girard-Bascou, J., Kück, U. Bennoun, P. and Rochaix, J.D. (1988) Cell 52, 903-913.

Debuchy, R., Purton, S. and Rochaix, J.D. (1989) EMBO J. 8, 2803-2809.

Deng, X. and Gruissem, W. (1987) Cell 49, 379-387.

Deng, X.W., Stern, D.B., Tonkya, J.C. and Gruissem, W. (1987) J. Biol. Chem. 262, 9641-9648.

Drapier, D., Girard-Bascou, J. and Wollman, F. (1992) Cell 4, 283-295.

Girard-Bascou, J., Choquet, Y., Schneider, M., Delorme, M. and Dron, M. (1987) Curr. Genet 12, 485-495.

Girard-Bascou, J., Pierre Y. and Drapier, D. (1982) Current Genet. 22, 47-52.

Goldschmidt-Clermont, M. (1991) Nucleic Acids Res. 19, 4083-4089.

Goldschmidt-Clermont, M., Girard-Bascou, J., Choquet, Y., and Rochaix, J.D. (1990) Mol. Gen. Genet. 223, 417-425.

Goldschmidt-Clermont, M., Choquet, Y., Girard-Bascou, J., Michel, F., Schirmer-Rahire, M. and Rochaix, J.D. (1991) 65, 135-143.

Harris, E. (1989). The Chlamydomonas Source Book.

Jensen, K.H., Harris, D., Plumley, G. and Schmidt, J.W. (1986) J. Cell Biol. 103, 1314-1325.

Kindle, K.L. (1990) Proc. Natl. Acad. Sci. USA 87, 1228-1232.

Kuchka, M., Mayfield, G.P. and Rochaix, J.-D. (1988) EMBO J. 7, 319-324.

Kuchka, M., Goldschmidt-Clermont, M., van Dillewijn, J. and Rochaix, J.D. (1989) Cell 58, 869-876.

Kück, U., Choquet, Y., Schneider, M., Dron, M. and Bennoun, P. (1987) EMBO J. 6, 2185-2195.

Miles, D. (1982) In Methods in chloroplast molecular biology. Edelman, M. et al. eds. pp. 85-107, Elsevier Biomedical Press.

Monod, C., Goldschmidt-Clermont, M. and Rochaix, J.D. (1992) Mol. Gen. Genet. 231, 449-459.

Mullet, J.E. and Klein, R.R. (1987) EMBO J. 6, 1571-1576.

Rochaix, J.D. (1992) Ann. Rev. Cell. Biol. 8, 1-28.

Rochaix, J.D., Kuchka, M., Mayfield, S.P. Schirmer-Rahire, M., Girard-Bascou, J. and Bennoun, P. (1989) EMBO J. 8, 1013-1021.

Sieburth, L.E., Berry-Lowe, S. and Schmidt, G.W. (1991) Plant Cell 3, 175-189.

Transcriptional Regulation of the Maize Anthocyanin Pathway

M.B. Sainz, S. A. Goff[1], J. M. Krahn, and V.L. Chandler
Institute of Molecular Biology
University of Oregon
Eugene, OR 97403 USA.

INTRODUCTION

The genetics of pigment production and its control in maize are well understood (Figure 1). Two classes of regulatory genes (*b/r* and *c1/pℓ*), required for the transcription of the genes encoding the anthocyanin biosynthetic enzymes (target genes), have been characterized (Paz-Ares et al. 1986; Dellaporta et al. 1988; Chandler et al. 1989; Cone and Burr 1989). The proteins encoded by *C1* and *Pℓ* alleles are 90% identical (K. Cone, personal communication). *C1* alleles function in the seed and embryo, while *Pℓ* alleles function in most seedling and plant tissues. The proteins encoded by the *B* and *R* genes are 80% identical (Radicella et al. 1991), and both genes have a large allelic diversity with respect to their tissue-specific expression (Styles et al. 1973). Six target genes have been cloned, *c2* (Wienand et al. 1986), *chi* (Grotewold and Peterson 1991), *a1* (O'Reilly et al. 1985), *a2* (Menssen et al. 1990), *bz1* (Fedoroff et al. 1984), and *bz2* (McLaughlin and Walbot 1987; Theres et al. 1987). Herein, we review recent data on each class of regulatory protein and discuss models for their mechanism of action.

The genetic requirement for a member of each class of regulatory protein has been recognized for many years (Coe et al. 1988). The mechanisms underlying the corequirement for *b/r* and *c1/pℓ* is unlikely to be transcriptional control of one gene by the other, as expression of *B* or *C1* cDNAs in transient transformation assays does not eliminate the requirement for the other gene (Goff et al. 1990). *Pℓ* mRNA accumulation is also not dependent on a functional *B* or *R* allele (Cone and Burr 1989), and *B* mRNA accumulation is not dependent on functional *Pℓ* alleles (V. Chandler, unpublished data).

Both B/R and C1 are required for activation of target gene promoter/luciferase fusions in transient transformation assays

---
1 Present address: Agricultural Biotechnology Research Unit, CIBA-Geigy Corporation, P.O. Box 12257, Research Triangle Park, NC 27709

NATO ASI Series, Vol. H 81
Plant Molecular Biology
Edited by G. Coruzzi and P. Puigdomènech

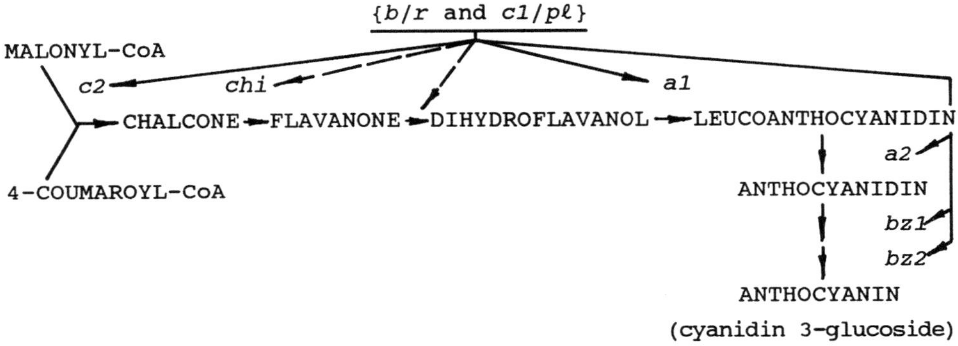

Figure 1. The target genes are indicated above the arrows, which symbolize the step catalyzed by the encoded enzyme. The regulatory genes stimulate the transcription of the target genes (solid lines). The dashed lines indicate target genes for which transcriptional regulation by B/R and C1/Pℓ has not been tested.

(Goff et al. 1990). Accumulation of target gene mRNA requires both B/R and C1 proteins (Cone et al. 1986; Ludwig et al. 1989; Chandler et al. 1989), and B is required for the transcription of *c2*, *a1*, and *bz1* in husk tissues (G. Patterson and V. Chandler, unpublished data). Sequence analyses of *R* and *B* cDNAs reveal that they encode proteins with 65% identity to the basic-helix-loop-helix (bHLH) domain found in several DNA-binding proteins in mammals, Drosophila and yeast (Ludwig et al. 1989; Perrot and Cone 1989; Radicella et al. 1991). The *C1* cDNA sequence revealed a motif with 55% identity to the DNA-binding domain in *myb* oncogenes (Paz-Ares et al. 1987).

FUNCTIONAL ANALYSES OF TARGET PROMOTER SEQUENCES

The sequences necessary and sufficient for regulation by B/R and C1/Pℓ proteins have been determined only for the *bz1* promoter. Deletion analysis demonstrated that sequences within 100 bp upstream of the transcription start site mediate significant induction by B and C1 (Klein et al. 1989). This 100 bp region contains consensus Myb-like and bHLH-like binding sites, and mutational analyses indicate that these sequences and the nucleotides between them are important for B and C1 activation (Goff et al. 1990; Roth et al. 1991). A 32 bp region spanning the putative consensus sites was sufficient for B and C1 induction of a heterologous promoter

(Roth et al. 1991). These results are intriguing given the homology between C1 and Myb and B/R and bHLH proteins, and demonstrate that these sequences are important. They do not, however, demonstrate that B and C1 are directly interacting with these sites.

While mutational analyses have not been published on the other promoters, the juxtaposition of adjacent Myb-like and bHLH-like consensus DNA sequences as in *bz1* is not observed in the other promoters. It is possible that the consensus sites in *bz1* are fortuitous and B and C1 activation requires nucleotides within or nearby, but not the consensus sites in particular. The mutational analysis substituted the 6 nucleotides of each consensus sequence simultaneously (Roth et al. 1991), so it is not clear which of the altered nucleotides are having the greatest effect.

The detailed mechanism of transcriptional induction by B and C1 may not be identical for each promoter. For example, B and C1 may interact directly with *bz1*, but indirectly with one or more of the other promoters through interaction with additional proteins that bind DNA. One possibile explanation for the limited sequence conservation between the promoters might be that while all require B and C1 for their transcription, some may also be regulated by additional proteins. For example, *a1* is required for both phlobaphene and anthocyanin synthesis (Dooner et al. 1991). These two pathways share the *c2*, *chi*, and *a1* genes, while *a2*, *bz1*, and *bz2* are specific for the anthocyanin pathway. The phlobaphene pathway is regulated by P, which shares 70% identity with the C1 Myb domain (Grotewold et al. 1991). Complex regulation of *c2*, *a1*, and *chi* may require differentially evolved sequences relative to *bz1*. Given that the B and C1 proteins interact (Goff et al. 1992), if one protein can bind to a promoter, the other may also be in the transcriptional complex, even if it does not directly bind to the DNA. Thus, every promoter need not contain both B and C1 binding sites.

FUNCTIONAL ANALYSES OF THE REGULATORY PROTEINS

In vitro mutagenesis and transformation assays in maize and yeast cells have identified regions of C1 and B important for transcriptional activation. C1 contains a transcriptional activation domain and a region that interacts with B. In addition, the N-terminus of C1 contains a potential DNA-binding domain, originally identified in Myb proteins (Lüscher and Eisenman 1990). In proteins of the Myb class, each of the 51-53 amino acid repeats has a predicted helix-turn-helix motif (Gabrielsen et al. 1991). C1 contains two of these repeats (Figure 2). The C-terminal acidic region of C1 functions as a transcriptional activation domain in maize and yeast cells when fused to the DNA-binding domain of the yeast regulatory protein GAL4 (Goff et al. 1991). When the acidic C-terminus of C1 is replaced by the GAL4 transcriptional activation domain, the resulting fusion protein activated both specific anthocyanin target gene promoters and the entire anthocyanin biosynthetic pathway in maize cells (Goff et al. 1991). Activation of the pathway and the *bz1* and *a1* promoters by these C1-GAL4 fusions remained dependent on an expressed *B* allele (Goff et al. 1991). Mutational analyses also suggest that the acidic C-terminus is important for C1 function. Sequence analysis of the naturally occurring *c1* dominant-inhibitory allele, *C1-I*, revealed that it contains, among other alterations, a frameshift that alters the last 21 amino acids of the acidic transcriptional activation domain (Paz-Ares et al. 1990). An *in vitro* generated frameshift mutation in C1 that altered the 15 most C-terminal amino acids is a dominant inhibitor of the wild-type C1 protein (Goff et al. 1991).

Functional analysis of the maize B protein demonstrated that its amino-terminus is essential for function. This region interacts with the amino terminal region of the C1 protein, enabling transcriptional activation in both maize and yeast cells (Goff et al. 1992). Using deletion derivatives, the regions sufficient for this interaction have been mapped to amino acids 1-245 of B and amino acids 1-117 of C1 (Figure 2; Goff et al. 1992). These studies also demonstrated that the highly conserved bHLH motif (Fig. 2) could be deleted without complete loss of transcriptional activity. We hypothesize that B proteins missing the putative DNA-binding domain still contribute to the transcriptional complex via protein/protein interactions with C1. Fusions of various regions of B, including the central acidic region, to the GAL4 DNA-binding domain failed to reveal a transcriptional activation domain (Goff et al. 1992). Our current hypothesis is that a B/C1 complex activates promoters via the C1 transcriptional activation domain. As the interaction occurs in yeast on a heterologous promoter, B and C1 do not require their normal DNA binding sites or other maize proteins to interact.

Figure 2: C1 and B proteins. Diagonal hatches indicate regions within the two proteins involved in their interaction.

MUTATIONAL ANALYSES OF THE C1 INTERACTION DOMAIN

The ability of the N-terminus of B, when fused to the GAL4 DNA-binding domain, to recruit C1 to a heterologous promoter and form a functional activation complex in yeast cells provides a convenient means to isolate mutants that have reduced interaction (Figure 3). In an initial PCR mutagenesis (Leung et al. 1989) of the C1 N-terminus, mutant C1 cDNA fragments were used to replace the wildtype sequence of a C1/GAL4 transcriptional activation domain fusion. Clones were tested for their ability to interact with the B-GAL4 DNA-binding fusion protein and activate transcription in yeast. The C1 variants were subsequently cloned into plant expression vectors and their ability to activate the *bz1* promoter in the presence of B was assessed in maize cells.

The yeast experiment assays the B/C1 interaction in the absence of the activators' target DNA. The plant assay, which uses the target gene promoters, is likely to require both DNA binding and the interaction between B and C1. The mutated region contains C1's putative DNA-binding domain. Thus, it is possible that mutants that are defective in both DNA binding and the interaction with B will be identified. Other mutants may interact poorly with B, but normally with their target DNA.

Eight variants have been analyzed in detail. Five mutants showed wildtype C1 protein levels in yeast, but reduced or background levels of β-galactosidase activity (Table 1). These mutants are candidates for altered interactions with B. Interpretations of the mutant phenotypes are presented based on a model in which C1 activity at *bz1* is dependent on both C1 DNA binding and C1/B interaction. Isolates 91 and 94 showed a reproducible 2-3 fold reduction in the yeast assay, and a more severe phenotype in maize. This phenotype can be explained by a mild interaction defect apparent in both the yeast and maize assays and a DNA binding defect

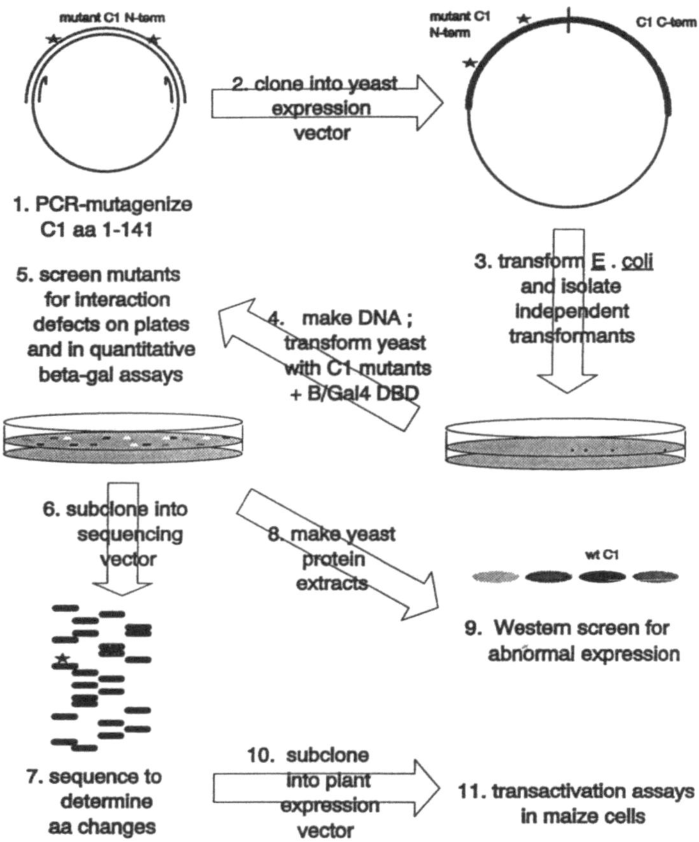

Figure 3. Protocol for Isolating C1 mutants defective in their interaction with B.

manifest only in maize. Isolate 6 showed no detectable interaction in yeast, but in maize the mutant phenotype was overcome by higher input DNA. This mutant can be explained as an interaction defect that decreases C1's affinity for B. Thus, at the low protein concentrations found in yeast and in the maize assays with low DNA, few mutant C1 molecules form a productive interaction with B. We suggest that at the 10 fold higher DNA concentrations, mass action drives the production of sufficient B/C1 complexes to give wildtype transactivation levels. Isolate #15 showed reduced activity in yeast, but was wild type in maize, even at the low DNA concentration. One interpretation for this mutation is that it has a

Table 1. *c1* mutants defective in their interaction with B.

c1 Clone	Amino Acid Substitutions	Yeast Interaction Phenotype	Maize Transactivation Phenotype LOW DNA	HIGH DNA
wt	none	100%	100%	100%
91	K_{105}→E	47%	0.4% +/- 0.06%	1% +/- 0.2%
94	V_{12}→A R_{14}→G V_{127}→A	34%	6%+ +/- 0.5%	5% +/- 0.4%
15	I_{78}→V N_{106}→S	9%	82% +/- 6%	100% +/- 6%
21	E_{10}→D A_{31}→T N_{109}→S	1%	1% +/- 0.1%	3% +/- 0.7%
6	N_{64}→H	1%	11% +/- 6%	121% +/- 9%

Yeast interaction assays (Fields and Song 1989), using cells containing a *Gal-lacZ* reporter gene, the B-GAL4 DNA-binding domain fusion protein, and the indicated C1 construct were performed. Wildtype C1 gives ≈ 500 Miller units (Miller 1972) of β-gal activity; background levels are ≤ 1% of wild-type levels. *b* and *c1* cDNA expression vectors were introduced into EG-5 suspension culture cells (Zurawski and Chourey 1981) by particle gun bombardment, together with 10 μg each of the pBz1Luc reporter (Klein et al. 1989) and a transformation efficiency control plasmid with a constitutively expressed CAT gene (Callis et al. 1987). C1 activity was calculated as a luciferase to CAT activity ratio. *c1* input DNA was either low (1μg) or high (10 μg). *b* DNA was present at twice the amount of *c1* DNA. Data represent 4 separate experiments normalized to wildtype C1 activity of 100%, +/- standard error of measurement.

reduced interaction with B in yeast, but this defect is compensated for in maize by an improved DNA-binding capacity or interaction with other maize proteins. Isolate 21 had background levels of activity in both yeast and maize. This mutant protein could be completely defective in its interaction with B. It is also possible that it is defective both in DNA binding and its interaction with B. Three mutants were not informative as Western blots showed they had low or undetectable amounts of C1 protein, resulting in low or no β-galactosidase activity. Additional C1 mutants are being examined, and hypotheses for the molecular basis of the mutant phenotypes will be tested with protein/protein and DNA/protein interaction assays.

CONCLUSIONS AND OPEN QUESTIONS

The B and C1 proteins can interact to form a functional transcription complex. This interaction requires the C1 Myb region, but does not require the B bHLH domain. The initial mutational analysis of the C1 Myb domain has identified several amino acids that may be important for the C1/B interaction. This functional interaction occurs both in maize and yeast cells demonstrating that other maize proteins are not required. However, it is possible that proteins in common to yeast and maize are mediating the interaction. The isolation of suppressor mutations in B that enable the C1 mutants to function would provide additional evidence that the B and C1 proteins directly interact.

B and C1 contain motifs similar to DNA-binding domains in other proteins, but there is not yet evidence that B and C1 directly interact with DNA. Various models for how B and C1 interact with the target gene promoters can be entertained. Either of the proteins could bind DNA, with the other protein becoming part of the transcriptional complex by protein/protein interactions. The DNA-binding protein could be made up of both B and C1 subunits. The B and C1 proteins may bind DNA efficiently only in the presence of a third protein. An alternative model is that B and C1 do not directly interact with the target DNA, but mediate their effects through interactions with other proteins that do bind the target gene promoters. While we think it less likely, it is also possible that B and C1 do not interact directly with the target gene promoters, but are required instead to activate another regulatory gene, whose product interacts directly with the target genes.

Biochemical assays for protein/DNA and protein/protein interactions together with further genetic experiments should begin to distinguish between these models. To investigate whether the various target genes are differentially regulated, it will be important to determine the minimum cis-acting sites in additional target gene promoters that are required for B and C1 regulation.

ACKNOWLEDGEMENTS

This research is supported by a NSF grant, DMB9004537 to VLC. MBS is a Howard Hughes Medical Institute Predoctoral Fellow. We thank Garth Patterson and Ken Kubo for comments on the manuscript.

LITERATURE CITED

Callis J, Fromm M, and Walbot V (1987) Introns increase gene expression in cultured maize cells. *Genes Dev*. **1**: 1183-1200.

Chandler VL, Radicella JP, Robbins TP, Chen J, and Turks D (1989) Two regulatory genes of the maize anthocyanin pathway are homologous: Isolation of *b* utilizing *r* genomic sequences. Plant Cell 1: 1175-1183.

Coe EH Jr, Hoisington DA and Neuffer MG (1988) The Genetics of Corn. In: Corn and Corn Improvements (Sprague, GF, Dudley, J, eds.) American Society of Agronomy, Madison, WI, pp. 88-258.

Cone KC, Burr FA and Burr B (1986) Molecular analysis of the maize anthocyanin regulatory locus *c1*. PNAS USA 83: 9631-9635.

Cone KC and Burr B (1989) Molecular and genetic analyses of the light requirement for anthocyanin synthesis in maize. In: The Genetics of Flavanoids. (DE Styles, GA Gavazzi, and ML Racchi, eds.) Edizioni Unicopli Milan. pp. 142-145.

Dellaporta S, Greenblatt I, Kermicle J, Hicks JB and Wessler S (1988) Molecúlar cloning of the maize *R-nj* allele by transposon-tagging with *Ac*. In: Chromosome structure and function: Impact of new concepts, 18th Stadler Genetics Symposium (ed. Gustafson JP and Appels R), pp. 263-282. Plenum Press, NY.

Dooner HK, Robbins TP and Jorgenson RA (1991) Genetic and developmental control of anthocyanin biosynthesis. Annu. Rev. Genet. 25: 173-199.

Fedoroff NV, Furtek DB and Nelson OE (1984) Cloning of the *bronze* locus in maize by a simple and generalizable procedure using the transposable element *Activator* (*Ac*). PNAS USA 81: 3825-3829.

Fields S and Song O (1989) A novel genetic system to detect protein-protein interactions. Nature 340: 245-246.

Gabrielsen OS, Sentenac A and Fromageot P (1991) Specific DNA binding by c-Myb: evidence for a double helix-turn-helix-related motif. Science 253: 1140-1143.

Goff SA, Klein TM, Roth BA, Fromm MF, Cone KC, Radicella JP and Chandler VL (1990) Transactivation of anthocyanin biosynthetic genes following transfer of *b* regulatory genes into maize tissues. EMBO J. 9: 2517-2522.

Goff SA, Cone KC, and Fromm MF (1991) Identification of functional domains in the maize transcriptional activator C1: comparison of wild-type and dominant inhibitor proteins. Genes Dev 5: 298-309.

Goff SA, Cone KC and Chandler VL (1992) Functional analysis of the transcriptional activator encoded by the maize *b* gene: Evidence for a direct functional interaction between two classes of regulatory proteins. Genes & Dev. 6: 864-875.

Grotewold E, Athma P and Peterson T (1991) Alternatively spliced products of the maize *p* gene encode proteins with homology to the DNA binding domain of *myb*-like transcription factors. Proc. Natl. Acad. Sci. USA 88:4587-91.

Grotewold E and Peterson T (1991) Cloning of a *P*-regulated chalcone-flavanone isomerase gene from maize. Maize Genet. Coop. Newslet. 65: 46-47.

Klein T, Roth BA and Fromm M (1989) Regulation of anthocyanin biosynthetic genes introduced into intact tissues by microprojectiles. PNAS USA 86: 6681-6685.

Leung DW, Chen E and Goeddel (1989) A method for random mutagenesis of a defined DNA segment using a modified polymerase chain

reaction. J. Methods Cell Mol. Bio 1: 11-15.
Ludwig SR, Habera LF, Dellaporta SL, and Wessler SR (1989) *Lc*, a member of the maize *r* gene family responsible for tissue-specific anthocyanin production, encodes a protein similar to transcriptional activators and contains the *myc*-homology region. Proc. Natl. Acad. Sci. 86: 7092-7096.
Lüscher B and Eisenman RN (1990) New light on Myc and Myb. Part II. Myb. Genes Dev. 4: 2235-2241.
McLaughlin M and Walbot V (1987) Cloning of a mutable *bz2* allele of maize by transposon tagging and differential hybridization. Genetics 117: 771-776.
Menssen A, Hohmann WM, Schnable PS, Peterson PA, and Saedler H (1990) The En/Spm transposable element of *Zea mays* contains splice sites at the termini generating a novel intron from a dSpm element in the *a2* gene. EMBO J. 9: 3051-3057.
Miller JH (1972) Assay of β-galactosidase. In: Experiments in molecular genetics, pp. 352-355. Cold Spring Harbor Laboratory, Cold Spring Harbor, N.Y.
O'Reilly C, Shepherd NS, Pereira A, Schwarz-Sommer Z, Bertram I, Robertson DS, Peterson PA, and Saedler H (1985) Molecular cloning of the *a1* locus of *Zea mays* using the transposable elements *En* and *Mu1*. The EMBO J. 4: 877-882.
Paz-Ares J, Wienand U, Peterson PA and Saedler H (1986) Molecular cloning of the *c* locus of *Zea mays*: a locus regulating the anthocyanin pathway. EMBO J. 5: 829-833.
Paz-Ares J, Ghosal D, Wienand U, Peterson P and Saedler H (1987) The regulatory *c1* locus of *Zea mays* encodes a protein with homology to *myb* proto-oncogene products and with structural similarities to transcriptional activators. EMBO J. 6: 3553-3558.
Paz-Ares J, Ghosal D, and Saedler H (1990) Molecular analysis of the *C1-I* allele from *Zea mays*: A dominant mutant of the regulatory *c1* locus. EMBO J. 9: 315-321.
Perrot GH and Cone KC (1989) Nucleotide sequence of the maize *R-S* gene. Nucleic Acids Res. 17: 8003.
Radicella JP, Turks D and Chandler VL (1991) Cloning and nucleotide sequence of a cDNA encoding *B-Peru*, a regulatory protein of the anthocyanin pathway in maize. Plant Mol. Biol. 17: 127-130.
Roth BA, Goff SA, Klein TM and Fromm ME (1991) *C1*- and *R*-dependent expression of the maize *bz1* gene requires sequences with homology to mammalian *myb* and *myc* binding sites. Plant Cell 3: 317-325.
Styles ED, Ceska O and Seah KT (1973) Developmental differences in action of *r* and *b* alleles in maize. Can. J. Genet. Cytol. 15: 59-72.
Theres N, Scheele T and Starlinger P (1987) Cloning of the *Bz2* locus of *Zea mays* using the transposable element *Ds* as a gene tag. Mol. Gen. Genet. 209: 193-197.
Wienand U, Weydemann U, Niesbach-Kloesgen U, Peterson PA and Saedler H (1986) Molecular cloning of the *c2* locus of *Zea mays*, the gene coding for chalcone synthase. Mol. Gen. Genet. 203: 202-207.
Zurawski DB and Chourey P (1981) Callus formation from protoplasts of a maize cell culture. Theor. & Appl. Genetics 59: 341-344.

THE *phy* GENE FAMILY: FUNCTION AND EXPRESSION

Peter H. Quail, Margaret T. Boylan, Katayoon Dehesh, Jorge Nieto-Sotelo*, Brian M. Parks†, James M. Tepperman, David E. Somers and Doris Wagner

U.C. Berkeley/USDA Plant Gene Expression Center
800 Buchanan Street
Albany, CA 94710 USA

Introduction

Plants constantly monitor and adapt to the light environment in which they find themselves (Kendrick and Kronenberg, 1986). To accomplish this task they employ a set of regulatory photoreceptors of which phytochrome is the best characterized (Quail, 1991, 1993). Although phytochrome had long been thought of by many as a single homogeneous entity, the complexities of environmental light signals and the diversity of plants' responses to them had over the years increasingly suggested to photophysiologists the possible involvement of more than one phytochrome in mediating these responses (Hillman, 1967; Smith and Whitelam, 1990; Smith, 1992). Indeed, parallel biochemical, immunochemical and spectroscopic studies provided steadily accumulating evidence for the existence of at least two molecular species of the photoreceptor (Furuya, 1989). Most recently, molecular evidence that phytochrome is in fact a family of photoreceptors encoded by five divergent genes, designated *phyA*, *phyB*, *phyC*, *phyD* and *phyE*, was obtained from studies with *Arabidopsis* (Sharrock and Quail, 1989; R. A. Sharrock, pers. comm.).

Cognizance of the existence of multiple molecular species of phytochrome has highlighted a number of basic questions regarding the photoreceptor: Do individual family members have distinct photobiological functions and/or molecular mechanisms of action? Do the different phytochromes exhibit similar, distinctive or overlapping patterns of expression and what are the molecular determinants of these patterns? This chapter examines recent studies directed at addressing some of these questions.

---

*Present address: Dept. de Biología Mol. de Plantas, Inst. de Biotecnologia/U.N.A.M., Cuernavaca, Mor 62271, Mexico.
†Present address: Dept. of Plant Biology, Ohio State University, Columbus, OH 43210.

NATO ASI Series, Vol. H 81
Plant Molecular Biology
Edited by G. Coruzzi and P. Puigdomènech

Specificity of Photosensory Function

Emerging evidence from studies with photomorphogenic mutants and transgenic overexpressors indicates that at least some individual members of the phytochrome family have discrete photosensory roles in regulating plant photomorphogenesis. The *hy*3 long hypocotyl mutants of *Arabidopsis* have been shown by western blot analysis with type-selective monoclonal antibodies to be deficient in phytochrome B (Somers *et al.*, 1991), and sequence analysis has shown that this deficiency results from mutations in the *phyB* gene (Reed *et al.*, 1993). Similarly, recently isolated *hy*8 *Arabidopsis* mutants have been shown to be deficient in functional phytochrome A (Parks and Quail, 1993) as a consequence of lesions in the *phyA* gene (K. Dehesh, unpublished). The *hy*3 mutant is unresponsive to continuous red light whereas the *hy*8 mutant exhibits wild-type responsiveness to these irradiation conditions (Koornneef *et al.* 1980; McCormac *et al.*, 1993; Parks and Quail, 1993). These data establish (a) that phytochrome B is necessary for continuous red-light perception, and (b) that phytochrome A is neither necessary nor sufficient for continuous red-light perception in seedling morphogenesis. Conversely, the *hy*8 mutant is unresponsive to continuous far-red light whereas the *hy*3 mutant exhibits wild-type responsiveness to this irradiation regime (Koornneef *et al.*, 1980; McCormac *et al.*, 1993; Parks and Quail, 1993). These data demonstrate (a) that phytochrome A is necessary for perception of continuous far-red light, and (b) that phytochrome B is neither necessary nor sufficient for continuous far-red light perception.

Data from transgenic *Arabidopsis* plants overexpressing either phytochromes A or B complement and support these results. Phytochrome B overexpressors are hypersensitive to red light, but retain wild-type responsiveness to continuous far-red (Wagner *et al.*, 1991; McCormac *et al.*, 1993). This result verifies that phytochrome B actively mediates red-light controlled responses but exhibits no intrinsic capacity to mediate responsiveness to continuous far-red light. By contrast, phytochrome A overexpressors are hypersensitive to continuous far-red (Boylan and Quail, 1991; Whitelam *et al.*, 1992; McCormac *et al.*, 1993) indicating that the capacity to mediate responsiveness to these irradiation conditions is an intrinsic property of the phytochrome A molecule. The assembled data indicate, therefore, that phytochromes A and B have reciprocal responsivities to continuous far-red and continuous red light, respectively, and therefore perform complementary photosensory functions in controlling photomorphogenesis in response to light in these regions of the spectrum. It can be concluded in addition that none of the other phytochromes C, D or E can substitute for phytochromes A or B in executing these respective functions.

The physiological consequences of deficiencies in phytochromes A or B enable definition of the functional role each plays in the wild type plant. The lack of responsiveness to red light in the *hy*3 mutant results in these plants lacking "shade avoidance" and "end-of-day far-red" responses (Nagatani *et al.*, 1991; Whitelam and Smith, 1991). These responses represent, effectively, the reversal of the normal inhibition of elongation growth imposed by red light. Phytochrome B therefore, appears to be principally, if not exclusively, responsible for the "proximity perception" function of phytochrome so important to adaptive development in fully green plants, as elegantly delineated by Smith and colleagues (Smith, 1982; Smith and Whitelam, 1990; Smith, 1992). By contrast, the lack of responsiveness of *hy*8 to continuous far-red light indicates that phytochrome A is principally, if not exclusively, responsible for the "far-red high irradiance response" of etiolated seedlings (Holmes and Schäfer; Smith and Whitelam, 1990). This response may provide a transient competitive advantage to newly emergent seedlings that penetrate the soil surface into a vegetatively shaded environment (McCormac *et al.*, 1992; Parks and Quail, 1993). Although the photosensory roles of phytochromes C, D and E are yet to be defined, it would not be unexpected if these family members also specialize in performing other discrete facets of the diverse spectrum of functions attributed to the photoreceptor, such as regulating seed germination or floral initiation.

phy Gene Expression

One factor that could contribute to potential differences in regulatory functions between individual phytochromes might be divergence in spatial, temporal and/or environmentally-regulated patterns of expression of the different family members. The evidence available for *phyA*, *phyB* and *phyC* indicates that indeed both level of expression and photoresponsiveness varies substantially among these genes. It is well-established that *phyA* is expressed strongly in darkness and is down-regulated in response to light in many plant species (Quail, 1991, 1993). In contrast, although the data are currently limited, *phyB* and *phyC* are expressed at relatively low levels which do not change in response to light. This pattern is illustrated for *Arabidopsis* in Figure 1A (Sharrock and Quail, 1989). In dark-grown seedlings *phyA* mRNA is present at considerably higher levels (~20-fold) than the *phyB* or *phyC* transcripts. Irradiation of seedlings with continuous white light leads to a strong reduction in the level of *phyA* mRNA, but does not significantly affect *phyB* or *phyC* transcript levels. Thus, expression of the *phyA* gene is negatively photoregulated whereas *phyB* and *phyC* are constitutively expressed. A similar pattern is observed at the protein level (Figure 1B; Somers *et al.*, 1991). The phytochrome A, B and C proteins are present at relative levels of approximately 500:10:1 in dark-grown seedlings. Exposure to light induces rapid degradation of phytochrome A as a result of the well known phenomenon of Pfr instability *in vivo* (Quail, 1991). By contrast phytochromes B and C are apparently stable in the Pfr form resulting in constant levels of these family members in light and dark. Similar behavior has been documented, at least for *phyA* and *phyB*, in the monocot rice (Dehesh *et al.*, 1991) suggesting that differential photoregulation of the *phy* gene family occurs generally in monocots and dicots.

Although the precise pattern of photoregulation of *phyA* varies between plant species (Quail, 1993), this response system has many features which make it an attractive paradigm for approaching the general question of the molecular basis of phytochrome-regulated gene expression. This system has been best characterized in monocots where there is clear evidence for strong, negative autoregulation of *phyA* expression by phytochrome A itself. As illustrated in Figure 2, pulse-irradiation of etiolated oat seedlings triggers a decrease in *phyA* transcription detectable between 2 and 5 min from Pfr formation (Lissemore and Quail, 1988). A decline in mRNA and protein levels follows as a result of the intrinsic rapid turnover rates of the *phyA* transcripts (Colbert *et al.*, 1985; Seeley *et al.*, 1991) and the Pfr form of the protein (Colbert *et al.*, 1983), respectively. The rapid repression of *phyA* transcription occurs in the absence of new protein synthesis (Lissemore and Quail, 1988). This result indicates that *phyA* responsiveness involves a primary transduction chain between photoreceptor and promoter where all necessary components are present in the cell before Pfr formation.

Using linker-scan mutagenesis and a microprojectile-based transient expression assay we have defined three functionally active DNA elements in the oat *phyA3* promoter

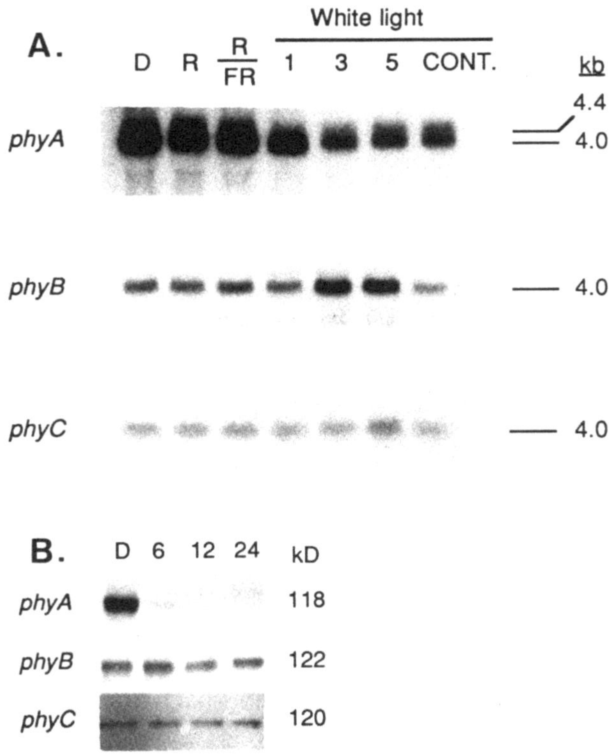

Fig. 1. Phytochrome mRNA and protein levels in *Arabidopsis* seedlings in response to light. A. Blot hybridization analysis of the *phy* mRNAs present in total RNA from *Arabidopsis* seedlings. RNA was isolated from seedlings that were grown for 5 days completely in the dark (D), given a pulse of red (R) or red followed by far red (R/FR) light 3 hr prior to harvest, transferred to white light for 1, 3, or 5 hr prior to harvest, or grown for 5 days in continuous white light (CONT). Blots were probed with ssDNA transcript-specific probes for *phyA*, *phyB*, and *phyC* mRNAs. (From Sharrock and Quail, 1989). **B.** Western blot analysis of phytochromes present in *Arabidopsis* seedling extracts. Phytochrome-enriched extracts from 7-day-old, dark-grown seedlings treated before harvest with 0 (D), 6, 12, or 24 hr of continuous red light. Immunoblot lanes were loaded with ammonium sulfate-precipitated fractions from 300 μg (rows *phyA* and *phyB*) and 1.2 mg (row *phyC*) of crude extract protein, and the blots were probed with the appropriate pool of monoclonal antibodies selective for the individual phytochrome species. (From Somers *et al.*, 1991).

involved in expression and photoregulation (Figure 3; Bruce *et al.*, 1991). Two of these elements, PE1 and PE3, function synergistically to support high transcription rates in the derepressed state, i.e., in dark-grown plants. The third element, RE1, mediates the transcriptional repression imposed by light-triggered Pfr formation.

To begin to understand this regulation we have isolated and characterized, from oats and rice, cDNAs encoding related proteins, designated oat PF1 (O-PF1) and rice PF1

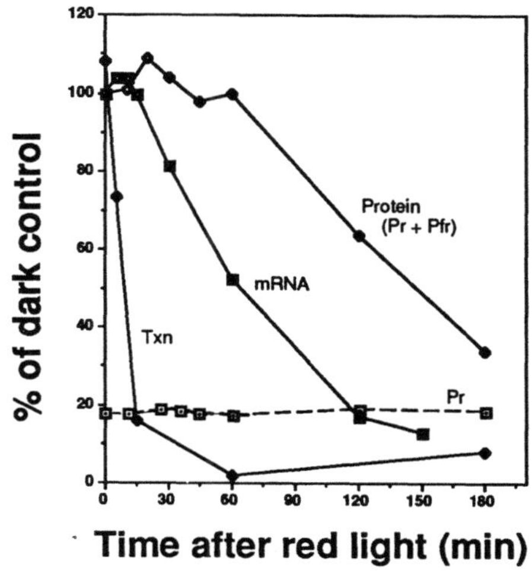

Fig. 2. Time-course of the changes in levels of *phy* gene transcription, *phy* mRNA, and phytochrome protein (presumably *phyA*; Quail, 1991) in etiolated oat seedlings in response to a saturating pulse of red light given at time zero. Transcription rates were measured by run-on transcription assay (Lissemore and Quail, 1988), *phy* mRNA levels by RNA blot analysis (Colbert *et al.*, 1985), and phytochrome protein (including levels of Pr and Pfr forms) by spectrophotometric assay (Colbert *et al.*, 1983).

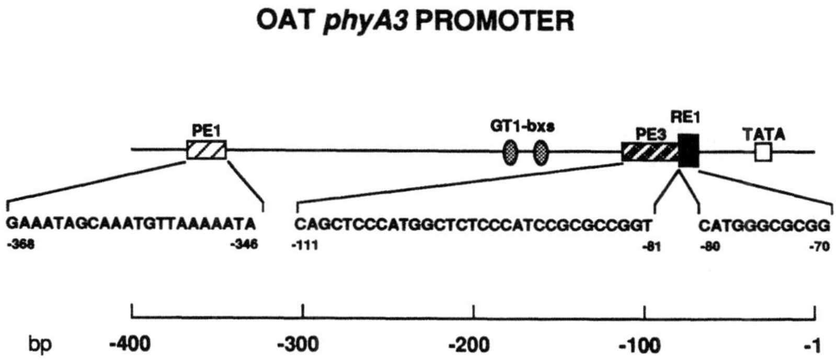

Fig. 3. Physical map of the oat *phyA3* promoter showing the locations and sequences of functionally active *cis*-elements, PE1 (positive element 1), PE3 (positive element 3), and RE1 (repressor element 1), defined by linker-scan mutagenesis in a microprojectile-mediated gene transfer assay (Bruce *et al.*, 1991). Also indicated are two GT1-boxes (GT1-bxs) related to GT-motifs in other promoters (see Fig. 5) but apparently functionally inactive in the oat *phyA3* promoter context (Bruce and Quail, 1990). Modified from Bruce *et al.* (1991).

(R-PF1), that bind in sequence-specific fashion to the PE1 element (Nieto-Sotelo and Quail, 1993; Nieto-Sotelo *et al.*, 1993). The structural features of these proteins are summarized in Figure 4. Although differing in length, the two proteins have extensive domains of sequence identity ranging between 60 and 71 percent (overall 67 percent).

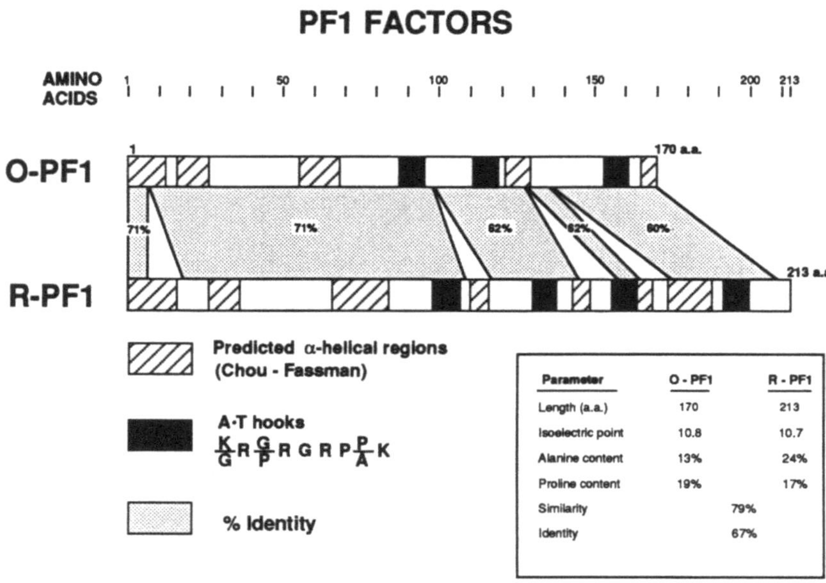

Parameter	O - PF1	R - PF1
Length (a.a.)	170	213
Isoelectric point	10.8	10.7
Alanine content	13%	24%
Proline content	19%	17%
Similarity	79%	
Identity	67%	

Fig. 4. Physical maps of recombinant PF1 factors cloned from oats (O-PF1) (Nieto-Sotelo and Quail, 1993) and rice (R-PF1) (Nieto-Sotelo *et al.*, 1993). Regions of sequence similarity between the two proteins are indicated (grey shading), as well as predicted α-helical regions and repeat nonapeptide sequences with strong similarity to A•T hook motifs in other proteins (Churchill and Travers, 1991). Consensus sequence for the A•T hook motifs in the PF1 factors is shown. Inset: selected features of the two proteins.

The most striking feature, however, is a series of 3 (O-PF1) or 4 (R-PF1) repeats of 9 amino acids previously defined as the "A·T hook" motif in other proteins, such as the HMGI/Y family (Eckner and Birnstiel, 1989; Johnson *et al.*, 1989) and the protein SB16 of soybean (Laux *et al.*, 1991). These other proteins bind to A-T-rich stretches of DNA, and the "A·T hook" nonapeptide repeat has been shown to be responsible for this DNA-binding activity (Churchill and Travers, 1991). The PE1 element of *phyA3* is also A-T-rich consistent with this pattern of DNA-protein interactions (Figure 3). Using COOH-terminal deletion derivatives of rice PF1 we have found that removal of the two A·T hooks closest to the COOH-terminus does not affect DNA-binding capacity, whereas removal of all three COOH-terminal A·T hooks eliminates detectable binding (Nieto-Sotelo, *et al.*, 1993). These results are consistent with the suggestion that one or more of the A·T hooks of PF1 are involved in PE1 recognition. To our knowledge, PF1 is the first cloned A·T-hook containing plant factor shown to bind to a functionally-defined *cis*-acting promoter element (Katagiri and Chua, 1992).

The general structure of the rice *phyA* promoter appears to be similar to that of the oat *phyA* gene, except that no AT-rich element equivalent to PE1 is present in the rice minimal promoter (Kay *et al.*, 1989). Instead, a complex of related GT-containing elements, designated GT1-box, GT2-box and GT3-box (Fig. 5), has been shown to function as a positive element in rice *phyA* expression (Dehesh *et al.*, 1990). We have obtained a recombinant clone from rice encoding a full-length protein, designated GT-2, that binds in a highly sequence-specific fashion to elements in the GT-box complex (Dehesh *et al.*, 1990, 1992). The DNA target-sequence specificity parallels that of a factor detected in rice nuclear extracts, namely, a high affinity for the GT2-box and GT3-box motifs, but relatively low affinity for the GT1-box. In addition, antibodies prepared against the recombinant protein recognize a polypeptide in rice nuclear extracts identical in size to that of recombinant GT-2. Together, these data support the conclusion that the cloned GT-2 protein is likely to represent the nuclear factor that interacts with the functionally defined GT-box *cis*-elements *in vivo* to facilitate rice *phyA* expression (Dehesh *et al.*, 1990, 1992).

A striking feature of the rice GT-2 protein is a repeated sequence of 75 amino acids present once in the NH_2-terminal and once in the COOH-terminal domain of the polypeptide (Fig. 6). This repeat, designated the trihelix motif, overlaps a strongly basic segment of polypeptide and is predicted to fold into three amphipathic α-helices separated from each other by two short loops (Dehesh *et al.*, 1992). This structure is reminiscent of that of the homeodomain proteins, but is unrelated in primary sequence to these proteins, or any others currently in the databases, with the single exception discussed below. Initial experiments to map the DNA-binding domains of GT-2 have shown that each of the NH_2-terminal and COOH-terminal domains containing the trihelix motif can bind independently to the GT-box complex (Dehesh *et al.*, 1992). More intriguing, however, is the evidence that the NH_2-terminal domain binds primarily to the GT3-box, the COOH-terminal domain primarily to the GT2-box, and neither domain to the GT1-box (Fig. 6). Extension of the NH_2-terminal domain to include a central region (that alone does not bind) results in weak binding to the GT1-box. Taken together, the data indicate that GT-2 represents a new class of factor with twin, autonomous DNA-binding domains that can discriminate with high specificity between closely related target sites. Because the trihelix motifs are the only sequences present in both domains, it appears likely that these motifs have a primary role in the DNA-binding activity of GT-2.

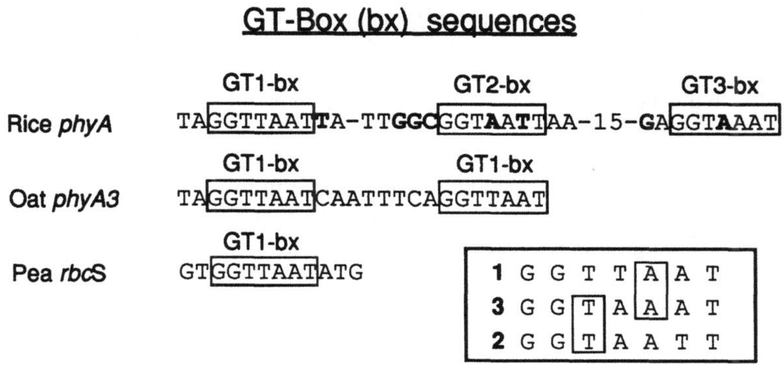

Fig. 5. Sequence comparisons of GT-motifs found in the promoters of the rice *phyA* (Kay *et al.*, 1989), oat *phyA3* (Hershey *et al.*, 1987), and pea *rbc*S-3A genes (Gilmartin *et al.* 1990). The core motifs are boxed, and the three different motifs in the rice *phyA* promoter are designated GT1-bx, GT2-bx, and GT3-bx. Fifteen base pairs (15) have been omitted between GT2-bx and GT3-bx. The bold letters indicate nucleotide differences between the GT2-bx and GT3-bx of rice *phyA* and the oat *phyA3* sequence. Inset shows a direct comparison of the GT1-bx, GT2-bx, and GT3-bx core sequences from the rice *phyA* promoter.

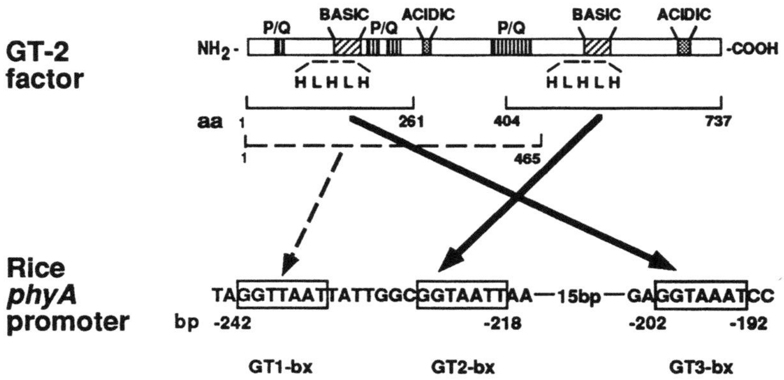

Fig. 6. Physical map of the rice GT-2 protein and its binding affinity for the three related GT-bx motifs in the rice *phyA* promoter (Dehesh *et al.*, 1992). Regions of the polypeptide rich in basic, acidic, or proline (P) + glutamine (Q) amino acid residues are indicated. The duplicated sequence predicted to fold into a helix-loop-helix-loop-helix (HLHLH) structure, designated the trihelix motif, is shown as occurring once each in the NH_2-terminal and COOH-terminal domains. Heavy arrows indicate the strong, autonomous binding preferences that the separately expressed NH_2-terminal (residues 1-261) and COOH-terminal (residues 404-737) domains of GT-2 exhibit toward the GT3-bx and GT2-bx motifs, respectively. Neither domain binds to the GT1-bx detectably, and the central domain (residues 262-403) exhibits no detectable, autonomous DNA binding activity. Dashed arrow indicates weak binding of extended NH_2-terminal domain (residues 1-465). Analogous extension of the COOH-terminal domain (residues 261-737) does not alter target sequence specificity.

A tobacco recombinant factor, designated GT-1a or B2F, containing a single region with sequence similarity to the twin trihelix motifs of GT-2, was recently reported by two groups simultaneously (Gilmartin *et al.*, 1992; Perisic and Lam, 1992). This tobacco factor has affinity for the GT1-box (see Fig. 5), originally identified as important to pea *rbcS* gene expression (Lam and Chua, 1990), but apparently does not recognize the GT2-box. These different factors do not appear to reflect a monocot-dicot difference because we have isolated a homolog of rice GT-2 from *Arabidopsis* (Kuhn *et al.*, 1993). Thus, plants may have a small family of related GT-factors that can select with high resolution among strongly similar sequence motifs. This consideration is consistent with the observation that the pair of GT1-boxes in the oat *phyA3* promoter (Figs. 3 and 5) were found by gene transfer assay to be functionally inactive in this promoter context (Bruce and Quail, 1990; Bruce *et al.*, 1991). The data imply that the weak recognition of the GT1-box by GT-2 (Fig. 6) is insufficient to support transcriptional activation from this motif, and that the interaction of any presumptive monocot homolog of GT-1a/B2F with the GT1-boxes is incapable of transactivating the promoter.

Acknowledgments

We are grateful to the many colleagues who contributed to the work from this laboratory cited here, and to R. A. Wells for preparing and editing this manuscript. Our research is supported by National Institutes of Health grant no. GM47475; Department of Energy grant no. PR03-92ER13742; USDA National Research Initiative Competitive Grants Program grant no. 92-37301-7678; National Science Foundation grant no. MCB 9220161; and USDA/ARS CRIS grant no. 5335-21000-006-00D.

References

Boylan MT, Quail PH (1991) Phytochrome A overexpression inhibits hypocotyl elongation in transgenic *Arabidopsis*. Proc. Natl. Acad. Sci. USA 88:10806-10810

Bruce WB, Quail PH (1990) Cis-acting elements involved in photoregulation of an oat phytochrome promoter in rice. Plant Cell 2:1081-1089

Bruce WB, Deng X-W, Quail PH (1991) A negatively acting DNA sequence element mediates phytochrome-directed repression of *phyA* gene transcription. EMBO J. 10:3015-3024

Churchill ME A., Travers AA (1991) Protein motifs that recognize structural features of DNA. Trends Biochem. Sci. 16:92-97

Colbert JT, Hershey HP, Quail PH (1983) Autoregulatory control of translatable phytochrome mRNA Levels. Proc. Natl. Acad. Sci. USA 80:2248-2252

Colbert JT, Hershey HP, Quail PH (1985) Phytochrome regulation of phytochrome mRNA abundance. Plant Mol. Biol. 5:91-102

Dehesh K, Bruce WB, Quail PH (1990) A trans-acting factor that binds to a GT-motif in a phytochrome gene promoter. Science 250: 1397-1399

Dehesh K, Tepperman J, Christensen AH, Quail PH (1991) *phyB* is evolutionarily conserved and constitutively expressed in rice-seedling shoots. Mol. Gen. Genetics 225:305-313

Dehesh K, Hung H, Tepperman JM, Quail PH (1992) GT-2: A transcription factor with twin autonomous DNA-binding domains of closely related but different target sequence specificity. EMBO J. 11:4131-4144

Eckner R, Birnstiel ML (1989) Cloning of cDNAs coding for human HMG I and HMG Y proteins: both are capable of binding to the octamer sequence motif. Nucl. Acids Res. 17:5947-5959

Furuya M (1989) Molecular properties and biogenesis of phytochrome I and II. Adv. Biophys. 25:133-167

Gilmartin PM, Sarokin L, Memelink J, Chua NH (1990) Molecular light switches for plant genes. Plant Cell 2:369-378

Gilmartin PM, Memelink J, Hiratsuka K, Kay SA, Chua NH (1992) Characterization of a gene encoding a DNA binding protein with specificity for a light-responsive element. Plant Cell 4:839-849

Hershey HP, Barker RF, Idler KB, Murray MG, Quail PH (1987) Nucleotide sequence and characterization of a gene encoding the phytochrome polypeptide from *Avena*. Gene 61:339-348

Hillman WS (1967) The physiology of phytochrome. Ann. Rev. Plant Physiol. 18:301-324

Holmes MG, Schäfer E (1981) Action spectra for changes in the "high irradiance reaction" in hypocotyls of *Sinapis alba* L. Planta 153:267-272

Johnson KR, Lehn DA, Reeves R (1989) Alternative processing of mRNAs encoding mammalian chromosomal high-mobility-group proteins HMG-I and HMG-Y. Mol. Cell. Biol. 9:2114-2123

Katagiri F, Chua N-H (1992) Plant transcription factors: present knowledge and future challenges. Trends in Genetics 8:22-27

Kay SA, Keith B, Shinozaki K, Chua NH (1989) The sequence of the rice phytochrome gene. Nucl. Acids Res. 17:2865-2866

Kendrick RE, Kronenberg GHM. (1986) "Photomorphogenesis in Plants". Dordrecht:Martinus Nijhoff

Koornneef M, Rolff E, Spruit C (1980) Genetic control of light-inhibited hypocotyl elongation in *Arabidopsis* thaliana (L.) Heynh. Z. Pflanzenphysiol. 100:147-160

Kuhn RM, Caspar T, Dehesh K, Quail PH (1993) DNA binding factor GT-2 from *Arabidopsis*. Plant Mol. Biol. submitted

Lam E, Chua NH (1990) GT-1 binding site confers light responsive expression in transgenic tobacco. Science 248:471-473

Laux T, Seurinck J, Goldberg RB (1991) A soybean embryo cDNA encodes a DNA binding protein with histone and HMG-protein-like domains. Nucl. Acids Res. 19:4768

Lissemore JL, Quail PH (1988) Rapid transcriptional regulation by phytochrome of the genes for phytochrome and chlorophyll a/b-binding protein in *Avena sativa*. Mol. Cell. Biol. 8:4840-4850

McCormac AC, Whitelam GC, Boylan MT, Quail PH, Smith H (1992) Contrasting responses of etiolated and light-adapted seedlings to red:far-red ratio:A comparison of wild type, mutant and transgenic plants has revealed differential functions of members of the phytochrome family. J. Plant Physiol. 140:707-714

McCormac AC, Wagner D, Boylan MT, Quail PH, Smith H, Whitelam GC (1993) Photoresponses of transgenic *Arabidopsis* seedlings expressing introduced phytochrome B-encoding cDNAs: Evidence that phytochrome A and phytochrome B have distinct photoregulatory functions. Plant J. in press

Nagatani A, Chory J, Furuya M (1991) Phytochrome B is not detectable in the *hy3* mutant of *Arabidopsis*, which is deficient in responding to end-of-day far-red light treatments. Plant Cell Physiol. 32:1119-1122

Nieto-Sotelo J, Quail PH (1993) Cloning and characterization of cDNAs encoding oat PF1: a protein that binds to the PE1 region in the oat *phyA3* gene promoter, In Bowles, DJ, Gilmartin, PM (eds.) "Molecular Botany: Signals and the Environment" London:Portland Press Ltd., in press

Nieto-Sotelo J, Ichida A, Quail PH (1993) PF1: an "A·T hook" containing DNA binding protein from rice that interacts with a dA-dT-rich functional element in the oat *phyA3* gene promoter. Mol. Cell. Biol., submitted

Parks BM, Quail PH (1993) *hy8*, a new class of *Arabidopsis* long-hypocotyl mutants deficient in functional phytochrome A. Plant Cell 5:39-48

Perisic O, Lam E (1992) A tobacco DNA binding protein that interacts with a light-responsive Box II element. Plant Cell 4: 831-838

Quail PH (1991) Phytochrome: A light-activated molecular switch that regulates plant gene expression. Ann. Rev. Genet. 25:389-409

Quail PH (1993) Phytochrome: *phy* genes and their expression, In Kendrick RE, Kronenberg GHM. (eds.) "Photomorphogenesis in Plants" Amsterdam:Martinus Nijhoff, in press

Reed JW, Nagpal P, Poole DS, Furuya M, Chory J (1993) Mutations in the gene for the red/far-red light receptor phytochrome B alter cell elongation and physiological responses throughout *Arabidopsis* development. Plant Cell 5:147-157

Seeley KA, Byrne DH, Colbert JT (1992) Red light-independent instability of oat phytochrome mRNA in vivo. Plant Cell 4:29-38

Sharrock RA, Quail PH (1989) Novel phytochrome sequences in *Arabidopsis* thaliana:Structure, evolution, and differential expression of a plant regulatory photoreceptor family. Genes Develop. 3:1745-1757

Smith H (1982) Light quality, photoreception, and plant strategy. Ann. Rev. Plant Physiol. 33:481-518

Smith H (1992) The ecological functions of the phytochrome family. Clues to a transgenic programme of crop improvement. Photochem. Photobiol. 56:815-822

Smith H, Whitelam GC (1990) Phytochrome, a family of photoreceptors with multiple physiological roles. Plant, Cell, Environ. 13:695-707

Somers DE, Sharrock RA, Tepperman JM, Quail PH (1991) The *hy3* long hypocotyl mutant of *Arabidopsis* is deficient in phytochrome B. Plant Cell 3:1263-1274

Wagner D, Tepperman JM, Quail PH (1991) Overexpression of phytochrome B induces a short hypocotyl phenotype in transgenic *Arabidopsis*. Plant Cell 3:1275-1288

Whitelam GC, Smith H (1991) Retention of phytochrome-mediated shade-avoidance responses in phytochrome-deficient mutants of *Arabidopsis*, cucumber, and tomato. J. Plant Physiol. 139:119-125

Whitelam GC, McCormac AC, Boylan MT, Quail PH (1992) Photoresponses of *Arabidopsis* seedlings expressing an introduced oat *phyA* cDNA:persistence of etiolated plant type repsonses in light-grown plants. Photochem. Photobiol. 56:617-621

CHARACTERIZATION OF PROTEIN FACTORS INTERACTING WITH THE rDNA PROMOTER REGION

Manuel Echeverria, Dominique Tremousaygue,Thomas Roscoe, Katia Wassmann, Paul Penon and Michel Delseny
University of Perpignan, Plant Physiology and Molecular Biology Lab, CNRS Unit 565,Av de Villeneuve, 66860 Perpignan, France

INTRODUCTION

The nuclear genes coding for ribosomal RNA are transcribed by the RNA-polymerase I system.This transcriptional machinery is,by far,much less documented than the RNA-polymerase II system which is used for protein coding genes. Before the various components of the RNA -polymerase I transcription system could be analyzed, it was necessary to elucidate the fine structure of ribosomal RNA genes in higher plants and to map the site(s) for initiation of transcription.This has been an important challenge from 1985 up to now. At the moment, rDNA spacer sequence is available for more than 15 higher plant species and the major site of transcription initiation has been determined for more than half of them.The next step in the analysis of RNA-polymerase I transcription is to identify the regulatory *cis* -acting sequences as well as the corresponding *trans* - activating factors. There are three classical and complementary methods to reach this goal .

a) make fusions between the studied promoter (or *in vitro* mutated promoter) and a reporter gene and re-introduce the chimeric construction into a cell. Analysis of the expression of the reporter gene under the control of the native or modified promoter usually allows one to identify important regulatory regions.

b) use an *in vitro* transcription system in which the various regulatory elements and purified or recombinant factors can be assayed.

c) identify, by biochemical methods, proteins which interact with a well defined region of the promoter.

Although these three approaches have been extensively used with animal RNA-polymerase II - transcribed genes, there are still many difficulties to overcome with plant genes.The situation is even worse with the ribosomal genes .So far, it has not been possible to overexpress a reporter gene under the control of a plant rDNA promoter and until very recently no *in vitro* plant RNA-polymerase I transcription system was available. Accordingly, only the last alternative was left and has been explored by a few groups.This paper summarizes and compares present results in plants with our knowledge of animal rDNA gene transcription.

NATO ASI Series, Vol. H 81
Plant Molecular Biology
Edited by G. Coruzzi and P. Puigdomènech

RESULTS

1)Organization of the higher plant rDNA intergenic spacers

The first plant rDNA spacers to be described were those of maize (McMullen et al. 1986; Toloczyki and Feix 1986) and wheat (Lassner and Dvorak 1986) for cereals and that of radish (Delcasso-Tremousaygue et al .1988) for dicots . These sequences revealed the presence of short tandem repeat sequences upstream of

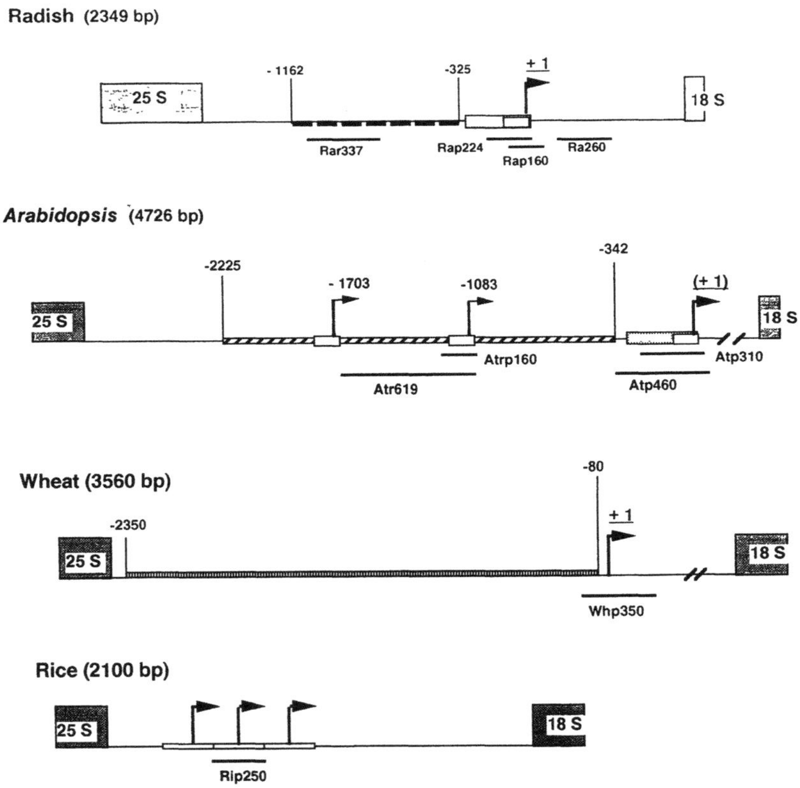

Fig.1. Structure of 4 plant rDNA spacers. (+1) indicates the initiation site when determined. Other arrows indicate putative initiation sites. Spacer repeats are indicated by thicker bars .The upstream box in radish is conserved in *Arabidopsis* at the same position, as well as a shorter sequence (open box) which is repeated twice upstream in the *Arabidopsis* spacer. The various probes used for gel shift assays are indicated although slightly different nomenclature is used in Fig.3

the transcription initiation site as was previously observed in animal rDNA spacers.These short repeats turned out to be highly specific for a given species and variation in their copy number accounts for a large part of the heterogeneity between rDNA genes among and within species (Cordesse et al.1993). By now, more than fifteen rDNA spacer sequences have been determined and, although there are some striking differences they all have in common a general organization with a set (or several sets) of repeats in the central part of the spacer as illustrated for in Fig. 1 . Additional repeats can be located elsewhere in the spacer. Reviews of the rDNA spacer organization have been published (Hemleben et al.,1988 ;Cordesse et al.,1993). Usually the sequence corresponding to the initiation site is well conserved as shown in Fig. 2. Indeed , a similar sequence is conserved at the initiation site in animal cells. From this limited number of rDNA sequence information , it is obvious

```
RICE 1                TGGCCCACGACTATAGTAGGGGGGAAGGGA
RICE 2                TGGCCCTAGACTATAGTAGGGGGGAAGGGA
RICE 3                TGGCCCTAGCCTATAGTAGGGGTGAGCGGA
RICE 4                GGCACGTAGCCTATAccgGGccGGgGGGaG
MAIZE                 TGCCCCTCaggTATAGTAGGGGGtAGGGaA
WHEAT                 aaaCCCTCGggTATAGTAGGGaGGAGGGGt
RYE                   aaaCCCTCGggTATAGTAGGGaGGAGGGGt

ARABIDOPSIS 1              TAAAGCTATA-TAGGGGTGGGT
ARABIDOPSIS 2              TTAAGCTATA-TAGGGGGGTGG
RADISH                     TAGTGTTATA-TAGGGGGTAGG
MUSTARD                    TAAGCATATATAAGGGGGGTAG
SPRING CABBAGE             TTTGTCTATATAAGGGGTAGGC
CUCUMBER                   TAAGCATATATAAGGGGGGTAG
SQUASH 1                   ATGTACTATA-TAGGGGGGTGA
SQUASH 2                   ATGTACTATA-TAGGGGGGTGA
PUMPKIN 1                  ATGTACTATA-TAGGGGGGTGA
PUMPKIN 2                  ATGTACTATA-TAGGGGGGTGA
TOMATO                     TAAGCATATATAAGGGGGGTAG
PEA                        CAAGCTTATA-TAGGGGGAGGC
CARROT 1                   TCCATTCGTA-TGGAGGGGACC
CARROT 2                   TCCATTCGTA-TGGAGGGGACC
CARROT 3                   TCCATTCGTA-TGGAGGGGACC
MUNG BEAN 1                TACCCCCTTA-TAGGGGGGGAG
MUNG BEAN 2                CGAATATATA-TGGGGGGGGAG
MUNG BEAN 3                ATATCGAATA-TGCAGGGGGAG
BROAD BEAN                 CAAACATATA-TAGGGGGAGGC
V. hirsuta                 TTGATATATA-TAGGGGGGGGG
V. angustifolia            TAGCCATATATATGGGGGGACA
```

Fig.2 Sequence alignments in the region of initiation of transcription. Initiation is usually on the last A upstream of the G stretch. (Cordesse et al. ,1993)

that most of the sequences of the spacer have extensively diverged with the exception of the initiation start region. Downstream sequences, which are transcribed are relatively conserved between closely related species but have diverged from one family to another.The function of the spacer repeats is not known but their rapid divergence has been correlated with the high species specificity of rDNA transcription and has led to the speculation that the transcription factors regulating rDNA transcription have co-evolved with the spacer subrepeats sequences (Dover and Flavell,1984) . Indeed, in animal cells, it has been demonstrated that some of the repeated elements act as enhancer elements (reviews :Sollner-Webb and Tower 1986 ; Reeder 1990).In higher plants, a similar role is assumed although very few indirect experimental data are available :in wheat, where different NOR (Nucleolar ORganizer) loci can be distinguished on the basis of the number of spacer repeats, the nucleolar activity is positively correlated with the number of repeats (Martini and Flavell 1985).

2)Scanning the radish rDNA spacer for sequence specific DNA-binding activity

As a first step to characterize putative *trans* - acting factors and identify their cognate *cis* -elements we subdivided the rDNA spacer in shorter fragments and we used these as probes in electrophoretic mobility shift assays.A radish nuclear extract was prepared from actively growing young leaves (Echeverria et al. 1992).

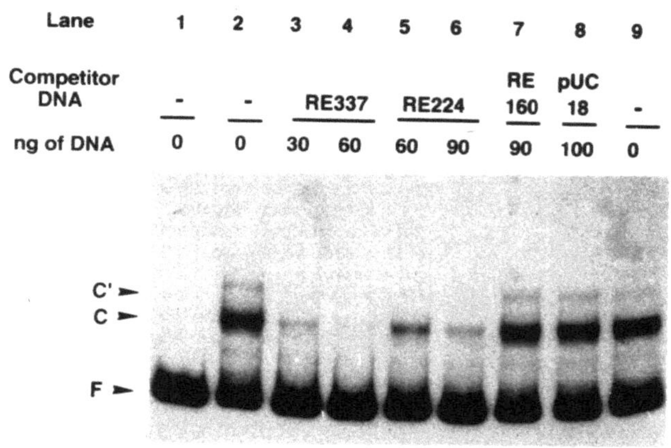

Fig 3 Gel retardation assay with radish nuclear extract and RE 337 probe. Lane 1: no extract, 2 and 9 :nuclear extract, 3-8 : same as 2 and 9 plus competitors as indicated

As shown in Fig. 3 several fragments ,RE 337 and RE 224 (identical to Rar 337 and Rap 224 in Fig.1) are retarded by proteins in the nuclear extract in the presence of an excess poly(dI-dC).RE 337 contains three spacer repeats and RE 224 overlaps the transcription initiation site. Another fragment, RE 260, overlaping a presumed processing site at + 185 also binds a nuclear protein (not shown).These retardation complexes are observed only if the extracts are prepared in the presence of protease inhibitors.The binding activity is lost by heating the extract at 65°C or by treating it with 2% trichloracetic acid. The specificity of the binding was confirmed by competition experiments with pUC18, as well as with the homologous and heterologous fragments.These experiments revealed that RE 337 and RE224 were binding the same protein because they can compete with each others. This protein was named NF A. On the other hand RE 260 binds another protein ,NFB,since this fragment can be competed only by itself in this preliminary survey.

3)Fractionnation of the nuclear extract

Since the gel gift assays and the competition experiments suggested that NF A anD NFB were different proteins we attempted to separate them by chromatography techniques.The nuclear extract was dialyzed against 50 mM Tris HCl pH8 , 6mM $MgCl_2$, 5 mM DTE ,15% glycerol and 100 mM KCl before being loaded on a DEAE- Sepharose column equilibrated in the same buffer. The flow -through and fractions eluted by increasing KCl concentration stepwise were assayed by gel retardation in the presence of RE 337 or RE 260 probes.This experiment demonstrated that NFA elutes in the flow-through fraction and NFB in the 300 mM KCl fraction. Each protein fraction was pooled , NFB was diluted to 200mM KCl and both fractions were further purified by Heparine Sepharose chromatography. In that way, partially enriched fractions were obtained for NF A and NF B.

4) Characterization of NFA binding site

In order to determine the precise location of NFA binding site the binding reaction was scaled up and the complex was partially digested with DNase I. The product of the digestion was run on a sequencing gel in parallel with a partial digestion of nake DNA.This experiment revealed clear footprints on both the coding and the non coding strand for RE 337 as well as for RE 224. The most conspicuous are the poly dA- polydT stretches which separate two consecutive repeat units. The other protected sequence is 13 nucleotide long and has also a 3-4 T stretch .The binding sequences detected byDNase I footprinting are shown in Fig 4. The similarity of the sequences explains why the two fragments can compete.

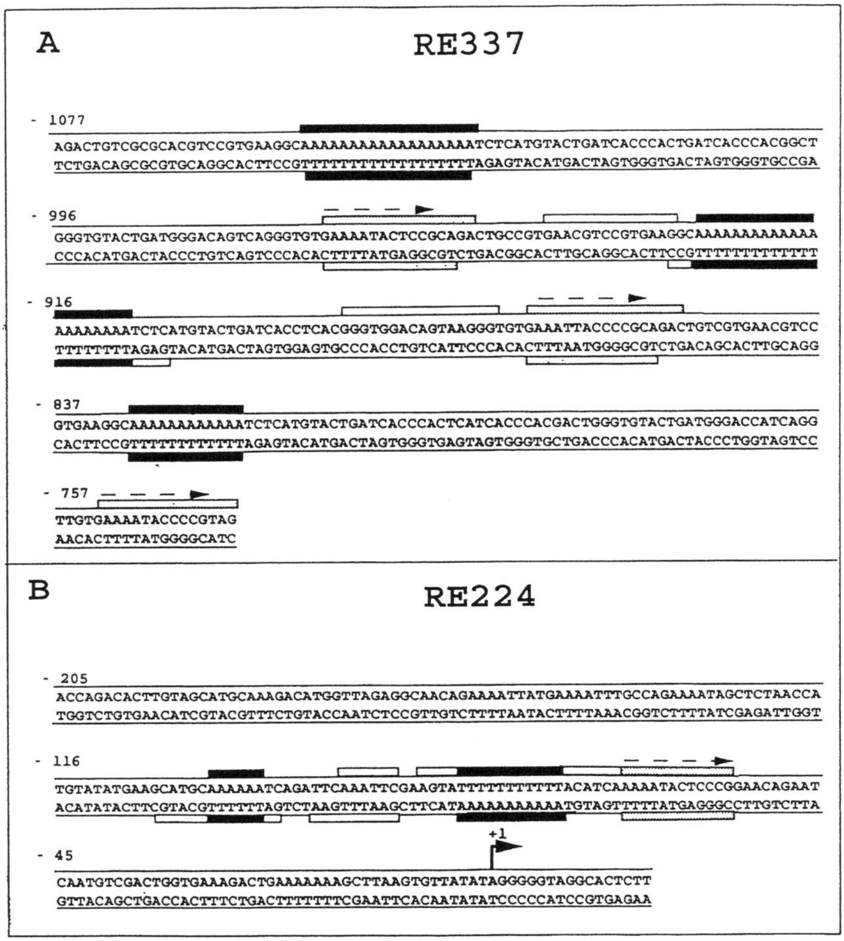

Fig.4 Sequences protected from DNase I by NF A. Boxes located above and below the sequence indicate the regions which are protected on each strand.

5)Characterization of NER B

Similar experiments were carried out with the NF B fraction . Indeed we could restrict the binding region to 57 nucleotides encompassing the processing site which is located at position + 185 .Footprint experiment identified a single region characterized by a G residue separating two pyrimidine stretches .The mapping was further refined by methylation interference studies which demonstrated that the central G as well as 4 others in the flanking sequences were important .The binding specificity for this sequence is stringent because several related sequences occur in the close vicinity. When oligonucleotides are derived from these related sequences and are challenged for complex formation in competition experiments, none of them can compete with the original sequence. Because a processing site for pre -rRNA occurs in this region, the possibility exists that NFB is an RNA- binding protein.This possibility was ruled out by incubating the purified extract with labelled sense and antisense RNA probes: no retardation was observed. In addition, the double stranded DNA-binding activity cannot be competed by either the upper or lower single strand. Interestingly, the core of the motif recognized by NFB is present in many plant rDNA spacer at a similar place.

6) Other proteins binding to the spacer

Other putative regulatory proteins are certainly binding to the rDNA promoter region.The two which are described here are the ones which have the highest affinity in radish. There are three lines of evidence for additional rDNA spacer binding proteins. Firstly a protein binding to the initiation site region as well as to the spacer repeats was described in maize (Smitz et al. 1989) .It binds to G-C rich regions .Jackson and Flavell (1992) recently described a protein binding to the wheat rDNA initiation site as well as to upstream sequences located within the repeats.The consensus sequence CATGG--GC-AAAAC was deduced from the different binding sites. Binding to spacer repeats as well as to the initiation site region and the presence of a short stretch of A residues are common features with NFA. Nakajima et al., 1992 described a protein which binds about 60 nucleotides upstream of the initiation site of *Vicia faba* rDNA. The target sequence seems to be different from that recognized in wheat and radish but shares a bipartite consensus sequence TATG(N)nCAGG with maize rDNA spacer. However the *Vicia* upstream repeats compete very poorly for binding suggesting that the protein which binds to this sequence is functionally different from those described in other species.In cucumber rDNA spacer ,at least 4 proteins were shown to have affinity for the core promoter (-41 to +16) and upstream (-164 to -105) sequences (Zentgraft and Hemleben, 1992). Some of these cucumber proteins seems to bind to single stranded DNA. Secondly our additional experiments in *Arabidopsis* also provided evidence for another protein binding very close to the presumed initiation site in addition to NFA . However we could not detect it in radish, presumably because NFA has a very strong affinity and masks binding of this new protein. Demonstration of an eventual relationship between these various DNA - binding proteins from different species awaits further work. Finally the last evidence comes from the sophisticated models that are presently available for animal rDNA transcription (Comai et al 1992).

7)Current model for rDNA transcriiption

In animal cells two major transcription factors are necessary to direct accurate and promoter -specific transcription of rRNA genes: the Upstream Binding Factor (UBF) and the selectivity factor SL1. UBF has been cloned from diferent animal species and it was possible to demonstrate in cross-species mixing experiments that both UBF and RNA - polymerase I are interchangeable and do not confer specificity in transcription (Bell et al 1990).UBF contains a DNA- binding domain with homology to the HMG proteins.All the attempts to clone UBF-like cDNA or gene from plants and yeast using heterologous probes or PCR have so far failed .The specificity of the transcription is due to SL 1, which is a complex factor which interacts with both UBF and RNA-poymerase I.For instance human SL 1 alone, does not bind by itself to the rDNA promoter, but in the presence of UBF and RNA-polymerase I a strong cooperative binding complex is formed with an extended binding region which is critical for accurate initiation on the human template.Careful analysis of the human SL 1 factor has recently identified a TATA-Binding factor as one of its major component (Comai et al. 1992).This TATA- binding factor is associated with other factors which confer species specificity to rRNA transcription and which recognize , as a complex , a sequence specific element .

DISCUSSION

At this stage of the experiments, we have partially characterized two proteins, NFA and NFB which bind to the radish rDNA spacer. We have also cloned a full length cDNA for the radish TATA- Binding factor and we are currently using this protein to identify and isolate associated factors (TAFs) in order to address the question of species specificity of transcription driven by RNA polymerase I . We know that there are probably many other proteins which specifically bind to the rDNA promoter region. We also have a model from the animal cells. The question is now to correlate these elements together.Before clarifying this situation it will be certainly necessary to characterize the various proteins at the sequence , structural and functional levels so that it may be established whether they belonged to the same family. NFA presents some properties that are reminiscent of UBF such as binding to promoter and upstream sequences . However UBF usually binds to GC -rich sequences in contrast to NFA. In Yeast, another protein, REB1 also presents many similarities with NFA and has been involved in the regulation of rRNA transcription and termination (Lang and Reeder 1993). NFB is a rather unusual protein in that it binds downstream the initiation site. No such protein has yet been described in animal cells. However, conservation of the NFB target site in many plant rDNA spacers at a similar location suggests an important role. Such a situation in which a regulatory factor interacts with a sequence downstream of the initiation site is evident in RNA- polymerase III transcription systems. An alternative posssibility is that NFB might be involved in coupling transcription with the first maturation step of the pre-rRNA.

Another difficult point will be to recognize whether a given protein is significant with respect to rDNA transcription or not. Indeed there are many indication that the spacer region is also involved in process other thantranscription of the rDNA unit. It has been shown that rDNA spacer can function as a replication origin (Van'tHof et al.,1987) or termination site (Hernandez et al., 1988).The rDNA spacer is also a hotspot for recombination (Voelkel-Meiman et al., 1988) and it certainly interacts to some extent with the chromosome scaffold and with the nucleolar matrix (Reeder 1990). These possible functions might overlap and interfere with the transcription itself. In order to discriminate between these possibilities much more effort is required to purify a number of these polypeptides, clone their genes and set up an accurate in vitro transcription system. Work is in progress in our lab, as well as in others to purify and clone putative RNA- polymerase I transcription factors and the recent report of a rDNA cell free transcription system (Yamashita et al., 1993), if reproducible, is a much needed development.

ACKNOWLEDGEMENTS

The authors wish to acknowledge the support from the CNRS to URA 565 .

REFERENCES

Bell SP , Jantzen HM and Tjian R (1990) Assembly of alternative multiprotein complexes directs RNA promoter selectivity. Genes Dev 4 : 943-954

Comai L,Tanese N and Tjian R (1992) The TATA -binding protein and associated factors are integral components of the RNA polymerase I transcription factor ,SL 1. Cell 68 :965-976

Cordesse F ,Cooke R, Tremousaygue D , Grellet F and Delseny M (1993) Fine structure and evolution of the rDNA intergenic spacer in rice and other cereals. J Mol Evol 36:369-379

Delcasso-Tremousaygue D, Grellet F,Panabières F, Ananiev E and Delseny M 1988) Structural and transcriptional characterization of the external spacer of the ribosomal RNA nuclear gene from a higher plant . Eur J Biochem 172 : 767-776

Dover G and Flavell RB (1984) Molecular coevolution: DNA divergence and the maintenance of the function . Cell 38 : 622-633

Echeverria M, Delcasso-Tremousaygue D and Delseny M (1992) A nuclear protein binding to dA/dT-rich sequences upstream from the radish rDNA promoter. Plant J 2 :211-219

Hemleben V,Ganal M ,Gerstner J,Schiebel K and Torres RA (1988) Organization and length heterogeneity of plant ribosomal RNA genes. In : Kahl G (ed).Architecture of eukaryotic genes. VCH Verlagsgesellschaft,Weinheim, Germany :371-383

Hernandez P, Lamm SS ,Bjerknes C A and Van't' Hof J (1988) Replication termini in the rDNA of synchronized pea root cells (*Pisum sativum*) .EMBO J 7 :303-308

Jackson SD and Flavell RB (1992) Protein-binding to reiterated motifs within the wheat rRNA gene promoter and upstream repeats. Plant Mol Biol 20: 911-919

Lang WH and Reeder RH (1993) The REB 1 site is an essential component of a terminator for RNA polymerase I in *Saccharomyces cerevisiae.* Mol Cell Biol 13 : 649-658

Lassner M and Dvorak J (1986) Preferential homogenization between adjacent and alternate subrepeats in wheat rDNA . Nucleic Acids Res 14 : 5499-5512

McMullen MD, Hunter B, Phillips RL and Rubinstein I (1986) The structure of the maize ribosomal DNA spacer region . Nucleic Acids Res 14 :4953-4968

Martini G and Flavell RB (1985) The control of nucleolus volume in wheat ; a genetic study at three developmental stages. Heredity 54 : 111-120

Nakajima T, Suzuki A, Tanifuji S and Kato A (1992) Characterization of nucleotide sequences that interact with a nuclear protein fraction in rRNA gene of *Vicia faba* . Plant Mol Biol 20 : 939-949

Reeder RH (1990) rRNA synthesis in the nucleolus . Trends Genet 6 :390-395

Sollner-Webb B and Tower J (1986) Transcription of cloned eukaryotic ribosomal RNA genes.Ann Rev Biochem 55 :801-830

Toloczyki C and Feix G (1986) Occurrence of 9 homologous repeat units in the external spacer region of a nuclear maize rRNA gene unit . Nucleic Acids Res 14 :4969-4986

Van'tHof J, Hernandez P, Bjerknes CA and Lamm SS (1987) Location of the replication origin of the 9 kb repeat size class of rDNA in pea (*Pisum sativum*).Plant Mol Biol 9 :87-95

Voelkel-Meiman K ,Veil RL and Roeder GS (1987) Recombination-stimulating sequences in yeast ribosomal DNA correspond to sequences regulating transcription by RNA- polymerase I . Cell 48 :1071-1079

Yamashita J, Nakajima T , Tanifuji S and Kato A (1993) Accurate transcription initiation of *Vicia faba* rDNA in a whole cell extract from embryonic axes. Plant J 3 :187-190

IDENTIFICATION OF DISTINCT FAMILIES OF HD-ZIP PROTEINS IN *ARABIDOPSIS THALIANA*

Giovanna Sessa, Monica Carabelli and Ida Ruberti,
Centro di Studio per gli Acidi Nucleici del CNR c/o Dip. di Genetica e Biologia Molecolare, Università di Roma La Sapienza
P.le Aldo Moro 5, 00185 Rome Italy

Sabrina Lucchetti, Simona Baima and Giorgio Morelli.
Unità di Nutrizione Sperimentale, Istituto Nazionale della Nutrizione
Via Ardeatina 546, 00178 Rome Italy

INTRODUCTION

Many of the regulatory genes involved in the control of development share a common sequence element of 180 bp, the homeobox (HB), which encodes a 60 amino acid motif, the homeodomain (HD; Scott *et al.*, 1989; Gehring *et al.*, 1990). The amino acid sequences of known HDs are conserved in evolution from yeast to higher vertebrates (Scott *et al.*, 1989). Recently HB genes have been identified in two plant species, maize (Vollbrecht *et al.*, 1991; Bellmann and Werr, 1992) and *Arabidopsis* (Ruberti *et al.*, 1991). Despite the differences in plant and animal development the discovery of homeobox genes in plants suggests that fundamental regulatory mechanisms that control development may be shared among all higher eukaryotes. The analysis of the maize *kn1* mutants has shown that ectopic expression of the *knotted* gene profoundly affects leaf development, suggesting that HD proteins in plants might be involved in differentiation and/or developmental control as they are in animals (Hake, 1992).

Several lines of evidence indicated that animal HD proteins are transcription factors in which the helix-3 and the N-terminal arm of the HD are responsible for

NATO ASI Series, Vol. H 81
Plant Molecular Biology
Edited by G. Coruzzi and P. Puigdomènech

sequence-specific recognition of DNA (Affolter *et al.*, 1990; Hayashi and Scott, 1990; Laughon, 1991; Furukubo-Tokunaga *et al.*, 1992). The homeodomain is sufficient for sequence-specific DNA binding activity, even without the context of a flanking polypeptide sequence. A detailed analysis of the DNA binding properties of the Ultrabithorax (Ekker *et al.*, 1991) and the fushi tarazu (Florence *et al.*, 1991) homeodomains has shown that the HD, as a monomer, recognizes its specific target site with a Kd lower than 10^{-10}. The ability of the HD to bind DNA with such high affinity and sequence discrimination suggests that the sequence specific DNA binding activity of HD proteins is a crucial determinant of their specificity of action *in vivo*. However, the question emerged of how the related regulators specifically recognize and regulate their appropriate target genes *in vivo* (Hayashi and Scott, 1990; Manak and Scott, 1993). For instance, although the Antennapedia and Ultrabithorax proteins elicit very different developmental programs, their HDs are very similar (54 of 61 residues identical). Moreover, the amino acid residues in the recognition helices of these proteins that make specific contacts with DNA are absolutely conserved, suggesting that specificity is only partially achieved by HD-DNA interaction. Insights into the ways by which specificity of HD proteins can be achieved have been provided by recent work on the yeast protein $\alpha 2$. These studies showed that the functional specificity of $\alpha 2$ is determined by both the sequence-specific DNA binding activity of the homeodomain and the specific interaction of $\alpha 2$ with an auxiliary protein factor (Smith and Johnson, 1992). Mechanistically similar strategies might be used by animal HD proteins to achieve their functional specificity (Schier and Gehring, 1993).

The *Arabidopsis* Athb-1 and -2 homeodomains exibit a strong homology with the helix-3 region and the highly conserved residues of the Antp class of HD proteins. The conservation of these residues has suggested that Athb-1 and -2 homeodomains may adopt three-dimensional structures similar to those of others. However, the Athb-1 and -2 proteins also contain a dimerization domain, a leucine zipper motif (Zip), located immediately COOH-terminal to the homeodomain. Therefore we have proposed that proteins with a contiguous homeodomain-leucine zipper architecture should be referred to as HD-Zip proteins (Ruberti *et al.*, 1991). Other members of the *Arabidopsis* HD-Zip class of proteins have been characterized: Athb-3 (Mattsson *et al.*, 1992), HAT 4, 5 and 22 (Schena and Davis, 1992) and Athb-4 (Carabelli *et al.*, 1993). Moreover the isolation of three new members (Athb-5, -6, -7) of this class of proteins has been recently reported (Mattsson *et al.*, 1993).

We have suggested that the HD-Zip proteins might use the dimerization domain to closely juxtapose a pair of DNA contacting surfaces, each of which fits into half of a dyad-symmetric recognition sequence (Ruberti *et al.*, 1991). It is clear that both specific protein-protein interactions between members of the HD-Zip class of proteins and sequence-specific DNA binding activity of the dimeric complexes might significantly contribute to the functional specificity of these HD proteins. Recently, the analysis of the DNA binding properties of Athb-1 *in vitro* showed indeed that the HD-Zip domain binds DNA as a dimer and it recognizes a pseudopalindrome of two 5 bp half sites that overlap on a central position (Sessa *et al.*, 1993). A preliminary analysis of the DNA binding properties of Athb-2 showed that it recognizes a distinct 9 bp dyad-symmetric sequence (Sessa *et al.*, 1993). Here we present a more detailed DNA binding study for the HD-Zip domain of Athb-2. In addition, we isolated and characterized other genes coding for HD proteins to evaluate the complexity of the HD-Zip class of proteins in *Arabidopsis thaliana.*

RESULTS

The Athb-1 and -2 HD-Zip domains recognize dyad-symmetric DNA sequences.

We have previously reported a detailed analysis of the DNA binding properties of the Athb-1 HD-Zip domain *in vitro* (Sessa *et al.*, 1993). DNA binding analysis performed using random-sequence DNA templates showed that the Athb-1 homeodomain-leucine zipper (HD-Zip-1) domain, but not the homeodomain alone, binds to DNA. The HD-Zip-1 domain recognizes a 9 bp dyad-symmetric sequence

(CAAT(A/T)ATTG), as determined by selecting high affinity binding sites from a DNA containing a 15 bp random core sequence. In addition, we determined that the Athb-2 HD-Zip (HD-Zip-2) recognizes a distinct 9 bp dyad-symmetric sequence. The HD-Zip-2 binding site was deduced by sequencing directly the pool of DNA molecules selected from a DNA (D1, see Fig. 2A) in which specific binding site positions are random in sequence (Sessa *et al.*, 1993).

Here, we report a more accurate analysis of the DNA binding properties of the Athb-2 HD-Zip domain *in vitro*. To express the HD and HD-Zip domains of Athb-2 in bacteria, we used the Glutathione-S-Transferase (GST) expression vector pGEX-2T (see Sessa *et al.*, 1993).

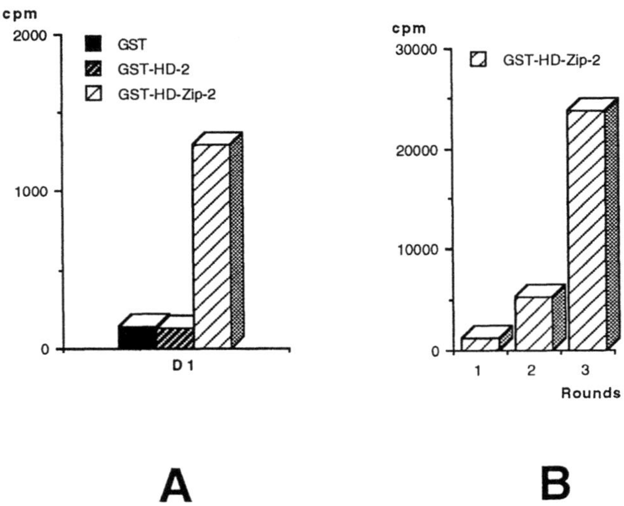

Fig.1. HD-Zip-2 binding site selection. **(A)** Schematic representation of the DNA binding analysis performed using the indicated proteins and the D1 template. **(B)** Representation of the HD-Zip-2 binding site selection. Results are given in cpm of DNA bound. The DNA binding assays were performed under the conditions previously described (Sessa *et al.* 1993) using 1.5 x 10^5 cpm of DNA template in each assay.

DNA binding assays performed using the D1 template showed that the GST-HD-Zip-2 fusion protein binds to D1 to a significant extent with respect to GST and GST-HD-2 (Fig. 1A). Selection of the DNA specifically bound by the GST-HD-Zip-2 was done by subsequent rounds of DNA binding, isolation of bound DNA by an affinity chromatography assay and PCR amplification.The DNA derived from two successive rounds of selection and amplification was bound by the GST-HD-Zip-2 fusion protein to a greater relative extent (Fig. 1B). Following three successive rounds of selection and amplification, the DNA was cloned into pBluescript II KS. 27 out of 1×10^5 colonies were picked and plasmid DNA prepared and sequenced.

D1	N	N	A	T	N	A	T	N	N
1	C	A	A	T	G	A	T	T	G
2	C	A	A	T	G	A	T	T	G
4rev	C	A	A	T	G	A	T	T	G
5	C	A	A	T	G	A	T	T	G
7rev	A	A	A	T	G	A	T	T	G
8	T	A	A	T	G	A	T	T	G
9	G	C	A	T	G	A	T	T	A
10	A	C	A	T	G	A	T	T	G
13rev	A	A	A	T	G	A	T	T	G
14	T	A	A	T	G	A	T	T	G
15	C	A	A	T	G	A	T	T	G
16	C	A	A	T	G	A	T	T	G
17rev	C	A	A	T	G	A	T	T	G
19	T	A	A	T	G	A	T	T	A
20	C	A	A	T	G	A	T	T	A
22rev	A	A	A	T	G	A	T	T	T
23	C	A	A	T	G	A	T	T	G
24	T	A	A	T	G	A	T	T	T
25	A	C	A	T	G	A	T	T	G
28rev	T	A	A	T	G	A	T	T	G
29	C	A	A	T	G	A	T	T	G
30	T	A	A	T	G	A	T	T	G
31	C	A	A	T	G	A	T	T	A
32rev	C	A	A	T	G	A	T	T	G
33rev	C	A	A	T	G	A	T	T	G
34	A	A	A	T	G	A	T	T	G
35	T	A	A	T	G	A	T	T	G
G	1	0			27			0	21
A	5	24			0			0	4
T	7	0			0			27	2
C	13	3			0			0	0

A

BS-1 CAAT(A/T)ATTG

BS-2 CAAT(G/C)ATTG

B

Fig. 2. The HD-Zip domain of Athb-2 recognizes a dyad-symmetric sequence. **(A)** Tabulation and analysis of the HD-Zip-2 binding site sequences present within affinity selected DNA molecules. The nucleotide sequences of DNA molecules cloned after the third round of selection are shown as aligned by the method described in the text. Rev indicates that the reverse complementary of the selected sequence is shown. The number of occurrences for the four nucleotides in positions ±3 and ±4 is shown. **(B)** The dyad-symmetric sequences recognized by the Athb-1 (BS-1) and Athb-2 (BS-2) HD-Zip domains are shown.

By using the alignment in which a purine residue in the central position has been chosen arbitrarily, the frequencies of each of the bases at positions ±3 and ±4 are shown in figure 2A. Although the most preferred nucleotides are the same on each side of the binding site, the frequencies of the preferred bases are affected significantly according to their side. The majority of deviations from CAATGATTG occurs to the left of the central position, suggesting that Athb-2 probably interacts more avidly with the right half site than with the left half site.

In order to prove that the CAAT(G/C)ATTG motif is specifically recognized by the

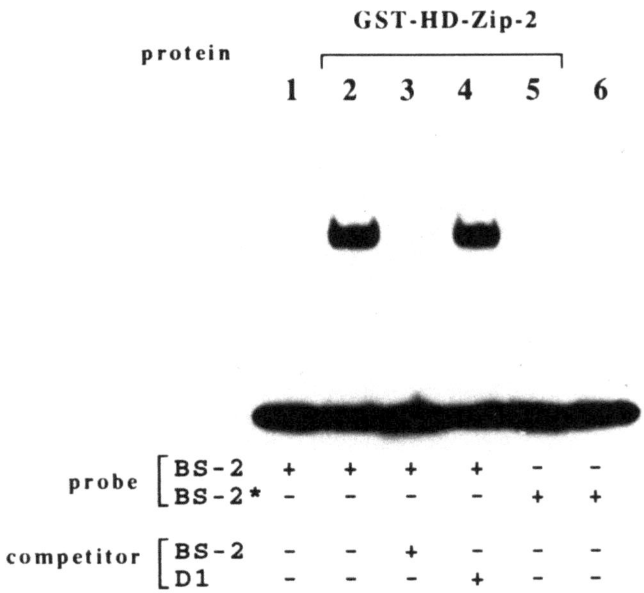

Fig. 3. The HD-Zip-2 domain binds specifically to DNA. Gel retardation analysis was performed as described in Materials and Methods with the oligonucleotide bearing the CAATGATTG motif (BS-2, lanes 1-4) and a derivative containing the CAATGGTTG sequence (BS-2*, lanes 5-6). To the reaction mixtures shown in lanes 3 and 4 a 100-fold molar excess of unlabelled competitor DNA (BS-2 in lane 3 and D1 in lane 4) was added.

GST-HD-Zip-2 fusion protein, gel retardation assays (Singh *et al.*, 1986) were performed using the GST-HD-Zip-2 fusion protein and BS-2, a DNA fragment containing the optimal HD-Zip-2 binding site.The incubation of GST-HD-Zip-2 and BS-2 resulted in a DNA-protein complex of reduced electrophoretic mobility relative to free DNA (Fig. 3, lanes 1 and 2). The formation of the retarded DNA-protein complex was specific: it was resistant against a 100-fold molar excess of unselected starting DNA template (D1, Fig. 3, lane 4), whereas a 100-fold molar excess of unlabelled BS-2 abolished the formation of the complex (Fig. 3, lane 3). We also performed retardation assays with a derivative of the BS-2 DNA (BS-2*, CAATGGTTG) containing a base substitution (G vs. A) in the right half site. No DNA-protein complex formation was seen performing a retardation assay with GST-HD-Zip-2 and BS-2* (Fig. 3, lane 5). This observation indicates that the ability of GST-HD-Zip-2 to recognize DNA is dependent on the presence of two intact half sites. The Athb-2 binding site deduced by the sequencing of affinity selected DNA molecules is shown in figure 2B. The Athb-2 binding site is a 9 bp pseudopalindrome in which the central position is likely to be alternatively recognized by one of the two monomers of the homodimeric complex. Comparison with the Athb-1 binding site (Sessa *et al.*, 1993) showed that the two recognition sequences differ only at the central position, Athb-1 recognizes an A/T and Athb-2 a G/C (Fig. 2B).

Isolation of genes encoding HD proteins in *Arabidopsis thaliana.*

To identify other genes coding for HD proteins, we took advantage of the strategy previously used to isolate Athb-1 and -2 (Ruberti *et al.*, 1991). For the new screening we designed HB-2 (see Materials and Methods), a degenerate oligonucleotide shorter than HB-1. HB-2 corresponds to a conserved six amino acid sequence from the helix-3 region of Athb-1 and -2. We selected 6 independent λ clones out of 75,000 clones screened from an *Arabidopis thaliana* genomic library (Clontech). A probe hybridizing fragment from each clone was subcloned and partially sequenced. To verify further the homeodomain coding regions the corresponding cDNAs were also obtained with a combination of cDNA cloning and PCR amplification with selected primers. DNA sequence analysis revealed that five clones, designated Athb-4 (Carabelli *et al.* 1993), -8, -9, -10 and -11, code for HD-Zip proteins, whereas the last one, named Athb-12, code for a HD protein. Sequence comparison of the genomic fragments with the respective cDNAs allowed the identification of exons and introns. All the Athb genes, but Athb-12, contain introns within the homeobox. Athb-1 has an intron positioned after amino acid 13. In Athb-8 and -9 genes the intron is localized in the same position after the amino acid 53. The other Athb genes have two introns: they are positioned after amino acids 28 and 55 in Athb-2 and -4, amino acids 18 and 57 in Athb-10 and -11. It is noteworthy that, with the exception of the Athb-1 gene, the homeodomain and the leucine zipper are encoded by different exons.

The deduced amino acid sequences of the Athb homeodomain-leucine zipper regions are shown in figure 4. Initial inspection of the Athb-8 and -9 sequences revealed a weak homology with the other HD-Zip proteins in the region upstream and downstream the helix-3. A better alignment can be obtained by looping out 4 amino acids between helix-2 and -3, and between helix-3 and the first leucine residue of the zipper. The first putative repeat of the Athb-10 and -11 leucine zipper motifs lacks the leucine residue at position d_1. Although we cannot exclude that the glutamic acid residue at position d_1 is active in dimerization, we preferred to align the Athb-10 and -11 to the other HD-Zip motifs by looping out 7 amino acids between helix-3 and the leucine residue of the second repeat.

The results of the amino acid comparison between either the entire HD-Zip regions (aas 1-75 in Fig. 4, upper panel), or the HDs (aas 1-57), or the Zips (aas 58-75) of the Athb proteins are shown in Table 1. This analysis shows that none of the new genes has a strong similarity to Athb-1, whereas Athb-4 is highly related to Athb-2 (Carabelli *et al.*, 1993). Moreover Athb-8 and -9, -10 and -11 are homologous to each other, respectively.

As already observed for Athb-1 and -2, the new Athb gene products diverge from previously characterized animal HD sequences. For this reason they cannot be

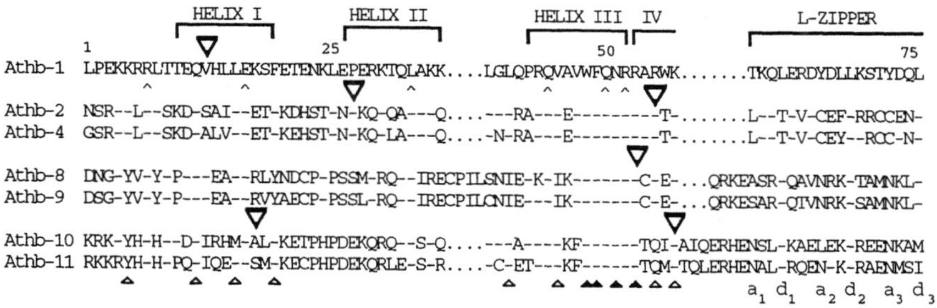

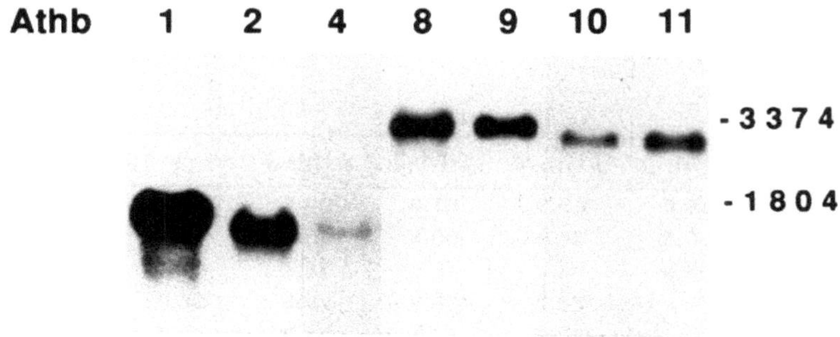

Fig. 4. Identification of four distinct HD-Zip protein families.
(Upper panel) Amino acid comparison of the Athb-1 HD-Zip domain with other HD-Zips. Single letter code designates amino acid residues. The invariant amino acids of the homeodomain are marked by black triangles, the highly conserved ones by open triangles (Gehring *et al.*, 1990). The first residue of each heptad repeat of the leucine zipper is marked by the letter **a**, with leucine residues at position **d** (Glover, 1989). The amino acids which are invariant in all the HD-Zip domains are marked by carets. The positions of the introns are indicated with white arrowheads.
(Lower panel) Northern blot analysis of the Athb genes. cDNA fragments were randomly labelled with ^{32}P and used to probe 1 µg of poly $(A)^+$ RNA from leaves of five weeks old *Arabidopsis* plants. The numbers indicate the size in bases of the ribosomal RNAs.

assigned to any of the previously identified classes of the HD proteins (Scott *et al.*, 1989). Nevertheless, residues that are 100% conserved in all homeodomains of higher eukaryotes (Scott *et al.*, 1989) are conserved in the seven *Arabidopsis* HD-Zip sequences. The invariant residues are tryptophan, phenylalanine, asparagine and arginine at positions 48, 49, 51 and 53, respectively. Most of the residues that are highly conserved in all homeodomains are also conserved in the *Arabidopsis* HD-Zip motifs. These are position 12 (glutamine), 16 (leucine/methionine), 20 (phenylalanine/ tyrosine), 40 (leucine/isoleucine), 45 (valine/isoleucine), 57 (lysine). At position 5 of Athb-1, -2 and -4 homeodomains there is a conservative substitution (lysine vs arginine); a tyrosine residue is present at this position in the other HD-Zip proteins. Finally, the last conserved amino acid (arginine at position

	Athb-2	Athb-4	Athb-8	Athb-9	Athb-10	Athb-11	
Athb-1	**48.0**	**49.3**	**33.3**	**34.7**	**34.7**	**30.6**	**HD-Zip**
	52.6	50.9	38.6	40.4	42.1	36.8	HD
	33.3	44.4	16.7	16.7	11.1	11.1	Zip
Athb-2		**86.7**	**26.7**	**28.0**	**38.7**	**33.3**	**HD-Zip**
		87.7	31.6	33.3	45.6	38.6	HD
		83.3	11.1	11.1	16.7	16.7	Zip
Athb-4			**25.3**	**29.3**	**36.0**	**34.7**	**HD-Zip**
			29.8	35.1	43.9	42.1	HD
			11.1	11.1	11.1	11.1	Zip
Athb-8				**86.7**	**40.0**	**36.0**	**HD-Zip**
				89.5	40.4	36.8	HD
				77.8	38.9	33.3	Zip
Athb-9					**38.7**	**40.0**	**HD-Zip**
					42.1	40.4	HD
					27.8	38.9	Zip
Athb-10						**66.7**	**HD-Zip**
						70.2	HD
						55.6	Zip

Table 1

55) is not present in Athb-10 and -11. In addition, other positions are invariant in all the *Arabidopsis* HD-Zip proteins identified so far. These are position 7 (arginine), 17 (glutamic acid), 34 (leucine), 44 (glutamine), 50 (glutamine), 52 (arginine).
It is interesting to note that the genes coding for the more related HDs show a very high degree of homology in the leucine zipper region (Table 1). We tentatively grouped these HD-Zip proteins in four different families, named HD-ZIP I, II, III and IV (Fig. 5). The members of each family are related not only by the homology in the HD-Zip region, but also by the conservation of the intron position within this region (Fig. 4). In addition, the related genes code for mRNA of similar length, as shown in figure 4. Finally, Athb-3 (Mattsson *et al.*, 1992) and HAT 4, 5 and 22 (Schena and

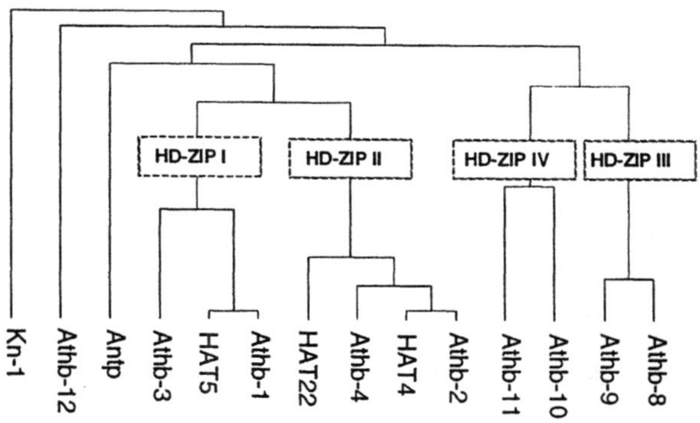

Fig. 5. Dendrogram of the HD-Zip domain sequences.The dendrogram shows the order of the pairwise alignment of the HD-Zip sequences characterized so far. Antp (McGinnis *et al.*, 1984), Kn-1 (Vollbrecht *et al.*, 1991) and Athb-12 (Sessa G., unpublished data) HD proteins were also included. Distance along the orizontal axis is proportional to the differences between sequences; distance along the vertical axis has no significance at all.

Davis, 1992) belong to either the HD-ZIP I or II protein families, as recently described by Carabelli *et al.* (1993).

CONCLUSION

Here we present the identification of four distinct families of HD-Zip proteins. Our DNA binding studies have established that the HD-Zip domain, unlike the classic HD, depends on dimerization for sequence-specific DNA binding activity. In this respect, it is noteworthy that a HD-Zip protein might be turned into a highly specific DNA binding factor by simply promoting the formation of a dimeric complex. Whereas the formation of heterocomplexes between proteins is an efficient way to increase the number of binding or regulatory properties, inappropriate heterodimerization is probably undesiderable. Amino acid comparison between HD-Zips showed that the leucine zipper motifs of members of different families are highly unrelated. This suggests that heterodimeric complexes may form only between members of each HD-Zip family. Supporting this hypothesis is the observation that the Athb-1 and -2 HD-Zip domains are unable to form heterodimeric complexes *in vitro* (Sessa *et al.,* 1993). In addition, the discovery of different HD-Zip architectures in the HD-Zip I and II with respect to the HD-Zip III family also suggests that an active binding complex might form only between members of each HD-Zip family.

Why the HD-Zip proteins have been found only in plants? It is tempting to speculate that these factors might control developmental pathways that are peculiar to plants. Of these, the light regulated pathways have an especially important role, not only in a complex process called photomorphogenesis, but also in other developmental processes, including floral induction and coordinate expression of several light-regulated nuclear and chloroplast-encoded genes. The observation that FAR-RED light rapidly induces the Athb-2 and -4 expression (Carabelli *et al.*, 1993) suggests that some of the HD-Zip gene products may be important components of the light signal trasduction pathways.

MATERIALS AND METHODS

Binding site selection

Binding site selection from the D1 templates (see Fig. 2A) was performed as previously described (Sessa *et al.*, 1993). Following three rounds of selection, double stranded 70-mer was digested with Xho I and Eco RI and ligated into similarly digested pBluescript II KS (Stratagene). 27 single colonies were picked and plasmid DNA prepared (Quiagen tip 100; Diagen). Sequencing was carried out on double stranded DNA using Sequenase 2 (USB) according to the manufacturer's instructions. DNA sequences were obtained using the M13 (20) primer.

Gel mobility shift assay

Double stranded ^{32}P labelled DNA (3 ng, $6x10^4$ cpm) was incubated with *E. Coli* protein extracts (0.2 µl containing about 2.5 ng of the fusion protein), in binding buffer (50 mM KCl; 20 mM Hepes, pH 7.4; 10% glycerol; 2 mM $MgCl_2$; 1 mM DTT; 0.5 mM EDTA, pH 7.5, 0.5% NP-40) which contained 1 µg poly (dI-dC) as non specific competitor DNA. Electrophoresis was performed as previously described (Sessa *et al.*, 1993).

Isolation of genes encoding homeodomain-containing proteins

75,000 clones from an *A. thaliana* genomic EMBL-3 library (Clontech) were screened with the degenerate oligonucleotide HB-2 (TGGTTYCARAAYMGNMG). Filters were prehybridized, hybridized and washed as previously described (Ruberti *et al.*, 1991), except that the washing temperature was 47°C.

DNA sequencing

Sequencing of the double stranded DNA was carried out with Sequenase 2 (USB). Initial sequences were obtained using HB-2 primer. Sequences were then extended by internal specific primers. Oligonucletides were synthetized on Beckman DNA-SM synthesizer and purified as previously described (Ruberti *et al.*, 1991)

Isolation of RNA from *Arabidopsis* and Northern analysis

Plants were grown in a light/dark cycle 16:8 h in growth chambers. Leaves were harvested from five weeks old *Arabidopsis* plants. Total RNA was isolated by the guanidinium thyocynate-phenol-chloroform method (Chirgwin *et al.*, 1979) and the polyA$^+$ RNA was extracted using PolyA Tract mRNA Isolation System (Promega) according to the manufacturer's instructions. Northern blot of *A. thaliana* RNA with a random priming labelled probe (Feinberg and Vogelstein, 1983) was performed essentially as described by Thomas, 1980.

Computer analysis

Comparative sequence analysis were performed with the PILEUP program from the Genetics Computer Group (Devereux, 1991).

AKNOWLEDGMENTS

This work was supported, in part, by Ministero Agricoltura e Foreste, Piano Nazionale "Sviluppo di Tecnologie Avanzate Applicate alle Piante" and Istituto Pasteur-Fondazione Cenci Bolognetti, Università di Roma La Sapienza.

REFERENCES

Affolter M, Schier A and Gehring WJ (1990) Homeodomain proteins and the regulation of gene expression. Current Opinion in Cell Biology 2: 485-495.

Bellmann R and Werr W (1992) Zmhox 1A, the product of a novel maize homeobox gene, interacts with the *Shrunken* 26 bp feedback control element. EMBO J 11: 3367-3374.

Carabelli M, Sessa G, Baima S, Morelli G and Ruberti I (1993) The *Arabidopsis Athb-2* and *-4* genes are strongly induced by far red rich light. Plant J in press.

Chirgwin J M, Pzrybyla A E, McDonald R J and Rutter W J (1979) Isolation of biologically active ribonucleic acid from sources enriched in ribonuclease. Biochemistry 18: 5294-5299.

Devereux J (1991) Program Manual, Sequence Analysis Software Package, Version 7, Genetics Computer Group, Madison.

Ekker S C, Young K E, von Kessler D P and Beachy P A (1991) Optimal DNA

sequence recognition by the Ultrabithorax homeodomain of *Drosophila*. EMBO J 11: 1179-1186.
Feinberg A P and Vogelstein B (1983) A technique for radiolabelling DNA restriction endonuclease fragments to high specific activity. Anal Biochem 132: 6-13.
Florence B, Handrow R and Laughon A (1991) DNA-binding specificity of the fushi tarazu homeodomain. Mol Cell Biol 11: 3613-3623.
Furukubo-Tokunaga K, Muller M, Affolter M, Pick L, Kloter U and Gehring W J (1992) *In vivo* analysis of the helix-turn-helix motif of the fushi tarazu homeodomain of *Drosophila melanogaster.* Genes & Dev 6: 1082-1096.
Gehring WJ, Muller M, Affolter M, Percival-Smith A, Billeter M, Qian Y Q, Otting G and Wuthrich K (1990) The structure of the homeodomain and its functional implications. Trends in Genetics 6: 323-329.
Glover C V C III (1989) Sequence-specific protein-DNA recognition by trancriptional regulatory proteins. Plant Mol Biol Reporter 7 (3): 183-208.
Hake S (1992) Unraveling the knots in plant development. Trends in Genetics 8: 109-114.
Hayashi S and Scott M P (1990) What determines the specificity of action of *Drosophila* homeodomains proteins? Cell 63: 883-894.
Laughon A (1991) DNA binding specificity of homeodomains. Biochemistry 30: 11357-11367.
Manak J R and Scott P (1993) Able assistants for homeodomain proteins. Current Biol 3: 318-320.
Mattsson J, Soderman E, Svenson M, Borkird C and Engstrom P (1992) A new homeobox-leucine zipper gene from *Arabidopsis thaliana.* Plant Mol Biol 18: 1019-1022.
Mattsson J, Soderman E, Svenson M and Engstrom P (1993) Tissue- and organ-specific expression of homeobox-containing genes from *Arabidopsis thaliana* and carrot. J Cell Biochem Sup 17B: 31.
McGinnis W, Levine M, Hafen E, Kuroiwa A and Gehring W J (1984) A conserved DNA sequence in homeotic genes of the *Drosophila Antennapedia* and *bithorax* complexes. Nature 308: 428-433.
Ruberti I, Sessa G, Lucchetti S and Morelli G (1991) A novel class of plant proteins containing a homeodomain with a closely linked leucine zipper motif. EMBO J 10: 1787-1791.
Schena M and Davis R W (1992) HD-Zip proteins: members of an *Arabidopsis* homeodomain protein superfamily. Proc Natl Acad Sci USA 89: 3894-3898.
Schier F A and Gehring W J (1993) Analysis of a *fushi tarazu* autoregolatory element: multiple sequence elements contribute to enhancer activity. EMBO J 12: 1111-1119.
Scott M P, Tamkun J W and Hartzell G W III (1989) The structure and function of the homeodomain. BBA Rev Cancer 989: 25-48.

Sessa G, Morelli G and Ruberti I (1993) The Athb-1 and -2 HD-Zip domains homodimerize forming complexes of different DNA binding specificity. EMBO J in press.
Smith D L and Johnson A D (1992) A molecular mechanism for combinatorial control in yeast: MCM1 protein sets the spacing and orientation of the homeodomain of an α2 dimer. Cell 68: 133-142.
Singh H, Sen R, Baltimore D and Sharp P A (1986) A nuclear factor that binds to a conserved sequence motif in transcriptional control elements of immunoglobulin genes. Nature (London) 319: 154-158.
Thomas P S (1980) Hybridization of denatured RNA and small DNA fragments transferred to nitrocellulose. Proc Natl Acad Sci USA 77: 5201-5205.
Vollbrecht E, Veit B, Sinha N and Hake S (1991) The developmental gene *Knotted-1* is a member of a maize homeobox gene family. Nature (London) 350: 241-243.

Determinants of mRNA Stability in Higher Plants

Crispin B. Taylor, Thomas C. Newman, Masaru Ohme-Takagi, Pauline A. Bariola, André B. Dandridge, and Pamela J. Green
Department of Energy Plant Research Laboratory
Michigan State University
East Lansing, MI 48824-1312, USA

Introduction

In eukaryotes, each of the steps between transcription of a gene and degradation of the corresponding protein are potential targets for regulation of expression. Of these, transcriptional control mechanisms are the best understood, both from the perspective of basic components of the transcriptional apparatus, and with respect to the mechanisms that regulate specific genes or sets of genes. However, it has become apparent that many eukaryotic genes are controlled to some extent by posttranscriptional events (Peltz *et al.*, 1991). The contribution of mRNA processing, transport, and stability, and of translational and post-translational mechanisms to the regulation of gene expression has therefore received increasing attention recently. Of these, regulation at the level of mRNA stability is among the most widely evoked, but the mechanisms by which such control is exerted remain poorly understood.

Unstable transcripts, with half-lives on the order of minutes, are the exception rather than the norm in higher eukaryotes. Average transcript half-lives are on the order of several hours (Siflow and Key, 1979; Brock and Shapiro, 1983), and it has therefore been suggested that the destabilization of specific transcripts is an active process (Hunt, 1988). The best studied unstable transcripts in mammalian systems often encode proteins involved in the regulation of cellular growth and differentiation, such as c-*myc* (Jones and Cole, 1987), granulocyte/macrophage - colony stimulating factor (GM-CSF) (Shaw and Kamen, 1986), and others. One compelling reason why transcripts such as these may be so unstable is that the corresponding proteins are required only transiently in the

NATO ASI Series, Vol. H 81
Plant Molecular Biology
Edited by G. Coruzzi and P. Puigdomènech

cell. The short half-lives of their transcripts are thought to facilitate the rapid attainment of new mRNA steady state levels (both higher and lower) following a change in the transcription rate of their genes (Peltz, *et al.*, 1991). Rapid changes in gene expression are likely to be particularly important in plants (eg. Theologis, 1986). As sessile organisms plants cannot move away from adverse stimuli but must respond by altering their metabolism, often quite quickly, if they are to survive. Plants are therefore likely to have a large number of genes whose expression must change quickly in response to environmental stimuli. Accordingly, mRNA stability may figure prominently in the regulation of these genes, in addition to genes involved in control of cellular growth and differentiation.

As a first step in towards elucidating the fundamental principles that govern mRNA stability in model plant species, we have begun to identify molecular components that may be involved in the process. Our initial efforts have concentrated on two distinct types of components: 1) sequences that target transcripts for rapid decay in tobacco, and 2) the analysis of the RNases of *Arabidopsis*. This report highlights our progress in both of these areas.

Identification of *cis*-acting sequences that mediate rapid mRNA decay

Messenger RNA decay rates are rarely measured directly in plants, so little is known about the sequences that trigger the degradation of highly unstable plant transcripts. To overcome this problem, we have developed a system to measure mRNA half-lives in stably transformed plant cells (Newman *et al.*, 1993). Our system makes use of a tobacco cell line, NT-1, that can be stably transformed with reporter constructs of interest, using *Agrobacterium* (An, 1985). To measure mRNA decay rates, transcription in the transformed cell lines is arrested using actinomycin D, and the subsequent disappearance of the reporter gene transcripts over time is monitored by northern blotting. The NT-1 cell system is modelled on those used to identify specific mRNA sequences that target transcripts for rapid decay in mammalian cells (reviewed in Peltz *et al.*, 1991). Among the first instability sequences to be described was an AU-rich element in the 3' untranslated region (UTR) of GM-CSF which, when placed in the 3'UTR of the otherwise stable β-globin transcript, caused the transcript to be rapidly degraded (Shaw and Kamen, 1986).

Recently the NT-1 system was used to test the function of a sequence element derived from the Small Auxin Up RNAs (*SAURs*), a family of auxin inducible transcripts initially characterized in soybean (McClure *et al.*, 1989a). The temporal expression of the *SAUR* genes in soybean hypocotyls and the distribution of their transcripts is consistent with a role for the SAUR proteins in auxin stimulated growth of plant cells, although this has not been directly demonstrated. When soybean hypocotyls are gravistimulated by being placed horizontally, the *SAUR* transcripts disappear very rapidly from the upper side of the seedlings (McClure and Guilfoyle, 1989). Their half-lives have been estimated in these and other experiments (Franco *et al.*, 1990) to be on the order of 10-50 minutes. As such, the *SAURs* appear to be among the most inherently unstable plant transcripts yet characterized. Conserved in the 3'UTRs of all the *SAUR* genes sequenced to date (McClure *et al.*, 1989a; Yamamoto *et al.*, 1992; Y. Liu, P. Gil and P.J.G., unpublished results) is a short (~40 bp) sequence termed DST (McClure *et al.*, 1989a). The DST element differs from mRNA instability sequences identified in mammalian cells, such as the 3'UTRs of GM-CSF (Shaw and Kamen, 1986) and c-*fos* (Shyu *et al.*, 1989). These sequences are AU-rich and contain multiple copies of an AUUUA motif. However, the conservation of the DST element in a non-coding region supports its functional significance; its location is similar to that of several mammalian instability determinants, suggesting that it may contribute to the instability of the *SAUR* transcripts (McClure *et al.*, 1989a).

To test this hypothesis, a synthetic DST element was multimerized and inserted into the 3'UTR of a β-glucuronidase (GUS) reporter transcript. As shown in Figure 1, a dimer of the DST sequence (DST x2) was sufficient to decrease markedly the half-life of the GUS transcript in NT-1 cells, relative to that of the no-insert control. An unmodified chloramphenicol acetyltransferase (CAT) transcript served as the internal reference for these experiments. The effect of the DST x2

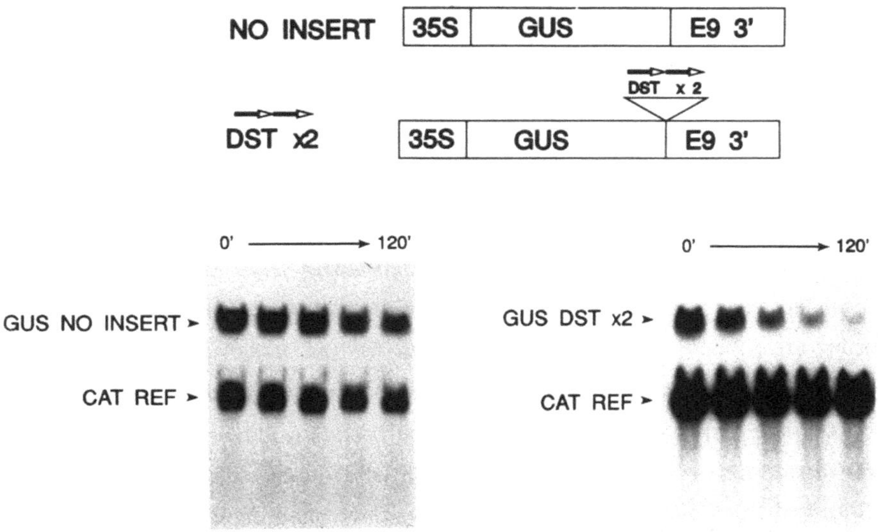

Figure 1. Destabilization of the GUS transcript by DST sequences in tobacco. Top: GUS genes containing no insert or a dimer of the DST sequence (DST x2) are shown. Each was introduced into NT-1 cells on the same vector as an unmodified CAT reference gene. Bottom: RNA was isolated from stably transformed cell lines every 30 min for a period of 2 hr after the addition of Actinomycin D. Northern blots containing 10 μg RNA per lane were probed for the GUS and CAT transcripts. Left: Northern blot of RNA from a cell line transformed with the GUS-no insert and the CAT reference genes. Right: Northern blot of RNA from a cell line transformed with the GUS-DST x2 and the CAT reference genes. The blot on the right was overexposed relative to the blot on the left to allow the GUS-DST x2 transcript, which was present in reduced amounts, to be visualized.

insert is not due to it's increasing the size of the GUS transcript because a spacer sequence of the same size (spacer x2) had little effect on GUS mRNA half-life (see Newman *et al.*, 1993). A parallel set of constructs, in which the GUS coding region was replaced with that of the human β-globin gene, yielded similar data. These results demonstrate that the DST element is a potent instability sequence that can function independent of the adjacent coding region in tobacco. The effect of the DST x2 element on transcript abundance in transgenic tobacco plants was also examined, using the *β*-globin constructs. Accumulation of the *β*-globin transcript containing the DST x2 insert was markedly lower than accumulation of the *β*-globin transcript containing the spacer x2 or no-insert controls (Newman *et al.*, 1993). Thus DST sequences are likely to target reporter transcripts for rapid decay in transgenic plants, because the effect of DST sequences on mRNA half-lives in NT-1 cells is mirrored by a corresponding decrease in transcript accumulation in leaves of tobacco.

Arabidopsis RNases and RNase Genes

RNases are likely to figure prominently in mRNA degradation pathways in plants. A large number of RNases have been described from various plant species, and have been classified into four major groups on the basis of characteristics such as molecular mass, pH optima, homopolymer specificity and cation requirements (Wilson, 1982). The levels of some of these RNases, or their activities, are known to respond to a variety of environmental stimuli and developmental cues, such as senescence (Matile and Winkenbach, 1971; Blank and McKeon, 1991). However, the RNases that play a role in mRNA degradation have not been differentiated from those with other functions. Without knowing which enzymes are involved, it is difficult to elucidate the mechanisms by which mRNAs are recognized and degraded.

One approach toward the identification of plant RNases that participate in the degradation of messenger RNA is to biochemically characterize as many RNases as possible from a model plant species. Using this work as a basis, it should then be possible to screen for mutants deficient in any of the RNases identified. These mutants can then be used to address the function of the corresponding RNase. Our initial studies have led to the identification of some 16 bands of RNase activity in extracts of *Arabidopsis* tissue using a gel-based RNase assay (Yen and Green, 1991). Among these are representatives of at least three of the four major classes of RNase previously described in plants (Wilson, 1982; Yen and Green, 1991). Most of these RNases are present in both stem and leaf extracts of *Arabidopsis*, but some are expressed in an organ specific fashion, either only in leaves, or only in stems (Yen and Green, 1991). The screening of M2 *Arabidopsis* plants for mutants affected in one or more of the RNase activities is now under way.

A complementary approach toward elucidating the function of RNases in *Arabidopsis* is to identify RNase genes in this species. Until recently, the only plant RNases that had been cloned were the S-RNases (McClure *et al.*, 1989b). The S-RNases, which are associated with self-incompatibility (SI) in the Solanaceae, were identified as RNases through their homology to a class of fungal RNases typified by RNase T_2 of *Aspergillus oryzae* (Kawata *et al.*, 1988; McClure *et al.*, 1989b). However, their role in SI is not clear (Haring *et al.*, 1990). Using primers directed to regions conserved between the S-RNases and the fungal RNases, we used PCR to amplify products corresponding to three *Arabidopsis* RNase genes, *RNS1*, *RNS2*, and *RNS3* (Taylor and Green, 1991). The identification of the *RNS* genes in *Arabidopsis*, which is self-compatible, demonstrates that not all RNases of this type are associated with SI. Therefore S-RNase homologs ("S-like" RNases) may perform more general roles in RNA catabolism in plants (Taylor and Green, 1991).

As an initial step in the identification of potential functions for the *RNS* gene products in *Arabidopsis*, we have studied the expression properties of the *RNS2* gene in this species. As shown in Figure 2, we have found that the expression of *RNS2* is induced during senescence in both leaves and petals of *Arabidopsis* (Taylor *et al.*, 1993). This result suggests that S-like RNases such as *RNS2* may be responsible for the marked increase in RNase activity that has been characterized during senescence in a number of plant species (Matile and Winkenbach, 1971; Blank and McKeon, 1991). The induction of catabolic enzymes, such as RNases, during senescence in plants is thought to be part of a mechanism for the remobilization of nutrients from senescing to non-senescing organs. Plants often grow under phosphate limiting conditions, so the RNases

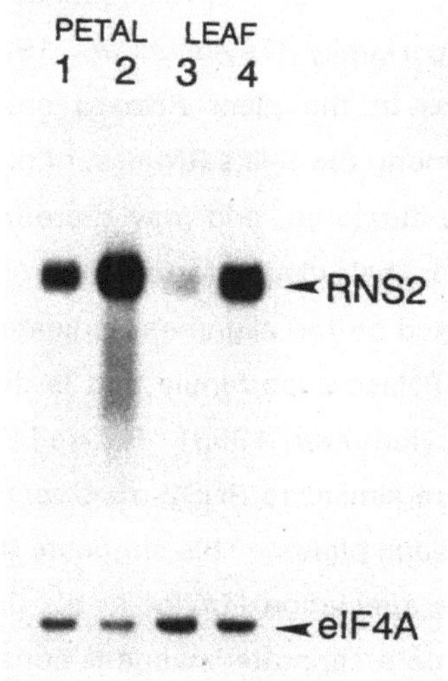

Figure 2. Induction of the *RNS2* transcript during senescence.
TOP: Samples of 5 µg of total RNA isolated from nonsenescing (lanes 1 and 3) and senescing (lanes 2 and 4) petals and leaves, as indicated above the lanes, were hybridized to the *RNS2* probe following northern blotting. The *RNS2* transcript is indicated. BOTTOM: As a control, the northern blot in the top panel was stripped and probed for the *eIF4A* transcript (see Taylor *et al.*, 1993). The arrowhead shows the *eIF4A* transcript.

induced during senescence may be involved in the recycling of phosphate. We found that the expression of *RNS2* in *Arabidopsis* is also induced under phosphate starvation conditions (Taylor *et al.*, 1993), a result that is consistent with a role for RNS2 in phosphate rescue.

In addition to studying the expression of the *RNS* genes in *Arabidopsis*, we were also interested in comparing the sequences of the S-like RNases to those of the S-RNases. To this end the *RNS2* PCR product was used as a probe to identify full-length *RNS2* cDNAs. The deduced amino acid sequence of RNS2 was aligned with those of fifteen S-RNases, four fungal RNases, and two S-like RNases for which protein sequence data are available [RNase LE of *Lycopersicon esculentum* (Jost *et al.*, 1991), and RNase MC of *Momordica charantia* (Ide *et al.*, 1991)].

This alignment indicates that there is extensive sequence conservation among the members of this RNase superfamily (Taylor *et al.*, 1993). More importantly, residues at certain positions in the plant RNases are conserved in mutually exclusive sets; either only among the S-like RNases, or only among the S-RNases. Some of these residues are clustered, and may therefore define regions of the proteins that are essential for their disparate functions.

A gene genealogy based on the alignment indicates that the S-like RNases form a subclass within the RNase superfamily that is distinct from the subclass formed by the S-RNases (Taylor *et al.*, 1993). RNase LE, an S-like RNase from a Solanaceous species, is more similar to RNS2 of *Arabidopsis* than it is to the S-RNases from other Solanaceous plants. This suggests that the S-RNases and S-like RNases diverged before speciation (Taylor *et al.*, 1993). Furthermore, the grouping of the S-RNases into a separate lineage is consistent with a specialized role for these proteins in the self-incompatible response.

Conclusions and Future Prospects

The work described here provides the foundation for future efforts aimed at elucidating mRNA decay pathways in plants. For example, DST sequences can now be used as tools in the identification of the specific components of the degradation machinery that recognize these elements. Further characterization of *Arabidopsis* RNases and RNase genes should facilitate the identification of those RNases with a specific role in mRNA degradation. Taken together, these approaches should provide fundamental information about the mechanisms that target transcripts for rapid degradation in plants. Moreover, the knowledge gained may suggest solutions to practical problems caused by the instability of some foreign transcripts in transgenic crop plants.

LITERATURE CITED

An, G. (1985) High efficiency transformation of cultured tobacco cells. Plant Physiol. 79: 568-570

Blank, A., and McKeon, T.A. (1991) Expression of three RNase activities during natural and dark-induced senescence of wheat leaves. Plant Physiol. 97: 1409-1413

Brock, M.L., and Shapiro, D.J. (1983) Estrogen stabilizes vitellogenin mRNA against cytoplasmic degradation. Cell 34: 207-214

Franco, A.R., Gee, M.A., and Guilfoyle, T.J. (1990) Induction and superinduction of auxin-responsive mRNAs with auxin and protein synthesis inhibitors. J. Biol. Chem. 265: 15845-15849

Haring, V., Gray, J.E., McClure, B.A., Anderson, M.A., and Clarke, A.E. (1990) Self-incompatibility: a self-recognition system in plants. Science 250: 937-941

Hunt, T. (1988) Controlling mRNA lifespan. Nature 38: 567-568

Ide, H., Kimura, M., Arai, M., and Funatsu, G. (1991) The complete amino acid sequence of ribonuclease from the seeds of bitter gourd (*Momordica charantia*). F.E.B.S. Lett. 284: 161-164

Jost, W., Bak, H., Glund, K., Terpstra, P., and Bientema, J.J. (1991) Amino acid sequence of an extracellular phosphate-starvation-induced ribonuclease from cultured tomato (*Lycopersicon esculentum*) cells. Eur. J. Biochem. 198: 1-6

Jones, T.R., and Cole, M.D. (1987) Rapid cytoplasmic turnover of c-myc mRNA: requirement of the 3' untranslated sequences. Mol. Cell. Biol. 7: 4513-4521

Kawata, Y., Sakiyama, F., and Tamaoki, H. (1988) Amino acid sequence of ribonuclease T_2 from *Aspergillus oryzae*. Eur. J. Biochem. 176: 683-697

Matile, P., and Winkenbach, F. (1971) Function of lysosomes and lysosomal enzymes in the senescing corolla of the Morning Glory (*Ipomoea purpurea*). J. Exp. Bot. 22: 759-771

McClure, B.A., and Guilfoyle, T. (1989) Rapid redistribution of auxin-regulated RNAs during gravitropism. Science 243: 91-93

McClure, B.A., Hagen, G., Brown, C.S., Gee, M.A., and Guilfoyle, T.J. (1989a) Transcription, organization, and sequence of an auxin-regulated gene cluster in soybean. Plant Cell 1: 229-239

McClure, B.A., Haring, V., Ebert, P., Anderson, M.A., Simpson, R.J., Sakiyama,

F., and Clarke, A.E. (1989b) Style self-incompatibility gene products of *Nicotiana alata* are ribonucleases. Nature 342: 955-957

Newman, T.C, Ohme-Takagi, M., Taylor, C.B. and Green, P.J. (1993) DST sequences, highly conserved among plant *SAUR* genes, target reporter transcripts for rapid decay in tobacco. Plant Cell: *in press*

Peltz, S.W., Brewer, G., Bernstein, P., Hart, P.A., and Ross, J. (1991) Regulation of mRNA turnover in eukaryotic cells. In Critical Reviews in Eukaryotic Gene Expression, G.S. Stein, J.L. Stein, and J.B. Lian, eds (Boca Raton: CRC Press), pp. 99-126

Shaw, G., and Kamen, R. (1986) A conserved AU sequence from the 3' untranslated region of GM-CSF mRNA mediates selective mRNA degradation. Cell 46: 659-667

Shyu, A.-B., Greenberg, M.E., and Belasco, J.G. (1989) The c-*fos* transcript is targeted for rapid decay by two distinct mRNA degradation pathways. Genes and Devel. 3: 60-72

Siflow, C.D. and Key, J.L. (1979) Stability of polysome-associated polyadenylated RNA from soybean suspension culture cells. Biochemistry 18: 1013-1018

Taylor, C.B., and Green, P.J. (1991) Genes with homology to fungal and S-gene ribonucleases are expressed in *Arabidopsis thaliana*. Plant Physiol 96: 980-984

Taylor. C.B., Bariola, P.A., delCardayré, S.B, Raines, R.T. and Green, P.J. (1993) RNS2: a senescence-associated RNase of *Arabidopsis* that diverged from the S-RNases before speciation. Proc. Natl. Acad. Sci. USA: *in press*

Theologis, A. (1986) Rapid gene regulation by auxin. Annu. Rev. Plant Physiol. 37: 407-438

Wilson, C.M. (1982) Plant nucleases: biochemistry and development of multiple molecular forms. *in* M.C. Rattazzi, S.C. Scandalios, G.S. Whitt, eds, Isozymes. Current Topics in Biological and Medical Research, Vol. 6. Alan R. Liss, Inc, New York, pp 33-54

Yamamoto, K.T., Mori, H., and Imaseki H. (1992) cDNA cloning of indole-3-acetic acid-regulated genes: Aux22 and SAUR from mung bean (*Vigna radiata*) hypocotyl tissue. Plant Cell Physiol. 33: 93-97

Yen, Y., and Green, P.J. (1991) Identification and properties of the major ribonucleases of *Arabidopsis thaliana*. Plant Physiol. 97: 1487-1493

POST-TRANSCRIPTIONAL GENE SILENCING IN TRANSGENIC PLANTS

Fernanda de Carvalho, Wout Boerjan, Ivan Ingelbrecht,
Ann Depicker, Dirk Inzé*, and Marc Van Montagu
Laboratorium voor Genetica and *Laboratoire associé de l'Institut National de la Recherche Agronomique (France),
Universiteit Gent, B-9000 Gent (Belgium)

INTRODUCTION

The recent successes of gene engineering in plants demonstrate the high potential of expressing transgenes in agronomically important crops. These approaches require that the introduced gene is stably expressed in subsequent plant generations. However, experiments involving transgenic plants show that the expression of chimeric genes can substantially vary among individual transformants.

Recently, attempts to overexpress certain genes in plants revealed the unexpected phenomenon of transgene-induced gene silencing (reviewed by Jorgensen, 1991; Kooter and Mol, 1993). Silencing occurs in some plants while others containing the same chimeric gene can show high levels of gene expression. Thus, silencing is not a direct consequence of the mere presence of the transgenes in plants. Accumulating evidence suggests that the number and configuration of transgene copies, levels of transgene expression as well as developmental and environmental factors may influence the occurrence of silencing of specific genes in plants. Furthermore, the variable frequency of silencing in plants might also reflect the different sites of transgene integration.

Transgenic plants with single copy insertions often show stable and high levels of gene expression (Hobbs *et al.*, 1990; Linn *et al.*, 1990; Mittelsen Scheid *et al.*, 1991). These authors also showed that reduced expression levels were associated with the presence of multiple transgene copies,

NATO ASI Series, Vol. H 81
Plant Molecular Biology
Edited by G. Coruzzi and P. Puigdomènech

suggesting a negative correlation between transgene insert number and gene expression. More recently, Hobbs *et al.* (1993) observed that an increased number of inserts can negatively affect gene expression but only when T-DNAs are in particular multimerized configurations (inverted or direct T-DNA repeats). It would be interesting to address whether in the suppression cases previously mentioned the transgenic plants also contain similar insert configurations as shown by Hobbs *et al.* (1993).

Increased DNA methylation has been shown to be associated with a number of transgene silencing events in plants (Linn *et al.*, 1990; Kilby *et al.*, 1992; Hobbs *et al.*, 1993; Matzke *et al.*, 1993). The first results showing a correlation between DNA methylation and transgene silencing were reported by Matzke *et al.* (1989). The introduction of a second T-DNA in transgenic tobacco plants already containing a T-DNA (T-DNA I), can lead to the inactivation of the first T-DNA encoded genes. The unidirectional suppression of T-DNA I genes is accompanied by an increase in methylation of specific sequences within these genes. In the mutual suppression phenomenon reported by Hobbs *et al.* (1993), simultaneous silencing and methylation of transgenes were also observed. However, it is currently not clear whether DNA methylation is the cause or a consequence of gene suppression.

Moreover, no correlation between DNA methylation and gene silencing was found in other related suppression cases (Goring *et al.*, 1991; Mittelsen Scheid *et al.*, 1991; Hart *et al.*, 1992). However, one can not exclude the possibility that potential methylated sites escaped the analyses performed. Alternatively, it may also suggest that different mechanisms of transgene-induced silencing exist in plants.

Introduction of chimeric genes in plants can also lead to suppression of endogenous plant genes. In almost all reported cases, the transgenes and homologous endogenous genes are simultaneously co-suppressed in the transgenic plants (Napoli *et al.*, 1990; Smith *et al.*, 1990; van der Krol *et al.*, 1990; Gottlob-McHugh *et al.*, 1992; Hart *et al.*, 1992). The only exception was reported by Elkind *et al.* (1990), where the introduction of a chimeric bean phenylalanine ammonia lyase (PAL) gene in tobacco results in the inactivation of only the endogenous PAL genes, probably due to feedback inhibition.

Other studies revealed that post-transcriptional mechanisms are responsible for suppression of gene expression (Mol *et al.*, 1991; de Carvalho *et al.*, 1992).

There might be mechanistic similarities between the numerous cases of suppression reported, but final conclusions whether they represent similar or distinct phenomena can still not be drawn. Therefore, further investigations should be done to unravel the mechanism(s) of gene silencing in plants.

In our laboratory we have identified and characterized in detail three suppression phenomena in transgenic plants. The results obtained from these analyses bring new, pertinent contributions to the understanding of co-suppression mechanisms in plants.

SUPPRESSION OF β-1,3-GLUCANASE GENES IN TRANSGENIC TOBACCO PLANTS

Plant β-1,3-glucanase genes constitute a small gene family coding for acidic and basic isoforms localized at different sub-cellular compartments. These hydrolytic enzymes are thought to be implicated in plant defense mechanisms and may also play a role in basic aspects of plant development (for review see Carr and Klessig, 1990; Eyal and Fluhr, 1991; Linthorst, 1991).

To address the potential physiological functions of β-1,3-glucanases in plants, transgenic tobacco plants producing constitutive high levels of this enzyme were generated. For this approach, the cauliflower mosaic virus (CaMV) 35S promoter was used to drive the expression of a β-1,3-glucanase gene (*gn1*) isolated from *Nicotiana plumbaginifolia* (De Loose *et al.*, 1988; Castresana *et al.*, 1990).

These studies revealed an interesting phenomenon of gene silencing in the progeny of the highest expressing transgenic plant (T17). While T17 hemizygous plants expressed high levels of GN1 ß-1,3-glucanase, expression of the chimeric *gn1* gene was completely blocked in T17 homozygous plants: no *gn1* steady state mRNA and GN1 protein could be detected. In addition, suppression was specific for *gn1* since the expression of the chimeric *npt*II marker gene present on the same T-DNA was not affected.

The characterization of this suppression phenomenon has been reported (de Carvalho *et al.*, 1992). Run-off transcription assays demonstrated that the p35S-*gn1* gene was transcribed in both hemizygous and homozygous plants; the rate of transcription was clearly higher in the homozygous plants. Therefore, the strong differences in *gn1* steady-state mRNA levels in T17 homozygous and hemizygous plants depend on post-transcriptional processes. Furthermore, *gn1* suppression occurs not only in homozygous plants but also in haploid tobacco plants derived from both homozygous and hemizygous T17 plants. Interestingly, the ratio between the transgene number and the genome copy is identical in the silenced homozygous and haploid plants (one transgene per haploid genome). Therefore, ß-1,3-glucanase transgene silencing is correlated with the transgene dose in the plant genome, which might reflect the levels of transgene expression.

To progress towards the elucidation of the mechanism of *gn1* suppression, further experiments were performed.

Induction of defense responses in T17 transgenic plants

We have previously reported that endogenous ß-1,3-glucanases were not co-suppressed together with GN1 in T17 homozygous plants (de Carvalho *et al.*, 1992). However, possible differences might have been overlooked due to the low ß-1,3-glucanase activity levels in healthy plants.

To clarify this point, the accumulation of ß-1,3-glucanases was induced in the plants by stress treatments such as 5 mM salicylic acid spraying and bacterial infection of plant leaves. By performing isoelectric focusing ß-1,3-glucanase assays, it was possible to distinguish the different isoforms in the plants. Those experiments revealed that a basic ß-1,3-glucanase isoform (pI≈ 8.3) from *Nicotiana tabacum* was co-suppressed together with GN1 in T17 homozygous plants. Interestingly, an acidic ß-1,3-glucanase isoform (pI ≈4.2) from *N. tabacum* was not affected, accumulating at similar levels as detected in untransformed plants. Northern blot analysis indicated that the inhibition of the endogenous basic ß-1,3-glucanase activity is correlated with the absence of the corresponding mRNA transcript. Since the *N. plumbaginifolia gn1* gene shows higher homology to the basic (80%) than to the acidic (60%) *N.tabacum* ß-1,3-glucanase genes, the suppression mechanism is likely to operate post-transcriptionally by recognizing specific sequences only present in the *gn1* and the basic *N. tabacum* ß-1,3-glucanase transcripts.

In contrast, acidic and basic *N.tabacum* ß-1,3-glucanases accumulated to high levels in stressed hemizygous plants. However, the induced levels of endogenous ß-1,3-glucanases were much lower than the constitutive GN1 levels. If increased expression levels are important to trigger the suppression phenomenon, one could postulate that the final ß-1,3-glucanase levels (endogenous+GN1) are still not high enough to trigger the suppression mechanism in stressed hemizygous plants.

Analysis of *gn1* expression during plant development

A detailed analysis of the developmental suppression of the p35S-*gn1* gene was performed. Northern blot analysis demonstrated that *gn1* is detected at high levels during the first weeks of plant development in the T17 homozygous plants. Then, it decreases reaching undetectable levels 7-8 weeks after seed germination in soil. *gn1* mRNA accumulates at similar levels in all the developmental stages examined in T17 hemizygous plants. In addition, the *N. tabacum* basic isoform is suppressed together with GN1 during plant development. These results suggest that disappearance of the ß-1,3-glucanase transcripts in homozygous plants is correlated with a certain stage of plant development.

Transformation of SR1 tobacco and T17 homozygous plants

Suppression of *gn1* expression was only observed in the homozygous plants derived from the highest expressing plant (T17). To test whether this phenomenon could be reproduced, new independent transgenic plants containing the p35S-*gn1* gene were generated.

Twelve independent SR1 transformants were examined. Two of them, containing multiple T-DNA copies, showed the co-suppression phenomenon. This was visualized by the absence of both GN1 and the endogenous basic ß-1,3-glucanase. These results demonstrate that this suppression is a more general phenomenon that can occur in independent transgenic plants. In addition, new transformants producing similar levels of *gn1* expression as T17 hemizygous plants were also obtained. Analysis of their progenies will be done to evaluate whether the *gn1* expression is also suppressed in the homozygous plants. In parallel, crosses between these plants and T17 homozygous plants will be performed to test whether the suppression can also be achieved with two independent T-DNA loci derived from two highly expressing plants.

In addition, T17 homozygous plants were retransformed with extra p35S-*gn1* gene copies to address whether the suppression phenomenon in homozygous plants was due to the high levels of *gn1* expression. In twelve out of seventeen T17 double-transformants examined, the newly introduced *gn1* copies were suppressed. These results indicate that the suppression phenomenon observed in T17 plants effectively inactivates other *gn1* copies, integrated at independent positions in the plant genome. However, five out of 17 plants had restored the expression of *gn1* as well as of the endogenous basic ß-1,3-glucanase to similar levels as observed in T17 hemizygous plants. Therefore, those plants are likely to represent revertants of the initial suppression. However, it is not know yet if the reversion is associated with the presence of the second T-DNA or if it was activated by tissue culture procedures. To address this question, we will examine whether retransformation of T17 homozygous plants with vector DNA could also generate revertants of the *gn1* suppression.

In conclusion, transgenes and homologous host genes coding for ß-1,3-glucanases can be co-suppressed in certain transgenic plants. The suppression mechanism occurs during plant development resulting in the disappearance of mRNA transcripts. This post-transcriptional suppression is likely to operate by recognizing specific sequences found in the *gn1* and the endogenous basic ß-1,3-glucanase gene transcripts. In addition, our results strongly suggest that the silencing of the ß-1,3-glucanase genes is triggered by the high levels of transgene expression in the homozygous plants.

OVEREXPRESSION AND SILENCING OF *S*-ADENOSYL-L-METHIONINE SYNTHETASE IN TOBACCO AND *ARABIDOPSIS*

S-adenosyl-L-methionine synthetase (SAM-S) catalyses the conversion of ATP and L-methionine into *S*-adenosylmethionine (SAM). SAM is a cofactor in numerous metabolic reactions and is the major methyl group donor for transmethylation reactions in both prokaryotes and eukaryotes. After decarboxylation, it serves as a propylamine group donor in the biosynthesis of polyamines. In plants, SAM is furthermore involved in the biosynthesis of the phytohormone ethylene and in the regulation of the biosynthesis of the amino acids derived from aspartate.

An *Arabidopsis* gene (*sam1*) encoding a SAM-S was isolated (Peleman *et al.*, 1989a). Analysis of a *sam1* promoter-reporter gene fusion in transgenic *Arabidopsis* and tobacco plants revealed that the *sam1* gene is preferentially expressed in vascular tissue (Peleman *et al.*, 1989a, 1989b). To investigate the consequence of overproducing SAM-S in plants, tobacco and

Arabidopsis were transformed with a T-DNA, containing the *Arabidopsis sam1* gene under the control of the constitutive CaMV 35S promoter.

Overexpression of *sam1* in tobacco

Progeny of five transgenic lines, having high *sam1* steady state mRNA levels as judged by Northern analysis, were selected on kanamycin sulphate and the resistant plants were grown in soil. After two months of growth in the greenhouse, a phenotypical analysis was made. These plants developed abnormal phenotypes: the leaves were yellow-green with dark-green veins and developed necrotic spots, that in the severely affected plants led to extensive death of the mesophyll cells. Severely affected plants were also reduced in height. Plants presenting this kind of phenotypes are referred to as T2 class OV plants (OVerexpression phenotypes; Figure 1).

T2 class OV plants with different degrees of phenotypic abnormalities were observed. SAM-S activity measurements indicated that the highest activity levels were detected in the plants displaying the most severe symptoms. Moreover, a gradient in SAM-S activity was observed in both wild-type and transgenic plants: the youngest leaves contain much higher SAM-S activities compared to mature leaves. Since similar *sam1* steady-state mRNA levels were detected in young and mature leaves, the gradient of SAM-S activity is established by (a) post-transcriptional mechanism(s).

Further analysis on the phenotype of the T2 plants, demonstrated that a number of plants developed one or more dark-green sectors on the leaves. Most of these sectors were circular, oval or irregular spots and in some cases they appeared to be associated with the veins. The development of the dark-green tissue bordering the leaf veins was often established at early stages of plant development (Figure 1). The relative surface area of this dark-green tissue increased with each successive leaf and as a result, most leaves of the mature plant were dark-green. Thus, the lower leaves of these plants have a patterned phenotype: part of the leaf is yellow-green and part is dark-green. Interestingly, these plants developed a morphology that was completely different compared to their T2 class OV siblings. Among the most striking symptoms were: thick, asymmetric and leather-like leaves (that tear as a result of tension), a reduction in height and, in the most severely affected plants, the inability to develop an inflorescence. These plants developed lateral shoots that were able to flower and set seed. Plants presenting this phenotype are referred to as T2 class S (Suppressed phenotype) and T2 class SN (Suppressed Normal) plants, depending on the severity of the phenotype (Figure 1).

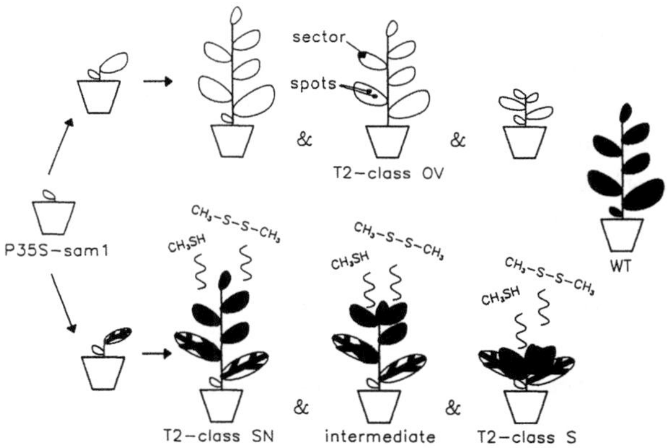

Figure 1. Schematic presentation of the different ways of development of T2 tobacco plants transformed with the chimeric P35S-*sam1* gene.
White corresponds to overexpressed tissue and black to suppressed tissue. T2 class OV plants exhibit different degrees of overexpression phenotypes (normal-sized plants and smaller plants). Suppressed tissue appears on T2 class OV plants as small spots or sectors. T2 seedlings can also develop into plants that have partially an overexpression and partially a suppressed phenotype. These plants further develop either into T2 class S plants that are stunted, into T2 class SN plants that exhibit a nearly wild-type phenotype or into intermediates between T2 class S and T2 class SN plants. Suppressed plants emit methanethiol and dimethyl disulfide. WT, wild-type.

SAM-S activity in the dark-green sectors was 8- to 18-fold lower compared to the neighboring pale-green tissue. In addition, in young as well as in mature leaves of T2 class S plants, SAM-S activities were extremely low compared to those detected in wild-type plants, suggesting co-suppression of the endogenous tobacco SAM-S genes and the transgene. These lower activity levels were correlated with undetectable levels of *sam1* mRNA.

The most remarkable phenotype of suppressed plants (T2 class S, T2 class SN plants and intermediates) is their cabbage-like smell which is caused, as shown by GC-MS analysis, by the production of methanethiol and its oxidation product dimethyl disulfide. Suppression of SAM-S leads to the accumulation of methionine that can be converted to methanethiol, which is volatile.

The developmental pattern of transgene silencing can be recognized due to the distinct phenotypes generated by the overexpression (yellow-green tissue) and the suppressed (dark-green tissue) states. As this can occur in one same leaf, this system will allow us to investigate different aspects of this gene silencing in detail.

Overexpression of *sam1* in *Arabidopsis*

Transgenic *Arabidopsis* plants overexpressing *sam1* have up to 16-fold higher SAM-S activity levels in rosette leaves compared to wild-type plants. Nevertheless, these plants are phenotypically wild-type. However, the progenies of three transformants having the highest SAM-S activity levels could be grouped in three main phenotypically distinct classes (MUT1, MUT2 and MUT3). MUT1 plants have a wild-type phenotype and show up to 16-fold higher SAM-S activity levels. MUT2 plants develop a high number of short stems and have a low fertility. MUT3 plants remain extremely small, do not develop an inflorescence or only a rudimentary one, and accumulate high amounts of anthocyanins. In addition, all intermediates between these classes were observed. Both MUT2 and MUT3 plants have extremely low SAM-S activity levels. Northern blot analysis of MUT2 stems indicates that the expression of both the transgene and the endogenous *sam1* genes is silenced. Northern blot analysis on a pool of MUT2 rosette leaves shows only reduced *sam1* steady-state mRNA levels compared to the levels found in MUT1 plants. Since stems develop later than rosette leaves, this suggests that a MUT2 plant initially develops as a plant that overexpresses *sam1*. At a certain stage of development, the expression of both *sam1* genes is silenced and this state is maintained throughout further development. The fact that intermediate phenotypes were observed, suggests that the switch from the overexpression state to the co-suppressed state can occur at any developmental stage. Segregation data indicate that the co-suppression phenomenon is not correlated with the zygosity of the transgene.

REVERSIBLE SILENCING OF REPORTER GENES IN TOBACCO

Chimeric genes consisting of the neomycin phosphotransferase II (*npt*II) coding sequence under the control of the CaMV 35S promoter were fused to different parts of the 3'-untranslated region of the *Anthirrinum majus* chalcone synthase gene, and introduced into SR1 tobacco plants using the *Agrobacterium* vector system. The hygromycin phosphotransferase (*hpt*) gene was used as a selectable marker. Irrespective of the construct used, transformed plants could be grouped into two classes: plants exhibiting high levels of *npt*II transcript, as expected for the expression directed by the strong 35S promoter, and plants with substantially lower *npt*II mRNA levels.

A comparative study of selected plants with normal (high) and reduced *npt*II gene expression was performed. Northern blot analysis indicated that the *npt*II mRNA levels were at least 20-fold lower in plants displaying low expression than in plants with normal levels of expression. In contrast, *hpt* mRNA levels were comparable in both sets of plants. Subsequently, run-off transcription analysis was performed to investigate whether reduced *npt*II mRNA levels correlated with reduced transcriptional activity of the *npt*II genes. Using an endogenous gene encoding the small subunit of ribulose-1,5-bisphosphate carboxylase (*rbcS*) and a ribosomal gene as internal controls, it was shown that *npt*II transcriptional activity was as high in plants with reduced *npt*II gene expression as in plants with normal expression levels, indicating that reduced expression levels are due to a post-transcriptional mechanism.

The T-DNA copy number and configuration was investigated by genetic and physical methods. This analysis showed that plants with reduced expression levels always contained multiple copies of the *npt*II gene. Plants with normal expression levels contained one or more T-DNA copies. Interestingly, the 5 plants showing reduced expression levels all contained at least one T-DNA inverted repeat centered about the right border.

Southern blot analyses reproducibly showed that DNA isolated from transgenic plants exhibiting reduced *npt*II mRNA levels was partially digested by the methylation sensitive enzyme *Bam*HI. The hybridization pattern suggested that in transgenic plants with reduced *npt*II gene expression a *Bam*HI site located in the 3′ portion of the *npt*II gene was (partially) modified.

This result was confirmed by analyzing a plant with low expression levels which contained three T-DNA copies: two of them linked as an inverted repeat (locus 1) and a third one (locus 2) which segregated independently. Analysis of progeny plants obtained after self-fertilization of this plant showed that progeny plants containing only T-DNA locus 2 had high levels of *npt*II transcript indicating that this co-suppression could be reversed. More importantly, the restoration of *npt*II gene expression was associated with loss of the DNA modification. This shows a correlation between transgene silencing and DNA modification and, conversely, between restoration of gene expression and loss of the DNA modification.

In plant cells, cytosine methylation is known to occur almost exclusively at CG or CXG sequences (Gruenbaum *et al.*, 1981). As *Bam*HI is sensitive to cytosine methylation it is obvious to conclude that the *Bam*HI sites in the silenced *npt*II genes are methylated. However, *Bam*HI does not contain CG or CXG in its recognition sequence and inspection of the neighboring DNA sequence indicated that in only four out of eight cases a CXG motif is present. This suggests that methylation in a sequence context different from CG or CXG is a frequent event in plants.

By using isoschizomeric restriction enzymes showing differential sensitivity to cytosine methylation, it was formally demonstrated that the *Bam*HI sites located in the 3' portion of the *npt*II gene were methylated. In addition, at least in one case, methylated cytosines also occurred in the promoter region. Even in this plant the silenced *npt*II genes were transcribed as efficiently as their non-methylated counterparts in plants showing normal *npt*II expression levels.

This latter finding might prove important with respect to different reports of transgene silencing in plants. Indeed, some gene silencing phenomena have been shown to be due to a post-transcriptional process, whereas in other cases a reversible correlation with DNA methylation has been observed. The fact that the gene silencing phenomenon described here is correlated with DNA methylation and at the same time is due to a post-transcriptional event might reveal a mechanistic relationship between some cases of transgene silencing in plants.

GENERAL CONCLUSIONS

The introduction of chimeric sense constructs in plants is certainly a very powerful tool for manipulating gene expression. This strategy allows the simultaneous analysis of the effects produced by overproduction and inhibition of host genes in plants. The variety of reported results indicate that the homology-based phenomena of gene silencing in plants arise from a combination of different factors. However, whether the distinct cases of suppression are governed by the same mechanism is still not known.

Many questions remain to be addressed: how are homologous transgenes and endogenous genes recognized and silenced simultaneously? Which factors control the developmental co-suppression phenomena? Which post-transcriptional mechanism is operating in silencing?

Following gene transcription, several post-transcriptional events can play important roles in establishing the final steady-state mRNA levels such as: formation and maturation of primary transcript, transport of mature mRNA to the cytoplasm from the nucleus, and control of mRNA stability. Identification of the post-transcriptional control mechanism will bring important information for the understanding of the silencing mechanisms. In addition, a yet unsolved problem for future study is how exactly DNA methylation can be associated to post-transcriptional processes. The elucidation of the mechanisms governing transgene-induced silencing in plants will bring a novel understanding of the functional relation between nuclei organization and gene expression. Furthermore, the wide occurrence of silencing in transgenic systems may indicate that similar processes are naturally happening in plants.

Acknowledgements

The authors thank M. De Cock and V. Vermaercke for help in preparing the manuscript and figures. This work was supported by grants from the Belgian Programme on Interuniversity Poles of Attraction (Prime Minister's Office, Science Policy Programming, #38) and the Vlaams Actieprogramma Biotechnologie (ETC 002). FdC is indebted to the Conselho Nacional de Desenvolvimento Científico e Tecnológico (CNPq 200353/90-0) for a predoctoral fellowship; DI is a Research Director of the Institut National de la Recherche Agronomique (France).

REFERENCES

Carr, J.P., and Klessig, D.F. (1990). The pathogenesis-related proteins of plants. In *Genetic engineering, Principles and Methods*, Vol. 11, J.K. Setlow (Ed.). New York, Plenum Press, 65-109.

Castresana, C., de Carvalho, F., Gheysen, G., Habets, M., Inzé, D., and Van Montagu, M. (1990). Tissue-specific and pathogen-induced regulation of a *Nicotiana plumbaginifolia* β-1,3-glucanase gene. *Plant Cell* **2:** 1131-1143.

de Carvalho, F., Gheysen, G., Kushnir, S., Van Montagu, M., Inzé, D., and Castresana, C. (1992). Suppression of β-1,3-glucanase transgene expression in homozygous plants. *EMBO J.* **11:** 2595-2602.

De Loose, M., Alliotte, T., Gheysen, G., Genetello, C., Gielen, J., Soetaert, P., Van Montagu, M., and Inzé, D. (1988). Primary structure of a hormonally regulated β-glucanase of *Nicotiana plumbaginifolia*. *Gene* **70:** 13-23.

Elkind, Y., Edwards, R., Mavandad, M., Hedrick, S.A., Ribak, O., Dixon, R.A., and Lamb, C.J. (1990). Abnormal plant development and down-regulation of phenylpropanoid biosynthesis in transgenic tobacco containing a heterologous phenylalanine ammonia-lyase gene. *Proc. Natl. Acad. Sci. USA* **87:** 9057-9061.

Eyal, Y., and Fluhr, R. (1991). Cellular and molecular biology of pathogenesis related proteins. *Oxford Surv. Plant Mol. Cell Biol.* **7:** 223-254.

Goring, D.R., Thomson, L., and Rothstein, S.J. (1991). Transformation of a partial nopaline synthase gene into tobacco suppresses the expression of a resident wild-type gene. *Proc. Natl. Acad. Sci. USA* **88:** 1770-1774.

Gottlob-McHugh, S.G., Sangwan, R.S., Blakeley, S.D., Vanlerberghe, G.C., Ko, K., Trupin, D.H., Plaxton, W.C., Miki, B.L., and Dennis, D.T. (1992). Normal growth of transgenic tobacco plants in the absence of cytosolic pyruvate kinase. *Plant Physiol.* **100:** 820-825.

Gruenbaum, Y., Naveh-Many, T., Cedar, H., and Razin, A. (1981). Sequence specificity of methylation in higher plant DNA. *Nature* (London) **292:** 860-862.

Hart, C.M., Fischer, B., Neuhaus, J.-M., and Meins, F. Jr (1992). Regulated inactivation of homologous gene expression in transgenic *Nicotiana sylvestris* plants containing a defense-related tobacco chitinase gene. *Mol. Gen. Genet.* **235:** 179-188.

Hobbs, S.L.A., Kpodar, P., and DeLong, C.M.O. (1990). The effect of T-DNA copy number, position and methylation on reporter gene expression in tobacco transformants. *Plant Mol. Biol.* **15:** 851-864.

Hobbs, S.L.A., Warkentin, T.D., and DeLong, C.M.O. (1993). Transgene copy number can be positively or negatively associated with transgene expression. *Plant Mol. Biol.* **21**: 17-26.

Jorgensen, R. (1991). Silencing of plant genes by homologous transgenes. *AgBiotech* **4**: 265N-273N.

Kilby, N.J., Leyser, H.M.O., and Furner, I.J. (1992). Promoter methylation and progressive transgene inactivation in *Arabidopsis*. *Plant Mol. Biol.* **20**: 103-112.

Kooter, J.M., and Mol, J.N.M. (1993). *Trans*-inactivation of gene expression in plants. *Curr. Opin. Biotechnol.* **4**: 166-171.

Linn, F., Heidmann, I., Saedler, H., and Meyer, P. (1990). Epigenetic changes in the expression of the maize A1 gene in *Petunia hybrida*: role of numbers of integrated gene copies and state of methylation. *Mol. Gen. Genet.* **222**: 329-336.

Linthorst, H.J.M. (1991). Pathogenesis-related proteins in plants. *Crit. Rev. Plant Sci.* **10**: 123-150.

Matzke, M.A., Primig, M., Trnovsky, J., and Matzke, A.J.M. (1989). Reversible methylation and inactivation of marker genes in sequentially transformed tobacco plants. *EMBO J.* **8**: 643-649.

Matzke, M.A., Neuhuber, F., and Matzke, A.J.M. (1993). A variety of epistatic interactions can occur between partially homologous transgene loci brought together by sexual crossing. *Mol. Gen. Genet.* **236**: 379-386.

Mittelsen Scheid, O., Paszkowski, J., and Potrykus, I. (1991). Reversible inactivation of a transgene in *Arabidopsis thaliana*. *Mol. Gen. Genet.* **228**: 104-112.

Mol, J., Van Blokland, R., and Kooter, J. (1991). More about co-suppression. *Trends Biotechnol.* **9**: 182-183.

Napoli, C., Lemieux, C., and Jorgensen, R. (1990). Introduction of a chimeric chalcone synthase gene into petunia results in reversible co-suppression of homologous genes *in trans*. *Plant Cell* **2**: 279-289.

Peleman, J., Saito, K., Cottyn, B., Engler, G., Seurinck, J., Van Montagu, M., and Inzé, D. (1989a). Structure and expression of the *S*-adenosylmethionine synthetase gene family in *Arabidopsis thaliana*. *Gene* **84**: 359-369.

Peleman, J., Boerjan, W., Engler, G., Seurinck, J., Botterman, J., Alliotte, T., Van Montagu, M., and Inzé, D. (1989b). Strong cellular preference in the expression of a housekeeping gene of *Arabidopsis thaliana* encoding *S*-adenosylmethionine synthetase. *Plant Cell* **1**: 81-93.

Smith, C.J.S., Watson, C.F., Bird, C.R., Ray, J., Schuch, W., and Grierson, D. (1990). Expression of a truncated tomato polygalacturonase gene inhibits expression of the endogenous gene in transgenic plants. *Mol. Gen. Genet.* **224**: 477-481.

van der Krol, A.R., Mur, L.A., Beld, M., Mol, J.N.M., and Stuitje, A.R. (1990). Flavonoid genes in petunia: addition of a limited number of gene copies may lead to a suppression of gene expression. *Plant Cell* **2**: 291-299.

PLANT GENES ENCODING HOMOLOGUES OF THE SNF1 AND SHAGGY PROTEIN KINASES

Martin Kreis[1], Michele W. Bianchi[1], Véronique Ferrant[1], Laurence Le Guen[1], Martine Thomas[1], Nigel G. Halford[2], Jacqueline H. A. Barker[2], Ulrich Hannappel[2], Jesus Vicente-Carbajosa[2] and Peter R. Shewry[2].
1) Université Paris-XI, Centre de Recherches sur les Plantes, URA 1128, Biologie du Développement des Plantes, F-91405, Orsay Cédex, Fr.
2) Department of Agricultural Sciences, University of Bristol, AFRC, Institute of Arable Crops Research, Long Ashton Research Station, Long Ashton, Bristol BS18 9AF, UK.

INTRODUCTION

Critical enzymes and proteins are often regulated by the reversible addition of phosphate. The family of enzymes catalyzing this reaction is huge and diverse. There are two major classes of protein kinases: those having specificity for serine and threonine and those enzymes that phosphorylate tyrosine residues. Phosphatases reverse the effects of protein kinases and also play important roles in cellular physiology.

All protein kinases share a conserved catalytic domain of about 260 to 300 residues. Throughout this core sequence are spread 11 conserved subdomains. Some interact directly with the substrate or Mg-ATP, while others appear to convey only structural stability. The eleven subdomains are separated by more variable regions where deletions and insertions are allowed. Based on sequence alignments, the protein kinases have been devided into subfamilies (see Hanks et al 1988; Hanks 1991). The relatedness of protein kinases implies that all Ser/Thr and Tyr protein kinases have evolved from one single archetypal gene.

Reversible phosphorylation is recognized as one of the most widely used mechanisms for signal transduction in eukaryotic cells.

NATO ASI Series, Vol. H 81
Plant Molecular Biology
Edited by G. Coruzzi and P. Puigdomènech

TABLE 1. PLANT GENES ENCODING PROTEIN KINASES*

Name	Origin	Homology group	SPECIFICITY** Auto-phosphorylation	SPECIFICITY** Substrate phosphorylation	Functional study	Reference
A. Homologues of evolutionary conserved proteins						
cdc2	Pea	cdc2/cdc28	-	-	-	Feiler and Jacobs (1990)
cdc2Ms	Alfalfa	cdc2/cdc28	-	-	**Yeast c.**	Hirt et al (1991)
cdc2 ZmA	Maize	cds2/cdc2	-	-	**Yeast c.**	Colasanti et al (1991)
R2	Rice	cdc2/cdc28	-	-	-	Hata (1991)
Aracdc2	Arabidopsis	cdc2/cdc28	-	-	**Yeast c.**	Ferreira et al (1991)
RKIN1	Rye	SNF1	-	-	**Yeast c.**	Alderson et al (1991)
AKin10	Arabidopsis	SNF1	-	-	-	Le Guen et al (1992)
cBKin12	Barley	SNF1	-	-	-	Halford et al (1992)
ASK	Arabidopsis	GSK-3	**Ser/Thr/Tyr**	**Ser/Thr**	-	Bianchi et al (1993)
MsK	Alfalfa	GSK-3	-	-	-	Pay et al (1993)
B. Other protein kinases						
PVRK-1	Bean	PKC-related	-	-	-	Lawton et al (1989)
G11A	Rice	PKC-related	-	-	-	Lawton et al (1989)
ZmPK1	Maize	Receptor protein kinase	-	-	-	Walker and Zhang (1990)
SRK-910	B. Oleracea	Receptor protein kinase	**Ser/Thr**	-	**S-locus**	Goring and Rothstein (1992)
TMK1	Arabidopsis	Receptor protein kinase	**Ser/Thr**	-	-	Chang et al (1992)
SRK6	Colza	Receptor protein kinase	-	-	**S-locus**	Stein et al (1991)
ATPK64	Arabidopsis		-	-	-	Mizoguchi et al (1992)
Phycer	Moss	-	**n. d.**	-	-	Algarra et al (1993)
pSK5	Soja	-	-	-	-	Harper et al (1991)
DcPK431	Carotte	-	-	-	-	Suen et al (1991)
MsERK1	Alfalfa	MAP kinase	**Tyr**	**n. d.**	-	Duerr et al (1993)
CTR1	Arabidopsis	Raf	-	-	**mutant**	Kieber et al (1993)
GmPK6	Soybean	-	-	-	-	Feng et al (1993)
ATPK7	Arabidopsis	-	-	-	-	Hayashida et al (1992)
APK1	Arabidopsis	-	**Ser/Thr**	**Ser/Thr**	-	Hirayama and Oka (1992)

Protein kinase cascades regulate a wide range of cellular processes in animals and yeast inluding the determination of cell identity, metabolic regulation and responses to external stimuli. Although only a few plant genes encoding protein kinases have been isolated and characterized (see table 1 of this communication), recent results show that reversible phosphorylation is equally important in higher plants.

The multiorganismal approach to signal transduction has revealed that most fundamental pathways have been conserved to different extents during evolution. Protein kinases acting in these pathways reveal structural and functional evolutionary conservation.

IDENTIFICATION AND CHARACTERIZATION OF *shaggy/GSK-3* HOMOLOGUES IN PLANTS

The homology cloning approach

Due to the recognition of the similarity between protein kinases a large number of genes have been isolated using "homology cloning". We used degenerate oligonucleotides (see also Lawton et al. 1989, 1990) targeting conserved motifs in protein kinases of the glycogen synthase kinase-3 (GSK-3) subfamily to PCR-amplify related plant sequences. Using this approach we have identified in *Arabidopsis thaliana* and *Petunia hybrida* L. (Bianchi et al 1993; Ferrant et al 1993) gene sequences encoding putative plant homologues of GSK-3. Related genes have also been identified in *Medicago sativa* by Pay et al (1993), using oligonucleotides deduced from the conserved sequences of the cdc2 kinases. The involvement of protein kinases of the GSK-3 subfamily in fundamental cellular processes, and their evolutionary

---

TABLE 1. PLANT GENES ENCODING PROTEIN KINASES

*Only complete genes and full length cDNA sequences are reported.

** In vitro analysis of proteins produced in *E. coli*.

n. d: the authors show auto or substrate phosphorylation but no phosphoamino acid analysis.

S-locus: S-locus linked.

Yeast c: yeast complementation; PKC: Protein kinase C.

mutant: ctr1 mutant of the ethylene signal transduction pathway.

conservation, makes it likely that their plant homologues will serve important biological functions.

Biological roles of GSK-3 and shaggy/zeste-white 3

The GSK-3 subfamily contains three protein kinases: shaggy/zeste-white 3 (sgg/zw3) (the gene is known as *shaggy* or *zeste-white 3*) from *Drosophila melanogaster* (Siegfried et al 1990), glycogen synthase kinase-3 (GSK-3) from mammals (Woodgett 1990), and MCK1 from *Saccharomyces cerevisiae* (Neigerborn and Mitchell 1991; Shero and Hieter 1991). These proteins share several distinct sequence motifs in the catalytic domain which clearly identifies them as members of an evolutionary conserved group.

Using genetic and molecular analyses it has been shown that the *sgg/zw3* gene specifies products that are required for the formation of the anterior-posterior compartment boundaries of the embryo, and other developmental processes. Recently Siegfried et al (1992) reported that *sgg/zw3* acts as a repressor of *engrailed* autoregulation. *engrailed* is a homeobox gene necessary for the specification of the fate of a subset of cells in each segment of the *Drosophila* embryo (Heemskerk et al 1993). Furthermore their experiments indicated that the gene product of *wingless* operates by inactivating sgg/zw3 repression of *engrailed*, hence also involved in position dependent cell differentiation (Perrimon and Smouse 1989; Limbourg-Bouchon et al 1991; Nusse and Varmus 1992; Siegfried et al 1990; Siegfried et al 1992; Perimon and Smouse 1989; Preat et al 1990).

GSK-3 was initially identified as one of the kinases capable of phosphorylating glycogen synthase. However, the enzyme has recently been shown to act *in vitro* on the product of c-*jun*, one of the components of a dimeric complex that is a member of the AP-1 transcription factor family. GSK-3 selectively phosphorylates *in vitro* several other proteins including the phosphatase inhibitor-2 (I-2), myelin basic protein and the c-*myb* and c-*myc* proto-oncogene products (Boyle et al 1991, for a review see Plyte et al 1992). The activity of GSK-3 has been implicated in the modulation of a wide variety of key regulatory elements all of which seem to be components of signaling pathways in the transduction of hormonal or growth factor stimuli. Results by Siegfried et al (1992) and Ruel et al (1993b) have shown that GSK-3 and sgg/zw3 are functionally homologous *in vivo*, since expression of *GSK-3* in *sgg/zw3* mutant embryos restores these embryos to wild type.

PCR identification of *Arabidopsis thaliana* and *Petunia hybrida sgg/zw3* related sequences

We have isolated plant sequences homologous to *sgg/zw3* using polymerase chain reaction and primers targeted to the conserved catalytic sub-domains VIb and VIII of the GSK-3 protein kinases. The strategy for the isolation of the plant *sgg/zw3* homologues was as follows. 160bp cDNA fragments amplified from cDNA libraries derived from *Arabidopsis* leaves and Petunia embryo sacs were cloned into pBS (Bianchi et al 1993; Ferrant et al 1993). Amplification of genomic DNA from *Arabidopsis thaliana* yielded several products ranging in size from 250 to 360bp, which were also cloned. The difference in size observed between the genomic and cDNA PCR products is due to the presence of introns. Screening of these "mini PCR-generated" libraries led to the identification of distinct amplification products. Nucleotide sequence analysis showed that these contained inserts encoding the amino acids targeted by the oligonucleotide primers. The deduced peptide sequences contained all the invariant residues, short conserved motifs and the characteristic subdomain VII (DFG) at the same relative position compared to other protein kinases (Hanks et al 1988; Hanks 1991). The cloned sequences, were termed ASK (*Arabidopsis* shaggy related protein kinase) and PSK (*Petunia* shaggy related protein kinase). Sequence comparisons of the *Arabidopsis* cDNA and genomic PCR products suggests the presence of a minimum of 5 genes. Hence, in contrast to *Drosophila* the plant transcripts do not originate from differential splicing.

Characterization of cDNA clones from *Arabidopsis*

Subsequently using the above "mini-inserts" as probes Bianchi et al (1993) isolated and characterized two full length *Arabidopsis* cDNA clones named, ASK-α and ASK-γ, encoding polypetides with a Mr of 45994 and 46557 repectively. Comparisons of the two *Arabidopsis* protein kinase sequences to other kinases in the data base showed as

```
1.  1  MASVGIAPNPGARDSTG----VDKLPEEMNDMKIRDDKEMEATVVDGNGTETGHII
2.  1 MM**G*V**AS*FI*KNASSVGVE*****************A**I************
3.  1                         MSGRPRTTSFAESC*PVQQPS-AFGSMKV-SRDKDGSK
6.  1                         MSGRPRTSSFAEGN*--QSPSLVLGGVKTCSRD--GSK

                          °         I                   II
1. 53 VTTIGGRNGQPK---QTISYMAERVVGHGSFGVVFQAKCLETGETVAIKKVLQDRRYKNR
2. 58 ******K*****---***************************************K*****
3. 37 ***VVATP**GPDRP*EV**TDTK*I*N******Y***LCDS**L*********K*F***
4. 35 I**VVATP**GTDRV*EV**TDTK*I*N**********LCD***L***********F***

      III         IV                   V
1.110 ELQTMRLLDHPNVVSLKHCFFSTTEK-DELYLNLVLEYVPETVHRVIKHYNKLNQRMPLI
2.115 **************************-****************S***R****M*****M*
3. 97 ***I**K***C*I*R*RYF*Y*SG**K**V******D******Y**AR**SRAK*TL*V*
4. 95 ***I**K*E*C*I*K*LYF*Y*SG**R**VF*******I****YK*ARQ*A*TK*TI*IN

                      VIa    VIb                    VII
1.169 YVKLYTYQIFRALSYIHRCIGVCHRDIKPQNLLVNPHTHQVKLCDFGSAKVLVKGEPNIS
2.174 *****S***C***A***NS*********************L*I*****************
3.157 *****M**L**S*A***-SF*I***********LD*D*AVL*********Q**R****V*
6.155 FIR**M**L**S*A***-SL*I***********LD*E*AVL*********Q*LH****V*
         VIII                IX
1.229 YICSRYYRAPELIFGATEYTTAIDVWSAGCVLAELLLGQPLFPGESGVDQLVHIIKVLGT
2.234 ************************I*******G*******************E*******
3.216 *****************D**SS**M***************I***D*******E*******
4.214 ****************IN***K******************I***D*******EV******

          X                                        XI
1.289 PTREEIKCMNPNYTEFKFPQIKAHPWHKIFHKRMPPEAVDLVSRLLQYSPNLRSAAIDTL
2.294 ******************************************************T**EA*
3.276 ****Q*RE******************T*V*RP*T****IA*C****E*T*TA*LTP*EAC
6.274 ****Q*RE**************S***Q*V*RI*T*T**IN***L**E*T*SA*ITP*KAC
          °
1.349 VHPFFDELRDPNAR-LPNGRFLPPAFHFKPHELKGVPLEMVAKLVPEHARKQCPWLGL
2.354 ****Y*DV****T*-*********L*N**VN******A**LV****P******ALF*SS
3.336 A*S*********VK-*****DT*AL*N*TTQ**SSN*PLATI-*I*P***I*AAASPPA+29
4.334 A********MEGNHT*****DM**L*N*TE***SIQ*SLVPQ-*L*K*--LQNASGPGG+124
```

Figure 1. Alignment of the Arabidopsis (1) ASK-α (Bianchi et al 1993). Alfalfa (2) MsK-3 (Pay et al 1993), rat GSK-3β (3)(Woodgett 1990) and Drosophila (4) zw3-A protein kinase sequences (Siegfried et al 1990). Residues conserved in all proteins are marked (*). Gaps (-) are introduced to optimize sequence similarity. The residues between ° are in the catalytic domain. The subdomains are indicated by the roman numbers and underlined (see Hanks et al 1988)

expected highest identity with the members of the GSK-3 subfamily of protein kinases. The catalytic domains of the plant proteins (Bianchi et al 1993; Pay et al 1993) are about 70 % identical to sgg/zw3 from *Drosophila* and GSK-3 from rat (see figure 1). The strikingly high sequence conservation between *Drosophila* and plant sgg/zw3 suggests functional homology. Transcripts hybridizing to ASK-α and ASK-γ were detected in roots, young shoots and fully extended leaves.

Protein kinase avctivity of the ASK proteins

ASK-α and ASK-γ were produced in *E.coli* as glutathione-S-transferase fusions (Smith and Johnson 1982). After purification on glutathione-agarose, both proteins autophosphorylated readily on threonine, serine and tyrosine residues (Bianchi et al 1993). The functional significance of ASK autophosphorylation is not yet determined, nor their in vivo phopshorylation states. However, GSK-3 shaggy and MCK1 all require phosphotyrosine for activity (Hughes et al 1993; G.S. Martin, personal communication). This is only the second reported example of a tyrosine phosphorylated protein from plants (see Duerr et al 1993). ASK-α and ASK-γ phosphorylated *in vitro* c-jun, myelin basic protein (MBP) and phosphatase inhibitor-2 (I-2) (Holmes et al 1986), which are also *in vitro* substrates of GSK-3 and sgg/zw3 (Plyte et al 1993). In the case of MBP and I-2, phosphorylation by ASK was restricted to serine and threonine residues respectively, suggesting a substrate specificty similar to that of the other members of the GSK-3 family (Plyte et al 1992; G.S. Martin, personal communication). The ASK protein represent therefore the first documented example, probably with the MsERK1 protein kinase from alfalfa (Duerr et al 1993) of plant dual specificity kinases, a novel group of enzymes able to act on all three hydroxyamino acids (Lindberg et al 1992).

ISOLATION AND CHARACTERIZATION OF THE PLANT HOMOLOGUE OF *SNF1*

Although about 20 plant genes encoding protein kinases have been isolated, the function of most of them remains to be elucidated (see table 1). However, we have recently isolated plant homologues of the yeast gene *SNF1* (Alderson et al 1991; Halford et al 1992; LeGuen et al 1992). In this case, a clear functional relationship between the yeast and the plant genes has been demonstrated by complementation.

Role of the *SNF1* gene product

SNF1 (Sucrose non fermenting) encodes a protein serine/threonine kinase (Celenza and Carlson 1986) which has been described as a global regulator of carbon metabolism in yeast. It is central to a pathway linking the perception of glucose levels with the derepression of a range of glucose-repressible genes. These include *Suc*, *Gal*, and *Mal* genes, including *Suc2* (encoding invertase) and genes encoding enzymes involved in the glyoxylate cycle (reviewed by Gancedo 1992). There is very little evidence for the regulation of gene expression by carbon metabolites in plants, but the isolation of plant *SNF1* homologues suggests that such a system might exist. If it does, it could be a prime target for genetic manipulation, with possible effects on carbohydrate sink/source relationships, carbon partitioning and storage compound deposition in the harvested organs of crop plants.

Molecular cloning of a rye *SNF1*-related cDNA

The first plant *SNF-1* related sequence to be characterized was cRKin1, a cDNA isolated from a rye endosperm cDNA library (Alderson et al 1991). cRKin1 encodes a protein of 520 amino acids with a Mr of 57710, containing all the invariant residues and conserved domains typical of eukaryotic protein kinases. The SNF1 protein kinase is larger with a Mr of 72045, but the searches of the Genbank and Embl nucleotide sequence databases matched the two sequences most closely with each other, the matchup being significantely higher than with any other sequence. The GAP program (Devreux et al 1984) calculates a similarity of 66.9 % between the two proteins. The most highly conserved regions are within the catalytic domain, which comprises about 240 amino acid residues in the N-terminal halves of the two proteins.

Complementation of a yeast *snf1* mutation

A clear functional relationship between the plant genes and the yeast gene, has been demonstrated by expressing cRKin1 in a yeast *snf1* mutant. *snf1* mutants are unable to utilise a number of carbon sources, including glycerol. Expression of the rye cDNA complemented the mutation, restoring *SNF1* function and enabling transformants to grow on glycerol minimal medium (Alderson et al 1991).

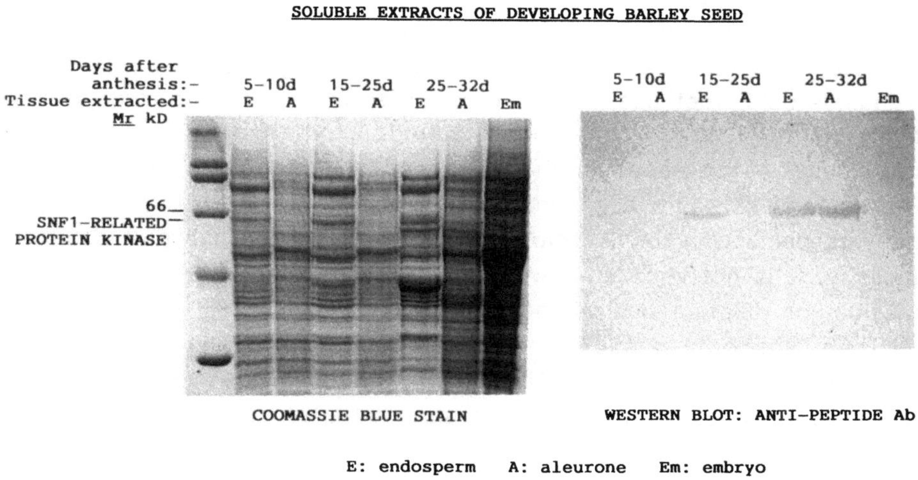

Figure 2. SDS-PAGE of proteins extracted from barley endosperms and aleurones at variuos stages of development and from developing embryos, and Western blot of the same reacted with an antibody raised to a peptide with a sequence characteristic of the plant SNF-1 related protein Kinases.

Structure and expression of the *SNF1*-related gene family in diverse plant species

SNF1-related genomic fragments, cDNAs and PCR products have been isolated from barley (Halford et al 1992), Potato (Hannappel et al in preparation) and *Arabidopsis thaliana* (Le Guen et al 1992). Southern blot and PCR analysis have shown the presence of similar sequences in wheat and sugar beet. A single copy of the *SNF1*-related gene is present in *Arabidopsis*, encoding a protein with a Mr of 58 909, with 67% similarity to the yeast protein. The gene is expressed in leaves, roots and shoots (other tissues have not been analysed). In contrast, barley, rye, wheat and potato all contain multigene families of *SNF1*-related genes. A detailed study of the barley multigene family has revealed that it can be subdivided into two subfamilies, encoding proteins with considerable differences in sequence in some regions, although both have an Mr of approximately 58 000 and show approximately 67% similarity to the yeast protein. These subfamilies are differently regulated, one being expressed in all tissues we have analysed, including roots, root tips, leaves coleoptiles and internodes, and at low levels in the seeds, the other being expressed at high levels specifically in the seeds in the early and mid stages of development. The subfamilies share a common gene structure with each other and in many respects with the *Arabidopsis* gene, clearly indicating a common ancestry. It is not yet known whether they encode isoforms of the same enzyme, or different but related protein kinases.

An antibody raised to a synthetic peptide with a sequence characteristic of the plant SNF1-related proteins has been shown to recognize a protein of the expected size in barley seed extracts (Figure 2), and is being used to monitor the purification of the protein.

CONCLUSION.

The striking evolutionary conservation of the protein kinases of the GSK-3 and SNF1 subfamilies implies an involvement in fundamental regulatory pathways. However, the ancient divergence of the plant lineage from the metazoa and yeast does not allow direct extrapolations. The manipulation of kinase activity in transgenic plants will allow us to probe for their *in vivo* functions.

Acknowledgment

This work was partly supported by a grant from the Ministère de la Recherche et de la Technologie. MB acknowledges the EEC for a training grant, EMBO for a short terme fellowship, and is grateful to JR Woodgett for advice and the hospitality while staying in his lab. UH was supported by EEC BRIDGE training grants (Biot-OO81-F(CH) and Seeds of Tomorrow Through Genetic Engineering).

REFERENCES

Algarra P, Linder S and Thümmler (1993) Biochemical evidence that phytochrome of the moss *Ceratodon purpureus* is a light-regulated protein kinase. FEBS Letters 315: 69-73

Alderson A, Sabelli PA, Dickinson JR, Cole D, Richardson M, Kreis M, Shewry PR and Halford NG (1991) Complementation of *snf1*, a mutation affecting global regulation of carbon metabolism in yeast, by a plant protein kinase cDNA. Proc Natl Acad Sci USA 88: 8602-8605

Bourouis M, Moore P, Ruel L, Grau Y, Heitzler P and Simpson P (1990) An early embryonic product of the gene *shaggy* encodes a serine/threonine protein kinase related to the CDC28/cdc2+ subfamily. EMBO J 9: 2877-2884

Bianchi M, Guivarc'h D, Thomas M, Woodgett JR, and Kreis M.(1993). *Arabidopsis* homologs of *shaggy* in *Drosophila* and *glycogen synthase kinase-3* in mammals: a gene family encoding dual-specificity protein kinases.(submitted)

Boyle WJ, Smeal T, Defize L , Angel P, Woodgett JR, Karin M and Hunter T (1991) Activation of protein kinase C decreases phosphorylation of c-jun at sites that negatively regulate its DNA-binding activity. Cell 64: 573-584

Budde RJA and Chollet R (1988) Regulation of enzyme activity in plants by reversible phosphorylation. Physiol Plant 72: 435-439

Budde RJA and Randall DD (1990) Light as a signal influencing the phosphorylation status of plant proteins. Plant Physiol 94: 1501-1504

Celenza JL, and Carlson M (1986) A yeast gene that is essential for the release from glucose repression encodes a protein kinase Science 233: 1175-1180

Chang C, Schaller GE, Patterson SE, Kwok SF, Meyerowitz EM and Bleecker AB (1992) The TMK1 gene from *Arabidopsis* codes for a protein with structural and biochemical characteristics of a receptor protein kinase. Plant Cell 4: 1263-1271

Colasanti J, Tyers M, and Sundaresan V (1991) Isolation and characterization of cDNA clones encoding a functional p34 cdc2 homologue from *Zea mays*. Proc Natl Acad Sci USA 88: 3377-3387

Devereux J, Haezbrelli M, and Smithies (1984) A comprehensive set of of sequence analysis programs for the Vax Nucl. Acids Res.12: 387-395

Duerr B, Gawienowski M, Ropp T and Jacobs T (1993) MsERK1: a mitogen-activated protein kinase from a flowering plant. Plant Cell 5: 87-96

Feiler HS and Jacobs TW (1990) Cell division in higher plants: A *cdc2* gene, its 34-kDa product, and histone H1 kinase activity in pea. Proc Natl Acad Sci USA 87: 5397-5401

Feng XH, Bottino PJ, Kung SD (1992) Molecular identification of a soybean protein kinase gene family by using PCR. Plant Mol. Biol. 18: 581-584

Ferrant F, Van Went J, Bianchi M, de Vries SC, and Kreis M (1993) Embryo sac and ovule soecific libraries of *Petunia hybrida* L. (submitted)

Ferreira PCG, Hemerly AS, Villarroel R, van Montagu M and Inze D (1991) The *Arabidopsis* functional homolog of the $p34^{cdc2}$ protein kinase. Plant Cell 3: 531-540

Gancedo JM (1992) Carbon catabolite repression in yeast. Eur J Biochem 206: 297-313

Goring DR and Rothstein SJ (1992) The S-Locus receptor kinase gene in a self-incompatible *Brassica napus* line encodes a functional serine threonine kinase. Plant Cell 4:1273-1281

Halford NG, Vicente-Carbajosa J, Sabelli PA, Shewry PR, Hannappel U, Kreis (1992) Molecular analyses to the yeast protein kinase gene *SNF1*. Plant J 2: 791-797

Hanks SK, Quinn AM and Hunter T (1988) The protein kinase family: conserved features and deduced phylogeny of the catalytic domains. Science 241: 42-52

Hanks SK (1991) Eukaryotic protein kinases. Curr Opin Struct Biol 1: 369-383

Harper JF, Sussman MR, Schaller GE, Putnam-Evans C, Charbonneau H, and Harmon AC (1991) Calcium dependent protein kinase with a regulatory domain similar to calmodulin. Science 252: 951-954

Hata S. (1991) cDNA cloning of a novel cdc2+/CDC28-related protein kinase from rice. FEBS 279: 149-152

Hayashida N, Mizoguchi T, Yamaguchi-Shinozaki K and Shinozakin K (1992) Characterization of a gene that encodes a homologue of protein kinase in *Arabidopsis_thaliana.* Gene 121: 325-330

Heemskerk J, DiNardoS., Kostriken and O'Farell P (1991) Multiple modes of engrailed regulation in the progression towards cell fate determination. Nature 352: 4043-410

Heitzler P and Simpson P (1991) The choice of cell fate in the epidermis of *Drosophila*. Cell 64: 1083-1092

Hirayama T and Oka A (1992) Novel protein kinases of *Arabidopsis thaliana* (APK1) that phosphorylates tyrosine, serine, and threonine Plant Mol Biol 20: 653-662

Hirt H, Pày A, Györgyey J, Bakó L, Németh K, Bögre L, Schweyen RJ, Heberle-Bors and Dudits D (1991) Complementation of a yeast cell cycle mutant by an alfalfa cDNA encoding a protein kinase homologous to $p34^{cdc2}$. Proc Natl Acad Sci USA 88: 1636-1640

Hunter T and Karin M (1992) The regulation of transcription by phosphorylation. Cell 70: 375-387

Holmes CFB, Campbell DG, Caudwell FB, Aitken A and Cohen P (1986) The protein phosphatases involved in cellular regulation-primary structure of inhibitor-2 from rabbit skeletal muscle. Eur J Biochem 155: 173-182

Hughes K, Nikolakaki E, Plyte SE, Totty NF and Woodgett JR (1993) Modulation of the glycogen synthase kinase family by tyrosine phosphorylation. EMBO J 12: 803-808

Kieber JJ, Rothenberg M, Roman G, Feldmann KA and Ecker JR (1993) CTR1, a negative regulator of the ethylene response pathway in *Arabidopsis*, encodes a member of the Raf family of protein kinases. Cell 72: 427-441

Lawton MA, Yamamoto RT, Hanks SK and Lamb CJ (1989) Molecular cloning of plant transcripts encoding protein kinase homologs. Proc Natl Acad Sci USA 86: 3140-3144.

Lawton MA (1990) Plant protein kinases: a molecular genetic approach. Curr Top Plant Biochem Physiol 9: 373-382

Le Guen L, Thomas M, Bianchi M, Halford NG and Kreis M (1992) Structure and expression of a gene from *Arabidopsis thaliana* encoding a protein related to SNF1 protein kinase. Gene 120: 249-254

Lindberg RA, Quinn AM and Hunter T (1992) Dual-specificity kinases: will any hydroxyl do? Trends Biochem Sci 17: 114-119

Limbourg-Bouchon B., Busson D., Lamour-Isnard (1991) Interactions between *fused*, a segment-polarity gene in *Drosophila*, and other segmentation genes. Development 112: 417-429

Neigeborn L and Mitchell AP (1991) The yeast *MCK1* gene encodes a protein kinase homolog that activates early meiotic gene expression. Genes Dev 5:533-548

Nusse R and Varmus HE (1992) *Wnt* genes. Cell 69:1073-1087

Mizoguchi T, Hayashida N, Shinozaki KY, Harada H and Shinozaki K (1992) Nucleotide sequence of a cDNA encoding a protein kinase homologue in *Arabidopsis thaliana* Plant Mol Biol 18: 809-812

Pay A Jonak.C, Bögre L, Meskiene I, Mairinger T, Szalay A, Heberle-Boros E, and Hirt H (1993) The *Msk* family of alfalfa protein kinase genes encodes homologues of *shaggy/glycogen synthase kinase-3* and shows differential expression patterns in plant organs and development. Plant J (in press)

Peifer M and Bejsovec A 1992. Knowing your neighbors: Cell interactions determine intrasegmental patterning in drosophila. Trends Gnet 8:243-249

Perrimon N and Smouse D (1989) Multiple functions of a *Drosophila* homeotic gene, *zeste-white 3*, during segmentation and neurogenesis. Dev Biol 135: 287-305

Perrimon N and Smouse D (1989) Multiple functions of a *Drosophila* homeotic gene, *zeste-white 3*, during segmentation and neurogenesis. Dev Biol 135, 287-305

Preat T, Thérond P, Lamour-Isnard C, Limbourg-Bouchon B, Tricoire H, Erk I, Mariol M.C and Busson D, 1990. A putative serine/threonine protein kinase encoded by the segment-polarity fused gene in *Drosophila*. Nature 347: 87-89

Plyte SE, Hughes K, Nikolakaki E, Pulverer BJ and Woodgett JR (1992) *Glycogen synthase kinase-3* - functions in oncogenesis and development. Biochim Biophys Acta 114: 147-162

Ruel L, Pantesco V, Lutz Y, Simpson P and Bourois M (1993a) Functional significance of a family of protein kinases encoded at the *shaggy* locus in *Drosophila*. EMBO J 12:1657-1669

Ruel L, Bourois M, Heitzler P, Pantesco V and Simpson P (1993b) The *Drosophila* shaggy kinase and rat glycogen synthase kinase-3 have conserved activities and act downstream of *Notch*. Nature, in press

Shero JH and Hieter P (1991) A suppressor of a centromere DNA mutation encofdes a putative protein Kinase (MCK1) Genes and Development 5: 549-560.

Siegfried E, Perkins LA, Capaci TM, and Perrimon N (1990) Putative protein kinase product of the drosophila segment-polarity gene *zeste-white 3*. Nature 345: 825-829

Siegfried E, Chou TB, Perrimon N, 1992. *wingless* signaling acts through *zeste-white 3*, the *Drosophila* homolog of *glycogen synthase kinase-3*, to regulate engrailed and establish cell fate. Cell 71:1167-1179

Smith DB and Johnson KS(1988) Single-step purification of polypeptides expressed in *Escherichia coli* as fusions with glutathione-S-transferase. Gene 67: 31-40

Stein JC, Howlett B, Boyes DC, Nasrallah ME and Nasrallah JB (1991) Molecular cloning of a putative receptor protein kinase gene encoded at the self-incompatibility locus of *Brassica oleracea*. Proc Natl Acad Sci USA 88: 8816-8820

Ranjeva R and Boudet AM (1987) Phosphorylation of proteins in plants: regulatory effects and potential involvement in stimulus/response coupling. Ann Rev Plant Physiol 38: 73-93

Suen KML and Choi JH (1991) Isolation and sequene analysis of a cDNA clone for a carrot calcium dependant protein kinase: homology to calcium/calmodulin-dependant protein kinases and to calmodulin. Plant Mol Biol 17: 581-590

Walker JC and Zhang R (1990). Relationship of a putative receptor protein kinase from maize to the S-locus glycoproteins of Brassica. Nature 345:743-746.
Woodgett JR. (1990) Molecular cloning and expression of glycogen synthase kinase-3/Factor A. EMBO 9: 2431-24538

Specific Nuclear Localization Sequence Binding to Plant Nuclei

Glenn R. Hicks, Valentina Kovaleva and Natasha Raikhel
MSU-DOE Plant Research Laboratory
Michigan State University
East Lansing, MI 48824-1312 USA

Introduction

The import of proteins into the nucleus is mediated by nuclear localization sequences (NLSs), which are short polypeptide regions having a high content of basic amino acids (for reviews, see Roberts, 1989; Garcia-Bustos et al., 1991; Raikhel, 1992). NLSs are found at many locations within proteins but are not removed following translocation as are signals for chloroplastic (Keegstra, 1989), mitochondrial (Hartl and Neupert, 1990) and vacuolar (Chrispeels and Raikhel, 1992) import. The best studied NLSs are those that resemble the import signal of the simian virus 40 (SV40) large T-antigen (PKKKRKV; Kalderon et al., 1984; Lanford and Butel, 1984). Two additional classes of NLSs have also been identified. These are exemplified by the yeast mating type α-2 (Mat α-2) NLS (Hall et al., 1984), which possesses hydrophobic and basic amino acids (KIPIK), and bipartite NLSs which are composed of two short regions of basic amino acids separated by approximately ten residues (for review, see Dingwall and Laskey, 1991).

While the precise mechanism is uncertain, it is clear from studies of animal and yeast cells that import occurs through the nuclear pores and involves two steps: an energy-independent binding step followed by translocation, which requires the hydrolysis of ATP (Newmeyer and Forbes, 1988; Richardson et al., 1988). In addition, import is likely to

NATO ASI Series, Vol. H 81
Plant Molecular Biology
Edited by G. Coruzzi and P. Puigdomènech

be receptor mediated, since the process is saturable and specific only for functional NLSs (Goldfarb et al., 1986). Recent effort has been applied toward the identification of components of the import apparatus, including receptors that recognize NLSs (for reviews, see Garcia-Bustos et al., 1991; Silver, 1991; Stochaj and Silver, 1992b; Forbes, 1992). In animals and yeast, such NLS-binding proteins have been found in the cytosol and at the nuclear envelope (Adam et al., 1989; Yamasaki et al., 1989; Stochaj and Silver, 1992a). Several additional studies have also demonstrated that cytosolic factors are essential for import in animal cells (Newmeyer and Forbes, 1990; Adam and Gerace, 1991; Moore and Blobel, 1992; Sterne-Marr et al., 1992).

Until recently, little was known about nuclear import in higher plants (for review, see Raikhel, 1992). Thus, as an initial step, it was necessary for our laboratory to characterize the NLSs of several nuclear proteins of plant origin. The transcription factor R, which is involved in the control of anthocyanin biosynthesis in maize, was found to possess three functional NLSs (Figure 1; Shieh et al., 1993): an SV40-type (NLS M), a Mat α-2 type (NLS C), and an unusual type containing three consecutive arginine residues found in some viruses (NLS A; Garcia-Bustos et al., 1991). Interestingly, specific combinations of two of the three NLSs are necessary for efficient nuclear localization of R:GUS (β-glucuronidase) fusion proteins in onion epidermal cells (Shieh et al., 1993). Another maize transcription factor, Opaque-2 (O2), which participates in the regulation of zein storage protein biosynthesis, was shown to localize to the nucleus of both monocot and dicot plant species (Varagona et al., 1991). The protein contains two NLSs (Figure 1): an SV40-type (NLS A) and a bipartite signal (NLS B), although the bipartite signal was found to be more efficient in directing import (Varagona et al., 1992). Based upon our extensive studies of these plant NLSs, we have begun to dissect the nuclear import apparatus of higher plants using biochemical and cell biology approaches.

Protein	Nuclear Localization Sequences	Suff/Nec
O2 (NLS B)	MPTEERV**RKRK**ESN**R**ES**ARR**S**R**Y**RK** AAHLKEL	+/+ with NLS A
O2 (NLS A)	MEEAVTMAPAAVSSAVVGDPMEYN AIL**RRK**LEEDLE	+/+ with NLS B
R (NLS A)	GN**RR**AAPARPAG	-/+ with NLS M
R (NLS M)	MSE**RKRR**EKL	+/+ with NLS A +/+ with NLS C
R (NLS C)	MISEAL**RK**AIG**KR**	+/+ with NLS M

Figure 1. Nuclear localization sequences (NLSs) of the R and Opaque-2 (O2) proteins from maize. Residues are in single letter code, and basic residues within functionally tested NLSs are in bold type. Suff, NLS is sufficient to redirect a reporter protein to the nucleus; Nec, NLS is necessary for import of intact protein.

Materials and Methods

Nicotiana tabacum (NT) suspension-cultured cells were maintained and protoplasts were prepared according to Bednarek et al. (1990). Maize Black Mexican Sweet (BMS) suspension-cultured cells were maintained and protoplasts prepared according to Fromm et al. (1987). Peptides were synthesized as previously described (Hicks and Raikhel, 1993). All other methodologies were according to Hicks and Raikhel (1993), with the following exception. For indirect immunofluorescence microscopy, purified tobacco nuclei were allowed to associate for 5 min on ice with 0.25 μM NLS-HSA or HSA in the absence or presence of SV40 T-antigen NLS or O2 NLS peptides as indicated and photographed using Kodak T-Max 100 film and fluorescence optics (Varagona et al., 1991).

Results and Discussion

To detect NLS receptors, we examined the specific association of plant NLSs with purified plant nuclei. A 31-residue synthetic peptide (Hicks and Raikhel, 1993) corresponding to the wild-type bipartite import signal of O2 (O2NLS) was ^{14}C-labeled and allowed to associate with purified tobacco nuclei (Figure 2A). To examine the specificity of NLS association, unlabeled O2NLS peptide was added as a competitor and resulted in a sigmoidal curve, indicating a single class of low-affinity binding sites (K_d=200 μM). In addition, several classes of NLSs compete for this site, since a ^{14}C-labeled peptide to the well-studied SV40 large T-antigen NLS (SV40NLS) was displaced by unlabeled O2NLS peptide (Figure 2B).

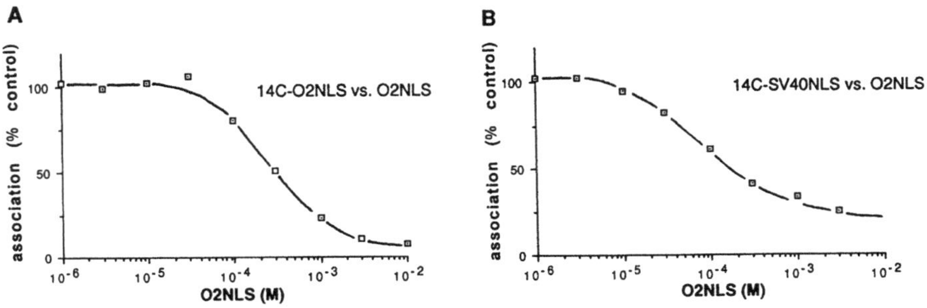

Figure 2. Binding of NLSs to plant nuclei is specific. Radiolabeled wild-type O2 and SV40 NLS peptides (5 μM) were incubated with tobacco nuclei, and pelleted cpms were plotted as percent binding of control versus concentration (M) of added competitor. (A) Association of ^{14}C-labeled O2NLS in the presence of O2NLS as a competitor, indicating specific binding. (B) Association of labeled SV40 T-antigen NLS in the presence of O2NLS, indicating competition for the same site.

This finding is consistent with several recent studies that indicate that the SV40 large T-antigen NLS is functional in plants (Lassner et al., 1991; van der Krol and Chua, 1991; Varagona and Raikhel, in preparation). In other studies, peptides to mutant NLSs of O2

(Varagona and Raikhel, in preparation) and SV40 large T-antigen (Kalderon et al., 1984; Lanford and Butel, 1984) that are defective in mediating nuclear import *in vivo* competed poorly with the wild-type NLSs for association with tobacco and maize nuclei; a reverse SV40 wild-type NLS competed weakly, whereas a non-NLS peptide displayed no ability to compete. In addition, the use of ^{125}I-labeled NLS peptides, which should have detected binding at nanomolar concentrations, did not reveal higher-affinity binding (Hicks and Raikhel, 1993). These results indicate that nuclei of dicot and monocot species possess a single low-affinity site capable of specifically recognizing functional NLSs. Furthermore, the recognition is not dependent solely upon the basic charge of the NLS, since the reverse wild-type SV40 peptide competed poorly for binding.

The observed NLS association represents specific binding - the first step of nuclear import - and not translocation, since the association of poorly diffusible ligands such as O2NLS conjugated to human serum albumin (HSA), which displays specific association similar to that of the unconjugated peptide, is not stimulated by ATP addition or inhibited by the animal cell translocation inhibitor, wheat germ agglutinin (Hicks and Raikhel, 1993). This indicates that cytosolic factors may be necessary for import, as has been demonstrated in amphibian and mammalian systems (Newmeyer and Forbes, 1990; Adam and Gerace, 1991; Moore and Blobel, 1992; Sterne-Marr et al., 1992). In addition, biochemical characterization indicates that the binding site is proteinaceous and tightly associated with the nucleus, since it is protease sensitive and resistant to extraction by Triton X-100, NaCl, and urea. We have found by electron microscopy that under these conditions, identifiable structures, including nuclear pores, are detectable (Hicks and Raikhel, 1993). Thus, it was possible that binding occurred at the nuclear pores.

To determine the location of NLS binding, indirect immunofluorescence and immunoelectron microscopy were used. For immunofluorescence, the binding of conjugated NLSs to isolated tobacco nuclei was detected via an anti-HSA antibody followed by a rhodamine-labeled secondary antibody. Wild-type O2NLS-HSA and SV40NLS-HSA were found to bind to the nuclear periphery (Figure 3A,C, panels 1). As observed for the binding of radiolabeled NLS peptides and NLS-HSA conjugates, the presence of 3 mM free wild-type O2NLS peptide (Figure 3A, panel 2) or 1 mM wild-type SV40NLS (Figure 3C, panel 2) greatly reduced detectable binding. The presence of mutant O2NLS (Figure 3A, panel 3) or mutant SV40NLS (Figure 3C, panel 3) was much less effective than the wild-type sequences in reducing conjugate binding.

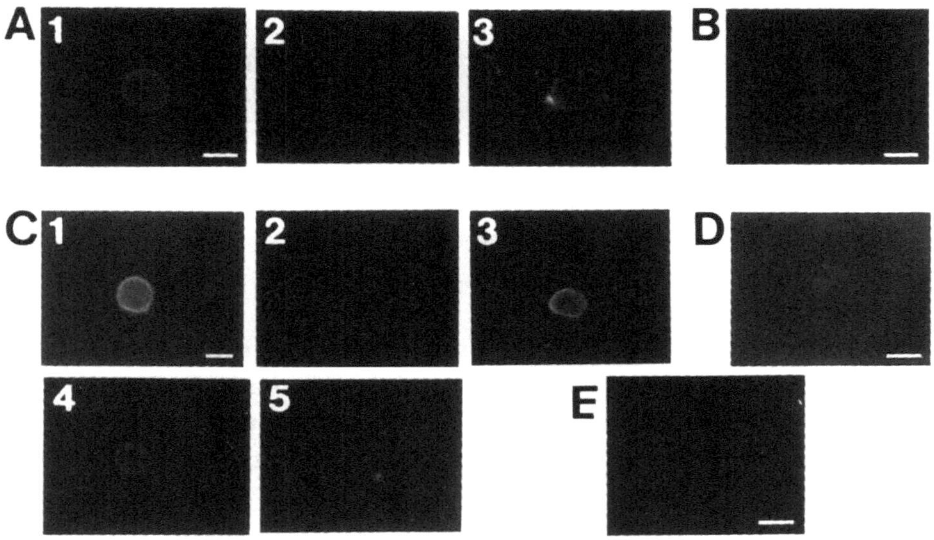

Figure 3. NLSs bind to the periphery of plant nuclei. Purified tobacco nuclei were incubated with 0.25 μM NLS-HSA conjugates or HSA, and binding was visualized with an anti-HSA antibody and a rhodamine-labeled secondary antibody. (A) Wild-type O2NLS-HSA in the absence (panel 1) or presence of 3 mM wild-type (panel 2) and 3 mM mutant O2NLS peptides (panel 3). (B) Binding of mutant O2NLS-HSA conjugate. (C) Wild-type SV40NLS-HSA binding in the absence (panel 1) or presence of 1 mM wild-type (panel 2), 1 mM mutant (panel 3) and 1 mM reverse wild-type (panel 4) SV40 NLS peptides or 1 mM wild-type O2NLS peptide (panel 5). (D) Binding of mutant SV40NLS-HSA conjugate. (E) Binding of unconjugated HSA as a control. Bars indicate 8 μm.

As in the *in vitro* binding studies (Hicks and Raikhel, 1993), the reverse wild-type SV40 sequence was only partially effective as a competitor of wild-type SV40NLS-HSA binding (Figure 3C, panel 4). Consistent with the displacement curve in Figure 2B, free wild-type O2NLS peptide effectively competed for SV40NLS-HSA binding. Mutant O2NLS-HSA (Figure 3B), mutant SV40NLS-HSA (Figure 3D) and HSA (Figure 3E) control ligands were found to yield little detectable binding to plant nuclei, as predicted for a functional NLS-specific interaction. The immunofluorescence binding was confirmed by laser-scanning microscopy to be at the nuclear periphery and not within the nucleoplasm (Hicks and Raikhel, 1993), which indicates binding to the nuclear pores or envelope.

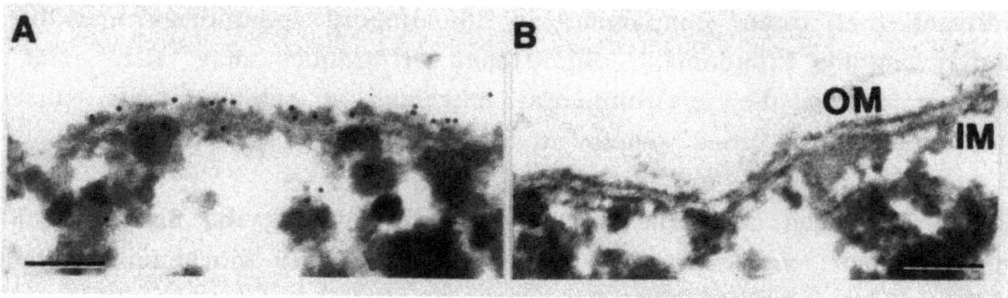

Figure 4. Import signals bind to nuclear pores. Tobacco nuclei were incubated with 5 µM wild-type O2NLS-HSA or HSA and binding was visualized with an anti-HSA antibody and Protein-A gold. (A) Nuclei incubated with O2NLS-HSA display gold label associated with pores and envelope in longitudinal section. (B) Control nuclei incubated with HSA show no specific labeling. Bars indicate 200 nm. OM, outer membrane; IM, inner membrane.

The actual nuclear location of O2NLS-HSA binding was determined by electron microscopy and immunogold labeling. O2NLS-HSA conjugate was allowed to associate with purified tobacco nuclei and fixed for electron microscopy. Binding was then detected by incubation with an anti-HSA antibody and Protein-A gold. Gold particles were found to be associated with nuclear pores (Figure 4A). However, binding may not be exclusive to this location, since label was also found in regions of the envelope apparently between pores. Again, the interaction was specific to the O2NLS because control nuclei treated with a similar concentration of HSA yielded essentially no labeling (Figure 4B).

Conclusions

Overall, our results indicate that dicot and monocot plant species possess low-affinity binding sites at the nuclear envelope that are specific for several classes of functional NLSs. The fact that NLSs of mammalian viral and plant origin bind to the same nuclear site indicates that some components of the import apparatuses may be similar among kingdoms. Significant differences may also exist, however, given the environmental extremes at which plants must survive and their broad genetic diversity.

Having developed a system to study specific binding, the first step of nuclear import, we are now in a position to identify biochemically the NLS-binding proteins by the use of chemical cross-linking or photoaffinity labeling. In addition, it will be possible by this approach to determine whether the binding site is exclusively nuclear or is also present in the cytosol, as has been shown for other NLS-binding factors (Adam et al., 1989, Yamasaki et al., 1989; Stochaj and Silver, 1992a). It will also be essential to develop higher plant *in vitro* nuclear import systems to test putative import components in a homologous system. Our *in vitro* binding system may provide one approach to this goal by the reconstitution of import via the addition of cytosolic extracts, as has been achieved in several non-plant systems (Newmeyer and Forbes, 1988; Richardson et al., 1988). Many exciting questions remain to be answered concerning nuclear import, and higher plants, which can be manipulated at the single-cell level and regenerated to a complex multicellular organism, may provide an excellent system for addressing the biology of this process.

Acknowledgments

The authors gratefully acknowledge Ms. Kathleen Kerekes for excellent technical assistance. We also thank Drs. Stephen Adam and Jose Garcia-Bustos for many helpful suggestions. Finally, we acknowledge other members of our nuclear targeting group: Dr. Marguerite Varagona, Ms. Antje Heese, and Mr. Mark Shieh for many stimulating discussions. This work was supported by grants from USDA (92-37301-7709) and DOE (DE-FG02-90ER20021) to NVR and an NSF postdoctoral fellowship (BIR-9203692) to GRH.

References

Adam, S.A., and Gerace, L. (1991) Cytosolic proteins that specifically bind nuclear location signals are receptors for nuclear import. Cell 66: 837-847.

Adam, S.A., Lobl, T.J., Mitchell, M.A., and Gerace, L. (1989) Identification of specific binding proteins for a nuclear location sequence. Nature 337: 276-279.

Bednarek, S.Y., Wilkins, T.A., Dombrowski, J.E., and Raikhel, N.V. (1990) A carboxyl-terminal propeptide is necessary for proper sorting of barley lectin to vacuoles of tobacco. Plant Cell 2: 1145-1155.

Chrispeels. M.J., and Raikhel, N.V. (1992) Short peptide domains target proteins to plant vacuoles. Cell 68: 613-616.

Dingwall, C., and Laskey, R. (1991) Nuclear targeting sequences - a consensus? Trends Biochem. Sci. 16: 478-481.

Forbes, D.J. (1992) Structure and function of the nuclear pore complex. Ann. Rev. Cell Biol. 8: 495-527.

Fromm, M., Callis, J., Taylor, L.P., and Walbot, V. (1987) Electroporation of DNA and RNA into plant protoplasts. Methods Enzymol. 153: 351-366.

Garcia-Bustos, J., Heitman, J., and Hall, M.N. (1991) Nuclear protein localization. Biochim. Biophys. Acta 1071: 83-101.

Goldfarb, D.S., Gariepy, J., Schoolnik, G., and Kornberg, R.D. (1986) Synthetic peptides as nuclear localization signals. Nature 322: 641-644.

Hall, M.N., Hereford, L., and Herskowitz, I. (1984) Targeting of *E. coli* β-galactosidase to the nucleus in yeast. Cell 36: 1057-1065.

Hartl, F.-U., and Neupert, W. (1990) Protein sorting to mitochondria: evolutionary conservations of folding and assembly. Science 247: 930-938.

Hicks, G.R. and Raikhel, N. (1993) Specific Binding of Nuclear Localization Sequences to Plant Nuclei. Plant Cell 5: 983-994.

Kalderon, D., Richardson, W.D., Markham, A.F., and Smith, A.E. (1984) Sequence requirements for nuclear location of simian virus 40 large-T antigen. Nature 311: 33-38.

Keegstra, K. (1989) Transport and routing of proteins into chloroplasts. Cell 56: 247-253.

Lanford, R.E., and Butel, J.S. (1984) Construction and characterization of an SV40 mutant defective in nuclear transport of T antigen. Cell 37: 801-813.

Lassner, M.W., Jones, A., Daubert, S., and Comai, L. (1991) Targeting of T7 RNA polymerase to tobacco nuclei mediated by an SV40 nuclear location signal. Plant Mol. Biol. 17: 229-234.

Moore, M.S., and Blobel, G. (1992) The two steps of nuclear import, targeting to the nuclear envelope and translocation through the nuclear pore, require different cytosolic factors. Cell 69: 939-950.

Newmeyer, D.D., and Forbes, D.J. (1988) Nuclear import can be separated into distinct steps *in vitro*: nuclear pore binding and translocation. Cell 52: 641-653.

Newmeyer, D.D., and Forbes, D.J. (1990) An N-ethylmaleimide-sensitive cytosolic factor necessary for nuclear protein import: requirement in signal-mediated binding to the nuclear pore. J. Cell Biol. 110: 547-557.

Raikhel, N.V. (1992) Nuclear targeting in plants. Plant Physiol. 100: 1627-1632.

Richardson, W.D., Mills, A.D., Dilworth, S.M., Laskey, R.A., and Dingwall, C. (1988) Nuclear protein migration involves two steps: rapid binding at the nuclear envelope followed by slower translocation through the nuclear pores. Cell 52: 655-664.

Roberts, B. (1989) Nuclear location signal-mediated protein transport. Biochim. Biophys. Acta 1008: 263-280.

Shieh, M., Wessler, S.R., and Raikhel, N.V. (1993) Nuclear targeting of the maize R protein requires two localization sequences. Plant Physiol. 101: 353-361.

Silver, P.A. (1991) How proteins enter the nucleus. Cell 64: 489-497.

Sterne-Marr, R., Blevitt, J.M., and Gerace, L. (1992) O-linked glycoproteins of the nuclear pore complex interact with a cytosolic factor required for nuclear protein import. J. Cell Biol. 116: 271-280.

Stochaj, U., and Silver, P.A. (1992a) A conserved phosphoprotein that specifically binds nuclear localization sequences is involved in nuclear import. J. Cell Biol. 117: 473-482.
Stochaj, U., and Silver, P.A. (1992b) Nucleocytoplasmic traffic of proteins. Eur. J. Cell Biol. 59: 1-11.
van der Krol, A.R., and Chua, N.-H. (1991) The basic domain of plant B-ZIP proteins facilitates import of a reporter protein into plant nuclei. Plant Cell 3: 667-675.
Varagona, M.J., Schmidt, R.J., and Raikhel, N.V. (1991) Monocot regulatory protein Opaque-2 is localized in the nucleus of maize endosperm and transformed tobacco plants. Plant Cell 3: 105-113.
Varagona, M.J., Schmidt, R.J., and Raikhel, N.V. (1992) Nuclear localization signal(s) required for nuclear targeting of the maize regulatory protein Opaque-2. Plant Cell 4: 1213-1227.
Yamasaki, L., Kanda, P., and Lanford, R.E. (1989) Identification of four nuclear transport signal-binding proteins that interact with diverse transport signals. Mol. Cell. Biol. 9: 3028-3036.

AQUAPORINS: WATER CHANNEL PROTEINS IN THE TONOPLAST OF PLANT CELLS

Christophe Maurel[1], Herman Höfte[2], Kenneth D. Johnson and Maarten J. Chrispeels

Department of Biology, University of California, San Diego, 9500 Gilman Drive, La Jolla, California 92093-0116 (CM, HH and MJC) and Department of Biology, San Diego State University, San Diego, California 92182-0057 (KDJ)

[1]Present address: Institut des Sciences Végétales, CNRS, 91198 Gif-Sur-Yvette, France
[2]Present address: Laboratoire de Biologie Cellulaire, INRA, Route de Versailles, 78026 Versailles Cedex, France

INTRODUCTION

The functional characterization of numerous transport systems across the vacuolar membrane (tonoplast) has provided a better understanding of the versatile role played by the vacuole in plant cells. The tonoplast H^+-ATPase and the H^+-pyrophosphatase pump protons into the vacuole making this organelle inside positive and acidic (Hedrich *et al.,* 1989; Taiz, 1992). Maintaining an electrochemical gradient of protons, in addition to its role in cytoplasmic pH regulation, energizes various transport processes across the tonoplast. Several types of ions channels have been identified by the patch clamp technique, but the selectivity of most of them and hence their physiological role remain elusive (Hedrich and Schroeder, 1989). They probably represent major transport pathways for ions such as K^+ and Cl^-, the main determinants of cell turgor, and NO_3^- which can be stored in and mobilized from the vacuole. Calcium-permeable channels as well as a Ca^{++}/H^+ antiporter occur in plant vacuoles (Alexandre *et al.*, 1990; Blackford *et al.*, 1990). They reflect the role that the vacuole as a major calcium store plays in signal transduction in plants. The tonoplast also carries Na^+/H^+ exchangers and Cd^+ transporters that can account for the detoxifying role of the vacuole during saline and heavy metal stress, respectively (Taiz, 1992). Finally,

NATO ASI Series, Vol. H 81
Plant Molecular Biology
Edited by G. Coruzzi and P. Puigdomènech

transporters of amino acids, malate or sucrose have been identified in various materials (Martinoia *et al.*, 1985; Dietz *et al.*, 1990; Preisser and Komor, 1991). They reflect the storage function of the vacuoles in these different tissues and cell types.

Protein storage vacuoles (protein bodies) are highly specialized vacuoles in seeds which are filled with proteins, phytin and complex sugars. In addition to the transport machinery common to all vacuoles, protein storage vacuoles likely possess specialized transport systems involved in seed maturation and germination. Several years ago, our group identified an abundant 26 kDa intrinsic protein in the tonoplast of protein storage vacuoles that was named α-TIP (for a-Tonoplast Intrinsic Protein) (Johnson *et al.*, 1989; Johnson *et al.*, 1990). Immunological characterization showed α-TIP to be seed specific but molecular cloning indicated the presence of other TIP proteins expressed in other parts of the plant or under specific physiological conditions (Höfte *et al.*, 1992). All these proteins exhibit a channel-like structure and are members of a large family of membrane proteins found in animals, bacteria and plants (Pao *et al.*, 1991). A few of these proteins have been shown to mediate transmembrane transport but no function had been reported for the plant homologs yet. Here, we report our molecular studies of tonoplast proteins and the functional characterization of one of the TIP homologs. We show that γ-TIP is a water channel and likely allows facilited diffusion of water across the tonoplast in plant cells. The name aquaporin designates this recently discovered class of proteins that form channels selective for water transport in plant and animal membranes (Preston *et al.*, 1992; Maurel *et al.*, 1993; Fushimi *et al.*, 1993; Agre *et al.*, submitted).

TIP PROTEINS BELONG TO A LARGE FAMILY OF MEMBRANE PROTEINS FOUND IN BACTERIA, ANIMALS AND PLANTS

Primary structure analysis revealed that TIP proteins belong to a large family of integral membrane proteins. The first identified and prototype member of this family is MIP, the Major Intrinsic Protein of fibers cells in the ocular lens of the rat (Pao *et al.*, 1991). Sequence homology among MIP family members occurs over a strech of 250-300 amino acids that encompasses in each case 6 putative transmembrane α-helices (Figure 1). In contrast, the C- and N-terminal tails of these proteins are highly divergent. All the proteins of the MIP family display a tandem sequence repeat that makes the membrane spanning domains 4-6 homologous to the first three domains. This and the perfectly conserved residues Asparagine-Proline-Valine in the loops between helices 2 and 3 and helices 5 and 6 constitute the most striking features common of all these proteins (Pao *et al.*, 1991; Wistow *et al.*, 1991).

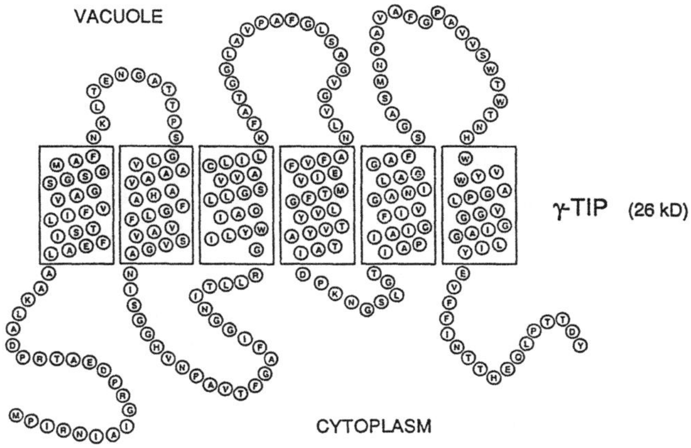

Figure 1. Proposed membrane topology of γ-TIP. Extensive studies performed on MIP (Pao *et al.*, 1991), CHIP28 (Smith and Agre, 1988) and NOD26 (Mia *et al.*, 1992) showed that for either of these proteins the amino and carboxy termini are located on the cytoplasmic side of the membrane.

The first plant protein of the MIP family to be identified was NOD26, a nodulin of legumes that is expressed in the peribacteroid membrane of symbiotic root nodules (Fortin *et al.*, 1987). Since then, 8 new members and their full cDNA sequences have been reported. They correspond to three TIP isoforms from bean and *Arabidopsis* (Johnson *et al.*, 1990; Höfte *et al.*, 1992), two root-specific proteins of tobacco and *Arabidopsis* (Yamamoto *et al.*, 1990), two proteins induced by water stress in pea and *Arabidopsis* (Guerrero *et al.*, 1990; Yamaguchi-Shinozaki *et al.*, 1992) and a protein specific to radiation-induced tumors tissue line of *Arabidopsis* (Campell and Town, 1992)

TISSUE SPECIFIC AND DEVELOPMENTALLY REGULATED EXPRESSION OF TIP PROTEINS

The two homologs α-TIP and γ-TIP were shown to be specifically located in the tonoplast (Johnson *et al.*, 1989; Höfte *et al.*, 1991; Höfte *et al.*, 1992). They constitute, with NOD26, the only members of the MIP family in plants for which the cellular location has been investigated. The α-TIP protein was characterized from protein storage vacuole and is strictly expressed in seeds (Johnson *et al.*, 1989; Höfte *et al.*, 1992). Immunocytochemical studies have shown that α-TIP accumulation occurs during late seed maturation and that α-TIP is catabolized occurs during early seedling growth (Melroy and Herman, 1991). Such a pattern is confered in part by the seed-specific transcription of the α-TIP gene (Ludevid *et al.*, 1992) but also by cellular mechanisms regulating its targeting to the vacuolar membrane and its specific removal from this membrane (Höfte and Chrispeels, 1992; Melroy and Herman, 1991).

The expression pattern of γ-TIP differs strongly from that of α-TIP since the protein could be detected in all parts of the plants except the seeds (Höfte *et al.*, 1992). The regulation of γ-TIP gene expression in *Arabidopsis* has been characterized using both plants containing a γ-TIP promoter-GUS gene fusion and *in situ* hybridization techniques (Ludevid *et al.*, 1992). These studies showed that γ-TIP gene expression occurs preferentially in elongating tissues of stems and roots. Such correlation between γ-TIP gene expression and cell elongation may correspond to a role for the protein during the formation of the large central vacuole and its subsequent enlargement.

More generally, the highly specific expression pattern of TIP proteins and their homologs in conditions where intense and specialized membrane processes occur suggests that these proteins may play a crucial role in adjusting membrane permeability to specific physiological requirements such as seed maturation and germination, *rhizobium*-legume symbiosis, cell elongation and drought adaptation.

γ-TIP FORMS CHANNELS SELECTIVE FOR WATER TRANSPORT

Although more than 15 genes in the MIP family have been identified, transport activity has been clearly demonstrated for only three classes of these proteins. Purified MIP protein was shown to form ion channels when incorporated in artificial membranes (Zampighi *et al.*, 1985; Ehring *et al.*, 1990). GlpF mediates the facilited diffusion of glycerol and other small molecules accross the cytoplasmic membrane of *E. coli* (Heller *et al.*, 1980). Very recently, a water transport activity was assigned to the CHIP28 protein from human red blood cells and kidney (Preston *et al.*, 1992, van Hoek and Verkman, 1992; Zeidel *et al.*, 1992).

Xenopus oocytes were used as an expression system to investigate the putative transport activity of TIP proteins. Oocytes were injected with *in vitro* transcribed γ-TIP mRNA or mRNA of two of its non-plant homologs, CHIP28 and GlpF. Water transport in oocytes expressing these proteins was investigated by transfering the oocytes to a hypoosmotic medium. Initial cell swelling during this osmotic challenge reflected water influx accross the plasma membrane of the hypertonic cell. Control oocytes injected with water instead of mRNA swelled slowly in response to the osmotic challenge. In contrast, oocytes expressing γ-TIP or CHIP28 displayed a fast swelling response and burst in less than

5 min (Figure 2). Oocyte osmotic water permeability was calculated from the rate of cell volume adjustment. This parameter increased by 8- to 10-fold in oocytes expressing γ-TIP or CHIP28 (Maurel *et al.*, 1993). We conclude that these proteins induced in the oocyte membrane a pathway for facilitated diffusion of water.

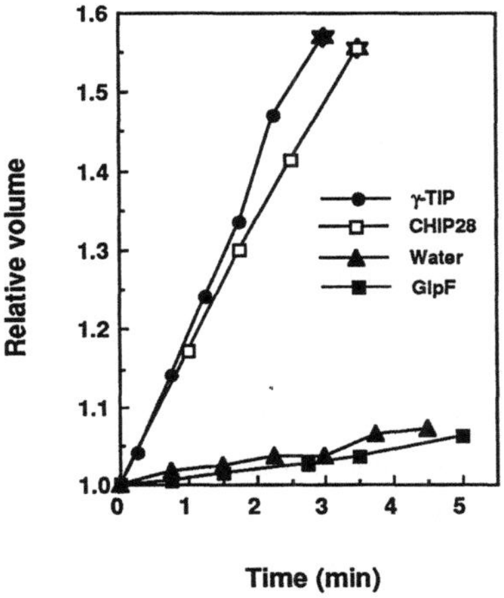

Figure 2. Osmotic swelling of *Xenopus* oocytes expressing γ-TIP, CHIP28 or GlpF. Oocytes were injected with water or *in vitro* transcribed mRNA specific for γ-TIP, CHIP28 or GlpF. After protein synthesis had occurred during the 2-3 d following mRNA injection, oocytes were transferred into a hypoosmotic (5 fold-diluted) medium and changes in cell volume were monitored (Maurel *et al.*, 1993). Stars indicate cell rupture. Representative data from experiments on the same lot of oocytes.

Uptake experiments with [^{14}C]glycerol showed that functional expression of GlpF occurred in oocytes injected with GlpF mRNA. In contrast to GlpF, γ-TIP did not enhance the glycerol uptake capacity of oocytes (Maurel *et al.*, 1993). A possible ion transport activity of γ-TIP was also investigated, using electrophysiological techniques with mRNA injected oocytes. Voltage clamping experiments showed that no electrogenic ion transport was induced by γ-TIP in oocytes, especially during osmotic challenge when massive transport of water occurs (Maurel *et al.*, 1993).

The functional expression of γ-TIP and CHIP28 in oocytes (Preston *et al.*, 1992; Maurel *et al.*, 1993) as well as reconstitution experiments with purified CHIP28 (van Hoek and Verkman, 1992; Zeidel *et al.*, 1992) have led to the proposal that this class of proteins could account for the so-called water channels that have first been described in the erythrocyte membrane (Macey, 1984; Finkelstein, 1987). Transport assays for small mineral ions or glycerol in oocytes expressing γ-TIP suggest that this protein can exclude these molecules. A high selectivity for water transport has also been demonstrated for the red blood cell water channels. This leads to the proposal that the two homologs γ-TIP and CHIP28 share the similar function of water selective channels. They are both distinct from bovine MIP and bacterial GlpF which transport ions and small polyols, respectively. This variety of transport functions among members of the MIP family suggests that all the TIP homologs in plants must possess membrane transport activity, but with possible differences in selectivity.

So far we have found a second member of the MIP family, δ-TIP that displays water channel activity similar to γ-TIP (Daniels *et al.*, in preparation). The overall amino acid identity between γ-TIP and δ-TIP is 61.5 %.

MOLECULAR MECHANISMS OF TRANSPORT

Water channels in mammalian cells are characteristically inhibited by mercury derivatives and this inhibition is reversed by reducing agents (Macey, 1984; Preston *et al.*, 1992). A similar feature is displayed by both γ-TIP and GlpF (Maurel *et al.*, 1993). This suggests the functional importance of cysteine residue(s) in the MIP family of proteins. A mercury sensitive residue of CHIP28 has recently been identified as Cys189 (Preston *et al.*, 1993). However, this residue is not conserved among other members of the MIP family. Substitution of the only cysteine of γ-TIP (Cys118) by an alanine residue was found to inhibit γ-TIP water channel activity by 60-85 % (Figure 3). Yet, the residual activity of the protein was insensitive to mercury action (Figure 3). These results confirm that Cys118 is the target of mercury effects on γ-TIP. They suggest that this residue plays a crucial role in water transport. Interestingly, Cys118 is located in the third transmembrane domain of the protein which forms a putative amphiphilic a-helix and may be part of the aqueous pore of the channel.

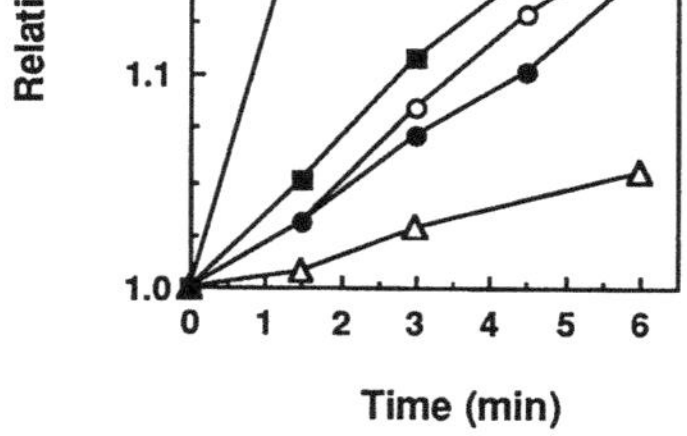

Figure 3. Effects of point mutation C118A on water channel activity and mercury sensitivity of γ-TIP. Osmotic swelling was performed as described in legends of figure 3 on oocytes either uninjected (uninj.) or injected with mRNA specific for γ-TIP (WT) or γ-TIP with the point mutation C118A (C118A). Mercury effects were tested in the presence of 3 mM $HgCl_2$ (Hg), with 10 min preincubation.

Phosphorylation, which is a common regulatory mechanism of membrane transport proteins, has been reported in the case of MIP, NOD26 and α-TIP (Ehring *et al.*, 1991; Weaver *et al.*, 1991; Johnson and Chrispeels, 1992). Phosphorylation of MIP was shown to modify the voltage sensitivity of the channels formed by the protein in lipid bilayers (Ehring *et al.*, 1991). No functional effect of NOD26 or α-TIP phosphorylation could be demonstrated. However, studies on α-TIP phosphorylation (Johnson and Chrispeels, 1992) clearly establish the presence of a calcium-dependent tonoplast-bound protein kinase that may be of major importance in the regulation of transport processes across the tonoplast.

DISCUSSION

We propose, on the basis of the present work, that γ-TIP and CHIP28 share not only amino acid sequence identity (32%) and a similar structure with six membrane spanning domains, but also the same function: the proteins form water selective channels in their respective membrane (γ-TIP in the tonoplast and CHIP28 in the red blood cell plasma membrane). Very recently, a new homolog of these proteins called WCH-CD has been characterized in rat kidney collecting tubules (Fushimi *et al.*, 1993). Its expression pattern and preliminary functional characterization suggest that WCH-CD very likely represents the vasopressin-regulated water channel of mammal kidney.

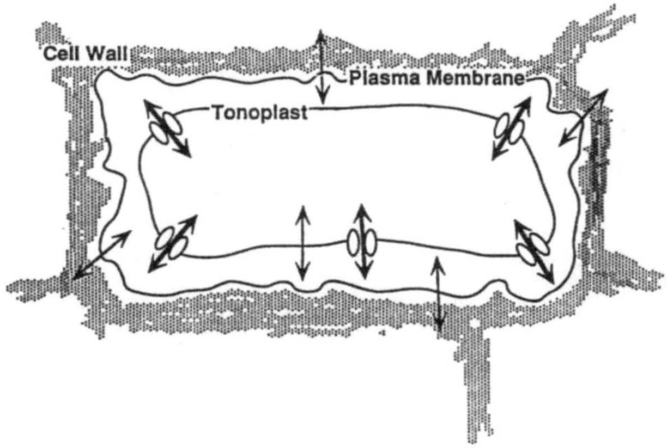

Figure 4. Putative role of aquaporins in cytoplasm osmoregulation in plant cells. The morphology of most plant cells as revealed by electron microscopy (Robinson, 1985) indicates that the plasma membrane may posses, due to its folding against the cell wall, a greater ratio of membrane surface to inner volume than the tonoplast. If only lipid mediated water transport occurred through these membranes, water exchange across the tonoplast may be limiting to rapidely compensate water exchange across the plasma membrane, exposing thus the cytoplasm to sudden osmotic change. Water channels in the tonoplast of plant cells can increase the rate of water transport accross this membrane. They may allow the cell to use the vacuolar space for rapid and homogenous osmoregulation of the cytoplasm.

Aquaporins has been proposed as a new name to designate this new class of proteins of the MIP family that transport water in a highly selective manner (Agre *et al.*, submitted).

Water slowly diffuses through membrane lipids and aquaporins can provide a pathway to rapidly equilibrate water concentration (i. e. osmotic) gradients accross membranes. The expression of γ-TIP mRNA preferentially in elongating tissues of roots and stems suggests a role for aquaporins in cell expansion. This process requires indeed massive water influx into the cells.

Most cellular processes in plants that require important water exchange, such as cell elongation, stomatal opening and closing, seed maturation and imbibition occur over periods of minutes or tens of minutes. This questions the need for a high rate of water transport accross plant membranes. The cytoplasm of plant cells forms a thin layer wedged between the vacuole and the plasma membrane and is exposed to rapid and drastic osmotic changes. We propose that a non-limiting water transport across the tonoplast may allow the cell to use the vacuolar space to buffer any osmotic change occuring in the cytoplasm after water or solute transport have occurred across the plasma membrane (Figure 4). Drastic osmotic change in the cytoplasm can also happen upon cell dessication or cell rehydration. Water transport activity of the water stress-induced (Yamaguchi-Shinozaki *et al.*, 1992) and other TIP homologs in *Arabidopsis* is presently under investigation in our group.

REFERENCES

Alexandre, J., Lassalles, J. P. and Kado, R. T. (1990) Opening of Ca^{2+} channels in isolated red beet root vacuole membrane by inositol 1,4,5-trisphosphate. *Nature,* **343,** 567-570.

Blackford, S., Rea, P. A. and Sanders, D. (1990) Voltage sensitivity of H^+/Ca^{2+} antiport in higher plant tonoplast suggests a role in vacuolar calcium accumulation. *J. Biol. Chem.,* **265,** 9617-9620.

Campell, B. R. and Town, C. D. (1992) Characterization of overexpressed cDNAs isolated from a hormone-autonomous, radiation-induced tumor tissue line of *Arabidopsis thaliana. Plant Physiol.,* **100,** 2018-2023.

Dietz, K.-J., Jäger, R., Kaiser, G. and Martinoia, E. (1990) Amino acid transport across the tonoplast of vacuoles isolated from barley mesophyll protoplasts. *Plant Physiol.,* **92,** 123-129.

Ehring, G. R., Zampighi, G., Horwitz, J., Bok, D. and Hall, J. E. (1990) Properties of channels reconstituted from major intrinsic protein of lens fiber membranes. *J. Gen. Physiol.,* **96,** 631-664.

Ehring, G. R., Lagos, N., Zampighi, G. A. and Hall, J. E. (1991) Phosphorylation modulates the voltage dependence of channels reconstituted from the major intrinsic protein of lens fiber membranes. *J. Membr. Biol.,* **126,** 75-88.

Fortin, M. G., Morrison, N. A. and Verma, D. P. S. (1987) Nodulin-26, a peribacteroid membrane nodulin is expressed independently of the development of the peribacteroid compartment. *Nucleic Acids Res.,* **17,** 813-824.

Fushimi, K., Uchida, S., Hara, Y., Hirata, Y., Marumo, F. and Sasaki, S. (1993) Cloning and expression of apical membrane water channel of rat kidney collecting tubule. *Nature,* **361,** 549-552.

Guerrero, F. D., Jones, J. T. and Mullet, J. E. (1990) Turgor-responsive gene transcription and RNA levels increase rapidly when pea shoots are wilted. Sequence and expression of three inducible genes. *Plant Mol. Biol.,* **15,** 11-26.

Hedrich, R., Kurkdjian, A., Guern, J. and Flügge, U. I. (1989) Comparative studies on the electrical properties of the H+ translocating ATPase and pyrophosphatase of the vacuolar-lysosomal compartment. *EMBO J.*, **8,** 2835-2841.

Hedrich, R. and Schroeder, J.K. (1989) The physiology of ion channels and electrogenic pumps in higher plants. *Ann. Rev. Plant Physiol.*, **40,** 539-569.

Heller, K. B., Lin, E. C. C. and Wilson, T. H. (1980) Substrate specificity and transport properties of glycerol facilitator of *Escherichia coli. J. Bacteriol.*, **144,** 274-278.

Höfte, H., Faye, L., Dickinson, C., Herman, E. M. and Chrispeels, M. J. (1991) The protein-body proteins phytohemagglutinin and tonoplast intrinsic protein are targeted to vacuoles in leaves of transgenic tobacco. *Planta,* **184,** 431-437.

Höfte, H. and Chrispeels, M. J. (1992) Protein sorting to the vacuolar membrane. *Plant Cell,* **4,** 995-1004.

Höfte, H., Hubbard, L., Reizer, J., Ludevid, D., Herman, E. M. and Chrispeels, M. J. (1992) Vegetative and seed-specific isoforms of a putative solute transporter in the tonoplast of *Arabidopsis thaliana. Plant Physiol.*, **99,** 561-570.

Johnson, K. D., Herman, E. M. and Chrispeels, M. J. (1989) An abundant, highly conserved tonoplast protein in seeds. *Plant Physiol.*, **91,** 1006-1013.

Johnson, K. D., Höfte, H. and Chrispeels, M. J. (1990) An intrinsic tonoplast protein of protein storage vacuoles in seeds is structurally related to a bacterial solute transporter (GlpF). *Plant Cell,* **2,** 525-532.

Johnson, K. D. and Chrispeels, M. J. (1992) Tonoplast-bound protein kinase phosphorylates tonoplast intrinsic protein. *Plant Physiol.*, **100,** 1787-1795.

Ludevid, D., Höfte, H., Himelblau, E. and Chrispeels, M. J. (1992) The expression pattern of the tonoplast intrinsic protein γ-TIP in *Arabidopsis thaliana* is correlated with cell enlargement. *Plant Physiol.*, **100,** 1633-1639.

Macey, R. I. (1984) Transport of water and urea in red blood cells. *Amer. J. Physiol.*, **246,** C195-C203.

Martinoia, E., Flügge, U. I., Kaiser, G., Heber, U. and Heldt, H. W. (1985) Energy-dependent uptake of malate into vacuoles isolated from barley mesophyll protoplasts. *Biochim. Biophys. Acta,* **806,** 311-319.

Maurel, C., Reizer, J., Schroeder, J. I. and Chrispeels, M. J. (1993) The vacuolar membrane protein γ-TIP creates water specific channels in *Xenopus* oocytes. *EMBO J.*, in press.

Melroy, D. L. and Herman, E. M. (1991) TIP, an integral membrane protein of the protein-storage vacuoles of the soybean cotyledon undergoes developmentally regulated membrane accumulation and removal. *Planta,* **184,** 113-122.

Miao, G.-H., Hong, Z. and Verma, D. P. S. (1992) Topology and phosphorylation of soybean Nodulin-26, an intrinsic protein of the peribacteroid membrane. *J. Cell Biol.*, **118,** 481-490.

Pao, G. M., Wu, L.-F., Johnson, K. D., Höfte, H., Chrispeels, M. J., Sweet, G., Sandal, N. N. and Saier, M. H., Jr. (1991) Evolution of the MIP family of integral membrane transport proteins. *Mol. Micro.*, **5,** 33-37.

Preisser, J. and Komor, E. (1991) Sucrose uptake into vacuoles of sugarcane suspension cells. *Planta,* **186,** 109-114.

Preston, G. M., Carroll, T. P., Guggino, W. B. and Agre, P. (1992) Appearance of water channels in *Xenopus* oocytes expressing red cell CHIP28 protein. *Science,* **256,** 385-387.

Preston, G. M., Jung, J. S., Guggino, W. B. and Agre, P. (1993) The mercury-sensitive residue at cysteine 189 in the CHIP28 water channel. *J. Biol. Chem.*, **268,** 17-20.

Smith, B. L. and Agre, P. (1991) Erythrocyte M_r 28,000 transmembrane protein exists as a multisubunit oligomer similar to channel proteins. *J. Biol. Chem.*, **266,** 6407-6415.

Taiz, L. (1992) The plant vacuole. *J. Exp. Biol.*, **172,** 113-122.

van Hoek, A. N. and Verkman, A. S. (1992) Functional reconstitution of the isolated erythrocyte water channel CHIP28. *J. Biol. Chem.*, **267,** 18267-18269.
Weaver, C. D., Crombie, B., Stacey, G. and Roberts, D. M. (1991) Calcium-dependent phosphorylation of symbiosome membrane proteins from nitrogen-fixing soybean nodules. *Plant Physiol.*, **95,** 222-227.
Wistow, G. J., Pisano, M. M. and Chepelinsky, A. B. (1991) Tandem sequence repeats in transmembrane channel proteins. *TIBS,* **16,** 170-171.
Yamaguchi-Shinozaki, K., Koizumi, M., Urao, S. and Shinozaki, K. (1992) Molecular cloning and characterization of 9 cDNAs for genes that are responsive to desiccation in *Arabidopsis thaliana*: Sequence analysis of one cDNA clone that encodes a putative transmembrane channel protein. *Plant Cell Physiol.*, **33,** 217-224.
Yamamoto, Y. T., Cheng, C.-L. and Conkling, M. A. (1990) Root-specific genes from tobacco and *Arabidopsis* homologous to an evolutionarily conserved gene family of membrane channel proteins. *Nucleic Acids Res.*, **18,** 7449.
Zampighi, G. A., Hall, J. E. and Kreman, m. (1985) Purified lens junctional protein forms channels in planar lipid films. *Proc. Natl. Acad. Sci. USA,* **82,** 8468-8472.
Zeidel, M. L., Ambudkar, S. V., Smith, B. L. and Agre, P. (1992) Reconstitution of functional water channels in liposomes containing purified red cell CHIP28 protein. *Biochemistry,* **31,** 7436-7440.

Genetic Analysis with RAPD Markers

Scott V. Tingey, J. Antoni Rafalski, and Michael K. Hanafey

Du Pont Agricultural Products, E.I. du Pont de Nemours & Company, Wilmington, Delaware 19880-0402

For many years the principles of genetics have been applied to crop variety improvement with great success. Several crop species, notably corn, wheat, and tomato, have been used as model genetic systems because of their central importance to food production. Until recently, virtually all progress in both breeding and model genetic systems has relied on a phenotypic assay of genotype. Since the efficiency of a selection scheme or genetic analysis based on phenotype is a function of the heritability of the trait, factors like the environment, multigenic and quantitative inheritance, or partial and complete dominance often confound the expression of a genetic trait (33). Many of the complications of a phenotype-based assay can be mitigated through direct identification of genotype with a DNA-based diagnostic assay (5, 7). For this reason, DNA-based genetic markers are being integrated into several genetic systems, and are expected to play an important role in the future of plant breeding.

The utility of DNA-based diagnostic markers is determined to a large extent by the technology that is used to reveal DNA-based polymorphisms. Currently, the technology of choice for many species is the RFLP or Restriction Fragment Length Polymorphism assay (6, 7). RFLP assays detect DNA polymorphisms through restriction endonuclease digestions, coupled with DNA blot hybridizations and are, in general, time-consuming and labor intensive. Over the last few years polymerase chain reaction (PCR) technology has led to the development of several novel genetic assays based on selective DNA amplification (17, 20). One of the strengths of these new assays is that they are more amenable to automation than conventional techniques. They are also simple to perform and are preferable in experiments where the genotype of a large number of individuals is to be determined at a few genetic loci. Unfortunately, because of a prerequisite for DNA sequence information, these assays are limited in their application.

Two years ago a new genetic assay was developed independently by two different laboratories (37, 34). This procedure, which we have called the RAPD assay, for Random Amplified Polymorphic DNA, detects nucleotide sequence polymorphisms in a DNA amplification based assay using only a single primer of arbitrary nucleotide sequence. In this reaction, a single species of primer binds to the genomic DNA at two different sites on opposite strands of the DNA template. If these priming sites are within an amplifiable distance of each other, a discrete DNA product is produced through thermocyclic amplification (Figure 1). The presence of each amplification product identifies complete or partial nucleotide sequence homology, between the genomic DNA and the oligonucleotide primer, at each end of the amplified product (Figure 2). On average, each primer will direct the amplification of DNA at several discrete loci in the genome, making the assay an efficient way to screen

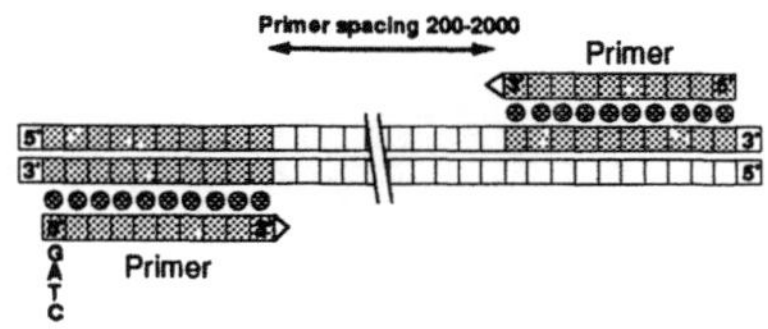

Fig. 1. Schematic diagram of random amplification of polymorphis DNA. A single species of primer binds to the genomic DNA at two different sites on opposite strands of the DNA template. If these priming sites are within an amplificble distance of each other, a discrete DNA product is produced through thermocyclic amplification. RAPD reaction conditions, except where noted, are those described by Williams et al. (1990).

NATO ASI Series, Vol. H 81
Plant Molecular Biology
Edited by G. Coruzzi and P. Puigdomènech

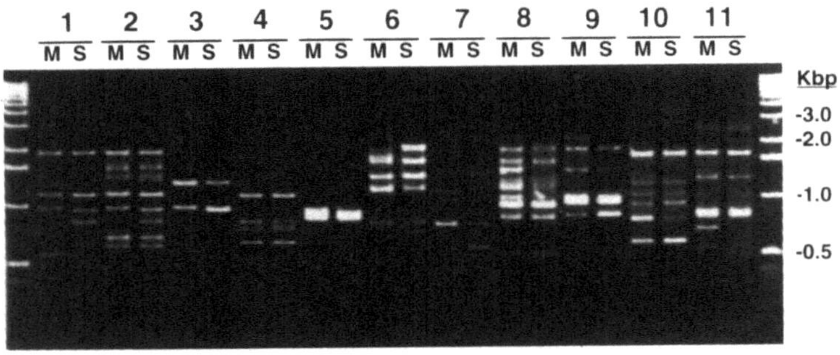

Fig. 2. Effect of nucleotide substitutions in the primer. Genomic DNA from soybean G. max cv Bonus ('M') and G. soja PI81762 ('S') was amplified with primers, which were derrived from a single primer sequence (primer 1= 5'-TGGTCACTGA), by changing only a single nucleotide relative to the original primer (primer 2=5'-AGGTCACTGA, 3=5'-TCGTCACTGA, 4=5'-TGCTCACTGA, 5=5'-TGGACACTGA, 6=5'-TGGTGACTGA, 7=5'-TGGTCTCTGA, 8=5'-TGGTCAGTGA, 9=5'-TGGTCACAGA, 10=5'-TGGTCACTCA, 11=5'-TGGTCACTGT), progressively from 5' to 3'. Molecular weight markers (kilobase pairs, Kbp) are as indicated.

for nucleotide sequence polymorphism between individuals (Figure 3). For example, the frequency of finding RAPD polymorphisms has been shown to be 0.3 per primer in Arabidopsis thaliana, 0.5 per primer in soybean, 1 per primer in corn, and 2.5 per primer in N. crassa . The major advantage of this assay is that there is no requirement for DNA sequence information. The protocol is also relatively quick and easy to perform, and uses fluorescence is used in lieu of radioactivity (38). Since the RAPD technique is an amplification-based assay, only nanogram quantities of DNA are required, and automation is feasible.

Applications of the RAPD assay

There are several applications of the RAPD assay, described below, that have been developed over the last two years. Each of these techniques exploits the efficiency of detection of DNA sequence-based polymorphisms in the RAPD assay; this efficiency permits genetic analysis in

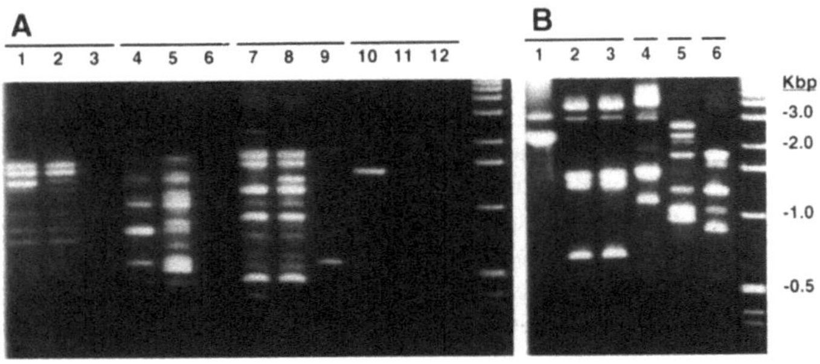

Figure 3. Amplification of DNA from different species. Panel A, amplification of eukaryotic DNA. DNA was amplified from a variety of species using primers of arbitrary nucleotide sequence (Methods). Lanes 1 and 2, human DNA Hu2 and Hu3, respectively, amplified with primer 5'-ACGGTACACT. Lanes 4 and 5, corn CM37 and T232, respectively, amplified with 5'-GCAAGTAGCT. Lanes 7 and 8, soybean G. max and G. soja,, respectively, amplified with 5'-CGGCCCCTGT. Lanes 10 and 11, N. crassa Oakridge and Mauriceville, respectively, amplified with 5'-CACATGCTTC. Genomic DNA was omitted in control reactions (lanes 3, 6, 9, and 12) to determine whether any of the bands seen with genomic DNA were primer artefacts; such an artefact is present in lane 9, but these bands were not seen when genomic DNA was present in the reaction mixture (lanes 7 and 8). Molecular weight markers (kilobase pairs, Kbp) are as indicated. Panel B, amplification of prokaryotic DNA. Lanes 1-3, Escherichia coli (strains 037, 641, 642, respectively). Lane 4, Listeria monocytogenes (strain 681). Lane 5, Staphylococcus aureus (strain 684). Lane 6, Salmonella typhimurium (strain 706). All genomic DNA samples were amplified with the primer 5'-TCACGATGCA. Molecular weight markers (kilobase pairs, Kbp) are as indicated.

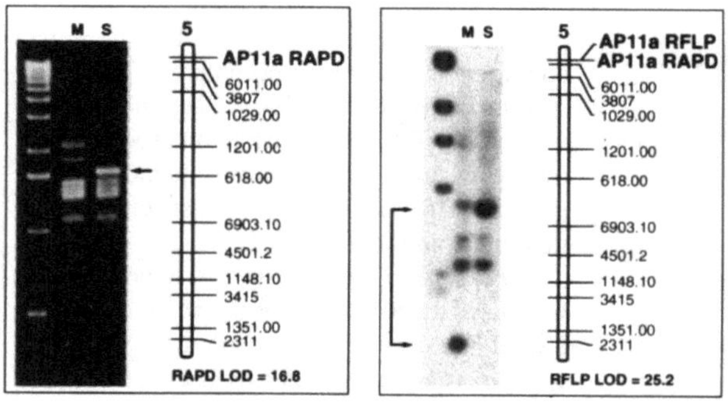

Figure 4. Map position of RAPD AP11a and its corresponding RFLP. Panel A, the segregation of RAPD AP11a (arrow) was scored and assigned a position on the soybean RFLP map (see text). The gel image shows the RAPD identified between the parents G. max cv Bonus and G. soja PI81762 (lanes 1 and 2, respectively). Molecular weight markers are in the unlabeled lane. A LOD score of 16.8 indicates that the mapped position of this RAPD is 1016.8 more likely than the probability of it being unlinked to this group of RFLP markers. Panel B, RAPD AP11a was used as a hybridization probe to detect the BclII RFLP AP11a. The RFLP maps to the same position as its corresponding RAPD probe, which confirms the equivalence of the two types of DNA marker.

biological systems which have not previously benefitted from the use of molecular markers. There is, however, a general caution that must applied when using single primer amplification technology. Because of the stochastic nature of DNA amplification with random sequence primers, it is important to maintain consistent reaction conditions that have been optimized for reproducible DNA amplification. Several factors including template DNA concentration, magnesium concentration, primer annealing temperature, primer length, and primer base composition all effect the reaction (38), and should be controlled carefully.

Development of Genetic Maps. One of the first practical uses of RAPD markers was in the creation of high density genetic maps (Figure 4). Reiter et al.(30) were able to place over 250 new genetic markers on a recombinant inbred population of A. thaliana in only four man-months, clearly demonstrating the utility of RAPD markers for quickly saturating both a global and local genetic map.

Historically, many important crop systems have suffered from a lack of genetic markers. For example, genetic linkage analysis in conifers has advanced slowly primarily due to the large size of the genome, and to the inherent difficulties involved in producing a segregating F2 population (10). Carlson et al. and Chaparro et al. (10, 11) have recently shown that the availability of RAPD technology has made mapping in conifers a reasonable endeavor. Chaparro et al. (11) were able to create a 191 marker RAPD map of loblolly pine in only 6 man months. Both the A. thaliana and loblolly pine genetic maps were developed with unprecedented speed, taking advantage of the unique reproductive biology of each system.

Because RAPD polymorphisms are the result of either a nucleotide base change that alters the primer binding site, or of an insertion or deletion within the amplified region (29, 37), polymorphisms are usually noted by the presence or absence of an amplification product from a single locus. This also means that the RAPD technique tends to provide only dominant markers. Individuals containing two copies of an allele are not distinguished quantitatively from those containing only one copy of the allele. The disadvantage of mapping with dominant markers is that markers linked in repulsion, for example markers residing on separate chromatids as can be found in a F2 population, provide little information for the estimate of genetic distance (1). Therefore when mapping with dominant markers it is necessary to work with markers that are only linked in coupling, that is markers residing on a single chromatid, as can be found in a backcross or recombinant inbred population, in haploid or gametophytic tissue, or alternately in an F2 population where only RAPD markers amplified from a single parent are mapped (38). Genetic simulations show that dominant markers linked in coupling are as efficient for mapping as codominant markers on a per gamete basis (See Appendix A). It is also possible to use RAPD

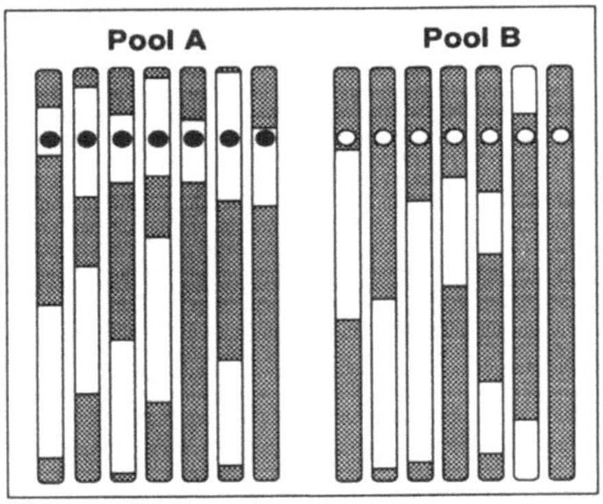

Fig. 5. Schematic diagram of bulk segregant analysis. Individuals are pooled from a segregating population based on a phenotypic or genotypic selection (open or closed circles). Individuals within a pool are at linkage disequilibrium at the selected locus. Genetic markers which show polymorphism between the pools identify loci that are linked to the selected locus.

products as hybridization probes to detect codominant RFLP polymorphisms. The utility of this approach, however, is dependent upon whether the RAPD amplification product is homologous to a single copy sequence in the genome (38), a characteristic specific to the genome being investigated.

Targeting Genetic Markers. Several groups have used the RAPD assay as an efficient tool to identify molecular markers that lie within regions of a genome which have been introgressed during the development of near isogenic lines (18, 25, 28). By definition, any region of the genome that is polymorphic between two near isogenic plants is potentially linked to the introgressed trait. Thus, Klein-Lankhorst et al. (18) were able to identify RAPD markers specific to chromosome 6 of tomato by screening an L. esculentum substitution line, while Martin et al. (25) were able to confirm linkage of RAPD markers to the Pto locus in tomato after screening two near isogenic lines. Paran et al. (28) used two different sets of near isogenic lettuce lines to identify RAPD markers linked to the Dm1, Dm3 and Dm11 loci. RAPD markers were 4 to 6 times more efficient, on a per assay basis, than were RFLP markers for identifyin these polymorphisms, and were over 10 fold more efficient in terms of time and labor.

Another advantage of this technology is that a genetic map of the entire genome is not required in order to identify markers linked to a trait of interest; instead, specific regions of the genome can be mapped. There are, however, two disadvantages to using near isogenic lines to identify markers linked to a genetic trait. The first is that several generations of backcrossing are needed to create near isogenic lines. The second is that there are frequently several regions of the donor genome which are inadvertently co-introgressed into the near isogenic line (41). This results in the identification of marker polymorphisms between near isogenic lines which are not necessarily linked to the trait that is being studied.

Pooling Strategies. Another technology has been developed which allows genetic markers linked to very specific regions of the genome to be identified. Arnheim et al. (2) outlined a genome pooling strategy that allows RFLP markers to be targeted to a region of the genome that is not in linkage equilibrium as a result of selection at a particular locus. The strategy requires pooling genomic DNA from individuals which are known to be genetically fixed at a particular locus. Markers linked to that locus are identified by their linkage disequilibrium with respect to the rest of the population. The limitation of this approach is that it relies on RFLP technology, which is relatively inefficient for identifying polymorphic regions of a genome, and more importantly, on the premise that linkage disequilibrium exists adjacent to the locus of interest within the source population.

Recently, Michelmore et al. (26) have described the use of RAPD markers to screen efficiently for markers linked to specific regions of the genome. This method, called Bulk Segregant Analysis (BSA), uses two bulked DNA samples gathered from individuals segregating in a single population. Each bulk is composed of individuals that differ for a specific phenotype or genotype, and/or are individuals at either extreme of a segregating population. For simple genetic traits, all loci in the genome should appear to be in linkage equilibrium except in the region of the genome linked to the selected locus. Markers linked to this locus should appear polymorphic between the two different pools. Since many segregating individuals are used to generate the pools, there is only a minimal chance that regions of the genome unlinked to the target locus will also be polymorphic between the pools. Random primers can then be used to amplify loci from each pool and to identify RAPD polymorphisms linked to the trait of interest.

Michelmore et al. (26) have used bulk segregant analysis successfully to target markers to the Dm5/8 locus in lettuce. The advantage of this technology (See Fig. 5) is that markers are targeted to a much smaller locus within the genome, and the likelihood of identifying false positive markers is small (26) compared to that for using near isogenic line analysis. Selections made from an F2 population will always be in linkage disequilibrium with respect to selected regions of the genome, and markers can be targeted to any locus where any form of selection can be applied, either phenotypic or genotypic.

Giovannoni et al. (13) demonstrated the use of a pooling strategy, based on known RFLP genotypes from existing mapping populations, to create pools of DNA from individuals homozygous for opposite parental alleles in a targeted chromosomal interval. RAPD markers were then targeted to regions of the tomato genome responsible for fruit ripening and pedicle abscission. Reiter et al. (30) used this pooling strategy to identify 100 RAPD markers specific to chromosome 1 of A. thaliana. As genetic maps approach saturation, pooling on phenotype or genotype will allow researchers to move away from a random approach to map saturation, and to focus more efficiently on specific regions of the genome.

Population Genetics. RAPD markers have now been used extensively to create DNA fingerprints for the study of individual identity and of taxonomic relationships in eukaryotic and procaryotic organisms (8, 35, 39, 14, 15, 16, 21, 36, 31, 24). An important question that has emerged from these studies is whether or not RAPD bands of equal molecular weight that are shared between individuals are homologous characters, which have been inherited from a common ancestor, or homoplastic characters, which have arisen independently within a population. It is more likely that closely related individuals would co-inherit a shared character state from a common ancestor rather than have acquired the same character independently. Williams et al. (38) demonstrated this to be the case by using single RAPD bands as hybridization probes to detect homologous characters on a DNA blot of RAPD products. Within the limits of the resolution of an agarose gel, RAPD bands that were amplified from different species of the genus Glycine, and that were scored as homologous by their relative mobility, were also shown to be homologous by hybridization.

Several groups have reported on the utility of RAPD markers as sources of phylogenetic information. Arnold et al. (3) were successful in using RAPD markers to test for interspecific nuclear gene flow between *Iris fulva* and *I. hexagona.*, and to study the presumed hybrid origin of I. nelsonii. Hu and Quiros (16) were able to show that the amplification products from only 4 random primers were sufficient to discriminate between 14 different broccoli and 12 different cauliflower cultivars (*Brassica oleracea* L.). RAPD markers have also been used to assess the amount of genetic diversity in germplasm collections. Using only 25 different decamer oligonucleotide primers, Kresovich et al. (21) were able to collect information on 140 different polymorphic characters in a array of individuals representing *Brassica oleracea* L. and *B. rapa* L. They showed the utility of the assay for discriminating between different individuals in a germplasm collection, and showed that the ability to distinguish between closely related individuals was simply a function of the number of RAPD bands that were observed. RAPD markers provided an efficient technology for discovering these polymorphic characters.

Using an arbitrary primer as short as 5 nucleotides, combined with silver staining to increase the sensitivity of DNA band detection, Caetano-Anolles et al. (8) were able to produced a detailed and relatively complex DNA fingerprint for several different species. This approach, termed DNA Amplification Fingerprinting, or DAF, has been reviewed recently (9), and promises to generate more genetic information from each amplification.

Conclusion

DNA-based diagnostics are now well established as a means to assay diversity at the locus, chromosome, and whole genome levels. As technology has advanced, DNA sequence-based assays have become easier to useand more efficient at screening for nucleotide sequence-based polymorphisms. The ultimate genetic assay would be based on the determination of the complete DNA sequence at any locus of interest. Increased genetic resolution would be obtained simply by sequencing a larger contiguous segment of DNA at the locus. As a prelude to this, there are now several good examples of DNA sequence-based diagnostic assays that are designed to identify the presence or absence of a specific nucleotide base pair at a discrete locus (23, 19, 27, 40, 32, 4, 12, 22). Today these assays are only limited in their application by the high cost of DNA sequence determination. As DNA sequencing technology becomes more cost efficient and automated, genetic assays may be based directly on DNA sequence analysis.

References

1. Allard, R.W. (1956). Formulas and tables to facilitate the calculation of recombination values in heredity. Hilgardia 24: 235-278
2. Arnheim, N., C. Strange, and H. Erlich. (1985). Use of pooled DNA samples to detect linkage disequilibrium of polymorphic restriction fragments and human disease: Studies of the HLA class II loci. Proc. Natl. Acad. Sci. USA 82: 6970-6974
3. Arnold, M.L., C.M. Buckner, and J.J. Robinson. (1991). Pollen-mediated introgression and hybrid speciation in louisiana irises. Proc. Natl. Acad. Sci. USA 88: 1398-1402
4. Barany, F. (1991). Genetic disease detection and DNA amplification using cloned thermostable ligase. Proc. Natl. Acad. Sci. USA 88: 189-193
5. Beckmann, J.S. and M. Soller. (1983). Restriction fragment length polymorphisms in genetic improvement: methodologies, mapping and costs. Theor. Appl. Genet. 67: 35-43
6. Botstein, D., R.L. White, M.H. Skolnick, and R.W. Davis. (1980). Construction of a genetic map in man using restriction fragment length polymorphisms. Am.J.Hum.Genet. 32: 314-331
7. Burr, B., S.V. Evola, and F.A. Burr. (1983). The application of restriction fragment length polymorphism to plant breeding, in: Genetic Engineering - Principles and Methods, J.K. Setlow and A. Hollander, Editor. Plenum Press: New York and London. p. 45-59.
8. Caetano-Anolles, G., B.J. Bassam, and P.M. Gresshoff. (1991). High resolution DNA amplification fingerprinting using very short arbitrary oligonucleotide primers. Biotechnology 9: 553-557
9. Caetano-Anollés, G., B.J. Bassam, and P.M. Gresshoff. (1991). DNA amplification fingerprinting: A strategy for genome analysis. Plant Mol. Biol. Rep. 4: 294-307
10. Carlson, J.E., L.K. Tulsieram, J.C. Glaubitz, V. Luk, C. Kauffeldt, and R. Rutledge. (1991). Segregation of random amplified DNA markers in f1 progeny of conifers. Theor. Appl. Genet. 83: 194-200

11. Chaparro, J., P. Wilcox, D. Grattapaglia, D. O'Malley, S. McCord, R. Sederoff, L. McIntyre, and R. Whetten. (1992). Genetic mapping of pine using RAPD markers: Construction of a 191 marker map and development of half-sib genetic analysis. in: Advances in gene technology: Feeding the world in the 21st century. Miami Winter Symposium, Miami, Florida.
12. Dockhorn-Dworniczak, B., B. Dworniczak, L. Brömmelkamp, J. Bülles, J. Horst, and W.W. Böcker. (1991). Non-isotopic detection of single strand conformation polymorphism (PCR-SSCR): A rapid and sensitive technique in diagnosis of phenylketonuria. Nuc. Acids Res. 19: 2500
13. Giovannoni, J.J., R.A. Wing, M.W. Ganal, and S.D. Tanksley. (1991). Isolation of molecular markers from specific chromosomal intervals using DNA pools from existing mapping populations. Nucleic Acids Res. 19: 6553-6558
14. Hadrys, H., M. Balick, and B. Schierwater. (1992). Applications of RAPD fingerprinting in molecular ecology. Mol. Ecol. In Press
15. Hedrick, P. (1992). Shooting the RAPDs. Nature 355: 679-680
16. Hu, J. and C.F. Quiros. (1991). Identification of broccoli and cauliflower cultivars with rapd markers. Plant Cell Reports 10: 505-511
17. Innis, M.A., D.H. Gelfand, J.J. Sninsky, and T.J. White, PCR Protocols. 1st ed. 1990, San Diego: Academic Press. 482.
18. Klein-Lankhorst, R.M., A. Vermunt, R. Weide, T. Liharska, and P. Zabel. (1991). Isolation of molecular markers for tomato (Lycopersicon esculentum) using random amplified polymorphic DNA (rapd). Theoretical And Applied Genetics 83: 108-114
19. Korner, J.S. and K.J. Livak. (1989). Mutation detection using nucleotide analogs that alter electrophoretic mobility. Nuc. Acids Res. 17: 7779-7783
20. Krawetz, S.A. (1989). The polymerase chain reaction: opportunities for agriculture. AgBiotech. News and Info. 1: 897-902
21. Kresovich, S., J.G.K. Williams, J.R. McFerson, E.J. Routman, and B.A. Schaal. (1992). Characterization of genetic identities and relationships of Brassica oleracea L. via random amplified polymorphic DNA assay. Theor. Appl. Genet. In Press
22. Kuppuswamy, M.N., J.W. Hoffman, C.K. Kasper, S.G. Spitzer, S.L. Groce, and S.P. Baja. (1991). Single nucleotide primer extension to detect genetic disease: Experimental application to hemophilia B (factor IX) and cystic fibrosis genes. Proc. Natl. Acad. Sci. USA 88: 1143-1147
23. Landegren, U., R. Kaiser, J. Sanders, and L. Hood. (1988). A Ligase-mediated gene detection technique. Science 241: 1077-1080
24. Lander, E.S., P. Green, J. Abrahamson, A. Barlow, M.J. Daly, S.E. Lincoln, and L. Newburg. 1987. Mapmaker: An interactive computer package for constructing primary linkage maps of experimental and natural populations. Genomics 1: 174-181
25. Lark, K.G., J. Evans, F. Basha, R. Bogden, R. Copeland, R. Ellison, D. Horne, K. Lee, K. McDonald, J. Pierson, W. Schuster, P. Wilhelm, and Y. Yu. (1992). Molecular phylogeny as a genetic tool for soybean breeding. Soybean Genetic Newsletter In Press
26. Martin, G.B., J.G.K. Williams, and S.D. Tanksley. (1991). Rapid identification of markers linked to Pseudomonas resistance gene in Tomato by using random primers and near isogenic lines. Proc. Natl. Acad. Sci. USA 88: 2336-2340
27. Michelmore, R.W., I. Paran, and R.V. Kesseli. (1991). Identification of markers linked to disease resistance genes by bulked segregant analysis: A rapid method to detect markers in

specific genomic regions using segregating populations. Proc. Natl. Acad. Sci. USA 88: 9828-9832

28. Newton, C.R., A. Graham, L.E. Heptinstall, S.J. Powell, C. Summers, N. Kalsheker, J.C. Smith, and A.F. Markham. (1989). Analysis of any point mutation in DNA. The amplification refractory mutation system (ARMS). Nuc. Acids Res. 17: 2503-2515
29. Paran, I., R. Kesseli, and R. Michelmore. (1991). Identification of restriction-fragment-length-polymorphism and random amplified polymorphic DNA markers linked to downy mildew resistance genes in lettuce, using near-isogenic lines. Genome 34: 1021-1027
30. Parks, C. ., L.-S. Chang, and T. Shenk. (1991). A polymerase chain reaction mediated by a single primer: cloning of genomic sequences adjacent to a serotonin receptor protein coding region. Nucleic Acids Res. 19: 7155-7160
31. Reiter, R.S., J. Williams, K.A. Feldmann, J.A. Rafalski, S.V. Tingey, and P.A. Scolnik. (1992). Global and local genome mapping in Arabidopsis thaliana by using recombinant inbred lines and random amplified polymorphic DNAs. Proc. Natl. Acad. Sci. USA 89: 1477-1481
32. Stiles, J.I., C. Lemme, S. Sondur, M.B. Morshidi, and R. Manshardt. (1992). Using randomly amplified polymorphic DNA for evaluating genetic relationships: Analysis of the relationship among 'Solo' and other papaya cultivars. Theor. Appl. Genet. In Press
33. Suzuki, Y., T. Sekiya, and K. Hayashi. (1991). Allele-specific polymerase chain reaction: A method for amplification and sequence determination of a single component amoung a mixture of sequence variants. Anal. Biochem. 192: 82-84
34. Tanksley, S.D., N.D. Young, A.H. Paterson, and M.W. Bonierbale. (1989). RFLP mapping in plant breeding: New tools for an old science. Biotechnology 7: 257-264
35. Welsh, J. and M. McClelland. (1990). Fingerprinting genomes using PCR with arbitrary primers. Nucleic Acids Res. 19: 303-306
36. Welsh, J., C. Petersen, and M. McClelland. (1991). Polymorphisms generated by arbitrarily primed PCR in the mouse: Application to strain identification and genetic mapping. Nucleic Acids Res. 19: 303-306
37. Wilde, J., R. Waugh, and W. Powell. (1992). Genetic fingerprinting of Theobroma clones using randomly amplified polymorphic DNA markers. Theor. Appl. Genet. 83: 871-877
38. Williams, J.G.K., A.R. Kubelik, K.J. Livak, J.A. Rafalski, and S.V. Tingey. (1990). DNA polymorphisms amplified by arbitrary primers are useful as genetic markers. Nucleic Acids Res. 18: 6531-6535
39. Williams, J.G.K., J.A. Rafalski, and S.V. Tingey. (1992). Genetic analysis using RAPD markers, in: Methods in Enzymology, R. Wu, Editor. Academic Press: Orlando, FL.
40. Wöstemeyer, J., C. Schäfer, M. Kellner, and M. Weisfeld. (1991). DNA polymorphisms detected by random primer dependent PCR as a powerful tool for molecular diagnostics of plant pathogenic fungi, in: Advances in Molecular Genetics. Hüthig-Verlag: Heidelberg.
41. Wu, D.Y., L. Ugozzoli, B.K. Pal, and R.B. Wallace. (1989). Allele-specific enzymatic amplification of ß-globin genomic DNA for diagnosis of sickle cell anemia. Proc. Natl. Acad. Sci. USA 86: 2757-2760
42. Young, N.D. and S.D. Tanksley. (1989). RFLP analysis

Appendix A
Genetic Mapping with Dominant Markers.

RAPD polymorphisms are the result of either a nucleotide base change that alters the primer binding site, or an insertion or deletion within the amplified region (29, 37), polymorphisms are usually noted by the presence or absence of an amplification product from a single locus. This also means that the RAPD technique tends to provide only dominant markers. Individuals containing two copies of an allele are not distinguished quantitatively from those containing only one copy of the allele. The disadvantage of mapping with dominant markers is that markers linked in repulsion, for example markers residing on separate chromatids as could be found in a F2 population, provide little information for the estimate of genetic distance (1). Computer simulation of meiosis and recombination in f2, backcross, and recombinant inbred populations was used to determine the error in the estimate of % recombination (R) between genetic markers linked at various distances, and markers in different linkage phase in each of these populations.

The average root mean square error (RMSE) for the estimate of R was determined for dominant and codominant markers, as a function of the genetic distance between each marker. This was done by simulating recombination based on a known genetic map with randomly spaced genetic markers at a series of expectation distances, simulating several different populations of different size and type (eg. f2, backcross, or recombinant inbred), scoring segregation in each population with dominant or codominant markers, and then estimating the distance between each marker using multipoint maximum likelihood techniques (24). The RMSE was computed using the squared deviation of the estimated value of R from the true genetic map. This set, iterating over 6 different average map distances, 6 different population sizes, and 5 different population/ marker types, was repeated 223 times. In each case the RMSE scaled approximately as the square root of 1/N with population size. This is expected in some cases, such as in a backcross population (1), but not necessarily in an F2 population scored with dominant markers.

The results for all population sizes were normalized to 100 individuals, the RMSE of R, relative to the codominant case, was calculated for each population type, and the results summarized in Table 1 and Figure 1. Table 1 presents the coeficients of the polynomial for computing the RMSE for various marker classes and population sizes as a function of genetic map distance Ø, and population size N. These equations are useful for comparing the relative error in the estimate of R using different marker classes, population types, or population sizes. Figure 1 shows that the

Table 1. Coeficients of the polynomial for computing the RMSE in the multipoint estimate or R, for various marker classes and population types, as a function of genetic map distance Ø, and population size N.

Population type and Marker class	Coeficients of the Polynomial to compute RMSE in %R.
F2 Codominant	$(0.93 + 0.127\text{Ø} - 0.00200\text{Ø}^2)\sqrt{100/N}$
F2 Dominant-Coupled	$(1.40 + 0.170\text{Ø} - 0.00262\text{Ø}^2)\sqrt{100/N}$
F2 Dominant-Random	$(1.65 + 0.229\text{Ø} - 0.00315\text{Ø}^2)\sqrt{100/N}$
BackcrossDominant	$(1.41 + 0.140\text{Ø} - 0.00235\text{Ø}^2)\sqrt{100/N}$
RI Dominant	$(0.83 + 0.207\text{Ø} - 0.00311\text{Ø}^2)\sqrt{100/N}$

relative RMSE of the estimate of R, for dominant markers linked in coupling in an F2 population, is roughly the square root of 2 times that for codominant markers in a F2 population. The higher relative RMSE can be explained simply by the difference in the effective number of gametes per individual, when only assaying markers linked in coupling phase, demonstrating that dominant markers linked in coupling are as efficient for mapping as codominant markers on a per gamete basis. Figure 1 also shows that the relative RMSE for dominant markers linked in coupling in an F2, is roughly equivalent to the relative RMSE for dominant markers in backcross and recombinant inbred populations. The differences that exist in the RMSE in these populations is a function of whether or not there is knowledge of linkage phase at any distance R. Of interest ment in the RMSE is due to the contribution of neighboring dominant markers, linked in coupling, to the multipoint estimate of R for a single interval flanked by markers in repulsion.

These genetic simulations demonstrate that dominant markers linked in coupling are as efficient for mapping as codominant markers on a per gamete basis. When mapping with dominant markers it is necessary to work with markers that are only linked in coupling, that is markers residing on a single chromatid, as can be found in a backcross or recombinant inbred population, in haploid, doubled haploid, or gametophytic tissue, or alternately in an F2 population where only RAPD markers amplified from a single parent are mapped (38).

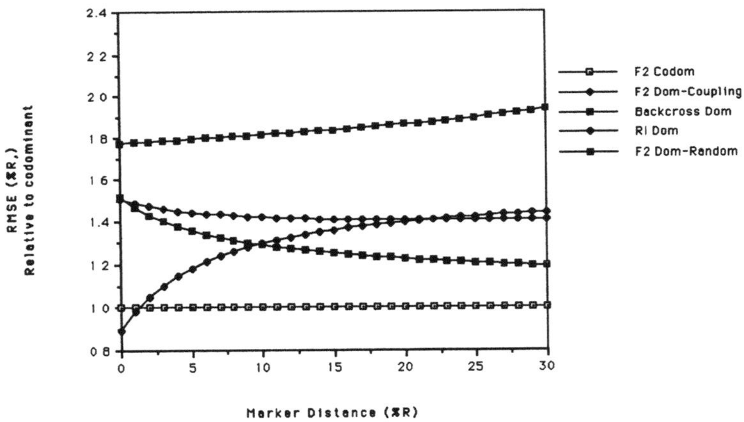

Figure 1. The RMSE for the estimate of R was determined at several marker distances for codominant markers in an F2, dominani markers linked in coupling in an F2, dominant markers linked in random phase in an F2, and dominant markers in a Backcross and recombinant inbred population. The RMSE of the estimate of R for each population and marker distance was then plotted relative the the RMSE of the estimate of R for codominant markers in an F2 population. This displays the RMSE in the estimate of R for domnant markers in each population, relative the the most efficient mapping scenario, that is, codominant markers in an F2 population.

Clusters Of Resistance Genes in Lettuce

(Map-based Cloning In Non-intensively Studied Species)

Richard Michelmore, Peter Anderson, Patricia Okubara, Hanneke Witsenboer
Department of Vegetable Crops
University of California
Davis
CA 95616
USA

Introduction

Strategies for map-based cloning of disease resistance genes have been previously reviewed (e.g. Michelmore *et al.*, 1993). In this chapter, we will discuss the recent progress and current limitations with particular emphasis on our efforts to clone and characterize genes for resistance to downy mildew in lettuce. The approaches discussed here are applicable to cloning the determinants of any trait for which only the genetics is known and the species has not been studied intensively at the molecular level.

Considerable progress has been made over the past two years towards cloning of disease resistance genes from several plant species. *Hm1* that confers resistance to *Cochliobolus carbonum* in corn has been cloned by transposon tagging (Johal and Briggs, 1992; Briggs and Johal, this volume). Markers have been identified that are genetically tightly linked to resistance genes in a wide variety of plants. A few resistance genes have been localized to small chromosomal regions in *Arabidopsis thaliana* (Katagiri and Ausubel, this volume) and tomato (Martin *et al.*, 1993); YAC clones thought to contain resistance genes have been isolated in several cases and map-based cloning seems imminent.

It is likely that there are several broad classes of resistance genes. Some resistance genes might encode receptors that detect either directly or indirectly the presence of the pathogen; activation of such receptors may initiate a signal transduction pathway that results in the induction of the generic resistance response genes (Lamb, this volume). Other resistance genes may encode products that detoxify/deactivate compounds required by the pathogen for pathogenesis; *Hm1* belongs to such a class. Yet other resistance genes may encode altered targets for pathogen molecules, such as toxins, that are required for disease. A fourth category of resistance genes may be a heterogeneous group some of which may have little or no generality in plant-pathogen interactions. To determine whether such classes of resistance genes exist and what are the characteristics of each class, it is necessary to

NATO ASI Series, Vol. H 81
Plant Molecular Biology
Edited by G. Coruzzi and P. Puigdomènech

characterize multiple genes conferring resistance to diverse pathogens, preferably from several plant species.

Lettuce Downy Mildew

Lettuce downy mildew, the interaction between *Lactuca sativa* and *Bremia lactucae*, has several characteristics that make it amenable to analysis at the classical and molecular level (Michelmore *et al.*, 1988). Although the causal fungus, *B. lactucae*, is an obligate biotroph, it can be readily cultured for most experimental purposes on young lettuce seedlings. The asexual cycle of the fungus is completed within one to two weeks. Large numbers of seedlings can be screened for their interaction phenotype. In an incompatible interaction, the penetrated epidermal cell undergoes rapid cell death and the fungus usually develops no further. In the compatible interaction, there is no rapid cell death, the fungus grows from the epidermal cell to form an intercellular mycelium, and there is minimal macroscopically-visible disturbance of the host until sporulation. *B. lactucae* is heterothallic; controlled crosses can be made to study the genetics of avirulence. We have constructed a partial genetic map of *B. lactucae* containing over 100 RFLP markers (Legg, 1991). We developed transformation vectors from *B. lactucae* sequences and used them to transform *Phytophthora infestans* (Judelson *et al.*, 1991); however, we have yet to obtain stable transformants of *B. lactucae*.

Lettuce (*Lactuca sativa*) is a diploid member of the Compositae (2n = 2x = 18) with a moderate genome size (n = approx. 2 pg). It has a fairly short life cycle (3 to 6 months depending on the genotype), crosses can be readily made and it produces large numbers of seeds. Therefore, extensive genetic studies and mutation experiments are feasible. Lettuce can be easily transformed using *Agrobacterium tumefaciens* (Michelmore *et al.*, 1987). We are developing a detailed genetic map of *L. sativa*. It currently comprises over 430 RFLP, RAPD, and other markers (Kesseli *et al.*, 1993 and unpublished).

Lettuce and *B. lactucae* interact in a precise gene-for-gene interaction (Crute and Johnson, 1976; Ilott *et al.*, 1989). To date, thirteen single dominant genes for resistance to *B. lactucae* (*Dm*) have been identified, matched by dominant avirulence genes in the pathogen. Several additional resistance factors have been identified but their single gene nature has yet to be confirmed by parallel pathogen genetics. Also, numerous additional resistant accessions have been identified that have yet to be characterized genetically. Therefore extensive genetic variation in resistance exists that awaits characterization at the molecular level.

Maximizing Genetics

Map-based cloning is comprised of several overlapping and integrated but discrete steps: classical genetics, genetic mapping, saturating the target region with markers, long-range restriction mapping, integrating genetic and physical distances, chromosome walking, gene identification, and gene complementation. While conceptually straightforward, many of the steps are technically demanding and labor intensive. When possible, it is necessary to utilize non-brute force approaches that minimize the effort at each stage. This is particularly true for species that are not intensively studied, such as lettuce, and where only small amounts of the genome are likely to be ever studied in detail. We have therefore adopted a strategy to maximize the genetics component and to minimize the extent of chromosome walking; we are also developing PCR-based methods for analyzing complex regions of the genome.

Maximization of genetics includes careful segregation analysis to ensure that only a single resistance gene specificity is being targeted. This is not trivial as resistance genes tend to be clustered and multiple tightly-linked genes exhibit monogenic segregation ratios. The number of active resistance genes can be determined using pathogen genetics. Avirulence genes in the pathogen tend to be unlinked; as there is a one-to-one correspondence between resistance and avirulence genes, segregation for avirulence on a particular plant genotype is indicative of the number of genes conferring resistance to that isolate.

Clusters of Resistance Genes

We are genetically characterizing many resistance genes in lettuce. Most of the resistance genes mapped so far are located in three major clusters. The majority of the genes characterized in detail confer resistance to downy mildew; however, these clusters also contain genes for resistance to turnip mosaic virus (*Tu*) and root aphid (*Ra*), that had been shown by previous workers to be linked to individual *Dm* genes (Fararra *et al.*, 1987; Zink and Duffus, 1973; Crute and Dunn, 1981). The largest group that includes 8 *Dm* genes and *Ra* spans 12 cM. The evolutionary and/or functional significance of this clustering has yet to be determined. However, it has led to the hypothesis that resistance genes are members of multigene families that have diverged to have different specificities (Pryor, 1987; Michelmore *et al.*, 1987). Therefore, the cloning of one member of the multigene family may provide access either by cross-hybridization or by functional implication to additional genes in that class of resistance genes.

With the help of numerous cooperators, we are continuing to map most of the known resistance genes in lettuce. Using bulked segregant analysis (Michelmore *et al.*, 1991), we have rapidly located two new genes for resistance to *B. lactucae*, one gene for resistance to

Plasmopara lactucae radicis, and one gene for resistance to lettuce mosaic virus (*mo*). The former three genes mapped to two of the previously characterized clusters of resistance genes (Maisonneuve *et al.*, unpublished; Kesseli *et al.*, 1993); *mo* was not linked to any known resistance gene (Irwin *et al.*, unpublished). We are currently mapping seven additional genes for resistance to five viruses and a bacterium and three additional *Dm* genes.

Saturation of Target Region with Markers

Before starting a chromosome walk, it is necessary to obtain markers as close as possible to the target gene to minimize the number of overlapping clones required and possibly to avoid using labor-intensive vector systems that are required for cloning large genomic fragments. We have used near-isogenic lines or bulked segregant analysis to identify RAPD markers linked to each of the clusters of *Dm* genes (Paran *et al.*, 1991; Michelmore *et al.*, 1991). This has resulted in many markers tightly linked to *Dm* genes. Currently we have 29 markers within 5 cM of a *Dm* gene (Table 1).

Table 1. Markers derived from RAPD fragments linked to *Dm* genes.

Resistance gene cluster	RAPDS within 5 cM	Number sequenced	SCARs dominant	SCARs codominant	Number with repeats	Hybridization probe
Dm1 et al.	14	9	7	2	7	6
Dm5/8 et al.	11	7	7	0	7	4
Dm4 et al.	4	2	2	0	-	-

RAPD markers have two weaknesses; they can be sensitive to reaction conditions and multiple loci are detected in each amplification. We have therefore developed reliable PCR-based markers from RAPD loci. RAPD fragments linked to *Dm* genes were cloned and the sequences at their termini determined. Longer primers (usually 24 bp) were designed and used with a high annealing temperature resulting in the reliable amplification of single loci. We designated these loci, SCARs (Sequence Characterized Amplified Regions; Paran and Michelmore, 1993). In some cases, SCAR primers amplify alleles from both genotypes under study, so that the marker may be scored as a codominant locus. Also, many of the RAPD fragments that we have cloned from the *Dm* regions, contain sequences that are repeated in the genome; therefore, the cloned sequences cannot be used directly as hybridization probes for single loci. However, it has proved possible in most cases to identify low-copy subfragments that could be used as hybridization probes (Table 1).

Mutagenesis of Resistance Genes

We have generated and characterized deletion mutations at *Dm* genes to increase the efficiency of marker identification and to help locate *Dm* genes on the physical map. Our ability to detect markers in a particular region using near-isogenic lines or bulked segregant analysis is limited by the level of polymorphism in the region between the two parents. In our most intensively studied mapping population, approximately 10% of all RFLP and RAPD loci are polymorphic. Identification of deletion mutants does not suffer this limitation. Comparison of a deletion mutant with its progenitor should detect all possible markers in the region.

We have generated *Dm* mutants following irradiation with fast neutrons or gamma rays. Seed of cv. Diana that carries four *Dm* genes was mutated and the M_1 plants selfed to generate several thousand M_2 families. These families were screened with a mixture of four isolates, each avirulent on a different *Dm* gene in cv. Diana. Sixteen mutant M_2 families were identified after screening 2,200 families. Most showed a loss of function of *Dm3*. As the closest markers are also at this locus, we are now focusing most of our efforts on cloning *Dm3*.

In order to screen for additional markers linked to *Dm* genes, we now utilize a standard panel of 12 DNA substrates: resistant and susceptible bulked DNA samples for each of the three major clusters of resistance genes plus six potential deletion mutants selected from each of the three clusters (Fig. 1). This has led to the identification of several RAPD markers that are missing in a deletion mutant but were not polymorphic between the bulked DNA samples.

Genetic versus Physical Distance

Prior to identifying genomic clones in the target region, it is necessary to determine the physical distance between the genetically tightly linked markers and the target gene. Due to variations in genetic versus physical distance, markers that are absolutely genetically linked might be within a few kilobases of the target or over a megabase away. Information on the physical distance indicates whether additional markers are required and what type of genomic library should be screened.

We have been developing a long range restriction map of the *Dm1 et al.* region. Four markers that spanned a 16 cM region, have been used as hybridization probes to blots of high molecular weight DNA restricted with rare-cutting endonucleases and separated by pulsed-field gel electrophoresis. The long-range restriction map detected with these probes did not coalesce; therefore, this region of the genome is not highly recombinagenic. Analysis of the potential deletion mutants for *Dm3* by pulsed-field gel electrophoresis confirmed that these were indeed deletion mutants lacking varying extents of DNA and localized *Dm3* to a 800 Kb interval.

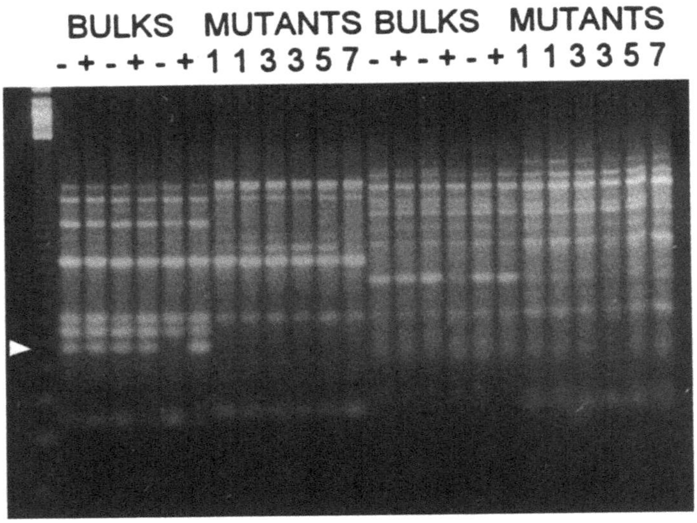

Figure 1. Standard panel of genotypes for screening for markers linked to *Dm* genes. The standard panel of 12 genotypes was used as substrate with two different RAPD primers. The first six lanes in each set show the results from pairs of bulked DNA samples for each of the three major clusters of *Dm* genes (- = pool from resistant individuals, + = pool from susceptible individuals; first pair for *Dm1* region, second pair for *Dm5/8* region, third pair for *Dm4* region) from a segregating population of cvs. Calmar x Kordaat. The next six lanes in each set show the reaction products from individual mutants of cv. Diana; the non-functional gene is shown above the lane. There is a clear polymorphism between the bulks for *Dm4* (arrowed).

Studies in Progress

We are currently identifying additional markers and recombinants flanking *Dm3* as well as EMS-induced and spontaneous mutants. These will be used to complete the physical map of the region and to confirm the relationship between physical and genetic distances.

We are utilizing two additional types of markers that have recently become available. A technique to identify anonymous cDNA fragments that differ between mRNA preparations has been developed by Liang and Pardee (1992). Essentially, this technique is similar to the RAPD protocol except that the substrate for the PCR reaction is first strand cDNA and an dT oligonucleotide with two discriminatory bases at the 3' end is used in addition to an arbitrary decamer as a primer. This reaction regularly yields at least 25 fragments per reaction. We have termed markers identified by this method, RATS (Random Amplified Transcribed Sequences). This method should identify transcribed sequences that are lacking in the deletion mutants. The second technique involves Amplified Fragment Length Polymorphisms (AFLPs) that have been recently developed by Keygene (Wageningen, Holland); these combine many of the advantages of RFLP analysis with the ease of PCR. A single PCR reaction can regularly assay over 50 to 100 loci. The details of this technique will be published shortly (M. Zabeau, pers. comm.). Both of these techniques provide access to over 100,000 loci using fast, PCR-based techniques. Therefore, we are no longer limited by the number of detectable loci and markers very close to any target gene should be readily obtainable (Table 2). We are presently screening for markers close to *Dm3*.

Table 2. Numbers of molecular markers detected by different marker technologies.

Marker		Loci detected per assay	Number of assays possible	Max. number of loci accessible
RFLP		1	130	150
RAPD	1 primer	8	1,500+	12,000
	2 primers	8+	1,000s	10,000s
AFLP		50+	1,000s	100,000s
RATS		20	1,000s	100,000s

Recombinants are being identified from a large F_2 population using SCAR markers that flank *Dm3*. PCR analysis with these markers allows the rapid identification of recombinant individuals as seedlings and allows all uninformative, non-recombinant individuals to be discarded. Numerous recombinants will be necessary to develop a fine resolution genetic map and locate *Dm3* relative to tightly linked markers.

We are developing several genomic libraries. The most labor intensive type is the YAC library; the production of a three genome equivalent library will require at least 20,000 clones with an average insert size of 300 Kb. The construction of this library is currently our rate limiting step.

Remaining Technical Challenges

There are several technical improvements that are necessary before map-based cloning is routine in plants. In particular, it would be useful to be able to clone specific large genomic fragments rather than having to make and screen a complete YAC library. Several technical advances may make this possible; RARE (*RecA*-Assisted Restriction Endonuclease) cleavage offers the potential to selectively release fragments from genomic DNA (Ferrin and Camrini-Otero, 1991; Koob *et al.*, 1992). For this to be routine however, the transformation rate for yeast will have to be improved. A second technical advancement would be the ability to conduct complementation experiments with large genomic fragments such as YAC clones. This would speed up the identification of YAC clones carrying the gene of interest and reduce the necessity for fine structure mapping of individual YAC clones. Also, more efficient methods for identification of genes contained on large fragments are required; this is particularly important in regions of high transcript density.

Summary

The major cluster of resistance genes in lettuce is comprised of at least nine members, eight conferring resistance to *B. lactucae* and one conferring resistance to root aphid. There is 12 cM between the most distant members. However, the region is not highly recombinagenic as it spans several megabases. The region is complex. Hybridization patterns of flanking markers indicate that there are several patterns of sequence duplication. Some sequences are duplicated locally, close to *Dm3*. Other sequences are duplicated throughout the region and in other clusters of resistance genes (Paran *et al.*, 1992). This supports the hypothesis that the clusters of resistance genes are related at the sequence level as well as functionally.

References

Crute IR and Johnson AG 1976 The genetic relationship between races of *Bremia lactucae* and cultivars of *Lactuca sativa*. Ann Appl Biol 83:125-137

Crute IR and Dunn JA 1981 An association between resistance to root aphid (*Pemphigus bursarius* L.) and downy mildew (*Bremia lactucae*) in lettuce. Euphytica 29:483-488

Farrara B, Ilott TW and Michelmore RW 1987 Genetic analysis of factors for resistance to downy mildew (*Bremia lactucae*) in species of lettuce (*Lactuca sativa* and *L. serriola*). Plant Pathol 36:499-514.

Ferrin LJ and Camrini-Otero RD 1991 Selective cleavage of human DNA: RecA-assisted restriction endonuclease (RARE) cleavage. Science 254:1494-1497

Ilott TW, Hulbert SH and Michelmore RW 1989 Genetic analysis of the gene-for-gene interaction between lettuce (*Lactuca sativa*) and *Bremia lactucae*. Phytopathology 79:888-897.

Johal G and Briggs SP 1992 Reductase activity encoded by the *Hm1* disease resistance gene in maize. Science 258:985-987

Judelson HS, Tyler B, and Michelmore RW 1991 Stable transformation of the potato late blight fungus, *Phytophthora infestans*. Molec Plant-Microbe Interact 4:602-607.

Kesseli RV, Paran I, Ochoa O, Wang W-C, and Michelmore RW 1990 Linkage map of lettuce (*Lactuca sativa*) 2N = 18. *In*: Genetic Maps, 6th Ed. S. J. O'Brien (ed.). Cold Spring Harbor Press. 6.229-233.

Koob M, Burkiewitcz A, Kur J, Szybalski W 1992 RecA-AC: Single site cleavage of plasmids and chromsomes at any predetermined restriction site. Nulc Acids Res 20:5831-5836

Legg EJ 1991 Genetic characterization of *Bremia lactucae* using molecular markers. M. Sc. thesis, University of California, Davis. pp101

Liang P and Pardee AB 1992 Differential display of eukaryotic messenger RNA by means of the polymerase chain reaction. Science 257:967-971

Martin G, Ganal M, and Tanksley SD 1993 Construction of a yeast artificial chromosome library of tomato and identification of clones linked to two disease resistance loci. Mol Gen Genet 233:25-32

Michelmore RW, Hulbert, Ilott TW and Farrara B 1988 Genetics of downy mildews. *In*: RH Williams and DS Ingram (eds.). Genetics of Pathogenic Fungi. Ed. GS Sidhu, Advances in Plant Pathology. Vol 6, pp. 53-79.

Michelmore RW, Marsh E, Seely S and Landry BS 1987 Transformation of lettuce mediated by *Agrobacterium tumefaciens*. Plant Cell Repts 6:439-442.

Michelmore RW, Paran I, and Kesseli RV 1991 Identification of markers linked to disease resistance genes by bulked segregant analysis: a rapid method to detect markers in specific genomic regions using segregating populations. Proc Natl Acad Sci 88:9828-9832.

Michelmore RW, Kesseli RV, Francis DM, Paran I, Fortin MG, and Yang C-H 1992 Strategies for cloning plant disease resistance genes. In: Molecular Plant Pathology - A Practical Approach. Ed. S Gurr. IRL Press. Vol 2, pp233-288.

Michelmore RW, Hulbert SH, Landry BS, and Leung H 1987 Towards a molecular understanding of lettuce downy mildew. In: Genetics and Plant Pathogenesis, Eds Day PR and Jellis GJ. Blackwell Scientific. p221-231.

Paran I., Kesseli RV, and Michelmore RW 1991 Identification of RFLP and RAPD markers linked to downy mildew resistance genes in lettuce using near-isogenic lines. Genome 34;1021-1027.

Paran I., Kesseli RV, and Michelmore RW 1992 Recent amplification of triose phosphate isomerase related sequences in lettuce. Genome 35:627-635.

Paran I. and Michelmore RW 1993 Development of reliable PCR-based markers linked to downy mildew resistance genes in lettuce. Theor Appl Genet 85:985-993.

Pryor AJ 1987 The origin and structure of fungal disease resistance genes in plants. Trend Genet 3:157-161

Zink FW and Duffus JE 1975 Inheritance and linkage of turnip mosaic virus susceptibility and downy mildew (*Bremia lactucae*) resistance in lettuce. J Amer Soc Hort Sci 98:49-51

Chromosome Microdissection and Megabase Technology in Plant Genome Analysis; Plant Chromosomes and Genes at High Resolution

Reinhold G. Herrmann, Regina Martin, Winfried Busch, Michael Kleine, Christian Eibl, Gerhard Wanner and Christian Jung
Botanisches Institut der Universität München
Menzinger Straße 67
D-8000 München 19

I. Introduction

The analysis of the complex eukaryotic genomes with their huge amounts of DNA and compartmentalized genomes has made substantial progress during the past two decades (e.g., Bernatzky and Tanskley, 1989; Tanskley et al., 1989; Jordan, 1988; Herrmann et al., 1992). The development began with the analysis of the subgenomes in the organelles, continued with expression cloning, which allowed the serological selection of cDNAs for nuclear genes from which products were known, and culminated in the eighthies with various approaches to manage and elucidate the complexity of the nucleo/cytosolic compartment with its estimated 50.000 to 100.000 genes. Milestones represent the development of a molecular marker system based on restriction fragment length polymorphisms (RFLPs), of procedures for tagging individual genes (e.g., Gierl and Saedler, 1992), and of the PCR (Saiki et al., 1985) as well as the megabase technologies (Jordan, 1988). The latter have filled a long lasting gap between genetic mapping procedures (linkage analysis) and conventional plasmid- or lambda-based cloning techniques. Basically three techniques, namely pulsed field gel electrophoresis, cloning into yeast artificial chromosomes (YACs), and chromosome jumping, allow to handle and/or physically map long coherent DNA segments.

RFLP markers possess a highly superior and significantly wider potential for both basic and applied research than customary phenotypic and isozyme markers used in the analysis of complex plant genomes, due to their codominant expression (environmental stability) and applicability in practically unlimited number per experiment (high resolution). The indispensible prerequisites for the application of this approach are the selection of recombinant DNA markers, which can be used to probe specific regions even in complex

NATO ASI Series, Vol. H 81
Plant Molecular Biology
Edited by G. Coruzzi and P. Puigdomènech

genomes for the presence of polymorphism at the DNA fragment level, and access to a genetic map of appropriate resolution (Tanksley et al., 1989). RFLPs provide the most efficient way for constructing detailed genetic linkage maps which can be established by conventional segregation analysis of F1 or BC1 populations. Until now maps have been constructed from various genomes, including *Arabidopsis* and, due to the enormous impact in plant breeding, predominantly from economically important crop species including maize (Coe et al., 1988), tomato (Tanksley, 1988), lettuce (Landry et al., 1987), rice (McCouch et al., 1988), potato (Gebhardt et al., 1989), rape seed (Slocum et al., 1990), in our laboratory for barley (Graner et al., 1991) and sugar beet (Pillen et al., 1992). Various strategies have been devised to maximize the benefits of RFLP-assisted selection in long-term breeding programs. RFLP markers represent not only useful tools for evolutionary and pedigree studies or for the characterization of germ plasm stocks. They can also serve for the isolation of developmentally crucial or economically important genes, for instance, genes conferring pathogen and drought resistance or quality characters which can usually be detected only by phenotype since their products are not known. The latter approach shares an access to *individual* genes with gene tagging. It differs in that it does not destroy the gene and is therefore independent of the viability of a mutant. RFLP markers offer in addition - and for the first time - the opportunity to study quantitative, *polygenic* traits (QTLs; e.g., Osborn et al., 1987; Paterson et al., 1988), that is, to dissect complex (developmental) processes into individual Mendelian loci and also determine the relative contribution of each locus even in dependence of environment.

In contrast to advances in our understanding of the "DNA thread", the dimensions between the macromolecule on the one hand and the cytological range on the other, and their consequences for processes of maintenance, evolution and expression of genetic information have largely escaped analysis. Their importance is illustrated by the enormous differences seen in the expression of a discrete transgene which are generally attributed to (ill-defined) "position effects". The recent development of the field emission scanning electron microscope (SEM) in combination with non-radioactive probe hybridization *in situ* (Lichter et al., 1990) offer unique perspectives and promises to bridge this (second) gap in gene and genome research (Wanner et al., 1990; 1991).

II. Chromosome-specific DNA libraries; high-resolution RFLP maps

Phenotypic and isozyme markers as well as the resolution of available linkage maps (in the order of 6 - 10 cM) are generally not sufficient to isolate genes for which products are not known with a reasonable input of labor. Apart from the fact that genetic distances expressed in recombination units and corresponding physical distances are not always consistent and may range from 0.5 - 5 Mb per cM, *chromosome walking* and *chromosome jumping* techniques present a critical hurdle in this DNA size range. There is therefore an obvious demand to select markers in the immediate proximity of a relevant gene in order to minimize the physical distance between marker loci and target gene, preferably to achieve *chromosome landing*, but usually to establish fine-scale physical maps of sets of overlapping cosmid or YAC clones.

A high marker saturation, i.e., 1 cM or less, is not readily attainable with customary techniques involving random selection of recombinant DNAs for RFLPs from shot-gun libraries. In principle, it can be substantially facilitated by directly cloning the DNA of individual chromosomes or chromosome segments, either by flow sorting (Conia et al., 1987) or microdissection of metaphase chromosomes via glass needles (Lüdecke et al., 1989) or laser beams (Hadano et al., 1991), but the cloning of subpicogram quantities of DNA has been generally hampered by the availability of sufficient chromosome numbers, the purity of (sorted) chromosomes, and the low cloning efficiency. Therefore, often more than 100 chromosomes or chromosome segments needed to be collected per cloning cycle (Edström et al., 1987). Furthermore, both approaches have almost exclusively been employed to human and animal chromosomes, probably due to technical problems caused by the rigid cell wall and the lack of sufficient synchronized material from plants. The first attempts for plant materials resulted in only a single or a few recombinant DNAs that could be investigated, or in cross-contamination (Sandery et al., 1991, Wang et al., 1992). The unequivocal identification of the desired chromosome or chromosome fragment is an indispensible prerequisite for the application of microdissection and sorting techniques. Unfortunately, the chromosome complements of many species lack pronounced morphological differences. The G-bandig technique, routine in human chromosome microdissection (Lüdecke et al., 1989), is of little use in plants. The HCl treatment, in turn, required for C-banding that is used in plant cytogenetics for the identification of individual chromosomes precludes the application of this technique for microdissection due to DNA damage. Aneuploid lines with well identifyable chromosome individuals in meiotic or mitotic metaphase plates can bypass these drawbacks, in principle (Jung et al., 1992b; Schondelmaier et al., 1993).

We have worked out a refined microdissection procedure and applied it to chromosomes of monosomic sugar beet material (Jung et al., 1992b; cf. also Section III) as well as to telosomic chromosomes in telotrisomic or ditelotetrasomic barley lines and telosomic wheat/barley addition lines (Schondelmaier et al., 1993) which are all well discernible by their morphology in mitotic metaphase spreads (Fig. 1). Microdissection and construction of chromosomal and subchromosomal libraries has been significantly improved by the use of (i) mitotic metaphase spreads which is time-saving compared to that of meiotic preparations since the plant material is available within one week, (ii) synchronized meristematic root tip tissue from which metaphase spreads of appropriate quality can be readily obtained, (iii) fixation times of less than three minutes that are possible with a novel drop-spread technique, (iv) a specific vector (Lüdecke et al., 1989) in combination with nl-scale PCR that allows to clone the DNA of 10 or less chromosomes or chromosome segments (Fig. 2), (v) mineral oil rather than a moist chamber to prevent evaporation from the collection drop, and (vi) an instrumental set-up which allows to select suitable metaphases, to store the coordinates of their positions, and to address sequentially individual metaphases or the collection drop automatically (Schondelmaier et al., 1993). The latter facility is particularly adventageous when dissection attempts the isolation of several fragments per chromosome arm (cf. Fig. 4), or chromosomes per cell.

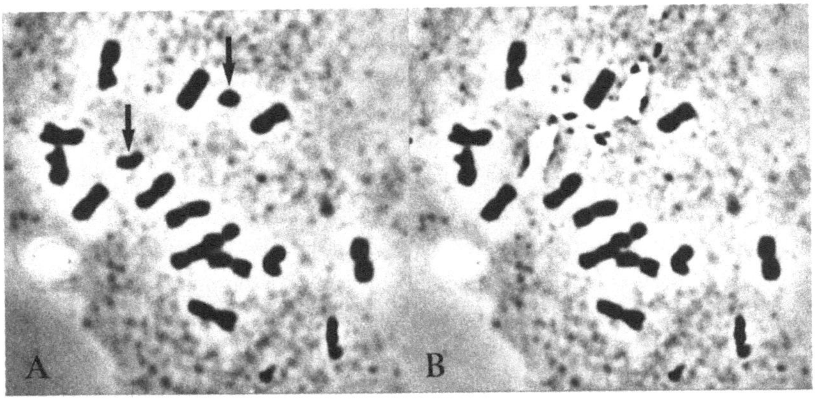

Figure 1: Mitotic metaphase plates of the barley ditelotetrasomic line 1HS (2n = 14 + 2t). A: The telocentric chromosomes are indicated by arrows. B: The telocentric chromosomes are removed by microsurgery from the same metaphase plate for microcloning.

Briefly, the DNA of the chromatin is collected in a microdroplet (1 - 2 nl), purified, restricted with *Rsa*I or *Pst*I, depending on the purpose, ligated into specifically designed vectors containing universal sequencing primers (Fig. 2), and amplified by PCR, all in nl-scale. *Pst*I digestion enriches for single- or low-copy sequences, that with *Rsa*I includes repetitive sequences as well. The amplified DNA is excised (*Eco*RI) in a way that eliminates largely linker sequences (Fig. 2) and is finally inserted into a standard plasmid vector for cloning. To date specific libraries are available for the *Beta patellaris* chromosome carrying the *bcn* gene for nematode resistance (Jung et al., 1992b; cf. Section III) and for the barley chromosome arms 1HS, 2HS, 2HL, 3HS, 3HL, 4HS and 4HL.

Chromosome 1HS bears the *hor1* and *hor2* (hordein) gene clusters as well as the economically important *mla* locus conferring resistance against powdery mildew. The size of the *Rsa*I-based library for this telocentric chromosome was estimated to be approximately 44.0000 recombinant plasmids, of which approximately 13% can be utilized for RFLP analysis. Monitoring approximately 600 randomly selected recombinant plasmids from this library established that (i) the insert sizes ranged between 70 and 1150 bp with a mean of 250 bp (somewhat larger average insert sizes are found in *Pst*I-based libraries), (ii) approximately 60% of the clones contained highly repetitive sequences, consistent with the fraction of repetitive DNA in the barley genome, and (iii) all single- or low-copy probes tested were chromosome-specific and originated from chromosome 1HS. Five probes were genetically mapped, using an interspecific *H. vulgare* x *H. spontaneum* F2 population, and shown to be well spread along the chromosome arm (Fig. 3). Similar results were obtained for chromosome 3HL (to be published). The outlined approach is therefore clearly superior to conventional microcloning techniques. It may serve to establish subchromosomal

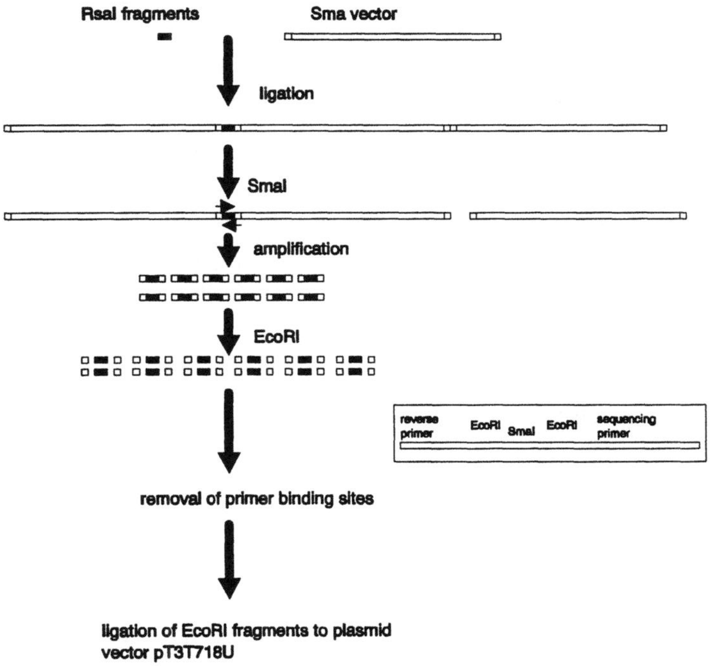

Figure 2: Vector, amplification and cloning strategy for the construction of chromosome-specific or subchromosomal DNA libraries.

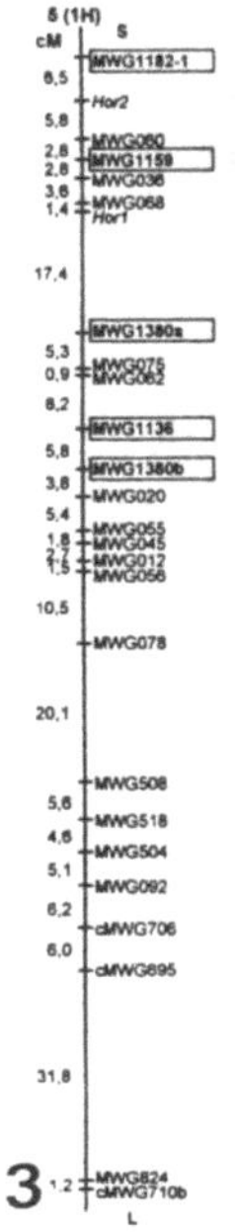

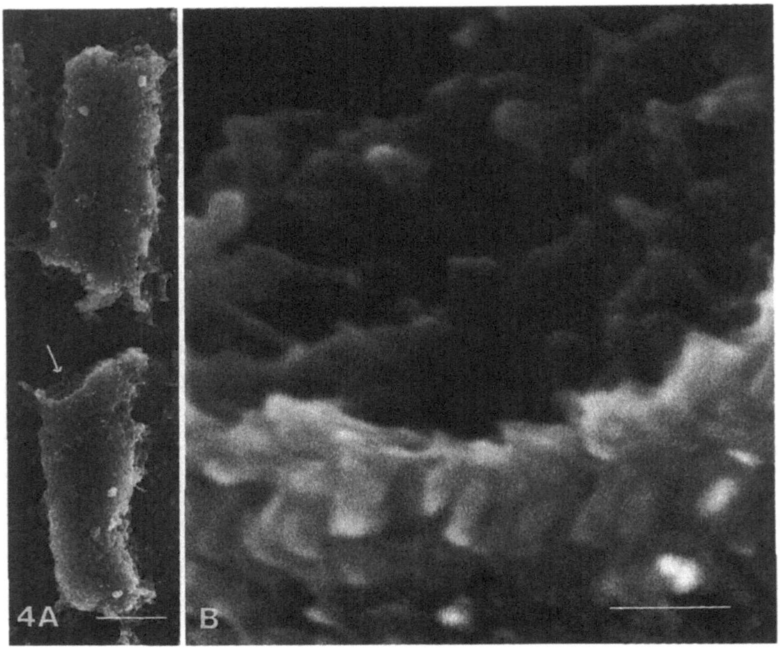

Figure 3: Chromosomal location of clones from the microdissection experiment on linkage group 5 of barley corresponding to chromsome 1H (boxed positions). The top part represents the short arm (S) of the chromosome (from Schondelmaier et al., 1993).

Figure 4: Scanning electron micrographs of a barley metaphase chromosome used in microsurgery. A: chromosome remnant after excision of a chromatin disk (bar length 1 μm). B: Detail of A (arrow) illustrating fine structural detail of the chromosome interior (bar length 100 nm).

libraries for the entire barley genome and to construct high-resolution maps from discrete genome regions (cf. also Fig. 4) which can greatly facilitate gene isolation and QTL studies.

III. Megabase technologies; isolation of genes

Four strategies are generally employed for tagging genes which can only be detected phenotypically, (i) the use of transposable elements for insertion mutagenesis (e.g., Gierl and Saedler, 1992), (ii) linkage analysis of populations cosegregating for RFLP markers and target gene (e.g., Yu et al., 1991), (iii) the use of RFLP markers on near-isogenic lines which are constructed by a backcross procedure with a recurrent parent and a donor parent homozygous for the gene in question (e.g., Sarfatti et al., 1989), and - relevant in this context - (iv) aneuploid stocks, translocation or deletion lines. Such material has proven useful to select markers for a monosomic wild beet chromosome in sugar beet that confers resistance to nematodes (Jung et al., 1990; 1992a).

The beet cyst nematode (*Heterodera schachtii*) causes severe damage in sugar beet (*Beta vulgaris*) and rape seed (*Brassica napus*). Due to the general absence of resistance in these genera, genes from resistant wild species, in *Beta* preferably from those of the section *Procumbentes*, have been introduced into sugar beet (summarized in Jung 1987), and the resulting species hybrids have been backcrossed with sugar beet to obtain monosomic, resistant addition lines and, subsequently, derived chromosome fragment addition lines as well as translocation lines carrying fragments or translocations of the wild beet chromosome (Fig. 5). In all these lines the gene(s) is (are) inherited in a dominant way since they display full resistance to the nematode. Since these resistant lines are economically poorly yielding and suffer from impaired recombination or transmission, it is highly desireable to isolate the corresponding gene.

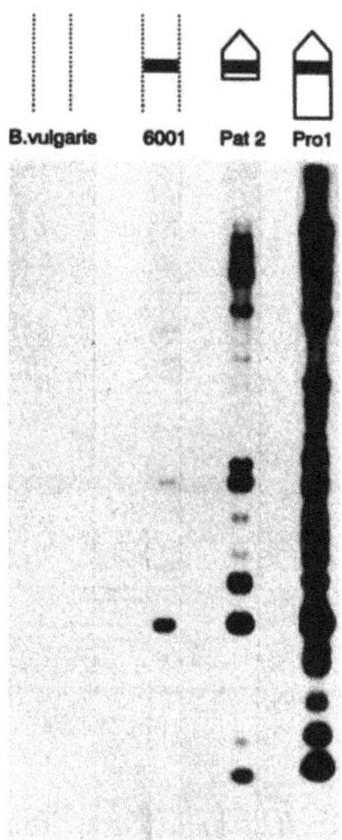

Figure 5: Characterization of the resistance-bearing region from cytogenetic mutants by hybridization with a genome-sepcific probe. Genomic DNA from resistant plants of chromosome fragment addition lines (Pro1, Pat2) and the translocation line 6001 were hybridized with a *B. procumbens*-specific dispersed repetitive probe. Susceptible sugar beet served as a control.

DNA markers, in particular dispersed repetitive elements, for which genomes of higher plants are highly enriched and which are often species-specific, offer unique advantages to select appropriate materials for breeding and for the physical isolation of the resistance gene. Using this strategy we have isolated dispersed repetitve DNA probes that are specific for wild beet genomes (Fig. 5) and applied them for the molecular characterization of fragment addition and translocation lines. Such lines offer two principal advantages: (i) wild beet-specific probes can be readily identified from shot-gun cloned libraries, and (ii) each of the probes isolated in this way must be physically linked to the resistance gene.

Using various wild beet-specific, repetitive DNA elements which are more or less evenly distributed throughout the *Procumbentes* genomes, we have not only found a strict cosegregation between the presence of these elements and the resistance gene, but also between the quantity of introduced wild beet chromatin and repeat copy numbers as judged from a decreasing number of bands with the decreasing fraction of wild beet chromatin in fragment addition vs. translocation lines (Fig. 5; Jung et al., 1990). Such probes are therefore not only superior as selectable markers for nematode resistance but also for the isolation of the gene. It is obvious that translocation lines which display a low signal number for dispersed repetitive elements represent the material of choice for that purpose. A translocation line which carries only five repeat copies (Fig. 5) was therefore chosen for further analysis.

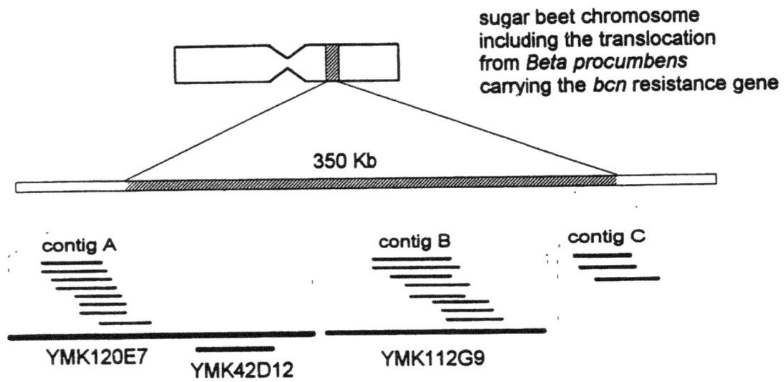

Figure 6: Contig cloning of the 350 kb *B. procumbens* translocation in sugar beet line 6001. YMK designate recombinant YACs, parallel lines lambda clones (Kleine et al., submitted)

The possibility of physically isolating plant genes from complex genomes based upon map position has been greatly enhanced by the development of yeast artificial chromosomes (Burke, 1990). Cloning large segments of genomic DNA into such vectors in combination with high-density RFLP maps and pulsed field gel electrophoresis provides access to such genes. We have constructed YAC libraries of approximately 15.000 clones from the described resistant sugar beet translocation line (and from barley; Kleine et al., 1993) representing approximately 2.2 haploid genome equivalents. Three translocation-specific recombinant YACs and various lambda clones have been selected from the corresponding libraries. They form a contig that covers approximately 70% of the translocation (Fig. 6). The translocated segment which bears the *bcn* gene has been estimated by pulse-field mapping to be in the order of 350 kb (Kleine et al., submitted). Work is in progress to identify this gene.

IV. Chromosomes at high resolution; gene mapping *in situ*

It has become obvious during the past years that genetic and physical maps are not directly comparable as crossing-overs are not evenly distributed along the chromosome axis. Genes which map genetically close to the centromere are often found in more distant regions (Curtis and Lakazewski, 1991). Furthermore the relationship between topographic, genetic and physical distances and its impact on genetic events (e.g., recombination frequency), gene expression, and perhaps evolution are largely unknown. It is therefore highly desireable to unravel details of chromosome structure and gene arrangement at the ultrastructural level.

Modern high resolution field emission scanning electron microscopy in combination with various detection systems (Peters, 1986) has now reached the "nanometer scale" of resolution. Image processing or stereo presentation is possible, and specimen can be continuously observed from the light microscopic to the SEM level. We have outlined relevant technical aspects of these approaches in three recent publications (Wanner et al., 1990; 1991; Lehfer et al., 1991). These approaches should provide access to yet largely unexplored topics such as (i) the arrangement of chromosomes or chromatin within the nuclear matrix, (ii) chromosomal fine structure at all stages of the cell cycle, (iii) the localization of individual genes including transgenes and, perhaps, (iv) to expression structures of genes. During the past few years satisfactory progress has been made in the preparation of plant specimen (Figs. 7 and 8; Wanner et al. 1990, 1991 and see below). The various structures seen or deduced include supercoiling of 12-nm fibres (nucleosomes as "beads on a string") into a 35-nm fibre (solenoid), other stages of packing in condensed (Fig. 7), condensing or decondensing chromatin or long fibrillar arrangement with little or no drilling. It is not yet possible to integrate all observations into a coherent picture nor to operate with *in situ* techniques for gene localization at the ultrastructural level. Although the detection of colloidal gold markers with diameters from 5 - 15 nm is now possible in SEM and the combination of backscattered electrons with the secondary electron image permits to localize the surface markers unequivocally, a critical step is the the denaturation/reassociation step required for the hybridization of the probe (Lehfer et al., 1991). However, satisfactory progress to visualize and map individual or even several plant genes was possible with an "intermediate" and complementing approach, the light microscopic hybridization *in situ* using fluorescent dyes.

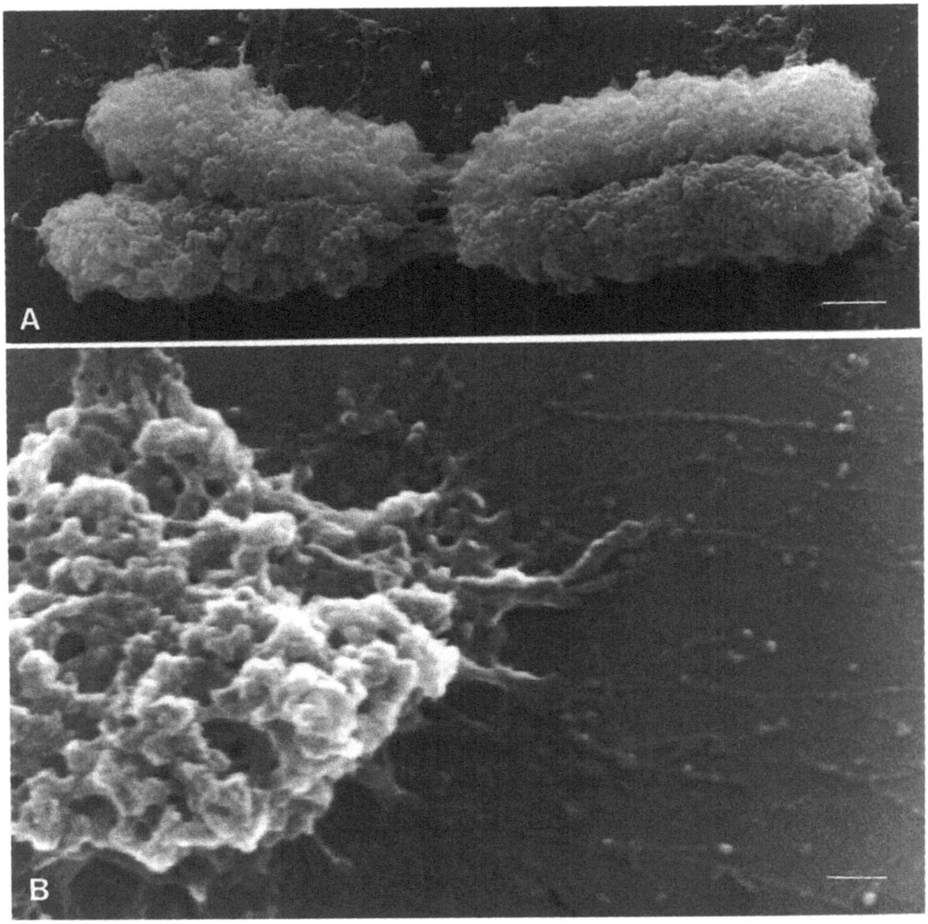

Figure 7: SE micrograph of (A) an untreated wheat chromosome exhibiting a knobby structure and chromatides. The centromeric region is characterized by parallel strands (bar length 400 nm). (B) Chromosomal detail after denaturation in 50% formamide at 70°C (bar length 100 nm). Note the release of DNA (10 nm fibre) from chromatin; the 35 nm fibre is apparently destroyed.

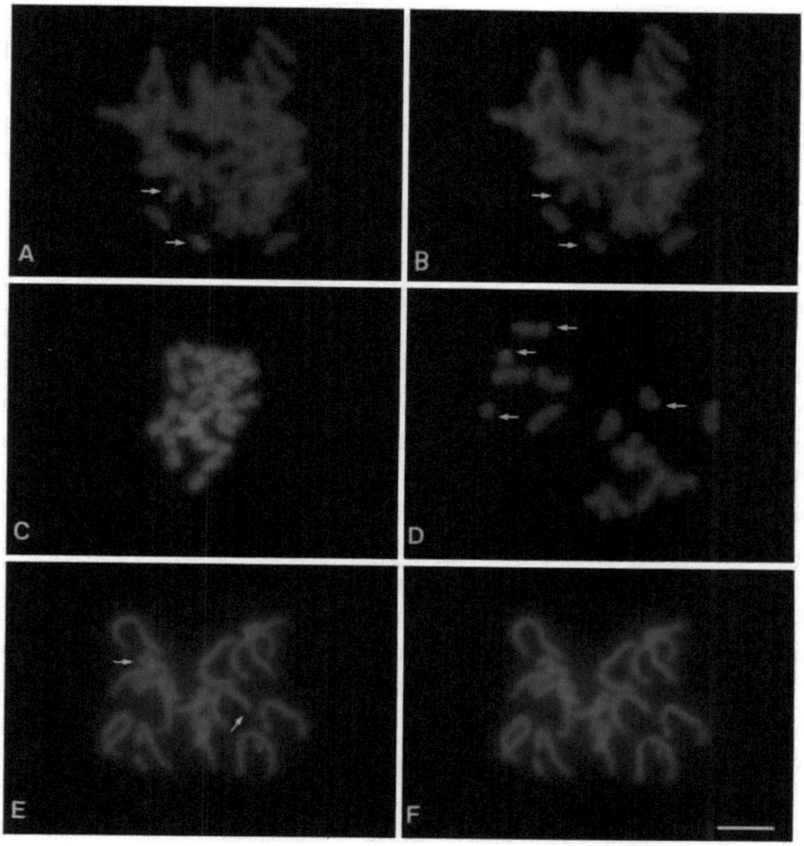

Figure 8: *In situ* hybridization with the wheat/barley addition line 3HS using probe MWG 1007. Visualization with Cy3. (A) Only barley chromosomes (arrows) display the signal. (B) Counterstain of the metaphase shown in (A). (C) A barley metaphase hybridized with a repetitive probe. Signal visualization with Cy3. All chromosomes show signals. (D) Hybridization of the barley ditelotetrasomic line 1HS with the B-hordein-specific probe pBSC21. The signal is seen on both telosomes and both chromosomes (arrow). (E) Hybridization to an early metaphase of barley with the same probe as in D (F counterstain). Bar length: 10 μm.

Since its first description by Gall and Pardue (1969), the development of improved techniques made *in situ* hybridization an increasingly powerful tool for detecting specific nucleic acid sequences directly on the chromosome (e.g., Lichter et al., 1990). This approach is well developed in the human field, even for single- and low-copy probes, and attempts are being made presently to map genes topographically. The use of different haptens for labeling DNA allows the detection of several probes on the same specimen by fluorescence techniques (Nederlof et al., 1990). The localization of specific sequences to plant chromosomes *in situ* has proven to be more difficult. While repetitive sequences could be detected quite readily (Gerlach and Peacock, 1980; Hutchinson and Lonsdale, 1982; Rayburn and Gill, 1985; Lapitan et al., 1989), only few attempts were successful in the demonstration of low- and single-copy sequences (e.g., Ambros et al., 1986; Mouras et al., 1987; Huang et al., 1988; Gustafson et al., 1990; Lehfer et al., 1991, 1993), but resolution was generally not satisfactorily.

Using a B-hordein-specific (i.e., low-copy) probe and FITC as fluorescence dye, we have recently mapped the *hor2* locus near the telomer of chromosome 1HS of barley (Fig. 8; Lehfer et al., 1991; 1993). An improved *in situ* hybridization procedure combining synchronization of root meristems, enzymatic removal of cell walls and the use of the advantageous fluorescent dye Cy3 has resulted in a significant increase of well spread metaphases and to an enhanced frequency and intensity of hybridization signals (Busch et al., 1993). When compared to FITC, Cy3 exhibits higher resistance to fading and a stronger signal intensity. Under appropriate conditions, practically all metaphases display signals, more than 60% of them on both chromatides of the gene-carrying chromosomes. This fulfils a crucial criterion for the construction of topographic maps (Lichter et al., 1990), preferably with mapped, chromosome-specific (RFLP) probes, and opens the perspective to tackle the relationship between the topographic position of a gene and its impact on genetic events. The technique is not restricted to metaphase chromosomes, but should also allow the detection of single-copy sequences, presumably at most stages of the cell cycle.

Acknowledgements: This work was supported by grants from the Bundesministerium für Forschung und Technologie, the Deutsche Forschungsgemeinschaft and the Fonds der Chemischen Industrie.

V. References

Ambros PF, Matzke MA, Matzke AJM (1986) Detection of a 17 kb unique sequence (T-DNA) in plant chromosomes by *in situ* hybridization. Chromosoma 94:11-18

Bernatzky R, Tanksley SD (1989) Restriction fragments as molecular markers for germplasm evaluation and utilization. Brown ADH, Frankel OH, Marshall DR, Williams JT (eds) The use of plant genetic resources. Cambridge University Press, Cambridge, pp 353-362

Burke DT (1990) Special section: Yeast cloning: Options and problems. GATA 7(5):94-99

Busch W, Martin R, Herrmann RG (1993) Sensitive enhancement of fluorescence *in situ* hybridization on plant chromosomes. Chromosome Res, submitted

Coe EH, Neuffer MG, Hoisington DA (1988) The genetics of corn. Sprague GF, Dudley JW (eds) Corn and corn improvement. American Society of Agronomy, Madison WI, pp 258

Conia J, Bergounioux C, Perennes C, Muller P, Brown S, Gadal P (1987) Flow cytometric analysis and sorting of plant chromosomes from *Petunia* hybrida protoplasts. Cytometry 8:500-508

Curtis CA, Lukazewski AJ (1992) Genetic linkage between C-bands and storage protein genes in chromosome 1B of tetraploid wheat. Theor Appl Genet 81:245-252

Edström JE, Kaiser R, Rohme D (1987) Microcloning of mammalian metaphase chromosomes. Methods Enzymol 151:503-516

Gall JG, Pardue ML (1969) Formation and detection of RNA-DNA hybrid molecules in cytological preparations. Proc Natl Acad Sci USA 63:378-383

Gebhardt C, Ritter E, Debener T, Schachtschnabel U, Walkemeier B, Uhrig H, Salamini F (1989) RFLP-analysis and linkage mapping in *Solanum tuberosum*. Theor Appl Genet 78:65-77

Gerlach WL, Peacock WJ (1980) Chromosomal localization of highly repeated DNA sequences in wheat. Heredity 44:269-276

Gierl A, Saedler H (1992) Plant transposable elements and gene tagging. Plant Mol Biol 19:39-49

Graner A, Jahoor A, Siedler H, Schondelmaier J, Pillen K, Fischbeck G, Wenzel G, Herrmann RG (1991) Construction of an RFLP map of barley. Theor Appl Genet 83:250-256

Gustafson JP, Butler E, McIntyre CL (1990) Physical mapping of a low copy DNA sequence in rye (*Secale cereale* L.). Proc Natl Acad Sci USA 87:1899-1902

Hadano S, Watanabe M, Yokoi H, Kogi M, Kondo I, Tsuchiya H, Kanazawa I, Wakasa K, Ikeda J-E (1991) Laser microdissection and single unique primer PCR allow generation of regional chromosome DNA clones from a single human chromosome. Genomics 11:364-373

Herrmann RG, Westhoff P, Link G (1992) Biogenesis of plastids in higher plants. Herrmann RG (ed) Plant Gene Research, vol.6. Springer, Wien, New York, pp 275-349

Huang P-L, Hahlbrock K, Somssich I (1988) Detection of a single-copy gene on plant chromosomes by *in situ* hybridization. Mol Gen Genet 211:143-147

Hutchinson J, Lonsdale DM (1982) The chromosomal distribution of cloned highly repetitive sequences from hexaploid wheat. Heredity 48:377-381

Jordan BR (1988) Megabase methods: a quantum jump in recombinant DNA techniques. Bio Essays 8:140-145

Jung C (1987) Breeding for nematode resistance in sugar beet. Ann Biol 3:15-25

Jung C, Kleine M, Fischer F, Herrmann RG (1990) Analysis of DNA from a Beta procumbens chromosome fragment in sugar beet carrying a gene for nematode resistance. Theor Appl Genet 79:663-672

Jung C, Koch R, Fischer F, Brandes A, Wricke G, Herrmann RG (1992a) DNA markers closely linked to nematode resistance genes in sugar beet (*Beta vulgaris* L.) using chromosome additions and translocations originating from wild beets of the Procumbentes species. Mol Gen Genet 232:271-278

Jung C, Claussen V, Horsthemke B, Fischer F, Herrmann RG (1992b) A DNA library from an individual *Beta patellaris* chromosome conferring nematode resistance obtained by microdissection of meiotic metaphase chromosomes. Plant Mol Biol 20:503-511

Kleine M, Michalek W, Graner A, Herrmann RG, Jung C (1993) Construction of a barley (*Hordeum vulgare* L.) YAC library and isolation of a *Hor 1*-specific clone. Theor Appl Genet, in press

Kleine M, Eibl C, Herrmann RG, Jung C (1993) YAC and lambda-contigs from a *Beta procumbens* translocation carrying the gene *bcn* for nematode resistance. Plant J, submitted

Landry BS, Kesseli RV, Farrara B, Michelmore RW (1987) A genetic map of lettuce (*Lactuca sativa* L.) with restriction fragment length polymorphism, isozyme, disease resistance and morphological markers. Genetics 116:331-337

Lapitan NVL, Ganal MW, Tanksley SD (1989) Somatic chromosome karyotype of tomato based on *in situ* hybridization of the TGRI satellite repeat. Genome 32:992-998

Lehfer H, Wanner G, Herrmann RG (1991) Physical mapping of DNA sequences on plant chromosomes by light microscopy and high resolution scanning electron microscopy. Herrmann RG, Larkins B (eds) Plant molecular biology 2. Plenum Publ.Corp., New York, pp 277-284

Lehfer H, Busch W, Martin R, Herrmann RG (1993) Localization of the B-hordein locus on barley chromosomes using fluorescence *in situ* hybridization. Chromosoma, in press

Lichter P, Chang Tang C, Call K (1990) High-resolution mapping of human chromosome 11 by *in situ* hybridization with cosmid clones. Science 24:64-69

Lüdecke HJ, Senger G, Claussen U, Horsthemke B (1989) Cloning defined regions of the human genome by microdissection of banded chromosomes and enzymatic amplification. Nature 338:348-350

McCouch SR, Kochert G, Yu ZH, Wang ZY, Khush GS, Coffman WR, Tanksley SD (1988) Molecular mapping of rice chromosomes. Theor Appl Genet 76:815-829

Mouras A, Saul MW, Essad S, Potrykus I (1987) Localization by *in situ* hybridization of a low copy chimaeric resistance gene introduced into plants by direct gene transfer. Mol Gen Genet 207:204-209

Nederlof PM, van der Flier S, Raap AK, Tanke HJ, Ploem JS, van der Ploeg M (1990) Multiple fluorescence *in situ* hybridization. Cytometry 11:126-131

Osborn TC, Alexander DC, Fobes JF (1987) Identification of restriction fragment length polymorphisms linked to genes controlling soluble solids content in tomato fruit. Theor Appl Genet 73:350-356

Paterson AH, Lander ES, Hewitt JD, Peterson S, Lincoln SE, Tanksley SD (1988) Resolution of quantitative traits into Mendelian factors by using a complete linkage map of restriction fragment polymorphisms. Nature 335:721-726

Peters KR (1986) Metal deposition by high-energy sputtering for high magnification electron microscopy. Koehler JK (ed) Advanced techniques in biological electron microscopy, vol. III. Springer, Berlin, pp 101-166

Pillen K, Steinrücken G, Wricke G, Herrmann RG, Jung C (1992) A linkage map of sugar beet (*Beta vulgaris* L.). Theor Appl Genet 84:129-135

Rayburn AL, Gill BS (1985) Use of biotin-labeled probes to map scientific DNA sequences on wheat chromosomes. J Hered 76:78-81

Saiki RK, Scharf S, Falcona F, Mullis KB, Horn GT, Erlich HA, Arnheim N (1985) Enzymatic amplification of ß-globulin genomic sequences and restriction site analysis for diagnosis of sickle cell anemia. Science 230:1350-1354

Sandery MJ, Forster JW, Macadam SR, Blunden R, Jones RN, Brown SDM (1991) Isolation of a sequence common to A- and B-type chromosomes of rye (*Secale cereale*) by microcloning. Plant Mol Biol Rep 9:21-30

Sarfatti M, Katan J, Fluhr R, Zamir D (1989) An RFLP marker in tomato linked to the *Fusarium oxysporum* rêsistance gene *I2*. Theor Appl Genet 78:755-759

Schondelmaier J, Martin R, Jahoor A, Houben A, Graner A, Koop HU, Herrmann RG, Jung C (1993) Microdisscetion and microcloning of the barley (*Hordeum vulgare* L.) chromosome 1HS. Theor Appl Genet, in press

Slocum MK, figdore SS, Kennard WS, Suzuki JY, Osborn TC (1990) Linkage arrangement of restriction fragment length polymorphism loci in *Brassica oleracea*. Theor Appl Genet 80:57-64

Tanksley SD (1988) Molecular mapping of plant chromosomes. Gustafson P, Appels R (eds) Chromosome structure and function. Plenum Publ.Corp., New York, pp 157-173

Tanksley SD, Young ND, Paterson AH, Bonierbale MW (1989) RFLP mapping in plant breeding: new tools for an old science. Biotechnology 7:257-264

Wang ML, Leitch AR, Schwarzacher T, Heslop-Harrison JS (1992) Construction of a chromosome-enriched HpaII library from flow-sorted wheat chromosomes. Nucleic Acids Res 20:1897-1901

Wanner G, Formanek H, Herrmann RG (1990) Ultrastructure of plant chromosomes by high-resolution scanning electron microscopy. Plant Mol Biol Rep 8:224-236

Wanner G, Formanek H, Martin R, Herrmann RG (1991) High-resolution scanning electron microscopy of plant chromosomes. Chromosoma 100:103-109

Yu ZH, Mackil DJ, Bonman JM, Tanksley SD (1991) Tagging genes for blast resistance in rice via linkage to RFLP markers. Theor Appl Genet 81:471-476

Development of an efficient transposon tagging system in *Arabidopsis thaliana*.

Caroline Dean and Ian Bancroft.
AFRC, IPSR Cambridge laboratory, John Innes Centre, Colney, Norwich NR4 7UH UK.

INTRODUCTION.

Insertional mutagenesis has a powerful advantage over other mutagens in that mutated genes are relatively easy to clone. In *Arabidopsis thaliana* (Arabidopsis), *Agrobacterium* T-DNA has been successfully used as an insertional mutagen and has provided many useful mutations (Feldmann et al 1989, Feldmann 1991). However, most laboratories have been unable to generate transformants using the seed transformation procedure used by Feldmann and there is a high ratio of non-tagged to tagged mutants in transformants generated from tissue culture. These factors have provided the incentive for the development of transposon tagging systems in Arabidopsis. The availability of such a system would bring a number of advantages. First, all interested laboratories could easily generate new transposition events. Second, modified transposons could be used to increase tagging efficiency and extend the range of genes that could be isolated. Third, the criteria for identifying a tagged mutation could be reversion to a wild-type phenotype rather than complementation experiments. Fourth, the various properties of the transposons could be taken advantage of, namely, transposition to linked sites could be used to target certain loci and, the presence of a "footprint" after the element has excised, could be used to generate an allelic series at the locus of interest.

The endogenous transposon systems in Arabidopsis have only recently been discovered and are relatively poorly characterized (Konieczny et al 1991, Peleman et al 1991, Tsay et al 1993). For this reason, a number of heterologous transposons from either maize or snapdragon (reviewed in Bhatt and Dean 1992) have been introduced into Arabidopsis using *Agrobacterium* transformation. Most effort has focused upon the autonomous maize transposable element *Ac* (*Activator*), and its non-autonomous derivative*Ds* (*Dissociation*) (van Sluys et al 1987, Schmidt and Willmitzer 1988, Dean et al 1992 Bancroft et al 1992, Swinburne et al 1992, Grevelding et al 1992, Keller et al 1992).

Here, the system in use in our laboratory will be described. The behaviour of the

NATO ASI Series, Vol. H 81
Plant Molecular Biology
Edited by G. Coruzzi and P. Puigdomènech

Ac/Ds elements in Arabidopsis will be summarized and the first transposon-induced mutants isolated using this system presented.

RESULTS.

Modified elements introduced into Arabidopsis.

Derivatives of *Ac* and *Ds* were constructed that would allow efficient monitoring of their excision, re-insertion and segregation. The essential features of the constructs are shown in Fig. 1

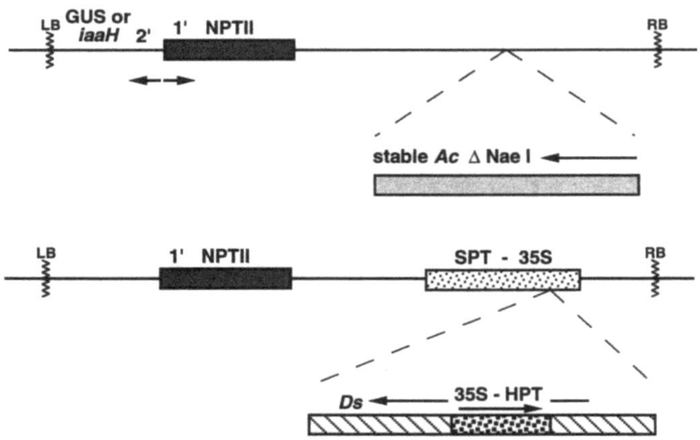

Fig.1. Schematic representation of the Hm^{R}*Ds* and *sAc* constructs introduced into Arabidopsis. LB and RB are the feft and right borders of the introduced T-DNAs. 1'NPTII is the fusion conferring kanamycin resistance. 35S-SPT confers streptomycin resistance and 35S-HPT confers hygromycin resistance. The *sAc* is an *Ac* element in which the 3' 180bp have been deleted.

All employ an NPTII fusion to provide resistance to kanamycin as the selection for primary transformants. The *Ds* element is cloned into the 5' untranslated leader region of a streptomycin phosphotransferase (SPT) fusion that confers resistance to streptomycin (Sm). Somatic excision of *Ds* is visualized as green sectors on bleached cotyledons after plating seeds on Sm-containing medium (Jones et al 1989, Dean et al 1992). Plants inheriting a germinal excision event are fully resistant to Sm and have a fully green (FG) phenotype on Sm-containing medium. The *Ds* contains a 35S-HPTII fusion conferring resistance to hygromycin (Hm), substituted for a portion of the transposase ORF. This allows the monitoring and selection for re-integration of the *Ds* element. The transposase source is provided by the *sAc* construct. This is an *Ac* element that has been stabilized by the deletion of the 3' terminal region (from bp 4385). It also carries a 537bp *NaeI* deletion in the 5' untranslated leader region of the transposase transcript. In Arabidopsis, this modification significantly increases the excision frequency of an autonomous *Ac* (E. Lawson, C. Dean unpublished) and *Ds* transactivation frequency (Bancroft et al 1992). The transposase source is linked to a β-glucuronidase (GUS) fusion so that progeny carrying the transposase source could be readily identified. The ability to easily identify individuals lacking a transposase source enables the isolation of mutations in genes which are not cell-autonomous, where somatic reversion would generate enough wild-type gene product to mask the mutant phenotype.

Somatic and germinal transactivation frequencies.

Four independent transformants carrying single copy insertions of the Hm^R*Ds* (B1 A3, E1, C12b) were crossed to a plant line carrying the ΔNaeI*sAc* construct linked to GUS. The results are summarized in Table I.

Table I. Somatic and germinal excision frequency of Hm^R *Ds* elements from different loci in the Arabidopsis genome.

Ds line	Av. no. of sectors per F1 seedling +/-SE	% fully Sm resistant seedlings
B1	3.9+/-0.75	0.58
A3	11.4+/-1.52	0.61
E1	6.8+/-1.35	0.74
C12b	32.2+/-2.12	3.6

Somatic excision, as assayed by the number of sectors on the cotyledons, varied approximately 10 fold and germinal excision, as assayed by the number of fully Sm resistant seedlings, varied approximately 6 fold (Bancroft and Dean 1993a). Thus, it appears that the efficiency of transactivation of *Ds* elements integrated into different regions of the genome can vary significantly, despite their immediate sequence context being identical (ie, the SPT fusion).

Effect of *Ds* and *sAc* copy number on levels of excision.

In maize, there is a negative dosage effect on transposition of *Ac* (McClintock 1948). In tobacco, there is a positive effect of increasing *Ac* dosage on excision (Jones et al 1989, Hehl and Baker 1990). Table II shows the analysis of varying *sAc* and *Ds* dosage on somatic activity as assayed by the average number of sectors on F1 seedlings. There is a low frequency of excision when both elements are heterozygous ie. present at a single copy. When a seedling is homozygous for the *Ds* element, ie contains 2 copies, but heterozygous for *sAc*, the number of sectors observed in the F1 seedlings is the same as that when both elements are heterozygous. When *sAc* is homozygous, with *Ds* either heterozygous or homozygous, the number of sectors per seedling increases at least 30 fold (Bancroft and Dean 1993a). There is, therefore, a strong positive dosage effect for the transposase source, but none for the excising element. This is a useful feature in a tagging experiment. High levels of *Ds* excision when the transposase source is homozygous facilitates obtaining germinal excision and reversion events, whilst lower levels when the transposase source is heterozygous reduce problems associated with high levels of somatic reversion.

Table II. Effect of copy number on excision frequency.

Genotype		Av. no. of sectors per F1 seedling +/-SE
sAc+	*+Ds*	1.4+/-0.31
sAc+	*DsDs*	1.3+/-0.44
sAcsAc	*+Ds*	>45+/-4.56
sAcsAc	*DsDs*	>69+/-2.74

Timing of germinal excision events.

An extensive analysis of the timing of excision events was undertaken by investigating the independence of transposition events in 45 F2 FG seedlings that had inherited a transposed *Ds*. These were progeny of 14 different F1 individuals, that had given rise to germinal excision frequencies in the F2 generation of between 0.4 and 5.9%. Five F2 seedlings, progeny of the same F1 plant that gave a high germinal excision frequency in the F2 (23%) were also included. Southern blot analysis was used to analyse the size of restriction fragment carrying the transposed *Ds* (*tDs*). The transposition events from the first 45 plants were independent, whilst the five plants from the population showing 23% germinal excision all carried the the same transposition event. The fact that the majority of individuals (isolated from populations showing a germinal excision frequency of less than 6%) carried independent events shows that the transposition must occur very late in the development of the plant, after the divergence of cell lines leading to pollen and ova. This characteristic of the system is very valuable as many FG individuals from the F2 progeny of one F1 plant can be used to generate families in which to screen for mutants. This is not the case when using a transposase source driven by the 35S promoter. Since the level of transposase expression is very high, the transposition events are extremely early resulting in many FG individuals all carrying the same transposition event (D. Long and G. Coupland pers. comm.). Thus only one FG seedling per F1 plant can be used to ensure independent transposition events are obtained.

Inheritance of re-integrated *Ds* elements.

The proportions of *tDs* elements inherited along with the excision marker was analysed by comparing the alleles present in the progeny as determined by the segregation of the antibiotic resistance markers. In progeny that were GUS negative (ie. lacked *sAc*), the segregation ratios of the markers, Sm resistance (to show whether active SPT is heterozygous or homozygous), Km resistance (the marker carried by both the *sAc* and *Ds* containing T-DNAs), and Hm resistance (the marker carried by the *Ds*) was determined. In the progeny that were GUS positive (ie. contained *sAc*), tests were performed for the segregation of Sm and Hm resistance and for the presence of variegation in their progeny (variegation would indicate the presence of a non-excised *Ds* element). Combining the data allowed the identification of the alleles present at the SPT locus, ie. whether there was a non-excised *Ds* present and hence whether there was a *tDs* present at another locus. Overall, of the 487 FG plants analysed, 24% carried *tDs* only, 30% carried a non-excised allele and 25% carried a transposed and a non-excised allele (Bancroft and Dean 1993b). There were 21% that showed anomalous segregation ratios and so could not be classified. There was little variation from locus to locus. Thus, the overall proportion of Hm-resistant (Hm^R), Sm-resistant (Sm^R) seedlings that contain a *tDs* is at least 49% (24% *tDs* only plus 25% *tDs* and non-excised *Ds*). As 68% of Sm^R seedlings were Hm^R (Bancroft and Dean 1993b), 33% (49% x 68%) of all *tDs* elements were inherited along with the excision marker at the donor locus. This compares with proportions of 42%, 58% and 53% for equivalent figures for *Ac* in maize (Dooner and Belachew 1989), tobacco (Dooner et al 1991) and Arabidopsis (Keller et al 1992), respectively.

Transposition pattern of the *Ds* elements.

In both maize and tobacco, *Ac* transposes preferentially to linked sites (61%, 59% in maize, Greenblatt 1984), Dooner and Belachew 1989, respectively; 72% in tobacco, Dooner et al 1991), but there are variable patterns of transposition from different loci. In order to define the patterns of *Ds* transposition in Arabidopsis, a detailed linkage analysis of 111 *tDs* elements to their original T-DNA donor loci in four independent transformants was undertaken. The proportions of *tDs* elements showing significant linkage to the donor locus is shown in Table III.

Table III. Proportions of transposed *Ds* elements showing significant linkage to the donor locus.

Locus	No. linked *tDs* elements	No. of unlinked *tDs* elements	% linked *tDs* elements
B1	30	14	68
A3	16	8	67
E1	16	10	62
C12b	14	3	82
Overall	76	35	68

The pattern of transposition from the four loci is shown in Fig.2. The patterns differ slightly. They form a range of distributions from that shown by C12b, where the transpositions are more tightly clustered to E1 where they are more dispersed.

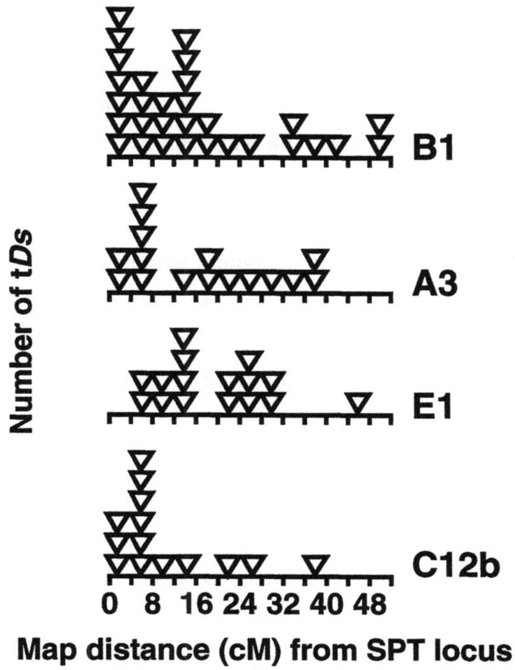

Fig. 2. Pattern of *Ds* transposition from four loci. Triangles represent independent *Ds* transposition events.

Identification of transposon-induced mutations.

487 F2 Sm^R, Hm^R plants were allowed to self-pollinate and their progeny were sown onto agar to test for the segregation of putatively tagged recessive mutations. Ten progeny from most of the 487 lines were subsequently transferred to soil and allowed to flower. From previous results, approximately 230 of the 487 plants would have contained a *tDs*, and would most likely represent independent transposition events. Five recessive mutations were identified and named as follows: *drl1* (*d*eformed *r*oots and *l*eaves, no inflorescence produced), *wlc1* (*w*avy *l*eaves, *c*otyledons furled back), *pcm1* (leaves of young plants wild-type, but turn *p*ale green after 3 to 4 weeks on *c*ulture *m*edium, wild-type phenotype maintained by transfer to fresh medium or soil), *dsr1* (leaves *d*arker green than wild-type, *s*tunted *r*oots) and *agamous* (Koornneef et al 1983, reiterating sepals and petals, no anthers or carpels). The mutant *wlc1* was derived from Hm^R Ds transformant C12b, the remainder from transformant B1 (Bancroft et al 1993).

If a gene has been transposon tagged, the mutated locus and transposon would be genetically inseparable by recombination. Preliminary investigations showed genetic linkage between the *tDs* and the mutations in *drl1 wlc1* and *pcm1*, but *dsr1* and *agamous* did not contain a *tDs*. The latter two mutations, therefore, are not tagged by *Ds*. Further investigations have so far been limited to the *drl1* mutation and these studies are summarized below.

The presence of *tDs* (marker by Hm resistance) was tested in *drl1* mutants and their wild-type siblings in populations segregating for both *drl1* and Hm resistance. After germination and scoring for *drl1* phenotype on agar, 179 *drl1* and 537 wild-type siblings were transferred to medium containing Hm. Seedlings were scored as resistant or sensitive after 14 days. All 179 *drl1* plants were resistant, while the wild-type plants segregated 358 resistamt to 179 sensitive. This perfect segregation of 1:0:2:1 (Hm^R mutants to Hm^S mutants to Hm^R wild-type to Hm^S wild-type) shows that *drl1* and the tDs insertion were genetically inseparable in the population used.

The linkage relationship between the *tDs* (and hence *drl1*) and the T-DNA was determined by analysing how many of the Sm^R seedlings were Hm^R and how many were Hm^S. The analysis indicated that the *tDs* was located approximately 4 centimorgans from the T-DNA. The *DRL1* locus therefore maps to chromosome 1, near T-DNA *Ds*-B1 (Bancroft and Dean 1993b).

The *drl1* mutation showed somatic and germinal reversion in the presence of a transposase source (Bancroft et al 1993). Southern analysis of plants carrying a germinal reversion event showed that they all contained a Ds excision event from the *DRL1* locus (Bancroft et al 1993). This correlation of reversion to wild-type morphology and excision of *Ds* demonstrated that the mutation had been caused by insertion of the *Ds* element.

Sequence analysis of the site of *Ds* excision from the *DRL1* locus showed that the footprints of all the wild-type revertants had restored a putative open reading frame. This suggests that the *Ds* had inserted into the translated region of the *DRL1* gene.

The *drl1* mutation severely disrupts the development of all major organs of Arabidopsis. We know relatively little about the expression or function of the DRL1 gene product. No hybridization was detected when a 155bp probe, that contained the sequence flanking the *Ds* element in the *drl1* mutant, was used to probe RNA gel blots. Either *DRL1* mRNA is present at a low levels in aerial tissues or the exon into which the *Ds* integrated is very small. When the nucleotide sequence of the 250bp HindIII fragment cloned from the *DRL1* locus was compared to entries in the EMBL database, no significant matches were found. Comparison of translation products (in all three reading frames and in both orientations), using BlastX software for analysis of the SwissProt protein library (version 24), revealed a 23 amino acid block with homology (10/23 identities, 17/23 positives, overall score 56, Poisson distribution score 0.15) to isopentenyl transferase. This enzyme catalyzes the initial step in cytokinin biosynthesis in Agrobacterium (Barry et al 1984). This is intriguing, given that there is a large amount of callus formation on the shoot apices of the stable mutants, but its relevance will only become clear once more extensive characterization of the gene product and its expression pattern have been conducted.

DISCUSSION

An efficient transposon tagging system is now available in Arabidopsis. An accurate assessment of the mutation rate is still not available as only the most obvious morphological or colour mutants were identified in the population screened. Three of the mutations identified in a population expected to contain approximately 230 transposition events were linked to the *tDs* and one has been shown to be definitely tagged. What caused the other two mutations has not been established. They were not the result of an insertion of the endogenous Arabidopsis transposon *Tag1* (Tsay et al 1993). They may have been caused by abortive transposition events that integrated little or none of the *Ds*, but resulted in a rearrangement at the receptor site. Further investigation of several such mutants will need to be conducted before a general cause can be identified.

The *drl1* mutation was linked to the T-DNA. The pattern of *Ds* transposition to linked sites in Arabidopsis will enable loci to be efficiently targeted. The limiting factor will be the number of mapped T-DNA integrations carrying *Ds* elements. Three of the T-DNA integrations described here have been mapped (Bancroft and Dean 1993b) and at least 20 more will be mapped very shortly in this laboratory (C. Lister, A. Wilson and C. Dean). Transactivation frequency between *Ds* elements integrated at different positions in the genome does vary and so all new integrations will need to be tested before initiating a targeted tagging experiment.

REFERENCES

Bancroft I, Bhatt AM, Sjodin C, Scofield S, Jones JDG and Dean C (1992) Development of an efficient two-element transposon tagging system in *Arabidopsis thaliana*. Molec. Gen. Genet. 233: 449-461.

Bancroft I. and Dean C. (1993a) Factors affecting the excision frequency of the maize transposable element *Ds* in *Arabidopsis thaliana*. Molec. Gen. Genet. in press.

Bancroft I. and Dean C. (1993b) Transposition pattern of the maize element *Ds* in *Arabidopsis thaliana*. Genetics in press.

Bancroft I, Jones J and Dean C (1993) Heterologous transposon tagging of the *DRL1* locus in Arabidopsis. Plant Cell in press.

Barry GF, Rogers SG, Fraley RT and Brand L (1984) Identification of a cloned cytokinin biosynthetic gene. Proc. Natl. Acad. Sci USA 81: 69-81.

Bhatt AM and Dean C 1992 Development of tagging systems in plants using heterologous transposons. Current Opinion in Biotechnology 3: 152-158.

Dean C, Sjodin C, Page T, Jones J and Lister C (1992) Behaviour of the maize transposable element *Ac* in *Arabidopsis thaliana*. Plant J. 2: 69-81.

Dooner HK and Belachew A (1989) Transposition pattern of the maize element *Ac* from the *bz-m2* (*Ac*) allele. Genetics 122: 447-457.

Dooner HK, Keller J, Harper E and Ralston E (1991) Variable patterns of transposition of the maize element *Activator* in tobacco. Plant Cell 3: 473-482.

Feldmann KA, Marks MD, Christianson ML and Quatrano RS (1989) A dwarf mutant of Arabidopsis generated by T-DNA insertional mutagenesis. Science 243: 1351-1354.

Feldmann, K.A. 1991. T-DNA insertion mutagenesis in Arabidopsis: Mutational spectrum. Plant J. 1: 71-82.

Greenblatt I (1984) A chromosome replication pattern deduced from pericarp phenotypes resulting from movements of the transposable element *Modulator* in maize. Genetics 108: 471-485.

Grevelding C, Becker D, Kunze R, von Menges A, Fantes V, Schell J and Masterson R (1992) High rates of *Ac*/*Ds* germinal transposition in Arabidopsis suitable for gene isolation by insertional mutagenesis. Proc. Natl. Acad. Sci. USA 89: 6085-6089.

Hehl R and Baker B (1990) properties of the maize transposable elemnet *Activator* in transgenic tobacco plants: a versatile inter-species genetic tool. Plant Cell 2: 709-721.

Jones JDG, Carland F, Maliga P and Dooner HK (1989) Visual detection of transposition of maize element *Activator* (*Ac*) in tobacco seedlings. Science 244: 204-207.

Keller J, Lim E, James DWJr and Dooner H (1992) Germinal and somatic activity of the maize element *Activator* (*Ac*) in Arabidopsis. Genetics 131: 449-459.

Koornneef M, van Eden J, Hanhart CJ, Stam P, Braaksma FJ and Feenstra FJ (1983) Linkage map of *Arabidopsis thaliana*. J. Hered.74: 265-272.

Konieczny A, Voytas DF, Cummings MP and Ausubel FM (1991) A superfamily of *Arabidopsis thaliana* retrotransposons. Genetics 127: 801-809.

McClintock B (1948) Mutable loci in maize Carnegie Inst. Washington Yearbook 47: 155-159.

Peleman J, Cottyn B, Van Camp W, van Montagu M and Inze D (1991) Transient occurrence of extrachromosomal DNA of an *Arabidopsis thaliana* transposon-like element, *Tat1*. Proc. Natl. Acad. Sci. USA 88: 3618-3622.

Schmidt R and Willmitzer L (1989) The maize autonomous element *Activator* (*Ac*) shows a minimal germinal excision frequency of 0.2%-0.5% in transgenic *Arabidopsis* plants. Mol. Gen. Genet. 220: 11-24.

Swinburne J, Balcells L, Scofield SR, Jones JDG and Coupland G (1992) Elevated levels of *Ac* transposase mRNA are associated with high frequencies of *Ds* excision in Arabidopsis. Plant Cell 4: 583-595.

Tsay Y-F, Frank MJ, Page T, Dean C and Crawford NM (1993) Identification of a mobile endogenous transposon in *Arabidopsis thaliana*. Science 260: 342-344.

van Sluys MA, Tempe J and Fedoroff N (1987) Studies on the introduction and mobility of the maize *Activator* element in *Arabidopsis thaliana* and *Daucus carota*. EMBO J. 6: 3881-3889.

Molecular Genetic Improvement of Wheat

Indra K. Vasil and Vimla Vasil
Laboratory of Plant Cell and Molecular Biology
1143 Fifield Hall
University of Florida
Gainesville, FL 32611-0692
U.S.A.

INTRODUCTION

Cereal grains like wheat, rice and maize were most likely amongst the first plants domesticated by prehistoric man, and ever since have exerted a dominant influence on the development of the human civilization. Even today they constitute the most important group of food plants by virtue of providing more than 50% of all food consumed by man in addition to being the principal source of calories and protein in the human diet. The yield as well as the quality of cereal crops have been greatly improved during the past several decades by the traditional methods of breeding and selection. Significant further improvement of cereal crops is, however, limited by the time consuming and labor intensive nature of plant breeding methods, the restricted gene pool, and most importantly by the severe natural barriers to hybridization with even closely related species. It is in this context that the novel methods of plant cell and molecular biology, which permit the introduction of selected and defined plant as well as bacterial, viral and other genes of interest into plants, offer new opportunities for supplementing as well as complementing plant breeding and selection for the improvement of crop plants.

Since the production of the first transgenic tobacco plants in 1983, a very productive partnership has evolved between plant cell and molecular biology. This has led to an impressive array of successes, including the production of improved transgenic crops such as wheat, rice, maize, sugarcane, etc., containing agronomically useful genes for resistance to insects, viruses and herbicides (Vasil 1991,

NATO ASI Series, Vol. H 81
Plant Molecular Biology
Edited by G. Coruzzi and P. Puigdomènech

Leemans 1993). This is particularly significant in light of the fact that world-wide nearly one-third of all crop production is lost to pests, pathogens and weeds. Wheat, the most important food crop in the world based on annual production (550 million metric tons) and cultivated area (226 million hectares), was the last of the major cereal grains to be genetically transformed (Bialy 1992).

Molecular genetic manipulation of plants requires the regeneration of plants from single cells in which foreign DNA has been delivered and stably integrated. In this paper we describe the experimental strategies that were successfully developed for the establishment of regenerable in vitro culture systems, the development of efficient methods for the delivery and integration of DNA, the selection of the rare transformation events, and the production of fertile transgenic wheat plants.

PLANT REGENERATION IN VITRO

Gramineous species in general, but wheat in particular, have proven to be very recalcitrant to in vitro manipulations. All attempts to establish actively dividing and regenerable tissue cultures were unsuccessful until about 1975, largely because in most instances tissue and organ explants comprised of differentiated cells were cultured on inadequate media (Yamada 1977). This resulted in the production of callus tissues which in many instances were made up almost entirely of proliferating root meristems. Such cultures could neither be maintained for long in vitro, nor did they form any plants. The difficulties faced in the establishment of regenerable cultures from differentiated tissues were attributed to the fact that in wheat, just as in other species of the Gramineae, the capacity for cell proliferation and totipotency is lost rather rapidly during cellular differentiation (Vasil 1988). Accordingly, explants containing undifferentiated and meristematic cells, still with a reversible developmental potential, were later used to initiate regenerable tissue cultures.

Regeneration from Callus Cultures. The first regenerable wheat cultures were thus established from immature embryos and young inflorescences on nutrient media containing 2,4-dichlorophenoxyacetic acid, a strong auxin with herbicide activity (Dudits et al. 1975, Chin and Scott 1977, Shimada 1978, Gosch-Wackerle et al. 1979). In these early reports which involved only a limited number of genotypes, the frequency of plant regeneration was often poor and sporadic, and the process of regeneration was either described as shoot morphogenesis or was not clearly documented. Further refinement of the methods and careful microscopic examination of the cultures showed them to be embryogenic in nature in which regeneration of plants took place via the formation of somatic embryos (Ozias-Akins and Vasil 1982, 1983a,b, Maddock et al. 1983). The best response was obtained on a nutrient medium containing double the concentration of Murashige and Skoog's (1962) salts (Ozias-Akins and Vasil 1983a). The embryogenic callus originated from divisions of the epithelial and sub-epithelial cells of the embryo scutellum, or from tissues of the rachis and glumes in young inflorescences (Ozias-Akins and Vasil 1982). The callus was comprised largely of small, isodiametric, richly cytoplasmic, non-vacuolated and starch-containing embryogenic cells. The identification and selective transfer of the embryogenic callus during subculture were shown to be important for the maintenance of regenerative ability, although long term maintenance was still rather difficult and unpredictable. This problem was later resolved by the identification of a special form of callus - Type C embryogenic callus - which proved to be useful not only for long term regeneration (Redway et al. 1990a) but also for the establishment of embryogenic suspension cultures (Redway et al. 1990b).

Regeneration from Cell Suspension Cultures. There are several early reports of the establishment of non-regenerable suspension cultures of wheat (Gamborg and Eveleigh 1968, Trione et al. 1968, Maddock 1987). Other reports have described suspension cultures consisting almost entirely of proliferating root meristems which continuously slough off senescing cells which are incapable of sustained growth. Finely dispersed and fast growing cell suspension cultures containing characteristic unorganized groups of small, highly cytoplasmic, non-vacuolated and starch-containing embryogenic cells, were established only recently (Redway et al. 1990b). The immature embryo-derived Type C embryogenic callus was found to be suitable for this purpose. It was necessary to plate the cells on agar nutrient media to induce the formation of somatic embryos which germinated upon maturity to give rise to rooted green plants.

Regeneration from Protoplast Cultures. Like other species of the Gramineae (Vasil 1988, Vasil and Vasil 1992), embryogenic cell suspension cultures of wheat have proven to be the only reliable source of totipotent protoplasts from which green plants have been regenerated (Vasil et al. 1990, Chang et al. 1991, He et al. 1992, Li et al. 1992, Qiao et al. 1992, Ahmed and Sagi 1993). In some of these studies, relatively high protoplast plating efficiencies (25% or higher) were obtained, and protoplast-derived plants were successfully grown in soil.

GENETIC TRANSFORMATION

The development of reliable protocols for the regeneration of plants from callus, cell suspension and protoplast cultures of wheat paved the way for the genetic transformation of this important food crop.

The Ti plasmid of *Agrobacterium tumefaciens* has been used as a very efficient vector for the transfer of genes to a wide variety of dicotyledonous plants. Unfortunately, except in a few rare instances, this method has not proven to be very suitable for the genetic transformation of most monocotyledonous plants - including cereal and grass species. This problem has, however, been successfully overcome by the development of methods for the direct delivery of DNA into regenerable cells.

Two methods of direct DNA delivery have proven useful in

wheat: (i) Osmotic (polyethylene glycol) or electric (electroporation) shock (Lorz et al. 1985, Hauptmann et al. 1987) which permeate protoplast membranes allowing entry of DNA, and (ii) high velocity bombardment (biolistics) of gold or tungsten microprojectiles coated with DNA into intact cells or tissues (Wang et al. 1988, Oard et al. 1989, Lonsdale et al. 1990, Chibbar et al. 1991, Vasil et al. 1991, Perl et al. 1992). In most of these reports only transient expression of the introduced gene/s was demonstrated. Only in a few instances, where selectable genes were introduced, stably transformed - but non-regenerable - cell lines were recovered (Lorz et al. 1985, Hauptmann et al. 1988, Vasil et al. 1991).

The CaMV 35S promoter is not very effective in inducing the expression of the introduced genes in cereal cells (Fromm et al. 1985, Hauptmann et al. 1987). Substantially increased levels of gene expression can, however, be attained by the use of the *Ubi1* promoter and/or *Adh1, Sh1* and *Ubi1* introns (Callis et al. 1987, Vasil et al. 1989, Christensen et al. 1992, Taylor et al. 1993). This strategy has proven particularly useful in the selection of stable transformants.

The plasmid pBARGUS was used to deliver the selectable *bar* gene coding for phosphinothricin acetyltransferase (PAT) into cells of Type C long-term regenerable callus cultures of wheat by microprojectile bombardment (Vasil et al. 1992). Putative transformed cell lines were selected on various concentrations of the broad spectrum herbicide basta, which contains the active ingredient phosphinothricin (PPT). PAT activity was demonstrated in the selected cell lines, which were then placed on regeneration media to obtain somatic embryos and rooted green plants that could be grown to maturity. Southern analyses confirmed the presence of the *bar* gene in the selected callus lines as well as in the regenerated plants. Transformed R0, R1 and R2 plants were resistant to topical applications of basta. There was a tight correlation between PAT activity, the physical presence of the gene as revealed by Southern analyses, and resistance to basta. The stable integration of the *bar* gene in germ line cells was demonstrated by its segregation as a dominant Mendelian trait in R1 and R2 plants. Homozygous R3 transgenic seed were obtained.

The production of fertile transgenic wheat plants as described above requires the identification, establishment and bulking of Type C callus cultures prior to bombardment. These are all difficult, labor intensive and time consuming processes. It takes anywhere from 15-18 months from the culture of immature embryos to the production of transgenic plants. We have, therefore, recently examined the possibility of reducing the time interval between the initial culture of the immature embryos and the recovery of transgenic plants. For this purpose we have introduced plasmid DNA directly either into immature embryos 4, 5, 6 or 9 days after culture or into 1-2 month old embryogenic callus. Several independent, putatively transformed lines were successfully selected and regenerated into fertile plants. PAT activity was demonstrated in the cell lines as well as in the regenerated plants. Mendelian segregation of basta resistance and/or PAT activity was shown in the sexual progenies of R0 plants. These results were obtained in three wheat varieties (RH770019, Bob White and Pavon), as compared to the earlier work in which transgenic progenies were obtained only from the variety Pavon (Vasil et al. 1992). In addition to pBARGUS, plasmids pAHC25 (containing the BAR and GUS coding regions driven by the maize ubiquitin promoter, first exon and first intron; Christensen et al. 1992), and pMON19526 (showing enhanced expression of the *bar* and GUS genes; M. Fromm, unpublished), were used to obtain transgenic plants.

ACKNOWLEDGMENTS

The work on wheat tranformation was supported by Monsanto Company (St. Louis, MO).

REFERENCES

Ahmed KZ, Sagi F (1993) Culture and fertile plant regeneration from regenerable embryogenic suspension cell-derived protoplasts of wheat *Triticum aestivum* L.). Plant Cell Rep 12:175-179

Bialy H (1992) Recalcitrant cereal: Transgenic wheat finally produced. Bio/Technology 10:630

Callis J, Fromm M, Walbot V (1987) Introns increase gene expression in cultured cells. Gens Develop 1:1183-1200

Chang Y, Wang W, Warfield CY, Nguyen HT, Wong JR (1991) Plant regeneration from protoplasts isolated from long-term cell cultures of wheat (*Triticum aestivum* L.). Plant Cell Rep 9:611-614

Chibbar RN, Kartha KK, Leung N, Qureshi J, Caswell K (1991) Transient expression of marker genes in immature zygotic embryos of spring wheat (*Triticum aestivum*) through microprojectile bombardment. Genome 34:453-460

Chin JC, Scott KJ (1977) Studies on the formation of roots and shoots in wheat callus cultures. Ann Bot 41:473-481

Christensen AH, Sharrock RA, Quail PH (1992) Maize polyubiquitin genes: structure, thermal perturbation of expression and transcript splicing, and promoter activity following transfer to protoplasts by electroporation. Plant Mol Biol 18:675-689

Dudits D, Nemet G, Haydu Z (1975) Study of callus growth and organ formation in wheat (*Triticum aestivum*) tissue cultures. Canad J Bot 53:957-963

Fromm M, Taylor LP, Walbot V (1985) Expression of genes transferred into monocot and dicot plant cells by electroporation. Proc Nat Acad Sci USA 82:5824-5828

Gamborg OL, Eveleigh DE (1968) Culture methods and detection of glucanases in suspension cultures of wheat and barley. Canad J Biochem 46:417-421

Gosch-Wackerle G, Avivi L, Galun E (1979) Induction, culture and differentiation of callus from immature rachises, seeds and embryos of *Triticum*. Z Pflanzenphysiol 91:267-278

Hauptmann RM, Ozias-Akins P, Vasil V, Tabaeizadeh Z, Rogers SG, Horsch RB, Vasil IK, Fraley RT (1987) Transient expression of electroporated DNA in monocotyledonous and dicotyledonous species. Plant Cell Rep 6:265-270

Hauptmann RM, Vasil V, Ozias-Akins P, Tabaeizadeh Z, Rogers SG, Fraley RT, Horsch RB, Vasil IK (1988) Evaluation of selectable markers for obtaining stable transformants in the Gramineae. Plant Physiol 86:602-606

He DG, Yang YM, Scott KJ (1992) Plant regeneration from protoplasts of wheat (*Triticum aestivum* cv. Hartog). Plant Cell Rep 11:16-19

Leemans J (1993) Ti to tomato, tomato to market: a decade of plant biotechnology. Bio/Technology 11:S22-S26

Li Z, Xia G, Chen H (1992) Somatic embryogenesis and plant regeneration from protoplasts isolated from embryogenic cell suspensions of wheat (*Triticum aestivum* L.). Plant

Cell Tiss Org Cult 28:79-85
Lonsdale D, Onde S, Cuming A (1990) Transient expression of exogenous DNA in intact, viable wheat embryos following particle bombardment. J Exp Bot 41:1161-1165
Lorz H, Baker B, Schell J (1985) Gene transfer to cereal cells mediated by protoplast transformation. Mol Gen Genet 199:178-182
Maddock SE (1987) Suspension and protoplast culture of hexaploid wheat (*Triticum aestivum* L.). Plant Cell Rep 6:23-26
Maddock SE, Lancaster VA, Risiott R, Franklin J (1983) Plant regeneration from cultured immature embryos and inflorescences of 25 cultivars of wheat (*Triticum aestivum*). J Exp Bot 34:915-926
Murashige T, Skoog F (1962) A revised medium for rapid growth and bioassays with tobacco tissue cultures. Physiol Plant 15:473-497
Oard JH, Paige D, Dvorak J (1989) Chimeric gene expression using maize intron in cultured cells of bread wheat. Plant Cell Rep 8:156-160
Ozias-Akins P, Vasil IK (1982) Plant regeneration from cultured immature embryos and inflorescences of *Triticum aestivum* L. (wheat): evidence for somatic embryogenesis. Protoplasma 110:95-105
Ozias-Akins P, Vasil IK (1983a) Improved efficiency and normalization of somatic embryogenesis in *Triticum aestivum* (wheat). Protoplasma 117:40-44
Ozias-Akins P, Vasil IK (1983b) Proliferation of and plant regeneration from the epiblast of *Triticum aestivum* (wheat; Gramineae) embryos. Amer J Bot 70:1092-1097
Perl A, Kless H, Blumenthal A, Galili G, Galun E (1992) Improvement of plant regeneration and GUS expression in cellular wheat calli by optimization of culture conditions and DNA-microprojectile delivery procedures. Mol Gen Genet 235:279-284
Qiao YM, Cattaneo M, Locatelli F, Luppoto E (1992) Plant regeneration from long term suspension culture-derived protoplasts of hexaploid wheat (*Triticum aestivum* L.). Plant Cell Rep 11:262-265
Redway FA, Vasil V, Lu D, Vasil IK (1990a) Identification of callus types for long-term maintenance and regeneration from commercial cultivars of wheat (*Triticum aestivum* L.). Theor Appl Genet 79:609-617
Redway FA, Vasil V, Vasil IK (1990b) Characterization and regeneration of wheat (*Triticum aestivum* L.) from embryogenic cell suspension cultures. Plant Cell Rep 8:714-717
Shimada T (1978) Plant regeneration from the callus induced from wheat embryo. Jap J Genet 53:371-374
Taylor MG, Vasil V, Vasil IK (1993) Enhanced GUS gene expression in cereal/grass cell suspensions and immature embryos using the maize ubiquitin-based plasmid pAHC25. Submitted
Trione EJ, Johnes LE, Metzger RJ (1968) In vitro culture of somatic wheat callus tissue. Amer J Bot 55:529-531

Vasil IK (1988) Progress in the regeneration and genetic manipulation of cereal crops. Bio/Technology 6:397-402
Vasil IK (1991) Plant tissue culture and molecular biology as tools in understanding plant development and in plant improvement. Curr Opin Biotech 2:158-163
Vasil IK, Vasil V (1992) Advances in cereal protoplast research. Physiol Plant 85:279-283
Vasil V, Brown SM, Re D, Fromm ME, Vasil IK (1991) Stably transformed callus lines from microprojectile bombardment of cell suspension cultures of wheat. Bio/Technology 9:743-747
Vasil V, Castillo AM, Fromm ME, Vasil IK (1992) Herbicide resistant fertile transgenic wheat plants obtained by microprojectile bombardment of regenerable embryogenic callus. Bio/Technology 10:667-674
Vasil V, Clancy M, Ferl RJ, Vasil IK, Hannah LC (1989) Increased gene expression by the first intron of maize *Shrunken-1* locus in grass species. Plant Physiol 91:1575-1579
Vasil V, Redway F, Vasil IK (1990) Regeneration of plants from embryogenic suspension culture protoplasts of wheat (*Triticum aestivum* L.). Bio/Technology 8:429-434
Wang YC, Klein TM, Fromm ME, Cao J, Sanford JC, Wu R (1988) Transient expression of foreign genes in rice, wheat and soybean cells following particle bombardment. Plant Mol Biol 11:433-439
Yamada Y (1977) Tissue culture studies in cereals. *In* Applied and Fundamental Aspects of Plant Cell, Tissue and Organ Culture (Reinert J, Bajaj YPS, eds), pp144-159. Springer-Verlag, Heidelberg

DIFFERENTIAL ACTIVITY OF WHEAT HISTONE H4 PROMOTER IN TRANSGENIC MAIZE

Serik Omirulleh[1], Ajnur Ismagulova[1], Attila Fehér, Metin Bilgin, Sándor Mórocz[2] and Dénes Dudits*
Institute of Plant Physiology, Biological Research Center,
Hungarian Academy of Sciences,
H-6701 Szeged, P.O. Box 521, HUNGARY

Introduction

The recent methodology of introducing foreign genes into the genome of monocot cereal plants, including maize, is based on a variety of experimental approaches. Several laboratories have demonstrated the potentials in the use of microprojectile bombardment to transform embryogenic maize tissue cultures (Fromm et al. 1990; Gordon-Kamm et al. 1990; Walters et al. 1992). Furthermore, the production of transgenic maize plants expressing agronomically significant traits was also achieved by this method (Koziel et al. 1993). Regeneration of fertile transgenic maize plants was recently reported by D'Halluin et al. (1992) too, who electroporated plasmid molecules into wounded tissues of immature zygotic embryos or of embryogenic callus cultures. Both of these methods are proposed to be widely applicable for transformation of different breeding stocks. However, the regeneration capability was shown to be genotype-dependent by several tissue culture studies (Tomes and Smith 1985; Lupotto 1986; Petersen et al. 1992).

The continuous improvement of tissue culture techniques has contributed to the progress in plant regeneration from cultured maize protoplasts (Rhodes et al 1988; Shillito et al 1989; Prioli and Söndahl 1990; Petersen et al. 1992). A unique maize tissue culture system has been developed by Mórocz et al. (1990), that allowed the maintenance of an efficient and reproducible plant regeneration from embryogenic maize protoplasts over a period as long as 4 years. This synthetic genotype (HE/89) combines several advantageous traits required for rapid release of protoplasts, considerable division frequency of protoplast-derived cells, embryogenic potential and regeneration of normal, fertile plants.

Permanent addresses: 1 Institute of Molecular Biology and Biochemistry, Alma-Ata, Michurina str. 86., Republic of Kazakhstan, 480012
2 Cereal Research Institute Szeged, P.O.Box 391, H-6701 HUNGARY

* Author for correspondence Tel/Fax: 36-62-433434

NATO ASI Series, Vol. H 81
Plant Molecular Biology
Edited by G. Coruzzi and P. Puigdomènech

Protoplasts, isolated from a suspension culture of the HE/89 genotype, have been successfully used for direct DNA uptake and for regeneration of fertile transformants (Omirulleh et al. 1993; Golovkin et al. 1993). This transformation system allowed the analysis of a chimeric promoter with the help of ß-glucuronidase (GUS) gene fusions (Omirulleh et al. 1993).

In the present work, we characterize the expression pattern of the GUS reporter gene under the control of different promoters. Special emphasis will be made on the activity of a wheat histone H4 promoter, previously isolated by Tabata et al. (1983). Regulation of wheat histone H3 gene was studied in transformed rice cells and several sequence motifs are suggested to be responsible for the control of cell cycle-dependent expression pattern (Terada et al. 1993, Nakayama and Iwabuchi 1993).

Selection of transformants and regeneration of transgenic maize plants

Earlier, we have showed that several resistance genes, including the neomycin-phosphotransferase (NPTII) and the phosphinotricin acetyl transferase (PAT) can be efficiently used for safe selection of maize transformants derived from plasmid treated protoplasts (Omirulleh et al. 1993). Here, we present an example for identification of putative transformed colonies, grown in the presence of 500 μM of L-phosphinotricin after incubation of pMGP1 plasmids (see Omirulleh et al. 1993) with HE/89 protoplasts in the presence of polyethylene glycol (Fig. 1/A). The selected resistant callus tissues were transferred onto 2,4-dichlorophenoxyacetic acid (2,4-D) free medium. During several subcultures, the so-called early-embryogenic tissues gradually switched to a late-embryogenic stage and differentiation of leaf meristems and shoots could be recognized in majority of the cultures. Fig. 1/B shows a regenerant, grown in *in vitro* culture. After the development of the root system, these plantlets could be transferred into soil (Fig. 1/C). Since the plasmid used here carries also the GUS reporter gene with a chimeric promoter, the regenerants were analyzed for GUS activity by fluorometric method. The expression of the reporter gene was confirmed in these selected plants (data not shown).

In experiments to characterize expression pattern of wheat histone H4 promoter - GUS fusions in transgenic maize, the co-transformation method was exploited by treatment of 10^6 maize protoplasts with mixture of 20 μg pHP23 (kindly provided by Dr. J. Paszkowszki; see Mendel et al. 1989) or pPAT (kindly provided by Hoechst AG, Germany) plasmids with selectable genes driven by CaMV 35S promoters and of 40 μg plasmid carrying promoter - GUS fusion. Schematic maps of the various promoter constructs used are shown by Fig. 2. The DNA treated protoplasts were cultured for 2 weeks, and microcolonies were plated onto the surface of selective medium solidified with Gelrite (0.2 %).

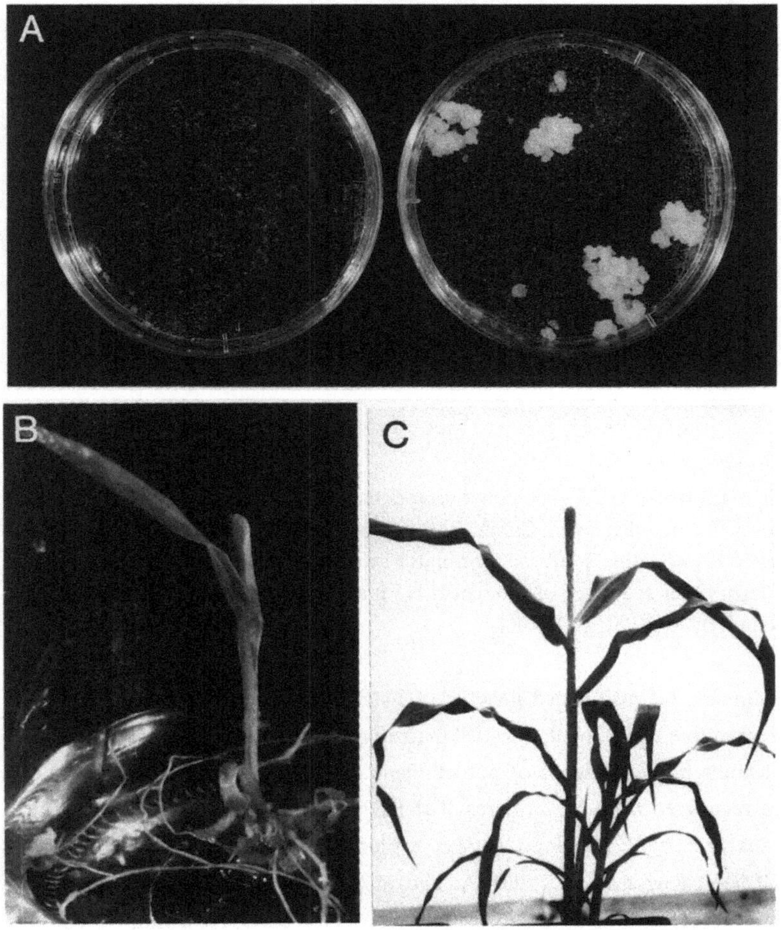

Figure 1. Selection of L-phosphinotricin resistant tissues and growth of a transformant up to flowering stage

A: Uptake of pMGP1 results in PPT resistant colonies (right); control culture under selective pressure (left)

B: Young plantlet regenerated from PPT resistant callus

C: Transformant grown in greenhouse

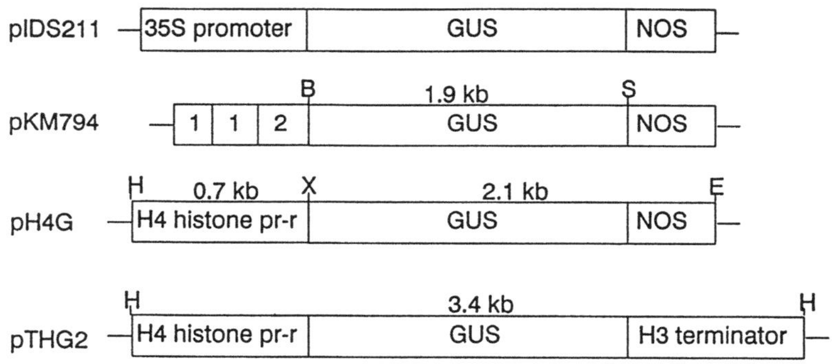

Figure 2. Schematic maps of the GUS gene containing constructs
1, enhancer sequences (-208 to -46) from CaMV 35S promoter; 2, truncated (-90 to +8) CaMV 35S promoter; NOS, nopaline synthase terminator sequence; H3 terminator, 850 bp long fragment of the terminator sequence of the wheat H3 histone gene; restriction sites: B, BamHI; S, SacI; H, HindIII; X, XhoI; E, EcoRI.

In these sets of experiments, we also tested geneticin (G418) antibiotic in replacement of kanamycin. In a representative experiment, 6 x 10^6 maize protoplasts were incubated with pH4G and pHP23 plasmids in the presence of polyethylene glycol and 184 colonies with G418 resistance were recovered from the cultures. Total DNA from randomly chosen 77 resistant clones was analyzed by PCR using CaMV 35S promoter specific primers (kindly provided by CIBA-GEIGY; Koziel et al. 1993). As shown by Fig. 3/A, a DNA fragment with the proper size could be amplified by PCR from the analyzed DNA samples. The estimation of co-transformation efficiency was based on the detection of GUS activity in the resistant transformants. According to these data, we observed 47% co-transformation frequency that is comparable with results from previous publications (Lyznik et al. 1989; Armstrong et al. 1990). Furthermore, these studies demonstrate that G418 can serve as a safe selective agent also in maize protoplast system. Previously, this antibiotic was successfully used in rice transformation experiments (Dekeyser et al. 1989).

Figure 3/A PCR amplification of 317 bp fragment from CaMV 35S promoter in DNA samples of G418-resistant callus lines.
a, molecular weight marker (PstI cut lambda DNA); b, negative control; c, positive control: plasmid pHP23.

Figure 3/B Southern blot analysis of the genomic DNA from GUS-positive transformants. 1.9 kb BamHI- SacI fragment of a coding region of the GUS gene from pKM794 construct was used for hybridization with 5 μg total plant DNA.
1, uncut total DNA from transformant HG4.15; 2, uncut total DNA from transformant THG.2; 3, DNA from control, untransformed HE/89 callus tissues after EcoRI digestion; 4, DNA of the HG4.15 line cut by XhoI-EcoRI; 5,6, DNA of the THG.2 line cut by HindIII and EcoRI respectively; 7, 7.5 pg plasmid DNA (pTHG2) after digestion with HindIII.

Figure 3/C-E Staining of various tissues with X-gluc.
C: Transverse section of a maize tissue with meristematic region and callus with enlarged cells;
D: Root differentiated from callus tissues;
E: Leaf pieces from control (2) and transformed plants (1).

Activity of 720 bp long promoter region from a wheat histone H4 gene in tissues of transformed maize

The structural organization and DNA sequence of a histone H4 gene from wheat were described by Tabata et al. (1983). The authors kindly provided us with the clone pTH011, from which 720 bp long, HindIII-XhoI promoter fragment was subcloned and linked to the GUS reporter gene. As shown by Fig. 2., we have constructed expression vectors with two types of termination signal. In pH4G vector the terminator was used from nopaline synthetase (NOS) gene, while in pTHG2 vector a 850 bp SauIII - HindIII fragment was cloned as termination signal from pTHO12 plasmid carrying the wheat histone H3 gene (Tabata et al. 1984). As reference vectors, we used pIDS211, with CaMV 35S promoter (Stefanov et al. 1992) and pKM794 with doubled enhancer sequence of CaMV 35S (-208 to -46 bp) inserted upstream to the domain A (-90 to +8 bp) of CaMV 35S promoter (Fejes and Nagy, unpublished). Based on the GUS activity values, presented in Table 1., we can suggest that in maize callus tissues, the wheat histone H4 promoter fragment with NOS terminator provides 2.5 times higher level of expression than the CaMV 35S promoters. Up to now, we have characterized one transgenic line originated from transformation with pTHG2.

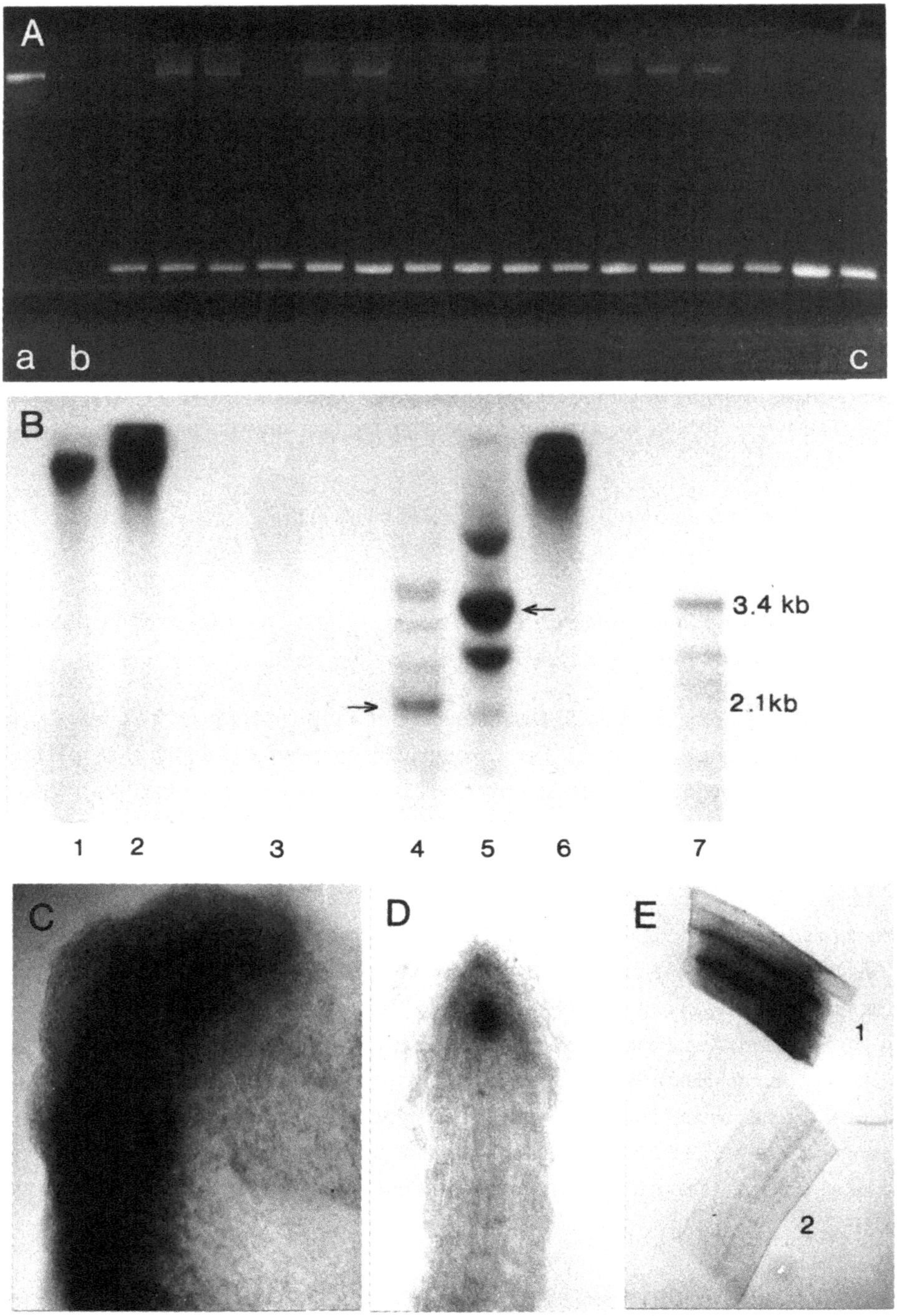
A
a
b
c
B
3.4 kb
2.1kb
1
2
3
4
5
6
7
C
D
E
1
2

Table 1. Comparison of various promoter activities in transformed maize callus tissues

Vector	Number of analyzed transgenic lines	GUS activity (nM MU/mg protein/hr)		
		minimum	average	maximum
pH4G	8	1.20	8.76	27.0
pIDS211	4	0.62	3.26	9.60
pKM794	8	0.42	3.86	10.50

In callus tissues of this line we found relatively high GUS activity values ranging from 33 to 200 nM MU/mg protein/hour. Further studies are in progress to analyze the influence of 3' end region from a histone H3 gene on mRNA accumulation.

Transgenic nature of two analyzed lines was also indicated by Southern analysis (Fig. 3/B). Integration of the GUS sequence into the high molecular weight genomic DNA can be concluded from the hybridization signal with uncut plant DNA. Digestion of plant DNA from clone THG2 with EcoRI which has no cutting site inside the vector, resulted in a hybridization also with a high molecular weight band. According to the comparison of the approximate copy numbers, there is a considerable difference between the two lines analyzed. In agreement with previous observations (Sanders et al. 1987; Klein et al. 1989), we could not find correlation between the number of integrated copies and the actual level of expression of GUS reporter gene (see Fig. 4.).

Expression pattern of histone H4 promoter - GUS gene fusions in maize tissues with different division activities.

The levels of GUS activity, under the control of various promoters, were analyzed in freshly transferred callus tissues and in old cultures, 6 weeks after transfer to new culture medium. As shown by Fig. 4., all transformants with wheat histone H4 promoter exhibited significantly higher expression levels in fresh callus tissues than in old ones. In contrast, the regulatory elements from CaMV 35S promoter resulted in an opposite order of GUS levels. The elevated histone promoter activity in callus with logarithmic growth phase, may be partially attributed to the increase in division rate. However, we have to consider the influence of other physiological factors on the stability of GUS enzyme in the analyzed tissues.

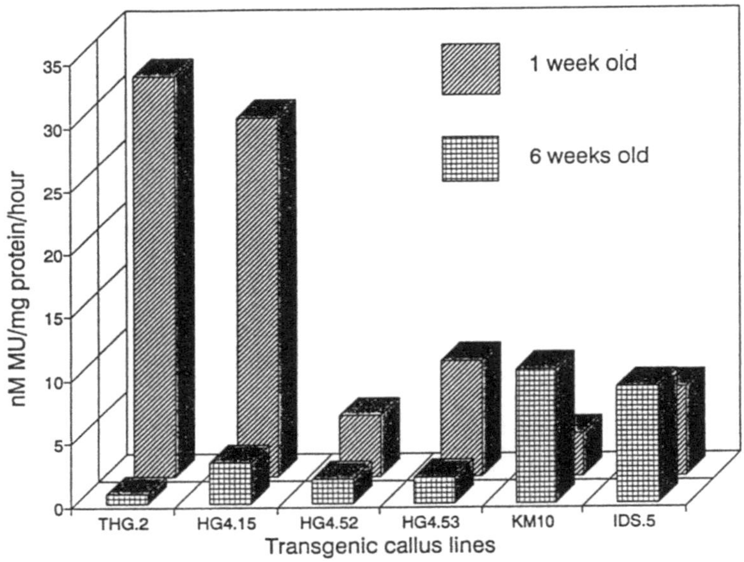

Figure 4. The level of the GUS gene expression in fresh and old callus tissues transformed by constructs with different promoters.
The callus lines were transformed: THG.2 by plasmid pTHG2; HG4.15, HG4.52 and HG4.53 by pH4G; KM10 by pKM794; IDS.5 by pIDS211.

The distribution of cells expressing the GUS gene was studied by histochemical staining of tissues from transformants. As shown by Fig. 3/C, the callus tissues consist of significantly different cell types. Intense staining was detected in small cells of the meristematic regions, while the enlarged, non-dividing cells failed to accumulate the indigo dye. Staining of roots revealed a very characteristic pattern of expression (Fig. 3/D). The GUS expressing cells were localized to the root meristematic region. Very similar expression patterns were described in roots of other transgenic plants carrying histone promoter regions linked to the GUS gene, either from monocot or dicot plants (Lepetit et al. 1992; Terada et al. 1993; Kapros et al. 1993). The basal regions of young leaves showed considerable GUS activities that could not be detected in mature leaves of regenerated transformants (Fig. 3/E). The described expression studies may support a conclusion that the analyzed wheat histone H4 promoter region specifies cell division-dependent expression, however, further experiments preferably based on synchronized cultures will be required to prove that this regulatory element is under cell cycle control.

References

Armstrong CL, Petersen WL, Buchholz WG, Bowen BA and Sulc SL (1990) Factors affecting PEG-mediated stable transformation of maize protoplasts. Plant Cell Rep 9:335-339

D'Halluin K, Bonne E, Bossut M, Beuckeleer M and Leemans J (1992) Transgenic maize plants by tissue electroporation. The Plant Cell 4:1495-1505

Dekeyser R, Claes B, Marichal M, Van Montagu M and Caplan A (1989) Evaluation of selectable markers for rice transformation. Plant Physiol 90:217-223

Fromm ME, Morrish F, Armstrong C, Williams R, Thomas J and Klein TM (1990) Inheritance and expression of chimeric genes in the progeny of transgenic maize plants. Biotechnol 8:833-839

Golovkin MV, Åbrahám M, Mórocz S, Bottka S, Fehér A and Dudits D (1993) Production of transgenic maize plants by direct DNA uptake into embryogenic protoplasts. Plant Sci (in press).

Gordon-Kamm WJ, Spencer TM, Mangano ML, Adams TR, Danies RJ, Start WG, O'Brien JV, Chambers SA, Adams WR, Jr.Willets NG, Rice TB, Mackey CJ, Krueger RW, Kausch AP and Lemaux, PG (1990) Transformation of maize cells and regeneration of fertile transgenic plants. The Plant Cell 2:603-618

Kapros T, Stefanov I, Magyar Z, Ocsovszky I and Dudits D (1993) A short histone H3 promoter from alfalfa specifies expression in S-phase cells and meristems. In Vitro Cell Dev Biol 29:27-32

Klein TM, Kornstein L, Sanford JC and Fromm ME (1989) Genetic transformation of maize cells by particle bombardment. Plant Physiol 91:440-444

Koziel MG, Beland GL, Bowman C, Carozzi NB, Crenshaw R, Crossland L, Dawson J, Desai N, Hill M, Kadwell S, Launis K, Lewis K, Maddox D, McPherson K, Meghiji MR, Merlin E, Rhodes R, Warren GW, Wright M and Evola SV (1993) Field performance of elite transgenic maize plants expressing an insecticidal protein derived from Bacillus thuringiensis. Biotechnol 11:194-200

Lepetit M, Ehling M, Chaubet N and Gigot C (1992) A plant histone gene promoter can direct both replication-dependent and -independent gene expression in transgenic plants. Mol Gen Genet 231:276-285

Lupotto E (1986) In vitro culture of isolated somatic embryos of maize (*Zea mays* L.) Madica 31:193-201

Lyznik LA, Ryan RD, Ritchie SR and Hodges TK (1989) Stable co-transformation of maize protoplasts with gusA and neo genes. Plant Mol Biol 13:151-161

Mendel RR, Müller B, Schulze J, Kolesnikov V and Zelenin A (1989) Delivery of foreign genes to intact barley cells by high-velocity microprojectiles. Theor Appl Genet 78:31-34

Mórocz S, Donn G, Németh J and Dudits D (1990) An improved system to obtain fertile regenerants via maize protoplasts isolated from a highly embryogenic suspension culture. Theor Appl Genet 80:721-726

Nakayama T and Iwabuchi M (1993) Regulation of wheat histone gene expression. Crit Rev Plant Sci 12:97-110

Omirulleh S, Åbrahám M, Golovkin M, Stefanov I, Karabaev IM, Mustárdy L, Mórocz S and Dudits D (1993) Activity of a chimeric promoter with the doubled CaMV 35S enhancer element in protoplast-derived cells and transgenic plants in maize. Plant Mol Biol 21:415-428

Petersen WL, Sulc S and Armstrong CL (1992) Effect of nurse cultures on the production of macro-calli and fertile plants from maize embryogenic suspension culture protoplasts. Plant Cell Rep 10:591-594

Prioli LM and Söndahl MR (1990) Plant regeneration and recovery of fertile plants from protoplasts of maize (*Zea mays* L.) Biotechnol 7:589-594

Rhodes CA, Pierce DA, Mettler IJ, Mascarenhas D and Detmer JJ (1988) Genetically transformed maize plants from protoplasts. Science 240:204-206

Rhodes CA, Lowe KS and Ruby KL (1988) Plant regeneration from protoplasts isolated from embryogenic maize cell cultures. Biotechnol 6:56-60

Shillito RD, Carswell GK, Johnson CM, DiMaio JJ and Harms D (1989) Regeneration of fertile plants from protoplasts of elite inbred maize. Biotechnol 7:581-587

Stefanov I, Ilubaev S, Fehér A, Margóczi K and Dudits D (1991) Promoter and genotype dependent transient expression of a reporter gene in plant protoplasts. Acta Biol Hung 42:323-331

Tabata T, Sasaki K and Iwabuchi M (1983) The structural organization and DNA sequence of a wheat histone H4 gene. Nucl Acids Res 11:5865-5875

Tabata T, Fukasawa M and Iwabuchi M (1984) Nucleotide sequence and genomic organization of a wheat histone H3 gene. Mol Gen Genet 196:397-400

Terada R Nakayama T, Iwabuchi M and Shimamoto K (1993) A wheat histone H3 promoter confers cell division-dependent and -independent expression of the gusA gene in transgenic rice plants. The Plant J 3:241-252

Tomes DT and Smith OS (1985) The effect of parental genotype on initiation of embryogenic callus from elite maize (*Zea mays* L.) germ-plasm. Theor Appl Genet 70:505-509

Walters DA, Vetsch CS, Potts DE and Lundguist RS (1992) Transformation and inheritance of a hygromycin phosphatase gene in maize plants. Plant Mol Biol 18:189-200

Transgenic Maize by Electroporation of Pectolyase-Treated Suspension Culture Cells

T. Michael Spencer, Cheryl Montain Laursen[1], Richard A. Krzyzek[2], Paul C. Anderson, Christopher E. Flick

Discovery Research
DEKALB Plant Genetics
62 Maritime Drive
Mystic, Connecticut 06355-1958
USA

Introduction

Recent progress in the development of direct gene transfer techniques has made genetic manipulation of important cereal crops possible. Direct DNA-delivery to protoplasts first yielded transgenic rice (Shimamoto et al., 1989) and ultimately transgenic maize (Omirulleh et al., 1993). Given the difficulties in developing protoplastable cell lines and problems associated with regenerating plants from protoplasts (Potrykus, 1990), other strategies have been pursued. Microprojectile bombardment (Klein et al., 1987) was the first direct DNA-delivery technique successfully employed for transformation of maize (Gordon-Kamm et al., 1990; Fromm et al., 1990; Walters et al, 1992) and wheat (Vasil et al., 1992). Alternatively, Dekeyser et al. (1990) demonstrated that it was possible to introduce DNA into intact monocot tissues using electroporation. Transient expression of reporter genes was demonstrated in leaf bases of rice, wheat, maize, and barley. Subsequently, D'Halliun et al. (1992) used electroporation in combination with enzymatic or mechanical wounding, to transform immature maize embryos and Type I callus, ultimately giving rise to transgenic plants. Here we describe a procedure for transformation of maize suspension culture cells by electroporation and present analysis of transgenic plants.

[1]Present Address: 6208 Hammersly Road, Madison WI 53711
[2]Present Address: R&D Systems Inc., 614 McKinley Place, Minneapolis MN 55413

NATO ASI Series, Vol. H 81
Plant Molecular Biology
Edited by G. Coruzzi and P. Puigdomènech

Enzyme Treatment and Electroporation

We have developed a system for maize transformation based on electroporation-mediated delivery of DNA to pectolyase-treated suspension culture cells (Laursen et al., submitted). DNA-delivery parameters were optimized using transient gene expression assays (Krzyzek et al., 1992). Numerous parameters were investigated using a plasmid encoding β-glucuronidase (GUS). These parameters included the type and concentration of cell wall degrading enzymes, the type and concentration of osmotic agent in the enzyme incubation and electroporation buffers, salt concentration in the electroporation buffer, the electroporation field strength and pulse duration, and the form and concentration of plasmid DNA used for transformation. Treatment of suspension culture cells with 0.5% pectolyase and electroporation with linear DNA using field strengths ranging from 280 to 560 volts/cm yielded the highest levels of transient GUS activity.

Two suspension cultures, SC716 and AT824, were used for stable transformation experiments. SC716 was initiated from A188 x B73 Type II callus (Armstrong and Green, 1985) and has been described in detail (Gordon-Kamm et al., 1990). Type II AT824 callus was initiated from immature embryos of an elite inbred line (Aves et al., 1992). A suspension culture was initiated as described for SC716. Both suspension cultures were cryopreserved and reinitiated (Gordon-Kamm et al., 1990) prior to electroporation.

Before electroporation, SC716 or AT824 suspension culture cells were sieved to remove large cell clusters and washed, 1.5 ml packed cells per 10 ml, in incubation buffer (0.2 M mannitol, 0.1% bovine serum albumin, 80 mM calcium chloride, and 20 mM 2-(N-morpholino)-ethane sulfonic acid (MES), pH 5.6). Cells were then treated for 60-90 minutes in incubation buffer containing 0.5% pectolyase Y-23 (Seishin Pharmaceutical, Tokyo, Japan) at a density of 1.5 ml packed cells per 5 ml of enzyme solution. During the enzyme treatment, cells were incubated in the dark at approximately 25° C on a rotary shaker at 60 rpm. Following pectolyase treatment, cells were washed once with 10 ml of incubation buffer followed by three washes with electroporation buffer (10 mM HEPES, 0.4 M mannitol, pH 7.5). Cells were resuspended in electroporation buffer at a density of 0.5 ml packed cells/ml. Cells were electroporated with 50-100 µg/ml *EcoRI* linearized pDPG165 alone or in combination with a second plasmid. Plasmid pDPG165 contains the Cauliflower Mosaic Virus 35S promoter, the *bar* gene from *Streptomyces hygroscopicus*, and the transcript 7 3'-end from *Agrobacterium tumefaciens* (Spencer et al., 1990). One ml of suspension culture cells in electroporation buffer (containing approximately 0.5

ml packed cells) was added to 1 ml aliquots of DNA in electroporation buffer. One-half ml aliquots of this mixture were transferred to the electroporation chamber (Puite et al., 1985) and electroporated with a 70, 100, or 140 volt pulse discharged from a 140 microfarad capacitor.

Selection of Transformants

Post-electroporation, cells were diluted with 2.5 ml suspension culture medium containing 0.3 M mannitol. The cells were dispensed onto a filter overlaying solid callus maintenance medium containing 0.3 M mannitol. After five days, the tissue and the supporting filters were transferred to callus maintenance medium containing 0.2 M mannitol. After seven days, tissue and supporting filters were transferred to callus maintenance medium without mannitol. Tissue growing on filters on callus maintenance medium was removed from the filter and transferred as clumps (approximately 0.5 cm in diameter) to the surface of solid selection medium. The selection medium consisted of callus maintenance medium supplemented with 1 mg/l bialaphos (Meiji Seika Kaisha, Ltd., Yokohama, Japan). Three weeks later, slowly growing tissue was transferred, as 0.5 cm clumps, to callus maintenance medium containing 3 mg/l bialaphos. Three to four weeks later, callus sectors that continued to grow were transferred to fresh medium containing 3 mg/l bialaphos. Callus lines that continued to grow after this subculture were transferred to fresh selection medium every two weeks. This selection protocol yielded approximately one transformant per one to two electroporations using either SC716 or AT824 (Laursen et al., submitted).

An alternate selection scheme involved transferring electroporated tissue, after recovery on solid mannitol medium, to liquid medium. Cells were cultured for four days in non-selective liquid medium, followed by transfer to liquid medium containing 2 mg/l bialaphos. After three days in liquid selection medium, cells were plated onto solid selection medium, containing 3 mg/l bialaphos, at a density of 0.1 ml packed cells per 100 x 15 mm petri dish of medium. Bialaphos resistant colonies were identified and transferred to fresh selection medium three to four weeks after plating on solid selection medium. Using this selection protocol and the AT824 suspension culture, over 10 transformants per electroporation were recovered in multiple experiments (Laursen et al., submitted).

Plant Regeneration and Analysis of Progeny

Approximately 50 plants were regenerated from 12 lines selected in our initial experiments (Laursen et al., submitted). Most of these plants have reached maturity and seed has been recovered from 11 of the 12 lines. Progeny have been recovered from crosses using the transgenic regenerants as either the male or the female parent in outcrosses to non-transformed inbred plants. Analysis of segregation of *bar* in four R_1 populations is shown in Table 1. Plants were assayed for expression of *bar* by localized foliar application of a 2% solution of the herbicide Basta^R (Gordon-Kamm et al., 1990). In most cases, inheritance of the Basta^R resistance phenotype approximated Mendelian segregation of a single gene.

Table 1. Segregation of *bar* in R_1 populations derived from suspension culture electroporation experiments.

transgenic plant	cross	R_1 plants painted with Basta^R	
		resistant	sensitive
EP1-3	female	7	10
EP1-6	female	18	14
EP1-9	female	5	8
EP3-137	male	5	5

Segregation of *bar* in several lines was analyzed by Southern blot hybridization; analysis of three lines is shown in Figure 1. Genomic DNA isolated from callus (C), regenerants (R_0), and progeny (R_1) was digested with *HindIII* and *BglII*, which release a 1.3 kb fragment from pDPG165 containing the 35S promoter and the *bar* gene. The Southern blots were hybridized with ^{32}P-labelled 0.6 kb *SmaI bar* fragment from pDPG165. In all three lines, multiple *bar*-hybridizing fragments segregated as a unit, indicating integration of multiple copies of *bar* at a single locus. Hybridization patterns observed in callus appeared to be conserved in R_0 and R_1 generations of all three lines. These results are consistent with most analyses of microprojectile bombardment-derived transgenic maize lines (Spencer et al., 1992).

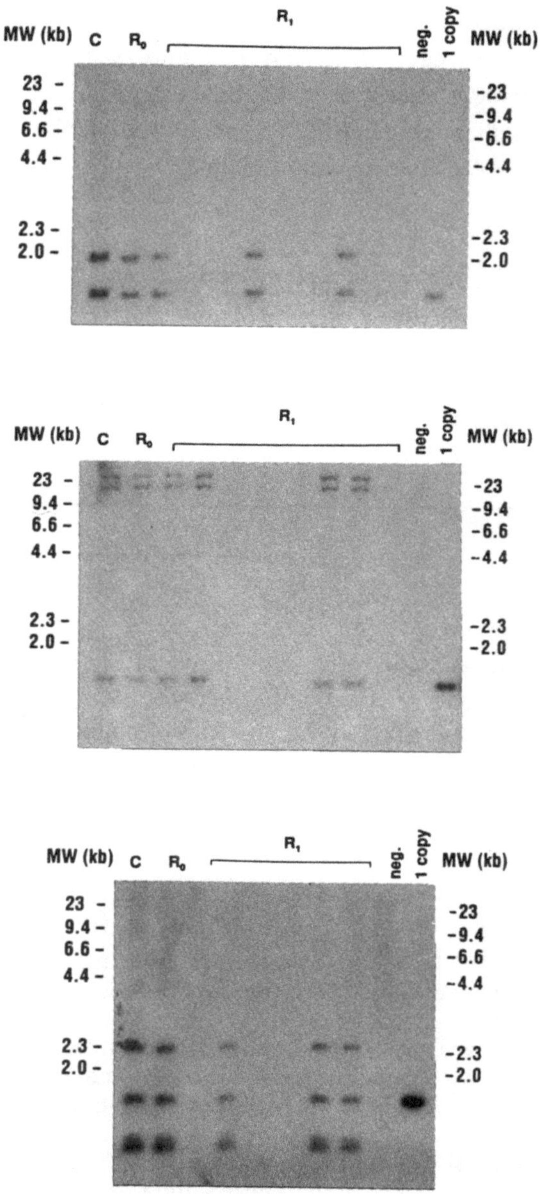

Figure 1. Southern blot analyses of *bar* segregation in R_1 populations from three transformed lines. *HindIII*/*BglII*-digested genomic DNA from callus (C), regenerants (R_0), and progeny (R_1) was probed with ^{32}P-labelled *bar*. Lanes labelled 1 copy contained *HindIII*/*BglII*-digested pDPG165 to represent 1 copy of the 1.3 kb *bar* fragment per maize genome.

One of the transgenic lines analyzed above, EP1-3, was perpetuated further by selfing R_1 plants. Segregation of *bar* was analyzed in the resulting R_2 population by spraying plants with a 1% Basta solution. Inheritance of the herbicide-resistance phenotype in EP1-3 R_2 population fit the expected ratio for Mendelian segregation of a single gene (Table 2).

Table 2. Segregation of *bar* in R_2 plants resulting from self pollination of an EP1-3 R_1 plant.

plants sprayed with Basta		expected	X^2	P
resistant	sensitive	ratio		
37	12	3:1	0.01	0.90

Conclusions

Our results show the feasibility of an alternative to the particle gun for maize transformation. Pectolyase treatment of embryogenic maize suspension culture cells, followed by electroporation in the presence of plasmid DNA has produced large numbers of stable transformants. Fertile transgenic plants were regenerated and transmission of introduced DNA to progeny was demonstrated.

References

Armstrong CL, Green CE: Establishment and maintenance of friable, embryogenic maize callus and the involvement of L-proline. Planta 164:207-214 (1985).

Aves K, Genovisi D, Willets N, Zachweija S, Mann M, Spencer T, Flick C, Gordon-Kamm W: Transformation of an elite maize inbred line through microprojectile bombardment of regenerable, embryogenic callus. In Vitro 28:124A (1992).

Dekeyser RA, Claes B, De Rycke RMU, Habets ME, Van Montagu MC, Caplan AB: Transient gene expression in intact and organized rice tissues. Plant Cell 2:591-602 (1990).

D'Halliun K, Bonne E, Bossut M, De Beuckeleer M, Leemans J: Transgenic maize plants by tissue electroporation. Plant Cell 4:1495-1505 (1992).

Fromm ME, Morrish F, Armstrong C, Williams R, Thomas J, Klein TM: Inheritance and expression of chimeric genes in the progeny of transgenic maize plants. Bio/Technology 8:833-839 (1990).

Gordon-Kamm WJ, Spencer TM, Mangano ML, Adams TR, Daines RJ, Start WG, O'Brien JV, Chambers SA, Adams WR, Willetts NG, Rice TB, Mackey CJ, Krueger RW, Kausch AP, Lemaux PG: Transformation of maize cells and regeneration of fertile transgenic plants. Plant Cell 2:603-618 (1990).

Klein TM, Wolf ED, Wu R, Sanford JC: High velocity microprojectiles for delivering nucleic acids into living cells. Nature 327:70-73 (1987).

Krzyzek RA, Laursen CR, Anderson PC: Stable transformation of maize cells by electroporation. European Patent Application No. WO92/12250 (1992).

Laursen CM, Krzyzek RA, Flick CE, Anderson PC, Spencer TM: Transformation of maize by electroporation of suspension culture cells. submitted.

Omirulleh S, Abraham M, Golovkin M, Stefanov I, Karabaev MK, Mustardy L, Morocz S, Dudits D: Activity of a chimeric promoter with the doubled CaMV 35S enhancer element in protoplast-derived cells and transgenic plants in maize. Plant Mol Biol 21:415-428 (1993).

Potrykus I: Gene transfer to cereals: an assessment. Bio/Technology 8:535-542 (1990).

Puite KJ, van Wikselaar P, Verhoeven H: Electrofusion: a simple and reproducible technique in somatic hybridization of *Nicotiana plumbaginifolia* mutants. Plant Cell Reports 4:274-276 (1985).

Shimamoto K, Terada R, Izawa T, Fujimoto H: Fertile transgenic rice plants regenerated from transformed protoplasts. Nature 338:274-276 (1989).

Spencer TM, Gordon-Kamm WJ, Daines RJ, Start WG, Lemaux PG: Bialaphos selection of stable transformants from maize cell culture. Theor Appl Genet 79:625-631 (1990).

Spencer TM, O'Brien JV, Start WG, Adams TR, Gordon-Kamm WJ, Lemaux PG: Segregation of transgenes in maize. Plant Mol Biol 18:201-210 (1992).

Vasil V, Castillo AM, Fromm ME, Vasil IK: Herbicide resistant fertile transgenic wheat plants obtained by microprojectile bombardment of regenerable embryogenic callus. Bio/Technology 10:667-674 (1992).

Walters DA, Vetsch CS, Potts DE, Lundquist RC: Transformation and inheritance of a hygromycin phosphotransferase gene in maize plants. Plant Mol Biol 18:189-200 (1992).

SUBJECT INDEX
(The first page of the respective article is listed)

Printing: Druckhaus Beltz, Hemsbach
Binding: Buchbinderei Schäffer, Grünstadt

NATO ASI Series H

Vol. 21: **Cellular and Molecular Basis of Synaptic Transmission.** Edited by H. Zimmermann. 547 pages. 1988.

Vol. 22: **Neural Development and Regeneration. Cellular and Molecular Aspects.** Edited by A. Gorio, J. R. Perez-Polo, J. de Vellis, and B. Haber. 711 pages. 1988.

Vol. 23: **The Semiotics of Cellular Communication in the Immune System.** Edited by E.E. Sercarz, F. Celada, N.A. Mitchison, and T. Tada. 326 pages. 1988.

Vol. 24: **Bacteria, Complement and the Phagocytic Cell.** Edited by F. C. Cabello und C. Pruzzo. 372 pages. 1988.

Vol. 25: **Nicotinic Acetylcholine Receptors in the Nervous System.** Edited by F. Clementi, C. Gotti, and E. Sher. 424 pages. 1988.

Vol. 26: **Cell to Cell Signals in Mammalian Development.** Edited by S.W. de Laat, J.G. Bluemink, and C.L. Mummery. 322 pages. 1989.

Vol. 27: **Phytotoxins and Plant Pathogenesis.** Edited by A. Graniti, R. D. Durbin, and A. Ballio. 508 pages. 1989.

Vol. 28: **Vascular Wilt Diseases of Plants. Basic Studies and Control.** Edited by E. C. Tjamos and C. H. Beckman. 590 pages. 1989.

Vol. 29: **Receptors, Membrane Transport and Signal Transduction.** Edited by A. E. Evangelopoulos, J. P. Changeux, L. Packer, T. G. Sotiroudis, and K.W.A. Wirtz. 387 pages. 1989.

Vol. 30: **Effects of Mineral Dusts on Cells.** Edited by B.T. Mossman and R.O. Begin. 470 pages. 1989.

Vol. 31: **Neurobiology of the Inner Retina.** Edited by R. Weiler and N.N. Osborne. 529 pages. 1989.

Vol. 32: **Molecular Biology of Neuroreceptors and Ion Channels.** Edited by A. Maelicke. 675 pages. 1989.

Vol. 33: **Regulatory Mechanisms of Neuron to Vessel Communication in Brain.** Edited by F. Battaini, S. Govoni, M.S. Magnoni, and M. Trabucchi. 416 pages. 1989.

Vol. 34: **Vectors asTools for the Study of Normal and Abnormal Growth and Differentiation.** Edited by H. Lother, R. Dernick, and W. Ostertag. 477 pages. 1989.

Vol. 35: **Cell Separation in Plants: Physiology, Biochemistry and Molecular Biology.** Edited by D. J. Osborne and M. B. Jackson. 449 pages. 1989.

Vol. 36: **Signal Molecules in Plants and Plant-Microbe Interactions.** Edited by B.J.J. Lugtenberg. 425 pages. 1989.

Vol. 37: **Tin-Based Antitumour Drugs.** Edited by M. Gielen. 226 pages. 1990.

Vol. 38: **The Molecular Biology of Autoimmune Disease.** Edited by A.G. Demaine, J-P. Banga, and A.M. McGregor. 404 pages. 1990.

NATO ASI Series H

Vol. 39: **Chemosensory Information Processing.**
Edited by D. Schild. 403 pages. 1990.

Vol. 40: **Dynamics and Biogenesis of Membranes.**
Edited by J. A. F. Op den Kamp. 367 pages. 1990.

Vol. 41: **Recognition and Response in Plant-Virus Interactions.**
Edited by R. S. S. Fraser. 467 pages. 1990.

Vol. 42: **Biomechanics of Active Movement and Deformation of Cells.**
Edited by N. Akkas. 524 pages. 1990.

Vol. 43: **Cellular and Molecular Biology of Myelination.**
Edited by G. Jeserich, H. H. Althaus, and T. V. Waehneldt. 565 pages. 1990.

Vol. 44: **Activation and Desensitization of Transducing Pathways.**
Edited by T. M. Konijn, M. D. Houslay, and P. J. M. Van Haastert. 336 pages. 1990.

Vol. 45: **Mechanism of Fertilization: Plants to Humans.**
Edited by B. Dale. 710 pages. 1990.

Vol .46: **Parallels in Cell to Cell Junctions in Plants and Animals.**
Edited by A. W Robards, W. J . Lucas, J . D. Pitts, H . J . Jongsma, and D. C. Spray. 296 pages. 1990.

Vol. 47: **Signal Perception and Transduction in Higher Plants.**
Edited by R. Ranjeva and A. M. Boudet. 357 pages. 1990.

Vol. 48: **Calcium Transport and Intracellular Calcium Homeostasis.**
Edited by D. Pansu and F. Bronner. 456 pages. 1990.

Vol. 49: **Post-Transcriptional Control of Gene Expression.**
Edited by J. E. G. McCarthy and M. F. Tuite. 671 pages. 1990.

Vol. 50: **Phytochrome Properties and Biological Action.**
Edited by B. Thomas and C. B. Johnson. 337 pages. 1991.

Vol. 51: **Cell to Cell Signals in Plants and Animals.**
Edited by V. Neuhoff and J. Friend. 404 pages. 1991.

Vol. 52: **Biological Signal Transduction.**
Edited by E. M . Ross and K . W. A. Wirtz. 560 pages. 1991.

Vol. 53: **Fungal Cell Wall and Immune Response.**
Edited by J. P. Latge and D. Boucias. 472 pages. 1991.

Vol. 54: **The Early Effects of Radiation on DNA.**
Edited by E. M. Fielden and P. O'Neill. 448 pages. 1991.

Vol. 55: **The Translational Apparatus of Photosynthetic Organelles.**
Edited by R. Mache, E. Stutz, and A. R. Subramanian. 260 pages. 1991.

Vol. 56: **Cellular Regulation by Protein Phosphorylation.**
Edited by L. M. G. Heilmeyer, Jr. 520 pages. 1991.

NATO ASI Series H

Vol. 57: **Molecular Techniques in Taxonomy.**
Edited by G . M . Hewitt, A. W. B. Johnston, and J. P. W. Young . 420 pages. 1991.

Vol. 58: **Neurocytochemical Methods.**
Edited by A. Calas and D. Eugene. 352 pages. 1991.

Vol. 59: **Molecular Evolution of the Major Histocompatibility Complex.**
Edited by J. Klein and D. Klein. 522 pages. 1991.

Vol. 60: **Intracellular Regulation of Ion Channels.**
Edited by M. Morad and Z. Agus. 261 pages. 1992.

Vol. 61: **Prader-Willi Syndrome and Other Chromosome 15q Deletion Disorders.**
Edited by S. B. Cassidy. 277 pages. 1992.

Vol. 62: **Endocytosis. From Cell Biology to Health, Disease and Therapie.**
Edited by P. J. Courtoy. 547 pages. 1992.

Vol. 63: **Dynamics of Membrane Assembly.**
Edited by J. A. F. Op den Kamp. 402 pages. 1992.

Vol. 64: **Mechanics of Swelling. From Clays to Living Cells and Tissues.**
Edited by T. K. Karalis. 802 pages. 1992.

Vol. 65: **Bacteriocins, Microcins and Lantibiotics.**
Edited by R. James, C. Lazdunski, and F. Pattus. 530 pages. 1992.

Vol. 66: **Theoretical and Experimental Insights into Immunology.**
Edited by A. S. Perelson and G. Weisbuch. 497 pages. 1992.

Vol. 67: **Flow Cytometry. New Developments.**
Edited by A. Jacquemin-Sablon. 1993.

Vol. 68: **Biomarkers. Research and Application in the Assessment of Environmental Health.** Edited by D. B. Peakall and L. R. Shugart. 138 pages. 1993.

Vol. 69: **Molecular Biology and its Application to Medical Mycology.**
Edited by B. Maresca, G. S. Kobayashi, and H. Yamaguchi. 271 pages. 1993.

Vol. 70: **Phospholipids and Signal Transmission.**
Edited by R. Massarelli, L. A. Horrocks, J. N. Kanfer, and K. Löffelholz. 448 pages. 1993.

Vol. 71: **Protein Synthesis and Targeting in Yeast.**
Edited by A. J. P. Brown, M. F. Tuite, and J. E. G. McCarthy. 425 pages. 1993.

Vol. 72: **Chromosome Segregation and Aneuploidy.**
Edited by B. K. Vig. 425 pages. 1993.

Vol. 73: **Human Apolipoprotein Mutants III. In Diagnosis and Treatment.**
Edited by C. R. Sirtori, G. Franceschini, B. H. Brewer Jr. 302 pages. 1993.

NATO ASI Series H

Vol. 74: **Molecular Mechanisms of Membrane Traffic.**
Edited by D. J. Morré, K. E. Howell, and J. J. M. Bergeron. 429 pages. 1993.

Vol. 75: **Cancer Therapy. Differentiation, Immunomodulation and Angiogenesis.**
Edited by N. D'Alessandro, E. Mihich, L. Rausa, H. Tapiero, and T. R. Tritton. 299 pages. 1993.

Vol. 76: **Tyrosine Phosphorylation/Dephosphorylation and Downstream Signalling.**
Edited by L. M. G. Heilmeyer Jr. 388 pages. 1993.

Vol. 77: **Ataxia-Telangiectasia.** Edited by R. A. Gatti, R. B. Painter. 306 pages. 1993.

Vol. 78: **Toxoplasmosis.** Edited by J. E. Smith. 272 pages. 1993.

Vol. 79: **Cellular Mechanisms of Sensory Processing. The Somatosensory System.**
Edited by L. Urban. 514 pages. 1994.

Vol. 80: **Autoimmunity: Experimental Aspects.**
Edited by M. Zouali. 318 pages. 1994.

Vol. 81: **Plant Molecular Biology. Molecular Genetic Analysis of Plant Development and Metabolism.**
Edited by G. Coruzzi, P. Puigdomènech. 579 pages. 1994.

Vol. 82: **Biological Membranes: Structure, Biogenesis and Dynamics.**
Edited by Jos A. F. Op den Kamp. 367 pages. 1994.

Vol. 83: **Molecular Biology of Mitochondrial Transport Systems.**
Edited by M. Forte, M. Colombini. 420 pages. 1994.

Vol. 84: **Biomechanics of Active Movement and Division of Cells.**
Edited by N. Akkaş. 587 pages. 1994.

MIX
Papier aus verantwortungsvollen Quellen
Paper from responsible sources
FSC® C105338

If you have any concerns about our products,
you can contact us on
ProductSafety@springernature.com

In case Publisher is established outside the EU,
the EU authorized representative is:
Springer Nature Customer Service Center GmbH
Europaplatz 3, 69115 Heidelberg, Germany

Printed by Libri Plureos GmbH
in Hamburg, Germany